Introduction to
NUMERICAL ANALYSIS

SECOND EDITION

CARL-ERIK FRÖBERG
Institute for Computer Sciences
University of Lund, Sweden

Introduction to

NUMERICAL ANALYSIS

SECOND EDITION

ADDISON-WESLEY PUBLISHING COMPANY
Reading, Massachusetts · Menlo Park, California · London · Don Mills, Ontario

This volume is an English translation of *Lärobok i numerisk analys* by Carl-Erik Fröberg, published and sold by permission of Svenska Bokförlaget Bonniers, the owner of all rights to publish and sell the same.

Preface

Numerical analysis is a science
—computation is an art.

The present text in numerical analysis was written primarily to meet the demand of elementary education in this field at universities and technical institutes. But it is also believed that the book will be useful as a handbook in connection with numerical work within natural and technical sciences. Compared with the first edition a thorough revision and modernization has been performed, and quite a few new topics have been introduced. Two completely new chapters, integral equations and special functions, have been appended. Major changes and additions have been made, especially in Chapters 6, 14, and 15, while minor changes appear in practically all the remaining chapters. In this way the volume of the book has increased by about 30%.

An introductory text must have a two-fold purpose: to describe different numerical methods technically, as a rule in connection with the derivation of the methods, but also to analyze the properties of the methods, particularly with respect to convergence and error propagation. It is certainly not an easy task to achieve a reasonable balance on this point. On one hand, the book must not degenerate into a collection of recipes; on the other, the description should not be flooded by complete mathematical analyses of every trivial method. Actually, the text is not written in a hard-boiled mathematical style; however, this does not necessarily imply a lack of stringency. For really important and typical methods a careful analysis of errors and of error propagation has been performed. This is true especially for the chapters on linear systems of equations, eigenvalue problems, and ordinary and partial differential equations. Any person who has worked on numerical problems within these sectors is familiar with the feeling of uncertainty as to the correctness of the result which often prevails in such cases, and realistic and reliable error estimations are naturally highly desirable. In many situations, however, there might also be other facts available indicating how trustworthy the results are.

Much effort has been spent in order to present methods which are efficient for use on a computer, and several older methods not meeting this requirement have been left out. As to the mathematical background only knowledge of elementary linear algebra, differential and integral calculus, and differential equations is presupposed. Exercises are provided at the end of each chapter. For those involving numerical computation access to a desk calculator and a suitable table (e.g. Chambers' shorter six-figure mathematical tables) is assumed.

Many suggestions for improvements have been presented, both in reviews and otherwise, and they have also been taken into account as far as possible. This helpful interest is hereby gratefully acknowledged.

Lund, December 1968   C-E. F.

Contents

Chapter 1

Numerical computations

A mathematician knows how to solve a
problem—but he can't do it.

W. E. MILNE.

1.0. Representation of numbers

As is well known, our usual number system is the decimal or decadic system. The number 10, which plays a fundamental role, is called the *base* of the system. The choice of 10 as the base of the system has evidently been made for historic reasons, and as a matter of fact any positive integer $N > 1$ could have been chosen instead. Those numbers N which have been discussed more seriously are displayed in the following table:

Base N	Number system
2	Binary
3	Ternary
8	Octal
10	Decimal
12	Duodecimal
16	Sedecimal*

When the base is given, N different symbols $0, 1, 2, \ldots, (N - 1)$ are necessary to represent an arbitrary number. A nonnegative number a can now be written in the form

$$a = a_m \cdot N^m + a_{m-1} \cdot N^{m-1} + \cdots + a_1 N + a_0 + a_{-1} N^{-1} + \cdots + a_{-n} N^{-n} .$$

Here $0 \leq a_k < N$ and, further, we have assumed that a can be represented by a finite number of digits. In the general case, an arbitrary real number can always be approximated in this manner. The representation is *unique*, for suppose that we have

$$a = a_m N^m + a_{m-1} N^{m-1} + \cdots = b_p N^p + b_{p-1} N^{p-1} + \cdots \qquad (a_m \neq 0 ; \; b_p \neq 0) .$$

Then we find

$$a_{m-1} N^{m-1} + \cdots + a_{-n} N^{-n} \leq (N - 1)(N^{m-1} + \cdots + N^{-n})$$

$$< (N - 1) \cdot \frac{N^{m-1}}{1 - 1/N} = N^m .$$

---

* Previously, the word *hexadecimal* was used in this case. However, since *hex* is Greek and *decimal* is Latin, the word *sedecimal*, with a pure Latin origin, is more satisfactory.

Putting $a - a_m N^m = R_m$, we get $R_m < N^m$. First we show that $m = p$. For suppose, for example, that $p < m$, then we would get $a = b_p N^p + \cdots < N^{p+1} \leq N^m$, or $a < N^m$, which is absurd.

Now we can write $(a_m - b_m)N^m + (a_{m-1} - b_{m-1})N^{m-1} + \cdots = 0$ or, using suitable notations, $\alpha_m N^m = \beta_{m-1} N^{m-1} + \cdots = S_{m-1}$. The signs should be chosen in such a way that $0 \leq \alpha_m < N$ and $-N < \beta_r < N$. From the preceding proof, we have directly $|S_{m-1}| < N^m$ and $\alpha_m = 0$, that is $a_m = b_m$, and so on.

The construction of high-speed electronic computers has greatly promoted the use of the binary number system. The octal and sedecimal systems are almost trivial variants, since 8 and 16 are powers of 2. Both systems have been widely used; in the latter case the numbers $10, 11, 12, 13, 14$, and 15 are usually represented by A, B, C, D, E, and F, respectively.

Very large or very small numbers are often expressed in a so-called *floating* representation using mantissa and exponent packed together. A certain number a can then be written

$$a = p \cdot N^q \qquad \text{(q positive or negative integer or zero)} .$$

This representation, of course, is not unique; we have, for example,

$$p \cdot N^q = (pN^2) \cdot N^{q-2} .$$

In the binary case ($N = 2$), p is often chosen in such a way that $-1 \leq p < -\frac{1}{2}$ or $\frac{1}{2} \leq p < 1$; the corresponding number is then said to be *normalized*.

1.1. Conversion

We will briefly treat the problem of translating a number from one representation to another. Suppose that the number a is given in the M-representation and that we want to find it in the N-representation. Thus we have the equation:

$$a = a_m M^m + a_{m-1} M^{m-1} + \cdots = x_r N^r + x_{r-1} N^{r-1} + \cdots ,$$

where the coefficients a_m, a_{m-1}, ... are known and the coefficients x_r, x_{r-1}, ... should be determined. Note that x_r, x_{r-1}, ... must be expressed as 1-digit symbols in the N-representation, $0 \leq x_s < N$.

We split a into an integer part b and a fractional part c, treating b first. Then we have $b = x_r N^r + x_{r-1} N^{r-1} + \cdots + x_1 N + x_0$, and dividing b by N, we get a quotient Q_1 and a remainder $R_0 = x_0$. Next, dividing Q_1 by N, we get the quotient Q_2 and the remainder $R_1 = x_1$, and it is obvious that in general $x_0, x_1, x_2, \ldots$ are the consecutive remainders when b is repeatedly divided by N. In an analogous way, we find the digits of the fractional part as the consecutive integer parts when c is repeatedly multiplied by N and the integer parts removed. The computations must be performed in the M-representation and N itself must also be given in this representation (in the N-representation $N = 10$).

EXAMPLE

Convert the decimal number 176.524 to ternary form. We find: $176 \div 3 = 58$, remainder 2; $58 \div 3 = 19$, remainder 1; $19 \div 3 = 6$, remainder 1; $6 \div 3 = 2$, remainder 0; last quotient $= 2$. Thus

$$176 = 20112 = 2 \cdot 3^4 + 1 \cdot 3^2 + 1 \cdot 3^1 + 2 \cdot 3^0.$$

Analogously,

$$0.524 = 0.112010222\ldots = 1 \cdot 3^{-1} + 1 \cdot 3^{-2} + 2 \cdot 3^{-3} + 1 \cdot 3^{-5} + \cdots.$$

1.2. On errors

In this section we shall discuss different kinds of errors, their sources, and the nature of their growth. Independently of the nature of the error, one can define an absolute and a relative error. Let x_0 be the exact number and x an approximation. Then the *absolute error* is defined by $\varepsilon = x - x_0$, while $|\varepsilon/x_0| = |x/x_0 - 1|$ is the *relative error*.

A number is rounded to position n by making all digits to the right of this position zero; the digit in position n is left unchanged or increased by one unit according as the truncated part is less or greater than half a unit in position n. If it is exactly equal to half a unit, the digit in position n is increased by one unit if it is odd; otherwise, it is left unchanged. Very often round-off to n decimals is made; in this case the digits beyond position n (which are made equal to zero) are simply left out.

Here we will briefly comment upon the concept of *significant digits*. We can say very roughly that the significant figures in a number are those which carry real information as to the size of the number apart from the exponential portion. It is obvious that a digit in a place farther to the left carries a larger amount of information than a digit to the right. When a number has been rounded to include only significant digits, these form a group which starts with the first nonzero digit and, as a rule, ends with the last nonzero digit. If the fractional part ends with one or several zeros, they are significant by definition. If the number is an integer ending with one or several zeros, it has to be decided from the context whether they are significant or not.

EXAMPLES

8632574 rounded to 4 significant figures ($4s$) is 8633000.
3.1415926 rounded to 5 decimals ($5d$) is 3.14159.
8.5250 rounded to $2d$ is 8.52.
1.6750 rounded to $2d$ is 1.68.

If a number is formed as a result of a physical measurement or of a numerical computation, it ought to be given with so many significant figures that the maximal error does not exceed a few units in the last significant digit. For example, it is rather meaningless to give the distance between two cities as 82.763 miles; this figure should be rounded to 83 miles.

We will now give an account of the round-off error bounds for the elementary operations, following von Neumann and Goldstine [2]. Let us suppose that we are working in a number system with base N, using n digits, and further that only such numbers x are allowed for which $-1 \leq x < 1$. The last condition is satisfied in many high-speed electronic computers. Then addition and subtraction will give no round-off errors (we suppose that the results do not grow out of range). Since the product of two n-digit numbers in general contains $2n$ digits and the quotient an infinite number, both results have to be rounded. We take this fact into account by introducing *pseudo operations* denoted by $\times$ for multiplication and $\div$ for division. Our basic inequalities then take the following form:

$$|a \times b - ab| \leq \tfrac{1}{2}N^{-n} \; ; \qquad |a \div b - a/b| \leq \tfrac{1}{2}N^{-n} \; .$$

It is evident that $a \times b = b \times a$. Hence the commutative law for multiplication is satisfied. The distributive law is no longer satisfied exactly:

$$
\begin{aligned}
|a \times (b &+ c) - a \times b - a \times c| \\
&= |a \times (b + c) - a \cdot (b + c) - (a \times b - ab) - (a \times c - ac)| \\
&\leq |a \times (b + c) - a \cdot (b + c)| + |a \times b - a \cdot b| + |a \times c - a \cdot c| \\
&\leq \tfrac{3}{2} \cdot N^{-n} \; .
\end{aligned}
$$

However, the initial expression contains only quantities rounded to n digits, and so we can strengthen the inequality replacing $\tfrac{3}{2}N^{-n}$ by N^{-n}. The associative law for multiplication is also modified:

$$
\begin{aligned}
|a \times (b \times c) &- (a \times b) \times c| \\
&= |\{a \times (b \times c) - a \cdot (b \times c)\} + \{a \cdot (b \times c) - a \cdot (b \cdot c)\} \\
&\quad - \{(a \times b) \times c - (a \times b) \cdot c\} - \{(a \times b) \cdot c - (a \cdot b) \cdot c\}| \\
&\leq \tfrac{1}{2}N^{-n}(2 + |a| + |c|) \; .
\end{aligned}
$$

The expression in the last parentheses becomes 4 only when $|a| = |c| = 1$, but in this special case the associative law is fulfilled. On the other hand, the difference must be a multiple of N^{-n}, and so we get:

$$|a \times (b \times c) - (a \times b) \times c| \leq N^{-n} \; .$$

Finally we consider

$$
\begin{aligned}
|(a \div b) \times b - a| &= |(a \div b) \times b - (a \div b) \cdot b + (a \div b) \cdot b - (a/b) \cdot b| \\
&\leq \tfrac{1}{2}N^{-n}(1 + |b|) \; .
\end{aligned}
$$

The difference must be a multiple of N^{-n} and is thus zero except when $|b| = 1$. In this case, however, the difference is zero trivially. Thus we have proved that

$$(a \div b) \times b = a \; .$$

On the other hand, we have

$$|(a \times b) \div b - a| \leq \tfrac{1}{2} N^{-n}(1 + |b|^{-1}) \, ,$$

which is a bad result when $|b|$ is small.

EXAMPLES

For $n = 2$ and $N = 10$, we have:

1. $0.56 \times 0.65 = 0.36$, $(0.56 \times 0.65) \times 0.54 = 0.19$,
   $0.65 \times 0.54 = 0.35$, $0.56 \times (0.65 \times 0.54) = 0.20$.

2. $(0.76 \times 0.06) \div 0.06 = 0.05 \div 0.06 = 0.83$.

When a floating representation is used instead, we obtain other results. In this case not even the associative law for addition is valid, as is shown in the following example.

$$(0.243875 \cdot 10^6 + 0.412648 \cdot 10^1) - 0.243826 \cdot 10^6$$
$$= 0.243879 \cdot 10^6 - 0.243826 \cdot 10^6 = 0.000053 \cdot 10^6$$
$$= \underline{0.530000 \cdot 10^2} \, ,$$
$$(0.243875 \cdot 10^6 - 0.243826 \cdot 10^6) + 0.412648 \cdot 10^1$$
$$= 0.000049 \cdot 10^6 + 0.412648 \cdot 10^1 = 0.490000 \cdot 10^2 + 0.412648 \cdot 10^1$$
$$= \underline{0.531265 \cdot 10^2} \, .$$

A detailed examination of the errors involved in floating-point operations has been performed by Wilkinson [3]. With the notations $\mathrm{fl}(x \pm y)$, $\mathrm{fl}(x \times y)$, and $\mathrm{fl}(x \div y)$ for the results of the corresponding floating-point operations on x and y, we have the basic inequalities:

$$\begin{cases} \mathrm{fl}(x + y) = (x + y)(1 + \varepsilon) \, , \\ \mathrm{fl}(x - y) = (x - y)(1 + \varepsilon) \, , \\ \mathrm{fl}(x \times y) = xy(1 + \varepsilon) \, , \\ \mathrm{fl}(x \div y) = (x/y)(1 + \varepsilon) \, , \end{cases}$$

with $|\varepsilon| \leq 2^{-t}$, t denoting the number of binary places in the mantissa. Thus for example, the computed sum of x and y is the exact sum of two modified numbers x' and y' which differ from x and y, respectively, by at most one part in 2^t. Further details in this matter can be obtained from Wilkinson's book.

In many cases one wants to estimate the error in a function $f(x_1, x_2, \ldots, x_n)$ when the individual errors $\Delta x_1, \Delta x_2, \ldots, \Delta x_n$ of the variables are known. We find directly that

$$\Delta f = \frac{\partial f}{\partial x_1} \Delta x_1 + \frac{\partial f}{\partial x_2} \Delta x_2 + \cdots + \frac{\partial f}{\partial x_n} \Delta x_n \, ,$$

where terms of second and higher orders have been neglected. The maximum error is given by

$$(\Delta f)_{\max} \leq \left| \frac{\partial f}{\partial x_1} \right| \cdot |\Delta x_1| + \cdots + \left| \frac{\partial f}{\partial x_n} \right| \cdot |\Delta x_n| .$$

In the special case $f = x_1 + x_2 + \cdots + x_n$, we have

$$(\Delta f)_{\max} \leq |\Delta x_1| + |\Delta x_2| + \cdots + |\Delta x_n| ,$$

while for $f = x_1^{m_1} x_2^{m_2} \cdots x_n^{m_n}$, we have instead

$$\left(\frac{\Delta f}{f} \right)_{\max} \leq |m_1| \cdot \left| \frac{\Delta x_1}{x_1} \right| + |m_2| \cdot \left| \frac{\Delta x_2}{x_2} \right| + \cdots + |m_n| \cdot \left| \frac{\Delta x_n}{x_n} \right| .$$

It is only fair to point out that in general the maximal error bounds are rather pessimistic, and in practical computations, the errors have a tendency to cancel. If, for example, 20,000 numbers, all of them rounded to four decimal places, are added together, the maximum error is $\frac{1}{2} \cdot 10^{-4} \cdot 20,000 = 1$. However, it is obvious that this case is extremely improbable. From a statistical point of view one would expect that in about 99% of all cases the total error will not exceed 0.005.

When one tries to classify the errors in numerical computations it might be fruitful to study the sources of the errors and the growth of the individual errors. The sources of the errors are essentially static, while the growth takes place dynamically. We shall here refrain from treating the "gross" errors in spite of the fact that they often play an important role in numerical computations and certainly do not lack interesting features. Then essentially three error sources remain:

1. Initial errors,

2. Local truncation errors,

3. Local round-off errors.

The *initial errors* are errors in initial data. A simple example is rendered when data are obtained from a physical or chemical apparatus. *Truncation errors* arise when an infinite process (in some sense) is replaced by a finite one. Well-known examples are computation of a definite integral through approximation with a sum, or integration of an ordinary or partial differential equation by some difference method. *Round-off errors* finally depend on the fact that practically each number in a numerical computation must be rounded to a certain number of digits.

Normally, a numerical computation proceeds in many steps. One such step means that from two approximations x' and y' of the exact numbers x and y we form the approximation z' of z by use of one of the four simple rules of arithmetic. Let us suppose that $x' = x + \delta$; $y' = y + \eta$, and $z = x/y$. Instead we compute $z' = (x'/y')_{\text{rounded}} = (x + \delta)/(y + \eta) + \varepsilon$, and hence we have $z' \simeq z + (1/y)\delta - (x/y^2)\eta + \varepsilon$. The error in z' is thus built up of propagated errors from x and y and further a new round-off error. This "compound-interest-effect" is very typical and plays a fundamental role in every error analysis.

In order to get a more detailed knowledge of error propagation we must isolate the different types of errors. Starting with the initial errors it might well be the case that they have a fatal effect on the solution. This means that small changes in initial data may produce large changes in the final results. A problem with this property is said to be *ill-conditioned*. Examples will be given in Chapters 2 and 4.

The truncation errors usually depend on a certain parameter, N say, such that $N \to \infty$ implies that the "approximate" solution approaches the "right" one. For example, in differential equations N often has the form $(b-a)/h$ where h is the interval length. As a rule, the local truncation error is $O(h^r)$ and the total truncation error is $O(h^s)$, where $s \leq r$. The truncation errors can be made arbitrarily small by choosing N sufficiently large or h sufficiently small.

In normal cases the round-off errors are accumulated completely at random, and this has the effect that the errors compensate each other to a large extent. For this reason, error estimates based on maximum errors in many cases are far too pessimistic. However, under special circumstances the round-off errors can grow like a rolling snowball. In particular this may happen when such an error can be understood as a small component of a parasitic solution which one really wants to suppress. This phenomenon is known as *instability* and will be treated in considerable detail in Chapters 14 and 15.

It can be of some interest to illustrate the different types of error a little more explicitly. Suppose that we want to compute $f(x)$ where x is a real number and f is a real function which we so far do not specify any closer. In practical computations the number x must be approximated by a rational number x' since no computer can store numbers with an infinite number of decimals. The difference $x' - x$ constitutes the *initial error* while the difference $\varepsilon_1 = f(x') - f(x)$ is the corresponding *propagated error*. In many cases f is such a function that it must be replaced by a simpler function f_1 (often a truncated power series expansion of f). The difference $\varepsilon_2 = f_1(x') - f(x')$ is then the *truncation error*. The calculations performed by the computer, however, are not exact but pseudo-operations of a type that has just been discussed. The result is that instead of $f_1(x')$ we get another value $f_2(x')$ which is then a wrongly computed value of a wrong function of a wrong argument. The difference $\varepsilon_3 = f_2(x') - f_1(x')$ could be termed the *propagated error from the roundings*. The total error is

$$\varepsilon = f_2(x') - f(x) = \varepsilon_1 + \varepsilon_2 + \varepsilon_3 .$$

We now choose the following specific example. Suppose that we want to determine $e^{1/3}$ and that all calculations are performed with 4 decimals. To start with, we try to compute $e^{0.3333}$ instead of $e^{1/3}$, and the propagated error becomes

$$\varepsilon_1 = e^{0.3333} - e^{1/3} = e^{0.3333}(1 - e^{0.0000333\ldots})$$
$$= -0.00004\,65196 .$$

Next, we do not compute e^x but instead

$$1 + \frac{x}{1!} + \frac{x^2}{2!} + \frac{x^3}{3!} + \frac{x^4}{4!}$$

for $x = 0.3333$. Hence, the truncation error is

$$\varepsilon_2 = -\left(\frac{0.3333^5}{5!} + \frac{0.3333^6}{6!} + \cdots\right) = -0.00003\,62750 .$$

Finally, the summation of the truncated series is done with rounded values giving the result

$$1 + 0.3333 + 0.0555 + 0.0062 + 0.0005 = 1.3955 ,$$

instead of $1.39552\,96304$ obtained with 10 decimals. Thus $\varepsilon_3 = -0.00002\,96304$ and the total error is

$$\varepsilon = 1.3955 - e^{1/3} = \varepsilon_1 + \varepsilon_2 + \varepsilon_3 = -0.00011\,24250 .$$

Investigations of error propagation are, of course, particularly important in connection with iterative processes and computations where each value depends on its predecessors. Examples of such problems are in first-hand linear systems of equations, eigenvalue computations, and ordinary and partial differential equations. In the corresponding chapters we shall return to these problems in more explicit formulations.

In error estimations one can speak about *a-priori* estimations and *a-posteriori* estimations. As can be understood from the name, the first case is concerned with estimations performed *without* knowledge of the results to be computed. In the latter case the obtained results are used in the error analysis. Further the notions *forward* and *backward analysis* should be mentioned. With forward analysis one follows the development of the errors from the initial values to the final result. In backward analysis one starts from a supposed error in the results tracing it backward to see between which limits the initial values must lie to produce such an error. This technique was introduced by Wilkinson, who used it with great success for error analysis of linear systems of equations.

When different numerical methods are compared one usually considers the truncation errors first. Then one investigates how the errors depend on some suitable parameter which in the ideal case tends toward 0 or ∞. Suppose that we consider the error ε as a function of h where it is assumed that $h \to 0$. The error analysis can now be performed on several different ambition levels. One might be content with showing that the method is convergent, i.e., $\varepsilon(h) \to 0$ when $h \to 0$. One might also derive results with respect to the convergence speed, e.g., $|\varepsilon(h)| \leq C\varphi(h)$, where C is a constant whose value, however, is not known. It might also happen that one can prove $\varepsilon(h)/\varphi(h) \to 1$ when $h \to 0$, i.e., an asymptotic formula for the error. Finally, one may also derive actual error estimates of the type $|\varepsilon(h)| \leq \varphi(h)$ for all $h < h_0$. In this case $\varphi(h)$ is an upper limit for the error giving us a possibility to guarantee a certain accuracy. A more detailed discussion of these problems can be found, for example, in [1].

> *Round numbers are always false.*
> SAMUEL JOHNSON.

1.3. Numerical cancellation

In the previous section it has been shown how several fundamental arithmetical laws must be modified in numerical applications. Against this background it is not surprising that expressions which are completely equivalent from a mathematical point of view may turn out to be quite different numerically. We will restrict ourselves to a few examples on this matter.

The second-degree equation $x^2 - 2ax + \varepsilon = 0$ has the two solutions

$$x_1 = a + \sqrt{a^2 - \varepsilon} \quad \text{and} \quad x_2 = a - \sqrt{a^2 - \varepsilon}.$$

If $a > 0$ and ε is small compared with a, the root x_2 is expressed as the difference between two almost equal numbers, and a considerable amount of significance is lost. Instead, if we write

$$x_2 = \frac{\varepsilon}{a + \sqrt{a^2 - \varepsilon}},$$

we obtain the root as approximately $\varepsilon/2a$ without loss of significance.

Next, suppose that for a fairly large value x, we know that $\cosh x = a$; $\sinh x = b$; and that we want to compute e^{-x}. Obviously,

$$e^{-x} = \cosh x - \sinh x = a - b,$$

leading to a dangerous cancellation. On the other hand,

$$e^{-x} = \frac{1}{\cosh x + \sinh x} = \frac{1}{a + b}$$

gives a very accurate result.

Finally, we present an example to show that one has to be careful when using mathematical formulas numerically.

The Bessel functions $J_n(x)$ are solutions of the differential equation (see Section 18.5)

$$\frac{d^2y}{dx^2} + \frac{1}{x}\frac{dy}{dx} + \left(1 - \frac{n^2}{x^2}\right)y = 0,$$

with

$$J_n(x) = \sum_{k=0}^{\infty} \frac{(-1)^k (x/2)^{n+2k}}{k!\,(n+k)!}.$$

It is easy to show that the recursion formula

$$J_{n+1}(x) = \frac{2n}{x} \cdot J_n(x) - J_{n-1}(x)$$

is valid.

We start with the following values, correctly rounded to six decimal places:

$$J_0(1) = 0.765198 \, ,$$
$$J_1(1) = 0.440051 \, .$$

Using the recursion formula, we try to compute $J_n(1)$ for higher values of n and obtain the following results (the correct figures are given in parentheses):

$$J_2(1) = 0.114904 \quad (0.114903) \, ,$$
$$J_3(1) = 0.019565 \quad (0.019563) \, ,$$
$$J_4(1) = 0.002486 \quad (0.002477) \, ,$$
$$J_5(1) = 0.000323 \quad (0.000250) \, ,$$
$$J_6(1) = 0.000744 \quad (0.000021) \, ,$$
$$J_7(1) = 0.008605 \quad (0.000002) \, .$$

It is obvious that this formula cannot be used here in this way.

On the other hand, putting $J_8(1) = 0$, $J_7(1) = k$, and applying the same formula in the other direction, we get:

$$J_6(1) = 14k \, ; \qquad J_5(1) = 167k \, ; \qquad J_4(1) = 1656k \, ; \qquad J_3(1) = 13081k \, ;$$
$$J_2(1) = 76830k \, ; \qquad J_1(1) = 294239k \, ; \qquad J_0(1) = 511648k \, .$$

The constant k can be obtained from the identity

$$J_0(x) + 2J_2(x) + 2J_4(x) + 2J_6(x) + \cdots = 1 \, .$$

We find that $k = 1/668648$, from which we obtain the correct values with an error of at most one digit in the sixth place. The explanation is that the former procedure is unstable but the latter is not. A detailed discussion of this phenomenon will be given in Chapter 14.

1.4. Computation of functions

We are not going to treat this extensive chapter systematically, but rather will point to some general methods which have proved efficient. We will restrict ourselves to real functions of one variable, even if a generalization to complex functions is straightforward in many cases.

First, we point out the importance of making use of computational schemes whenever possible. The arrangement becomes clear, the computational work is facilitated, and finally the computations are easy to check. Suppose that the function $y = \exp(-x + \arctan \sqrt{x^2 + 1})$ has to be tabulated from $x = 0$ to

$x = 1$ in steps of 0.2 [this is conveniently written $x = 0(0.2)1$]. The following outline is then suitable.

x	$z = \sqrt{x^2 + 1}$	$u = \arctan z$	$v = u - x$	$y = e^v$
0	1	0.7854	0.7854	2.1933
0.2	1.0198	0.7952	0.5952	1.8134
0.4	1.0770	0.8225	0.4225	1.5258
0.6	1.1662	0.8620	0.2620	1.2995
0.8	1.2806	0.9078	0.1078	1.1138
1.0	1.4142	0.9553	-0.0447	0.9563

Note that it is advisable to perform the operations *vertically*, so far as this is possible, i.e., all square roots are computed at the same time, next all arctangents are looked up in the table, and so on.

The function in this example was defined by an explicit formula. It has become more and more common to give a function by a recursive process. In the usual mathematical language, such formulas would often be clumsy. The advent of ALGOL, which at the same time is a language for describing computational procedures and an autocode system, has introduced considerable advantages.

We will give a few examples of this recursion technique and start with the polynomial

$$P(x) = x^n + a_1 x^{n-1} + \cdots + a_n$$

and its derivative,

$$P'(x) = nx^{n-1} + (n - 1)a_1 x^{n-2} + \cdots + a_{n-1}.$$

We put

$$\begin{cases} p_0 = 1 \\ p_0' = 0 \end{cases} \quad \text{and} \quad \begin{cases} p_{r+1} = p_r x + a_{r+1} \\ p_{r+1}' = p_r' x + p_r \end{cases} \quad r = 0, 1, \ldots, (n - 1).$$

It is then easy to prove that $p_n = P(x)$ and $p_n' = P'(x)$. These formulas can easily be generalized to the mth derivative:

$$p_{r+1}^{(m)} = p_r^{(m)} x + m p_r^{(m-1)}.$$

Many functions are easily obtained from their power series expansions. In the domain of convergence, the function can be computed with arbitrary accuracy, at least in principle. Also for such expansions a recursive technique is often suitable. As an example, we take the Bessel function of zero order:

$$J_0(x) = \sum_{k=0}^{\infty} \frac{(-1)^k x^{2k}}{2^{2k}(k!)^2}.$$

Putting $u_0 = s_0 = 1$; $u_k = -u_{k-1}x^2/4k^2$; $s_k = s_{k-1} + u_k$, we see that the partial sums s_k tend toward $J_0(x)$ when $k \rightarrow \infty$. The expansion is conveniently truncated when $|u_k|$ is less than some given tolerance; the remainder is then less than the absolute value of the first neglected term.

1.5. Computation of functions using binary representation

In some cases it is possible to use the special properties of a function to compute one binary digit at a time. We will demonstrate some of these procedures and start with the function $y = x^\alpha$, where $0 < \alpha < 1$. Suppose that α is written in binary form. If, for example, $\alpha = 0.101100111\ldots$, then

$$x^\alpha = \sqrt{x} \cdot \sqrt[8]{x} \cdot \sqrt[16]{x} \cdot \sqrt[128]{x} \cdots ,$$

and the problem has been reduced to computing a series of square roots. In order to formulate the general method, we first define the integral part entier (z) of a number z as the largest integer $\leq z$.

Now put $\beta_0 = \alpha$, $x_0 = x$, and $y_0 = 1$, and form $\alpha_k = $ entier $(2\beta_{k-1})$, $\beta_k = 2\beta_{k-1} - \alpha_k$, $x_k = (x_{k-1})^{1/2}$, and $y_k = y_{k-1}(1 + (x_k - 1)\alpha_k)$. It is not hard to infer that $\lim_{k\to\infty} y_k = x^\alpha$. The method is not too fast but can easily be programmed for a computer.

As our second example, we choose $y = \log_2 x$ and suppose that $1 < x < 2$. Putting $y = y_1 \cdot 2^{-1} + y_2 \cdot 2^{-2} + \cdots$, we have $x = 2^{y_1 \cdot 2^{-1} + y_2 \cdot 2^{-2} + \cdots}$ and, after squaring, $x^2 = 2^{y_1 + y_2 \cdot 2^{-1} + \cdots}$. Consequently, we get $y_1 = 1$ if $x^2 \geq 2$, while $y_1 = 0$ if $x^2 < 2$. From this we obtain the following recursion: Starting with $x_0 = x$, we set

$$
\begin{array}{llllll}
y_k = 0 & \text{and} & x_k = x_{k-1}^2 & \text{if} & x_{k-1}^2 < 2 ; \\
y_k = 1 & \text{and} & x_k = \tfrac{1}{2}x_{k-1}^2 & \text{if} & x_{k-1}^2 \geq 2 .
\end{array}
$$

In this way $y_1, y_2, \ldots$ are defined, and finally we obtain

$$\log_2 x = y = y_1 \cdot 2^{-1} + y_2 \cdot 2^{-2} + y_3 \cdot 2^{-3} + \cdots$$

An analogous technique can be defined also for bases other than 2, but the computations will not be so simple any longer.

Last, we also consider the function $y = (2/\pi) \arccos x$ (cf. Fig. 1.5.). We use the notations

$$y = y_0 + y_1 \cdot 2^{-1} + y_2 \cdot 2^{-2} + \cdots$$

and

$$x_k = \cos\left[\frac{\pi}{2}(y_k + y_{k+1} \cdot 2^{-1} + \cdots)\right]$$

with $x_0 = x$. Two different cases can be distinguished: $x_k > 0$ and $x_k \leq 0$. In the first case, we have $0 \leq (2/\pi) \arccos x_k < 1$ and, consequently, $y_k = 0$, which leads to

$$x_k = \cos\left[\frac{\pi}{2}(y_{k+1} \cdot 2^{-1} + y_{k+2} \cdot 2^{-2} + \cdots)\right].$$

Using the formula, $\cos 2z = 2 \cos^2 z - 1$, we obtain

$$x_{k+1} = 2x_k^2 - 1 .$$

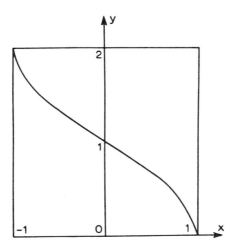

Figure 1.5

In the second case we have $y_k = 1$ and, consequently,

$$y_{k+1} \cdot 2^{-1} + y_{k+2} \cdot 2^{-2} + \cdots = \frac{2}{\pi} \left(\arccos x_k - \frac{\pi}{2} \right).$$

Hence

$$x_{k+1} = \cos \left[\frac{\pi}{2} \left(y_{k+1} + y_{k+2} \cdot 2^{-1} + \cdots \right) \right] = \cos \left(2 \arccos x_k - \pi \right)$$

$$= -\cos \left(2 \arccos x_k \right) = 1 - 2x_k^2 \, .$$

Starting with $x_0 = x$, we have the recursion:

$$\begin{cases} x_{k+1} = 2x_k^2 - 1 & \text{and} \quad y_k = 0 \quad \text{if} \quad x_k > 0 \, , \\ x_{k+1} = 1 - 2x_k^2 & \text{and} \quad y_k = 1 \quad \text{if} \quad x_k \leq 0 \, . \end{cases}$$

The function $\arcsin x$ can then be obtained from

$$\frac{2}{\pi} \arcsin x = 1 - \frac{2}{\pi} \arccos x \, .$$

EXAMPLE

$$x = \tfrac{1}{2}$$
$$x_0 = \tfrac{1}{2} \qquad\qquad\qquad y_0 = 0$$
$$x_1 = -\tfrac{1}{2} \qquad\qquad\qquad y_1 = 1$$
$$x_2 = \tfrac{1}{2} \qquad\qquad\qquad y_2 = 0$$
$$\vdots \qquad\qquad\qquad\qquad \vdots$$

$$y = \frac{1}{2} + \frac{1}{8} + \frac{1}{32} + \cdots = \frac{2}{3} = \frac{2}{\pi} \arccos \frac{1}{2} \, .$$

1.6. Asymptotic series

Important tools for computation of functions are the asymptotic series. They are defined in the following way. Suppose that $s_n(z) = \sum_{k=0}^{n} a_k z^k$ and further that

$$\lim_{z \to 0} \frac{|f(z) - s_n(z)|}{|z|^n} = 0$$

for an arbitrary fixed value of n; then the following notation is used: $f(z) \sim \sum_{n=0}^{\infty} a_n z^n$ when $z \to 0$. In a certain sense, the series expansion represents the function $f(z)$ in such a way that the nth partial sum approximates $f(z)$ better than $|z|^n$ approximates 0. The remarkable fact is that we do not claim that the series is convergent, and as a matter of fact it is in general *divergent*. In this case we say that the series is a *semiconvergent* expansion of the function $f(z)$.

An asymptotic series expansion around the point z_0 is written in the form $f(z) \sim \sum_{n=0}^{\infty} a_n (z - z_0)^n$, with the meaning

$$\lim_{z \to z_0} \frac{|f(z) - s_n(z)|}{|z - z_0|^n} = 0$$

for a fixed value of n when $s_n(z) = \sum_{k=0}^{n} a_k (z - z_0)^k$.

Asymptotic expansions at infinity are of special interest. The following notation is used: $f(z) \sim \sum_{n=0}^{\infty} a_n z^{-n}$ when $z \to \infty$ if

$$\lim_{z \to \infty} |z|^n |f(z) - s_n(z)| = 0$$

for a fixed value of n when $s_n(z) = \sum_{k=0}^{n} a_k z^{-k}$. In many cases $f(z)$ has no asymptotic series expansion, but it might be possible to obtain such an expansion if a suitable function is subtracted or if we divide by an appropriate function. The following notation should present no difficulties:

$$f(z) \sim g(z) + h(z) \cdot \sum_{n=0}^{\infty} a_n z^{-n} .$$

It is obvious that $g(z)$ and $h(z)$ must be independent of n.

EXAMPLE

$$f(x) = \int_x^{\infty} \frac{e^{-t}}{t} \, dt = \frac{e^{-x}}{x} - \int_x^{\infty} \frac{e^{-t}}{t^2} \, dt = \frac{e^{-x}}{x} - \frac{e^{-x}}{x^2} + 2! \int_x^{\infty} \frac{e^{-t}}{t^3} \, dt = \cdots$$

$$= \frac{e^{-x}}{x} \left[1 - \frac{1}{x} + \frac{2!}{x^2} - \frac{3!}{x^3} + \cdots + (-1)^{n+1} \cdot \frac{(n-1)!}{x^{n-1}} \right] + R_n ,$$

with

$$R_n = (-1)^n \cdot n! \cdot \int_x^{\infty} \frac{e^{-t}}{t^{n+1}} \, dt .$$

Thus we have $|R_n| < e^{-x} n! / x^{n+1}$, and by definition,

$$f(x) \sim \frac{e^{-x}}{x} \left[1 - \frac{1}{x} + \frac{2!}{x^2} - \cdots + (-1)^n \frac{n!}{x^n} + \cdots \right].$$

In general, this expression first decreases but diverges when $n \to \infty$, since x has a fixed value.

Thus $f(x)$ is expressed as a series expansion with a remainder term, the absolute value of which first decreases but later on increases to infinity from a certain value of n. Hence the series is divergent, but if it is truncated after n terms where n is chosen so that $|R_n| < \varepsilon$, the truncated series can nevertheless be used, giving a maximum error less than ε.

In our special example we denote the partial sums with S_n, and for $x = 10$, we obtain the following table:

n	$10^7 \cdot e^x \cdot S_n$	$10^7 \cdot e^x \cdot R_n$	n	$10^7 \cdot e^x \cdot S_n$	$10^7 \cdot e^x \cdot R_n$
1	1000000	− 84367	11	915891	− 186
2	900000	+ 15633	12	915420	+ 213
3	920000	− 4367	13	915899	− 266
4	914000	+ 1633	14	915276	+ 357
5	916400	− 767	15	916148	− 515
6	915200	+ 433	16	914840	+ 793
7	915920	− 287	17	916933	− 1300
8	915416	+ 217	18	913376	+ 2257
9	915819	− 186	19	919778	− 4145
10	915456	+ 177	20	907614	+ 8019

The correct value is

$$e^{10} \int_{10}^{\infty} \frac{e^{-t}}{t} \, dt = 0.0915633 .$$

In general, the error is of the same order of magnitude as the first neglected term.

REFERENCES

[1] Henrici: *Discrete Variable Methods in Ordinary Differential Equations* (Wiley, New York, 1962).

[2] von Neumann-Goldstine: *Bull. Am. Math. Soc.*, **53**, (11), 1038.

[3] Wilkinson: *Rounding Errors in Algebraic Processes* (Her Majesty's Stationery Office, London, 1963).

[4] Rainville: *Special Functions* (MacMillan, New York, 1960).

[5] de Bruijn: *Asymptotic Methods in Analysis* (Amsterdam, 1958).

EXERCISES

1. Convert the sedecimal number *ABCDEF* to decimal form.

2. Convert the decimal fraction 0.31416 to octal form.

3. Determine the maximum relative error when p_1 is calculated from the relation $p_1v_1^\kappa = p_2v_2^\kappa$ ($\kappa = 1.4$). The maximum relative errors of v_1, v_2, and p_2 are 0.7, 0.75, and 1%, respectively.

4. One wants to determine $1/(2 + \sqrt{3})^4$ having access to an approximate value of $\sqrt{3}$. Compare the relative errors on direct computation and on using the equivalent expression $97 - 56\sqrt{3}$.

5. Obtain an asymptotic series expansion for the function $y = e^{-x^2} \cdot \int_0^x e^{t^2}\,dt$ by first proving that y satisfies the differential equation $y' = 1 - 2xy$.

6. Find the coefficients in the asymptotic formula

$$\int_x^\infty \left(\frac{\sin t}{t}\right) dt = \cos x \left(\frac{a_1}{x} + \frac{a_2}{x^2} + \cdots\right) + \sin x \left(\frac{b_1}{x} + \frac{b_2}{x^2} + \cdots\right).$$

Chapter 2

Equations

When it is known that x is the same as 6 (which by the way is understood from the pronunciation) all algebraic equations with 1 or 19 unknowns are easily solved by inserting, x, substituting 6, elimination of 6 by x, and so on.

FALSTAFF, FAKIR.*

2.0. Introduction

Solutions of equations and systems of equations represent important tasks in numerical analysis. For this reason it seems appropriate to point out some difficulties arising in connection with this problem. If no special assumptions are given for the function $f(x)$, we shall in this chapter suppose that $f(x)$ is continuously differentiable of sufficiently high order to make the operations performed legitimate.

We will first consider the conception of a root of an equation $f(x) = 0$. By definition, a is a root if $f(a) = 0$. However, in numerical applications it must be understood that the equation usually cannot be satisfied *exactly*, due to round-off errors and limited capacity. Therefore, we want to modify the mathematical definition of a root, and to start with, we could think of the condition $|f(a)| < \varepsilon$, where ε is a given tolerance. The inequality defines an interval instead of a point, but this can hardly be avoided. Another consequence is that the equations $f(x) = 0$ and $M \cdot f(x) = 0$, where M is a constant, do not any longer have the same roots.

In the case of simple roots, the following procedure might be conceivable. In some suitable way, we construct two sequences of numbers $s_1, s_2, s_3, \ldots$ and $t_1, t_2, t_3, \ldots$, where $s_1 < s_2 < s_3 < \cdots$ and $t_1 > t_2 > t_3 > \cdots$. The numbers in both sequences are supposed to be successive approximations to an exact root of the equation. Further, we assume that $s_i < t_k$, $f(s_i)f(s_k) > 0$, $f(t_i)f(t_k) > 0$, $f(s_i)f(t_k) < 0$ for $i, k = 1, 2, 3, \ldots$ If now $t_n/s_n - 1 < \varepsilon$, where ε is a given tolerance, then $\frac{1}{2}(s_n + t_n)$ is defined as a root. Only the case of $f(0) = 0$ has to be treated separately.

---

* Famous Swedish humorist-author (1865–1896), influenced, for one, by Mark Twain. His real name was Axel Wallengren, and the quotation above comes from the book, *Everyone His Own Professor.*

17

Another difficulty occurs when one or several roots of an algebraic equation are extremely sensitive to changes in the coefficients. Consider the polynomial

$$f(z) = z^n + a_1 z^{n-1} + a_2 z^{n-2} + \cdots + a_n \, ,$$

and let r be a root of the equation $f(z) = 0$. Differentiating we get:

$$\left(\frac{\partial f}{\partial a_k} \right)_{z=r} = f'(r) \frac{\partial r}{\partial a_k} + r^{n-k} = 0 \, . \tag{2.0.1}$$

Hence $\partial r / \partial a_k = -r^{n-k}/f'(r)$, and this relation is written in the following form:

$$\frac{\partial r}{r} = - \frac{a_k \cdot r^{n-k-1}}{f'(r)} \cdot \frac{\partial a_k}{a_k} \, . \tag{2.0.2}$$

Now put $A_k = |a_k \cdot r^{n-k-1}/f'(r)|$, and we can summarize as follows. Large values of A_k have the effect that small changes in the coefficient a_k cause large changes in the root r. Large values of A_k occur when r is large and $f'(r)$ is small; the latter is the case, for example, when some of the roots lie close together.

A well-known example has been given by Wilkinson [1]:

$$(x + 1)(x + 2) \cdots (x + 20) = 0 \qquad \text{or} \qquad x^{20} + 210 x^{19} + \cdots + 20! = 0 \, .$$

We choose $k = 1$ and $\delta a_1 = 2^{-23}$, which means that the coefficient 210 is changed to 210.0000001192. Then the roots $-1, -2, \ldots, -8$ are shifted only slightly; among the remaining roots we find, for example, $-14 \pm 2.5i$, $-16.73 \pm 2.81i$, and -20.85. For $r = -16$ we obtain

$$A_1 = \frac{210 \cdot 16^{18}}{15! \, 4!} = 3.2 \cdot 10^{10} \, .$$

This result indicates that we must have 10 guard digits, apart from those corresponding to the wanted accuracy. It should also be noted that the value of A_k cannot be used for computation of the exact root, since in (2.0.2) we have given only the first order terms; here higher terms are of decisive importance.

Equations where small changes in the coefficients cause large changes in the roots are said to be *ill-conditioned*. If one wants to determine the roots of such an equation, one has to work with a suitable number of guard digits. It may be difficult to tell offhand whether a given equation is ill-conditioned or not. If the computations are performed on an automatic computer, the following process could be used.

A number is replaced by an *interval* in such a way that the limits of the interval represent the upper and lower limits of the number in question. The limits are always rational numbers chosen to fit the computer used. Now we get more complicated calculation rules which will always give new intervals as results. For example, if $a_1 < x_1 < b_1$, and $a_2 < x_2 < b_2$, then $a_1 + a_2 < x_1 + x_2 < b_1 + b_2$. By using such "range operations," we can keep numerical errors under control all the time. If the final result is represented by a large interval, the problem is ill-conditioned. As mentioned before, however, estimates of maximal errors are often far too pessimistic.

2.1. Cubic and quartic equations. Horner's scheme

As early as in the sixteenth century, it was known that algebraic equations of first, second, third, and fourth degree could be solved by square and cube roots. In 1824 Abel proved that the roots of algebraic equations of fifth and higher degrees could not in general be represented as algebraic expressions containing the coefficients of the equation, where only addition, subtraction, multiplication, division, and root extraction were allowed. We will now examine cubic and quartic equations a little more closely.

Equations of third and fourth degree can, of course, be solved numerically by some of the methods which will be explained later. However, it is not too easy to design a general technique in such a way that no root possibly can escape detection, but the following procedures seem to be satisfactory from this point of view.

First, consider the equation $x^3 + ax^2 + bx + c = 0$. If $c > 0$, there is a root between $-\infty$ and 0, and if $c < 0$, there is one between 0 and ∞. There exist general theorems regarding the localization of the roots of a given equation. Suppose that $f(z) = z^n + a_1 z^{n-1} + a_2 z^{n-2} + \cdots + a_n$. Putting $\lambda = \max |a_i|$, it can be proved that for any root z of the equation $f(z) = 0$, we have $|z| \leq 1 + \lambda$. In most cases this is a very crude estimate, but here we have at least one real root localized in a finite interval. By repeated interval halvings, the lower limit is moved upward or the upper limit downward. The upper and lower limits can be identified with our previous sequences s_k and t_k. When the root has been obtained with sufficient accuracy, it is removed, and we are left with a second-degree equation.

Next we consider an equation of the fourth degree, $x^4 + ax^3 + bx^2 + cx + d = 0$. The cubic term is removed through the transformation $x = y - a/4$, which gives the equation $y^4 + qy^2 + ry + s = 0$, where $q = b - 3a^2/8$, $r = c - ab/2 + a^3/8$, $s = d - ac/4 + a^2 b/16 - 3a^4/256$. Since the coefficients are real, the equation may have 4, 2, or no real roots. In any case there exists a partition into two quadratic factors:

$$(y^2 + 2ky + l)(y^2 - 2ky + m) \equiv y^4 + qy^2 + ry + s$$

with real k, l, and m. Comparing the coefficients we get:

$$\begin{cases} l + m - 4k^2 = q, \\ 2k(m - l) = r, \\ lm = s. \end{cases}$$

Eliminating l and m and putting $k^2 = z$, we obtain $z^3 + \alpha z^2 + \beta z + \gamma = 0$, with $\alpha = q/2$, $\beta = (q^2 - 4s)/16$, and $\gamma = -r^2/64$. Since $\gamma \leq 0$, there is always at least one positive root z, and hence k, l, and m can be determined one after another. In this way the problem is reduced to the solution of two second-degree equations. The method just described is originally due to Descartes.

We touched upon the problem of determining upper and lower bounds for the roots of an algebraic equation. It is interesting to note that this problem is best treated with matrix-theoretical methods, and we will return to it in Section 3.3.

We will now give a brief account of how roots can be removed from algebraic equations by means of so-called synthetic division (Horner's scheme). Suppose that the equation $f(z) = z^n + a_1 z^{n-1} + \cdots + a_n = 0$ has a root $z = \alpha$. Then there exists a polynomial $z^{n-1} + b_1 z^{n-2} + \cdots + b_{n-1} = 0$ such that

$$(z - \alpha)(z^{n-1} + b_1 z^{n-2} + \cdots + b_{r-1} z^{n-r} + b_r z^{n-r-1} + \cdots + b_{n-1})$$
$$\equiv z^n + a_1 z^{n-1} + \cdots + a_r z^{n-r} + \cdots + a_n .$$

Comparing the coefficients of the z^{n-r}-term, we find that $b_r - \alpha b_{r-1} = a_r$, and the coefficients b_r can be computed recursively from $b_r = a_r + \alpha b_{r-1}$, with $b_0 = 1$. Without difficulty we obtain $b_1 = a_1 + \alpha$; $b_2 = a_2 + a_1\alpha + \alpha^2$; $\ldots b_n = a_n + a_{n-1}\alpha + \cdots + \alpha^n = f(\alpha) = 0$. Conveniently, the following scheme is used:

$$
\begin{array}{c c c c c}
1 & a_1 & a_2 & \cdots\, a_n & \alpha \\
& \alpha & a_1\alpha + \alpha^2 & \\
\hline
1 & a_1 + \alpha & a_2 + a_1\alpha + \alpha^2 & \cdots\, b_n &
\end{array}
$$

Every term in the second row is obtained by multiplication with α of the preceding term in the third row, and every term in the third row is obtained by addition of the terms above. If α is not a root, we get $b_n = f(\alpha)$, and hence the scheme can also be used for computation of the numerical value of a polynomial for a given value of the variable.

EXAMPLE

The equation $x^5 - 3x^4 + 4x^3 + 2x^2 - 10x - 4 = 0$ has one root $x = 2$ and another $x = -1$, and both should be removed. We obtain the following scheme.

$$
\begin{array}{r r r r r r | r}
1 & -3 & 4 & 2 & -10 & -4 & 2 \\
& 2 & -2 & 4 & 12 & 4 & \\
\hline
1 & -1 & 2 & 6 & 2 & 0 & -1 \\
& -1 & 2 & -4 & -2 & & \\
\hline
1 & -2 & 4 & 2 & 0 & &
\end{array}
$$

After the first division, we are left with $x^4 - x^3 + 2x^2 + 6x + 2$, and after the second, with $x^3 - 2x^2 + 4x + 2$.

Obviously, there are no difficulties in generalizing this technique to division with polynomials of higher degree. Division by $z^2 + \alpha z + \beta$ leads to the following scheme.

$$
\begin{array}{c|cccccc}
1 & a_1 & a_2 & a_3 & \cdots & a_n & \\
& -\alpha & -c_1\alpha & -c_2\alpha\cdots & & & -\alpha \\
& & -\beta & -c_1\beta\cdots & & & -\beta \\
\hline
1 & c_1 & c_2 & c_3 & \cdots & &
\end{array}
$$

2.2. Newton-Raphson's method

We are now going to establish general methods for computing a root of the equation $f(x) = 0$, where $f(x)$ can be an algebraic or transcendental function. We intend to start with a trial value and then construct better and better approximations.

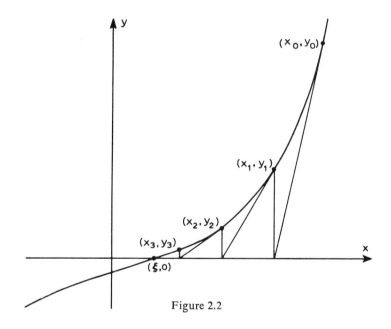

Figure 2.2

If we represent the function $y = f(x)$ graphically, the problem can be so formulated that we are looking for the intersection between the curve and the x-axis. The basic idea is now to replace the curve by a suitable straight line whose intersection with the x-axis can easily be computed. The line passes through the last approximation point (x_n, y_n), but its direction can be chosen in many different ways. One can prescribe either a fixed direction, e.g., parallel to the secant through the first points, (x_0, y_0) and (x_1, y_1), or to the tangent through the first point, or such a variable direction that the line simply coincides with the secant through (x_{n-1}, y_{n-1}) and (x_n, y_n), or with the tangent through (x_n, y_n). If the slope of the line is denoted by k, we have the following four possibilities:

1. $k = (y_1 - y_0)/(x_1 - x_0)$ (fixed secant).
2. $k = f'(x_0)$ (fixed tangent).
3. $k = (y_n - y_{n-1})/(x_n - x_{n-1})$ (variable secant).
4. $k = f'(x_n)$ (variable tangent).

In all cases we obtain an iteration formula of the following form:

$$x_{n+1} = x_n - f(x_n)/k .$$

We shall now examine the convergence of the different methods. Let ξ be the exact value of the simple root we are looking for: $f(\xi) = 0, f'(\xi) \neq 0$. Further we set $x_n = \xi + \varepsilon_n$, and for the first two cases we get

$$y - y_n = k(x - x_n) ,$$

and, putting $y = 0$,

$$x = x_{n+1} = x_n - y_n/k .$$

Hence

$$\xi + \varepsilon_{n+1} = \xi + \varepsilon_n - f(\xi + \varepsilon_n)/k = \xi + \varepsilon_n - [f(\xi) + \varepsilon_n f'(\xi) + \cdots]/k$$

and

$$\varepsilon_{n+1} = [1 - f'(\xi)/k]\varepsilon_n + \cdots .$$

We see that ε_{n+1} will be considerably less than ε_n, if k does not deviate too much from $f'(\xi)$. In this case the convergence is said to be geometric.

In case 3 we find

$$x_{n+1} = x_n - y_n \cdot \frac{x_n - x_{n-1}}{y_n - y_{n-1}} = \frac{x_{n-1} y_n - x_n y_{n-1}}{y_n - y_{n-1}}$$

and

$$\varepsilon_{n+1} = \frac{\varepsilon_{n-1} f(\xi + \varepsilon_n) - \varepsilon_n f(\xi + \varepsilon_{n-1})}{f(\xi + \varepsilon_n) - f(\xi + \varepsilon_{n-1})} = \varepsilon_{n-1} \varepsilon_n \cdot \frac{f''(\xi)}{2f'(\xi)} + \cdots$$

If the remaining terms are neglected, we get $\varepsilon_{n+1} = A\varepsilon_n \varepsilon_{n-1}$. We will now try to determine a number m such that $\varepsilon_{n+1} = K \cdot \varepsilon_n^m$. Then also $\varepsilon_n = K\varepsilon_{n-1}^m$, that is, $\varepsilon_{n-1} = K^{-1/m} \varepsilon_n^{1/m}$, and hence

$$\varepsilon_{n+1} = A \cdot \varepsilon_n \cdot K^{-1/m} \cdot \varepsilon_n^{1/m} = K \cdot \varepsilon_n^m .$$

From this we get

$$1 + \frac{1}{m} = m \quad \text{and} \quad m = \frac{1 \pm \sqrt{5}}{2} ,$$

where the plus sign should be chosen. Thus we have found

$$\varepsilon_{n+1} = K \cdot \varepsilon_n^{1.618} \, . \qquad (2.2.1)$$

The method with variable secant which has just been examined is known as *Regula falsi*.

In case 4 we find:

$$\xi + \varepsilon_{n+1} = \xi + \varepsilon_n - \frac{f(\xi + \varepsilon_n)}{f'(\xi + \varepsilon_n)} \, ,$$

or

$$\varepsilon_{n+1} = \frac{\varepsilon_n f'(\xi + \varepsilon_n) - f(\xi + \varepsilon_n)}{f'(\xi + \varepsilon_n)}$$

$$= \frac{\varepsilon_n f'(\xi) + \varepsilon_n^2 f''(\xi) + \cdots - f(\xi) - \varepsilon_n f'(\xi) - (\varepsilon_n^2/2) f''(\xi) - \cdots}{f'(\xi + \varepsilon_n)} \, .$$

Hence the leading term is

$$\varepsilon_{n+1} = \frac{f''(\xi)}{2f'(\xi)} \, \varepsilon_n^2 \, , \qquad (2.2.2)$$

and the convergence is said to be quadratic. This fact also means that the number of correct decimals is approximately doubled at every iteration, at least if the factor $f''(\xi)/2f'(\xi)$ is not too large.* The process which has just been described is known as the Newton-Raphson method. The iteration formula has the form

$$x_{n+1} = x_n - f(x_n)/f'(x_n) \, . \qquad (2.2.3)$$

As has already been mentioned, the geometrical interpretation defines x_{n+1} as the intersection between the tangent in (x_n, y_n) and the x-axis.

Newton-Raphson's formula can also be obtained analytically from the condition

$$f(x_n + h) = f(x_n) + h \cdot f'(x_n) + \frac{h^2}{2!} f''(x_n) + \cdots = 0 \, ,$$

where x_n is an approximation of the root. If all terms of second and higher degree in h are neglected, we get $h = -f(x_n)/f'(x_n)$ and $x_{n+1} = x_n - f(x_n)/f'(x_n)$ exactly as in (2.2.3).

The method can be used for both algebraic and transcendental equations, and it also works when coefficients or roots are complex. It should be noted, however, that in the case of an algebraic equation with real coefficients, a complex root cannot be reached with a real starting value.

The derivation presented here assumes that the desired root is simple. If the root is of multiplicity $p > 1$ the convergence speed is given by $\varepsilon_{n+1} \simeq [(p-1)/p]\varepsilon_n$

---

* See also Exercise 19.

(linear convergence, cf. Section 2.5). The modified formula

$$x_{n+1} = x_n - pf(x_n)/f'(x_n)$$

restores quadratic convergence. If the multiplicity is not known we can instead search for a zero of the function $f(x)/f'(x)$. Assuming that $f'(x)$ is finite everywhere this function has the same zeros as $f(x)$ with the only difference that all zeros are simple. This leads to the formula

$$x_{n+1} = x_n - f_n f_n'/(f_n'^2 - f_n f_n''),$$

where again we have quadratic convergence.

If $f(x)$ is twice continuously differentiable and $f(a)f(b) < 0$, $a < b$, it is easy to show that Newton-Raphson's method converges from an arbitrary starting value in the interval $[a, b]$ provided that $f'(x) \neq 0$ and $f''(x)$ has the same sign everywhere in the interval, and further

$$\max\{|f(a)/f'(a)|, |f(b)/f'(b)|\} \leq b - a$$

(global convergence). The geometrical interpretation of the conditions are left to the reader.

Finally, we give an error estimate which can be used for any method. We suppose that $x_n = \xi + \varepsilon_n$ is an approximation to the root ξ of the equation $f(x) = 0$, where $f(x)$ is analytic. We also suppose that x_n is so close to ξ that $\varphi(x) = f(x)/f'(x)$ varies monotonically in the interval between ξ and x_n, and further that $f'(x) \neq 0$ in this same interval. Since $\varphi(\xi) = 0$, by using the mean value theorem we get

$$\varphi(x_n) = \varphi(x_n) - \varphi(x_n - \varepsilon_n) = \varepsilon_n \varphi'(x_n - \theta \varepsilon_n),$$

where $0 < \theta < 1$. Hence $\varepsilon_n = \varphi(x_n)/\varphi'(x_n - \theta \varepsilon_n)$ and

$$\varepsilon_n = \frac{(f/f')_{x_n}}{1 - [(f/f') \cdot (f''/f')]_{x_n - \theta \varepsilon_n}}.$$

Putting $K = \sup |f''/f'|$ and $h = f(x_n)/f'(x_n)$, we find

$$|\varepsilon_n| < \frac{|h|}{1 - K|h|};\qquad\qquad(2.2.4)$$

where, of course, we have supposed that $|h| < 1/K$.

For the error estimate (2.2.4), we must know x_n, $f(x_n)$, and $f'(x_n)$. Hence, if we use Newton-Raphson's method, we could as well compute x_{n+1}, and we would then be interested in an estimate of the error ε_{n+1} expressed in known quantities. Let ξ be the exact root, as before. Expanding in Taylor series and using Lagrange's remainder term, we get

$$f(\xi) = f(x_n - \varepsilon_n) = f(x_n) - \varepsilon_n f'(x_n) + \tfrac{1}{2}\varepsilon_n^2 f''(x_n - \theta \varepsilon_n) = 0$$

with $0 < \theta < 1$. Putting

$$h = f(x_n)/f'(x_n) \qquad \text{and} \qquad Q = f''(x_n - \theta \varepsilon_n)/f'(x_n),$$

we find
$$2h - 2\varepsilon_n + Q\varepsilon_n^2 = 0 \qquad \text{or} \qquad \varepsilon_n = \frac{2h}{1 + \sqrt{1 - 2Qh}}.$$

(The other root is discarded since we are looking for a value of ε_n close to h.)
We put $K = \sup |Q|$ and suppose that we are close enough to the root ξ to be
sure that $K|h| < s < \frac{1}{2}$. But $\varepsilon_{n+1} = \varepsilon_n - h$ and hence

$$|\varepsilon_{n+1}| = |h| \cdot \left| \frac{2}{1 + \sqrt{1 - 2Qh}} - 1 \right| < |h| \left(\frac{1}{1 - \alpha K|h|} - 1 \right) = \frac{\alpha K h^2}{1 - \alpha K|h|}.$$

where $\alpha \geq (1 - \sqrt{1 - 2s})/2s$. Naturally, we must assume that $|h| < (2K)^{-1}$.
For $s = \frac{1}{2}$ we get

$$|\varepsilon_{n+1}| < \frac{Kh^2}{1 - K|h|}; \qquad\qquad (2.2.5)$$

however, in normal cases s is much smaller, and when $s \to 0$ we have $\alpha \to \frac{1}{2}$.
If we neglect the term $\alpha K|h|$ in the denominator, we are essentially back at
(2.2.2).

EXAMPLES

1. $x^4 - x = 10$.
By Newton-Raphson's formula, setting $f(x) = x^4 - x - 10$, we get

$$x_{n+1} = x_n - \frac{x_n^4 - x_n - 10}{4x_n^3 - 1} = \frac{3x_n^4 + 10}{4x_n^3 - 1},$$

$$x_0 = 2,$$
$$x_1 = 1.871,$$
$$x_2 = 1.85578,$$
$$x_3 = 1.855585,$$
$$x_4 = 1.85558452522,$$
$$\vdots$$

2. $f(z) = z^5 + (7 - 2i)z^4 + (20 - 12i)z^3 + (20 - 28i)z^2 + (19 - 12i)z + (13 - 26i)$
$= 0$.
Choosing $z_0 = 3i$, we find

$$z_1 = -0.293133 + 2.505945i,$$
$$z_2 = -0.548506 + 2.131282i,$$
$$z_3 = -0.819358 + 1.902395i,$$
$$z_4 = -1.038654 + 1.965626i,$$
$$z_5 = -0.997344 + 1.999548i,$$
$$z_6 = -1.000002 + 1.999993i,$$
$$z_7 = -1.000000 + 2.000000i.$$

3. $e^{-x} = \sin x$.

$$x_{n+1} = x_n + \frac{e^{-x_n} - \sin x_n}{e^{-x_n} + \cos x_n} \, .$$

$$x_0 = 0.6 \, ,$$
$$x_1 = 0.5885 \, ,$$
$$x_2 = 0.58853274 \, ,$$
$$\vdots$$

This converges toward the smallest root; the equation has an infinite number of roots lying close to π, 2π, 3π, ...

Newton-Raphson's formula is the first approximation of a more general expression, which will now be derived. Let ξ be the exact root and x_0 the starting value with $x_0 = \xi + h$. Further we suppose that $f'(x_0) \neq 0$ and put $f/f' = \alpha$, $f''/f' = a_2$, $f'''/f' = a_3$, $f^{IV}/f' = a_4$, ... Hence

$$f(\xi) = f(x_0 - h) = f - hf' + \frac{h^2}{2}f'' - \cdots = 0$$

or after division by $f'(x_0)$,

$$\alpha - h + a_2 \frac{h^2}{2} - a_3 \frac{h^3}{6} + a_4 \frac{h^4}{24} - \cdots = 0 \, ,$$

Now we write h as a power-series expansion in α,

$$h = \alpha + c_2\alpha^2 + c_3\alpha^3 + c_4\alpha^4 + \cdots \, ,$$

and inserting this value in the preceding equation, we have

$$c_2 = a_2/2 \, ; \qquad c_3 = \tfrac{1}{6}(3a_2^2 - a_3) \, ; \qquad c_4 = \tfrac{1}{24}(15a_2^3 - 10a_2a_3 + a_4) \, ; \cdots$$

Thus we find

$$\xi = x_0 - [\alpha + \tfrac{1}{2}a_2\alpha^2 + \tfrac{1}{6}(3a_2^2 - a_3)\alpha^3 + \tfrac{1}{24}(15a_2^3 - 10a_2a_3 + a_4)\alpha^4 + \cdots] \, .$$

This formula is particularly useful if one is looking for zeros of a function defined by a differential equation (usually of second order).

EXAMPLES

Find the least positive zero of the Bessel function $J_0(x)$ satisfying the differential equation $xy'' + y' + xy = 0$. The zero is close to $x = 2.4$, where we have

$$y(2.4) = 0.00250\ 76833 \, , \qquad y'(2.4) = -0.52018\ 52682 \, .$$

From the differential equation, we easily get

$$y''(2.4) = 0.21423\ 61785 \, ; \qquad y'''(2.4) = 0.34061\ 02514 \, ;$$
$$^{VI}y(2.4) = -0.20651\ 12692$$

and further

$$a_2 = -0.4118459\,, \qquad a_3 = -0.6547864\,, \qquad a_4 = 0.3969956\,;$$
$$\alpha = -0.00482\ 07503\,, \qquad \alpha^2 = 0.00002\ 32396\,, \qquad \alpha^3 = -0.00000\ 01120\,,$$
$$\alpha^4 = 0.00000\ 00005\,.$$

Hence

$$
\begin{aligned}
h = {}&-0.00482\ 07503 \\
&-\quad 47856 \\
&-\quad\ 217 \\
&-\quad\ \ \ 1 \\
\hline
&-0.00482\ 55577
\end{aligned}
$$

and $\xi = x_0 - h = 2.40482\ 55577$.

The Newton-Raphson technique is widely used on automatic computers for calculation of various simple functions (inverse, square root, cube root). Older computers sometimes did not have built-in division, and this operation had to be programmed. The quantity a^{-1} can be interpreted as a root of the equation $1/x - a = 0$. From this we obtain the simple recursion formula

$$x_{n+1} = x_n + \frac{1/x_n - a}{1/x_n^2}$$

or

$$x_{n+1} = x_n(2 - ax_n)\,. \tag{2.2.6}$$

This relation can also be written $1 - ax_{n+1} = (1 - ax_n)^2$, showing the quadratic convergence clearly. It is easy to construct formulas which converge still faster, for example, $1 - ax_{n+1} = (1 - ax_n)^3$, or $x_{n+1} = x_n(3 - ax_n + a^2x_n^2)$, but the improved convergence has to be bought at the price of a more complicated formula, and there is no real advantage.

If one wants to compute $\sqrt{a}$, one starts from the equation $x^2 - a = 0$ to obtain

$$x_{n+1} = \frac{1}{2}\left(x_n + \frac{a}{x_n}\right)\,. \tag{2.2.7}$$

This formula is used almost universally for automatic computation of square roots. A corresponding formula can easily be deduced for Nth roots. From $y = x^N - a$, we find

$$x_{n+1} = x_n - (x_n^N - a)/Nx_n^{N-1} = [(N-1)x_n^N + a]/Nx_n^{N-1}\,. \tag{2.2.8}$$

Especially for $\sqrt[3]{a}$ we have

$$x_{n+1} = \frac{1}{3}\left(2x_n + \frac{a}{x_n^2}\right) \tag{2.2.9}$$

and for $1/\sqrt{a}$

$$x_{n+1} = \tfrac{1}{2}x_n(3 - ax_n^2)\,. \tag{2.2.10}$$

The last formula is remarkable because it does not make use of division (apart from the factor $\frac{1}{2}$); $\sqrt{a}$ can then be obtained through multiplication by a.

An alternative method used to compute square roots is the following: Putting $r = 1 - x_0^2/a$ and $s = a/x_0^2 - 1$ when x_0 is an approximation of $\sqrt{a}$, we get

$$\sqrt{a} = x_0(1 - r)^{-1/2} = x_0\left(1 + \frac{r}{2} + \frac{3r^2}{8} + \frac{5r^3}{16} + \frac{35r^4}{128} + \frac{63r^5}{256} + \cdots\right)$$

and

$$\sqrt{a} = x_0(1 + s)^{1/2} = x_0\left(1 + \frac{s}{2} - \frac{s^2}{8} + \frac{s^3}{16} - \frac{5s^4}{128} + \frac{7s^5}{256} - \cdots\right).$$

Analogously for the cube root with $r = 1 - x_0^3/a$ and $s = a/x_0^3 - 1$, where x_0 is an approximation of $\sqrt[3]{a}$:

$$\sqrt[3]{a} = x_0\left(1 + \frac{r}{3} + \frac{2r^2}{9} + \frac{14r^3}{81} + \frac{35r^4}{243} + \frac{91r^5}{729} + \cdots\right);$$

$$\sqrt[3]{a} = x_0\left(1 + \frac{s}{3} - \frac{s^2}{9} + \frac{5s^3}{81} - \frac{10s^4}{243} + \frac{22s^5}{729} - \cdots\right).$$

2.3. Bairstow's method

As mentioned before, Newton-Raphson's method can also be used for complex roots. However, if we are working with algebraic equations with real coefficients, all complex roots appear in pairs $a \pm ib$. Each such pair corresponds to a quadratic factor $x^2 + px + q$ with real coefficients. Let the given polynomial be $f(x) = x^n + a_1x^{n-1} + \cdots + a_n$. If we divide by $x^2 + px + q$, we obtain a quotient $x^{n-2} + b_1x^{n-3} + \cdots + b_{n-2}$ and a remainder $Rx + S$. Our problem is then to find p and q, so that

$$R(p, q) = 0; \qquad S(p, q) = 0.$$

For arbitrary values p and q, these relations are not satisfied in general, and we try to find corrections Δp and Δq, so that

$$R(p + \Delta p, q + \Delta q) = 0; \qquad S(p + \Delta p, q + \Delta q) = 0.$$

Expanding in Taylor series and truncating after the first-order terms, we get:

$$\begin{cases} R(p, q) + \dfrac{\partial R}{\partial p}\,\Delta p + \dfrac{\partial R}{\partial q}\,\Delta q = 0, \\[2mm] S(p, q) + \dfrac{\partial S}{\partial p}\,\Delta p + \dfrac{\partial S}{\partial q}\,\Delta q = 0. \end{cases} \qquad (2.3.1)$$

We regard this as a linear system of equations for Δp and Δq, and when the system has been solved, the procedure is repeated with the corrected values for p and q.

In order to compute the coefficients $b_1, b_2, \ldots, b_{n-2}, R$, and S, we use the identity

$$x^n + a_1 x^{n-1} + \cdots + a_n \equiv (x^2 + px + q)(x^{n-2} + b_1 x^{n-3} + \cdots + b_{n-2}) + Rx + S,$$

from which we obtain:

$$
\begin{cases}
a_1 = b_1 + p, \\
a_2 = b_2 + pb_1 + q, \\
a_3 = b_3 + pb_2 + qb_1, \\
\quad \vdots \\
a_k = b_k + pb_{k-1} + qb_{k-2}, \\
\quad \vdots \\
a_{n-2} = b_{n-2} + pb_{n-3} + qb_{n-4}, \\
a_{n-1} = R + pb_{n-2} + qb_{n-3}, \\
a_n = S + qb_{n-2}.
\end{cases}
\tag{2.3.2}
$$

The quantities $b_1, b_2, \ldots, b_{n-2}$, R, and S can be found recursively. Conveniently, we introduce two other quantities b_{n-1} and b_n, and define:

$$b_k = a_k - pb_{k-1} - qb_{k-2} \qquad (k = 1, 2, \ldots, n), \tag{2.3.3}$$

with $b_0 = 1$ and $b_{-1} = 0$. Then we get $b_{n-1} = a_{n-1} - pb_{n-2} - qb_{n-3} = R$ and $b_n = a_n - pb_{n-1} - qb_{n-2} = S - pb_{n-1}$. Hence

$$
\begin{cases}
R = b_{n-1}, \\
S = b_n + pb_{n-1}.
\end{cases}
\tag{2.3.4}
$$

These values are inserted into (2.3.1) to give

$$\frac{\partial b_{n-1}}{\partial p} \Delta p + \frac{\partial b_{n-1}}{\partial q} \Delta q + b_{n-1} = 0,$$

$$\left(\frac{\partial b_n}{\partial p} + p\frac{\partial b_{n-1}}{\partial p} + b_{n-1}\right) \Delta p + \left(\frac{\partial b_n}{\partial q} + p\frac{\partial b_{n-1}}{\partial q}\right) \Delta q + b_n + pb_{n-1} = 0.$$

If the first of these equations is multiplied by p and subtracted from the second, we obtain

$$
\begin{cases}
\dfrac{\partial b_{n-1}}{\partial p} \Delta p + \dfrac{\partial b_{n-1}}{\partial q} \Delta q + b_{n-1} = 0, \\
\left(\dfrac{\partial b_n}{\partial p} + b_{n-1}\right) \Delta p + \dfrac{\partial b_n}{\partial q} \Delta q + b_n = 0.
\end{cases}
\tag{2.3.5}
$$

Differentiating (2.3.3), we find

$$
\begin{cases}
-\dfrac{\partial b_k}{\partial p} = b_{k-1} + p\dfrac{\partial b_{k-1}}{\partial p} + q\dfrac{\partial b_{k-2}}{\partial p}; \qquad \dfrac{\partial b_0}{\partial p} = \dfrac{\partial b_{-1}}{\partial p} = 0; \\
-\dfrac{\partial b_k}{\partial q} = b_{k-2} + p\dfrac{\partial b_{k-1}}{\partial q} + q\dfrac{\partial b_{k-2}}{\partial q}; \qquad \dfrac{\partial b_0}{\partial q} = \dfrac{\partial b_{-1}}{\partial q} = 0.
\end{cases}
\tag{2.3.6}
$$

Putting $\partial b_k/\partial p = -c_{k-1}$ $(k = 1, 2, \ldots, n)$, we obtain by induction

$$\frac{\partial b_k}{\partial q} = \frac{\partial b_{k-1}}{\partial p} = -c_{k-2}.$$

Equations (2.3.6) then pass into the following equations from which the quantities c can be determined:

$$c_{k-1} = b_{k-1} - pc_{k-2} - qc_{k-3}, \qquad c_{k-2} = b_{k-2} - pc_{k-3} - qc_{k-4}.$$

These can be comprised in one single equation:

$$c_k = b_k - pc_{k-1} - qc_{k-2}; \qquad c_0 = 1; \qquad c_{-1} = 0;$$
$$(k = 1, 2, \ldots, n-1). \qquad (2.3.7)$$

Hence c_k is computed from b_k in exactly the same way as b_k is from a_k. Equations (2.3.5) for determination of Δp and Δq can now be written

$$\begin{cases} c_{n-2} \cdot \Delta p + c_{n-3} \cdot \Delta q = b_{n-1}, \\ (c_{n-1} - b_{n-1}) \cdot \Delta p + c_{n-2} \cdot \Delta q = b_n. \end{cases} \qquad (2.3.8)$$

From equation (2.3.1), it is easily understood that the convergence of Bairstow's method is quadratic, and, further, that convergence always occurs if the starting value is not too bad.

EXAMPLES

The equation $x^4 + 5x^3 + 3x^2 - 5x - 9 = 0$ is to be solved, and we shall try to find two quadratic factors. Starting with $p_0 = 3$, $q_0 = -5$, we obtain the following scheme:

(a_k)	1	5	3	-5		-9		
		-3	-6	-6		3	-3	
			5	10		10	5	
(b_k)	1	2	2	-1		4		
		-3	3	-30				
			5	-5				
(c_k)	1	-1	10	-35				
		$\downarrow$	$\downarrow$	$\downarrow$	$\downarrow$	$\downarrow$		
		c_{n-3}	c_{n-2}	$c_{n-1} - b_{n-1}$	b_{n-1}	b_n		

Note that in the last addition of the computation of c_k, the term b_{n-1} (in this example, -1) is omitted. This gives directly $c_{n-1} - b_{n-1}$. Introducing u and v instead of Δp and Δq, we obtain the system:

$$\begin{cases} 10u - v = -1, \\ -35u + 10v = 4, \end{cases}$$

which gives $\begin{cases} u = -0.09, \\ v = 0.08, \end{cases}$ and $\begin{cases} p_1 = 2.91, \\ q_1 = -4.92. \end{cases}$

Next, the computation is repeated with the new values for p and q:

1	5	3	-5	-9	
	-2.91	-6.08	-5.35	0.20	-2.91
		4.92	10.28	9.05	4.92
1	2.09	1.84	-0.07	0.25	
	-2.91	2.37	-26.57		
		4.92	-4.03		
1	-0.82	9.13	-30.60		

$\begin{cases} 9.13u - 0.82v = -0.07, \\ -30.60u + 9.13v = 0.25, \end{cases}$ $\begin{cases} u = -0.00745, \\ v = 0.00241, \end{cases}$ $\begin{cases} p_2 = 2.90255, \\ q_2 = -4.91759; \end{cases}$

1	5	3	-5	-9	
	-2.90255	-6.08795	-5.31062	-0.01097	-2.90255
		4.91759	10.31440	8.99742	4.91759
1	2.09745	1.82964	0.00378	-0.01355	
	-2.90255	2.33684	-26.36697		
		4.91759	-3.95915		
1	-0.80510	9.08407	-30.32612		

$\begin{cases} 9.08407u - 0.80510v = 0.00378, \\ -30.32612u + 9.08407v = -0.01355, \end{cases}$ $\begin{cases} u = 0.000403, \\ v = -0.000146, \end{cases}$

$\begin{cases} p_3 = 2.902953, \\ q_3 = -4.917736; \end{cases}$

1	5	3	-5	-9	
	-2.902953	-6.087629	-5.312715	-0.000026	-2.902953
		4.917736	10.312724	8.999983	4.917736
1	2.097047	1.830107	0.000009	-0.000043	
	-2.902953	2.339507	-26.380150		
		4.917736	-3.963233		
1	-0.805906	9.087350	-30.343383		

$\begin{cases} 9.087350u - 0.805906v = 0.000009, \\ -30.343383u + 9.087350v = -0.000043, \end{cases}$ $\begin{cases} u = 0.00000081, \\ v = -0.00000202, \end{cases}$

$\begin{cases} p_4 = 2.902954, \\ q_4 = -4.917738. \end{cases}$

The errors in p_4 and q_4 are probably less than one unit in the last place. One more synthetic division finally gives the approximate factorization:

$$x^4 + 5x^3 + 3x^2 - 5x - 9$$
$$\simeq (x^2 + 2.902954x - 4.917738)(x^2 + 2.097046x + 1.830110).$$

2.4. Graeffe's root-squaring method

Consider the equation $x^n + a_1 x^{n-1} + \cdots + a_n = 0$. For the sake of simplicity, we suppose all roots to be real and different from each other. Collecting all even terms on one side and all odd terms on the other, we get

$$(x^n + a_2 x^{n-2} + a_4 x^{n-4} + \cdots)^2 = (a_1 x^{n-1} + a_3 x^{n-3} + a_5 x^{n-5} + \cdots)^2 .$$

Putting $x^2 = y$, we obtain the new equation

$$y^n + b_1 y^{n-1} + b_2 y^{n-2} + \cdots + b_n = 0$$

with

$$\begin{cases} b_1 = -a_1^2 + 2a_2 , \\ b_2 = a_2^2 - 2a_1 a_3 + 2a_4 , \\ b_3 = -a_3^2 + 2a_2 a_4 - 2a_1 a_5 + 2a_6 , \\ \vdots \\ b_n = (-1)^n a_n^2 , \end{cases}$$

or

$$(-1)^k b_k = a_k^2 - 2a_{k-1} a_{k+1} + 2a_{k-2} a_{k+2} - \cdots . \tag{2.4.1}$$

The procedure can then be repeated and is finally interrupted when the double products can be neglected, compared with the quadratic terms on formation of new coefficients. Suppose that, after m squarings, we have obtained the equation $x^n + A_1 x^{n-1} + \cdots + A_n = 0$ with the roots $q_1, q_2, \ldots, q_n$, while the original equation has the roots $p_1, p_2, \ldots, p_n$. Then $q_i = p_i^{2^m}$, $i = 1, 2, \ldots, n$. Further, suppose that $|p_1| > |p_2| > \cdots > |p_n|$ and $|q_1| \gg |q_2| \gg \cdots \gg |q_n|$.* Hence

$$\begin{cases} A_1 = -\sum q_k \simeq -q_1 , \\ A_2 = \sum q_i q_k \simeq q_1 q_2 , \\ A_3 = -\sum q_i q_k q_l \simeq -q_1 q_2 q_3 , \end{cases} \tag{2.4.2}$$

and consequently,

$$\begin{cases} q_1 \simeq -A_1 , \\ q_2 \simeq -A_2/A_1 , \\ q_3 \simeq -A_3/A_2 , \\ \vdots \end{cases} \tag{2.4.3}$$

---

* The sign $\gg$ means "much larger than."

Finally, we get p_i by m successive square-root extractions of q_i; the sign has to be determined by insertion of the root into the equation.

EXAMPLE

$$x^3 - 8x^2 + 17x - 10 = 0 ,$$
$$(x^3 + 17x)^2 = (8x^2 + 10)^2 ,$$

or, putting $x^2 = y$,

$$y^3 - 30y^2 + 129y - 100 = 0 .$$

Squaring twice again, we find

$$z^3 - 642z^2 + 10641z - 10^4 = 0 ,$$

and

$$u^3 - 390882u^2 + 100390881u - 10^8 = 0 .$$

Hence

$$|p_1| = \sqrt[8]{390882} = 5.00041 ,$$
$$|p_2| = \sqrt[8]{100390881/390882} = 2.00081 ,$$
$$|p_3| = \sqrt[8]{10^8/100390881} = 0.999512 .$$

The exact roots are 5, 2, and 1.

A more detailed description of the method can be found in [2]; the case of complex roots has been treated particularly by Brodetsky and Smeal [3]. As has been shown by Wilkinson [9] a well-conditioned polynomial may become ill-conditioned after a root-squaring procedure; in particular this seems to affect the complex roots.

2.5. Iterative methods

Strictly speaking, Newton-Raphson's and Bairstow's methods, which have been treated above, could be considered as iterative methods. The latter can only be used in a special case, while the former is general. It is also characteristic for Newton-Raphson's method that the derivative must be computed at each step. We will now consider some methods which do not make use of the derivative.

The first of these methods is due to Muller [4], and it is primarily supposed to be useful for solution of algebraic equations of high degree with complex coefficients. The general principle can be described as follows: Let

$$f(x) = a_0 x^n + a_1 x^{n-1} + \cdots + a_n = 0 ,$$

be the equation to be solved. Suppose that (x_{i-2}, f_{i-2}), (x_{i-1}, f_{i-1}), and (x_i, f_i) are three points on the curve. We can then find a parabola $y = ax^2 + bx + c$ passing through these three points. The equation of the parabola can be written down directly:

$$y = \frac{(x - x_{i-1})(x - x_i)}{(x_{i-2} - x_{i-1})(x_{i-2} - x_i)} f_{i-2} + \frac{(x - x_{i-2})(x - x_i)}{(x_{i-1} - x_{i-2})(x_{i-1} - x_i)} f_{i-1}$$
$$+ \frac{(x - x_{i-2})(x - x_{i-1})}{(x_i - x_{i-2})(x_i - x_{i-1})} f_i . \tag{2.5.1}$$

This equation is obviously of the second degree and, as is easily found, it is satisfied by the coordinates of the three points (cf. also Section 8.1). Putting $h = x - x_i$, $h_i = x_i - x_{i-1}$, $h_{i-1} = x_{i-1} - x_{i-2}$, we obtain

$$y = \frac{(h + h_i)h}{-h_{i-1}(-h_{i-1} - h_i)} f_{i-2} + \frac{(h + h_i + h_{i-1})h}{h_{i-1} \cdot (-h_i)} f_{i-1}$$
$$+ \frac{(h + h_i + h_{i-1})(h + h_i)}{(h_i + h_{i-1})h_i} f_i . \tag{2.5.2}$$

Further, introducing $\lambda = h/h_i$, $\lambda_i = h_i/h_{i-1}$, and $\delta_i = 1 + \lambda_i$, we get:

$$y = \frac{1}{\delta_i} [\lambda(\lambda + 1) \cdot \lambda_i^2 f_{i-2} - \lambda(\lambda + 1 + \lambda_i^{-1})\lambda_i \delta_i f_{i-1}$$
$$+ (\lambda + 1)(\lambda + 1 + \lambda_i^{-1})\lambda_i f_i]$$
$$= \lambda^2 \cdot \delta_i^{-1}[f_{i-2}\lambda_i^2 - f_{i-1}\lambda_i \delta_i + f_i \lambda_i]$$
$$+ \lambda \delta_i^{-1}[f_{i-2}\lambda_i^2 - f_{i-1}\delta_i^2 + f_i(\lambda_i + \delta_i)] + f_i . \tag{2.5.3}$$

This expression is equated to zero and divided by $f_i \lambda^2$, and we next solve for $1/\lambda$. With $g_i = f_{i-2}\lambda_i^2 - f_{i-1}\delta_i^2 + f_i(\lambda_i + \delta_i)$, we get

$$\lambda = \lambda_{i+1} = \frac{-2f_i \delta_i}{g_i \pm \sqrt{g_i^2 - 4f_i \delta_i \lambda_i [f_{i-2}\lambda_i - f_{i-1}\delta_i + f_i]}} . \tag{2.5.4}$$

Since $\lambda = h/h_i = (x - x_i)/(x_i - x_{i-1})$, a small value for λ will give a value of x close to x_i. For this reason we should try to make the denominator of (2.5.4) as large as possible by choosing the sign accordingly. Hence the result of the extrapolation is

$$x_{i+1} = x_i + \lambda_{i+1}h_i = x_i + h_{i+1} . \tag{2.5.5}$$

The process is conveniently started by making $x_0 = -1$, $x_1 = 1$, and $x_2 = 0$, and further by using $a_n - a_{n-1} + a_{n-2}$ for f_0, $a_n + a_{n-1} + a_{n-2}$ for f_1, and a_n for f_2, that is, $\lambda_2 = -\frac{1}{2}$ and $h_2 = -1$. This corresponds to the approximation $f \simeq a_n + a_{n-1}x + a_{n-2}x^2$ close to the origin.

Muller's method can be applied in principle to arbitrary equations, but so far it seems to have been tried only on algebraic equations of high degree. The experiences must be considered as quite good. The speed of convergence is given by $\varepsilon_{n+1} \simeq A\varepsilon_{n-2}\varepsilon_{n-1}\varepsilon_n$, or approximately $\varepsilon_{n+1} \simeq K \cdot \varepsilon_n^{1.84}$ (where 1.84 is the largest root of the equation $m^3 = m^2 + m + 1$).

We now take up a method designed for equations of the type

$$x = g(x) .$$ (2.5.6)

Starting with a suitable value x_0, we form $x_{i+1} = g(x_i)$ for $i = 0, 1, 2, \ldots$ If the sequence x_i has a limit ξ, it is obvious that ξ is a root of (2.5.6). Any

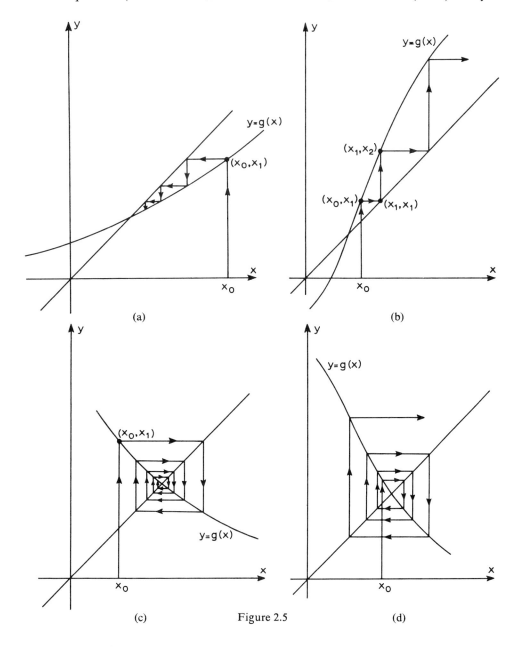

(a)   (b)

(c)   Figure 2.5   (d) equation $f(x) = 0$ can always be brought to this form, in fact, in an infinite number of ways. In order to show that this choice is not insignificant, we consider the equation $x^3 - 2 = 0$. It is rewritten on one hand as (a) $x = x^3 + x - 2$ and on the other as (b) $x = (2 + 5x - x^3)/5$. Starting with $x_0 = 1.2$, we get the results:

	(a)	(b)
$x_1 =$	0.928	1.2544
$x_2 =$	-0.273	1.2596
$x_3 =$	-2.293	1.2599
$x_4 =$	-16.349	1.25992

The correct value $\sqrt[3]{2} \simeq 1.259921$ was completely lost in (a) but rapidly attained in (b).

We will now examine the convergence analytically. Starting with $x_1 = g(x_0)$, we find that $x - x_1 = g(x) - g(x_0) = (x - x_0)g'(\xi_0)$, with $x_0 < \xi_0 < x$. Analogously $x - x_2 = (x - x_1)g'(\xi_1); \ldots; |x - x_n| = (x - x_{n-1})g'(\xi_{n-1})$. Here x is the wanted root of the equation $x = g(x)$. Multiplying, we obtain: $x - x_n = (x - x_0)g'(\xi_0)g'(\xi_1) \cdots g'(\xi_{n-1})$.

Now suppose that $|g'(\xi_k)| \leq m$. Then we get $|x - x_n| \leq m^n \cdot |x - x_0|$, and hence we have convergence if $m < 1$, that is, if $|g'(x)| < 1$ in the interval of interest. We attain the same result by simple geometric considerations (see Fig. 2.5).

It is also easy to give an error estimate in a more useful form. From $x - x_{n-1} = x - x_n + x_n - x_{n-1}$, we get, using $x - x_n = g(x) - g(x_{n-1})$,

$$|x - x_{n-1}| \leq |x - x_n| + |x_n - x_{n-1}| \leq |x - x_{n-1}| \cdot |g'(\xi_{n-1})|$$
$$+ |x_n - x_{n-1}| \leq m \cdot |x - x_{n-1}| + |x_n - x_{n-1}|.$$

Hence

$$|x - x_{n-1}| \leq \frac{|x_n - x_{n-1}|}{1 - m},$$

and

$$|x - x_n| \leq \frac{m}{1 - m} |x_n - x_{n-1}|. \tag{2.5.7}$$

If in practical computation it is difficult to estimate m from the derivative, we can use

$$m \simeq \frac{|x_{n+1} - x_n|}{|x_n - x_{n-1}|}.$$

In particular, choosing $g(x) = x - f(x)/f'(x)$ (Newton-Raphson's method) we find:

$$x_{n+1} = x_n - \frac{f(x_n)}{f'(x_n)} = x_n - h \quad \text{and} \quad g'(x) = \frac{f(x)}{f'(x)} \cdot \frac{f''(x)}{f'(x)}.$$

Using the same arguments as before we get $|x - x_n| \leq |h|/(1 - M)$ where $M = |g'(\xi_n)|$ and ξ_n lies between x and x_n. But $|x - x_{n+1}| = M|x - x_n|$ and hence

$$|x - x_{n+1}| < \frac{M|h|}{1 - M}.$$

If we now assume that $f(x)/f'(x)$ is monotonic we have

$$\left|\frac{f(\xi_n)}{f'(\xi_n)}\right| < \left|\frac{f(x_n)}{f'(x_n)}\right| = |h|.$$

Further, $|f''(\xi_n)/f'(\xi_n)| \leq K$, where $K = \sup |f''(x)/f'(x)|$ taken over the interval in question. Hence we have $M/(1 - M) \leq K|h|/(1 - K|h|)$ and we finally obtain

$$|x - x_{n+1}| \leq \frac{Kh^2}{1 - K|h|}$$

in accordance with previous results.

EXAMPLE

$x = \frac{1}{2} + \sin x$.
The iteration $x_{n+1} = \frac{1}{2} + \sin x_n$, with $x_0 = 1$, gives the following values:

$$x_1 = 1.34, \qquad x_2 = 1.47, \qquad x_3 = 1.495,$$
$$x_4 = 1.4971, \qquad x_5 = 1.49729.$$

For m we can choose the value $m = \cos 1.4971 \cong 0.074$, and hence $|x - x_5| \leq (0.074/0.926) \cdot 0.00019 = 0.000015$. In this case it is obvious that $x > x_5$, and this leads to the estimate

$$1.497290 < x < 1.497305.$$

The correct value to six decimal places is $x = 1.497300$.

Previously we have worked under the tacit assumption that a solution exists and that it is unique. We shall now discuss these questions to some extent and also treat the problems of convergence speed and computation volume. An equation $f(x) = 0$ can be brought to the form $x = \varphi(x)$, for example, by defining $\varphi(x) = x - f(x)g(x)$. For solving the equation $x = \varphi(x)$ we consider the iteration

$$x_{n+1} = \varphi(x_n). \tag{2.5.8}$$

If the relation $\xi = \varphi(\xi)$ is satisfied, ξ is said to be a *fix-point* of φ. If φ is regarded as a mapping, it is possible to prove the existence and uniqueness of a fix-point under various assumptions.

We now assume that the function φ is defined on a closed interval $[a, b]$ and that the function values belong to the same interval. We also suppose the function φ to be continuous in the Lipschitz sense. i.e., satisfying the condition

$$|\varphi(s) - \varphi(t)| \leq L|s - t| , \qquad 0 \leq L < 1 , \qquad (2.5.9)$$

for arbitrary s and t. Obviously, this implies usual continuity, while the reverse is not true (consider, for example, $\varphi(x) = \sqrt{x}$ in $[0, 1]$). First we show the existence of a fix-point. From the conditions above we have $\varphi(a) \geq a, \varphi(b) \leq b$. The continuous function $\varphi(x) - x$ is then ≥ 0 for $x = a$, ≤ 0 for $x = b$, and consequently it must vanish in some point in the interval. To show the uniqueness we suppose that there are two solutions ξ_1 and ξ_2. Then we would have

$$|\xi_1 - \xi_2| = |\varphi(\xi_1) - \varphi(\xi_2)| \leq L|\xi_1 - \xi_2| < |\xi_1 - \xi_2|$$

leading to a contradiction.

It is now easy to show that the sequence $\{x_n\}$, defined by (2.5.8), converges to ξ. For

$$|x_{n+1} - \xi| = |\varphi(x_n) - \varphi(\xi)| \leq L|x_n - \xi| , \quad \text{i.e.,} \quad |x_n - \xi| \leq L^n|x_0 - \xi| ,$$

and $\lim_{n \to \infty} x_n = \xi$ since $L^n \to 0$. We now turn to the problem of convergence speed and related to this the question of effectivity for different methods. If we can find constants $p \geq 1$ and $C > 0$ such that

$$\lim_{n \to \infty} \frac{|x_{n+1} - \xi|}{|x_n - \xi|^p} = C , \qquad (2.5.10)$$

the convergence is said to be of *order p* while C is called the *asymptotic error constant*. In particular, the convergence is *linear* if $p = 1$ (in this case C must be <1) and *quadratic* if $p = 2$. The effectivity of a method naturally depends on not only how fast the convergence is but also how many *new* evaluations s of the function f and its derivatives are needed for each step (the unit of s is often called *Horner*). Usually one neglects the fact that sometimes several extra evaluations are needed to start the process. The *effectivity index E* is defined through

$$E = p^{1/s} , \qquad (2.5.11)$$

and the method is better the larger E is (cf. [8]). We present a survey in the following table.

Method	s(Horners)	p	E
Fixed secant	1	1	1
Regula falsi	1	1.62	1.62
Muller	1	1.84	1.84
Newton-Raphson	2	2	1.41
Chebyshev	3	3	1.44
Multiplicity-indep. N.-R.	3	2	1.26

Chebyshev's formula has the form

$$x_{n+1} = x_n - f(x_n)/f'(x_n) - [f(x_n)]^2 f''(x_n)/[2f'(x_n)^3] ,$$

while the multiplicity-independent formula is

$$x_{n+1} = x_n - f(x_n)f'(x_n)/[f'(x_n)^2 - f(x_n)f''(x_n)] .$$

It seems fair to emphasize methods that do not require the derivatives, and among these *Regula falsi* and Muller's methods should be mentioned first. In order to attain reasonably fast convergence, it is important that the starting values are not too far out.

2.6. The Q-D method

This is a very general method, constructed by Stiefel, Rutishauser, and Henrici. Its full name is the Quotient-Difference Algorithm, and it can be used for determination of eigenvalues of matrices and for finding roots of polynomials. In this last case the method works as follows. One constructs a rhombic pattern of numbers $e_i^{(k)}$ and $q_i^{(k)}$:

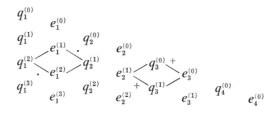

Rules

In a rhomb with an e-element on top, the product of two elements is equal to the product of the others, as indicated in the scheme.

In a rhomb with a q-element on top, the sum of two elements is equal to the sum of the others, as indicated in the scheme.

Formulas

$$\begin{cases} q_{k+1}^{(n)} \cdot e_k^{(n)} = q_k^{(n+1)} \cdot e_k^{(n+1)} , \\ q_k^{(n)} + e_k^{(n)} = q_k^{(n+1)} + e_{k-1}^{(n+1)} . \end{cases} \tag{2.6.1}$$

We will give at least some background to the Q-D method when used for algebraic equations. We take a polynomial $P(x)$ of degree n with the simple roots $x_1, x_2, \ldots, x_n$ and $0 < |x_1| < |x_2| < \cdots < |x_n|$. Then $1/P(x)$ can be expressed as a sum of partial fractions:

$$\frac{1}{P(x)} = \frac{A_1}{x - x_1} + \frac{A_2}{x - x_2} + \cdots + \frac{A_n}{x - x_n} . \tag{2.6.2}$$

The terms on the right-hand side can be expanded in a geometric series:

$$\frac{A_r}{x - x_r} = -\frac{A_r}{x_r}\frac{1}{1 - x/x_r} = -A_r\left(\frac{1}{x_r} + \frac{x}{x_r^2} + \frac{x^2}{x_r^3} + \cdots\right).$$

Now put $\alpha_i = -\sum_{r=1}^{n} A_r/x_r^{i+1}$ to obtain:

$$\frac{1}{P(x)} = \sum_{i=0}^{\infty} \alpha_i x^i . \tag{2.6.3}$$

We compute the quotient between two consecutive coefficients:

$$
\begin{aligned}
q^{(i)} = \frac{\alpha_i}{\alpha_{i-1}} &= \frac{(A_1/x_1^{i+1}) + (A_2/x_2^{i+1}) + \cdots + (A_n/x_n^{i+1})}{(A_1/x_1^i) + (A_2/x_2^i) + \cdots + (A_n/x_n^i)} \\
&= \frac{1}{x_1}\frac{1 + (A_2/A_1)(x_1/x_2)^{i+1} + \cdots + (A_n/A_1)(x_1/x_n)^{i+1}}{1 + (A_2/A_1)(x_1/x_2)^i + \cdots + (A_n/A_1)(x_1/x_n)^i}.
\end{aligned}
$$

But since $|x_1/x_k| < 1$, $k = 2, 3, \ldots, n$, we have for $i \to \infty$:

$$\lim_{i \to \infty} q^{(i)} = \frac{1}{x_1} . \tag{2.6.4}$$

In the same way it is easily shown that

$$\lim_{i \to \infty} \frac{(1/x_1) - q^{(i)}}{(x_1/x_2)^i} = \frac{1}{x_1}\frac{A_2}{A_1}\left(1 - \frac{x_1}{x_2}\right). \tag{2.6.5}$$

Replacing i by $i + 1$, we find:

$$\lim_{i \to \infty} \frac{(1/x_1) - q^{(i+1)}}{(x_1/x_2)^i} = \frac{1}{x_2}\frac{A_2}{A_1}\left(1 - \frac{x_1}{x_2}\right);$$

and subtracting:

$$\lim_{i \to \infty} \frac{q^{(i+1)} - q^{(i)}}{(x_1/x_2)^i} = \left(\frac{1}{x_1} - \frac{1}{x_2}\right)\frac{A_2}{A_1}\left(1 - \frac{x_1}{x_2}\right).$$

Introducing $e^{(i)} = q^{(i+1)} - q^{(i)}$, we obtain:

$$\lim_{i \to \infty} \frac{e^{(i+1)}}{e^{(i)}} = \frac{x_1}{x_2} . \tag{2.6.6}$$

Equation (2.6.4) gives the first root x_1, while the second root x_2 is obtained by eliminating x_1 from (2.6.6):

$$\lim_{i \to \infty} \frac{e^{(i+1)}}{e^{(i)}} q^{(i+1)} = \frac{1}{x_2} . \tag{2.6.7}$$

It is now convenient to change notations so that the quantities q and e get a lower index 1: $q_1^{(i)}$, $e_1^{(i)}$, and so forth. Further, we introduce:

$$\frac{e_1^{(i+1)}}{e_1^{(i)}} q_1^{(i+1)} = q_2^{(i)} .$$

Hence $\lim_{i \to \infty} q_1^{(i)} = 1/x_1$; $\lim_{i \to \infty} q_2^{(i)} = 1/x_2$.

Further, a new e-quantity is introduced

$$e_2^{(i)} = q_2^{(i+1)} - q_2^{(i)} + e_1^{(i+1)} ,$$

and as before, it is shown that

$$\lim_{i \to \infty} \frac{e_2^{(i+1)}}{e_2^{(i)}} = \frac{x_2}{x_3} .$$

Next $q_3^{(i)}$ is calculated, and it is now clear that the so-called rhomb rules have been attained.

The initial values for solution of an algebraic equation by the Q-D method are established by using a fairly complicated argument which cannot be reproduced here. When we are looking for the roots of the equation

$$a_0 x^n + a_1 x^{n-1} + \cdots + a_n = 0,$$

the scheme ought to start in the following way:

e_0	q_1	e_1	q_2	$e_2 \cdots q_{n-1}$	e_{n-1}	q_n	e_n
	$-\dfrac{a_1}{a_0}$		0	$\cdots$ 0		0	
0		$\dfrac{a_2}{a_1}$		$\dfrac{a_3}{a_2}$	$\dfrac{a_n}{a_{n-1}}$		0
0							0
$\vdots$							$\vdots$

All coefficients a_k are supposed to be nonzero. If all roots are real and simple, then with $|x_1| > |x_2| > \cdots > |x_n|$, we have

$$x_k = \lim_{n \to \infty} q_k^{(n)} .$$

EXAMPLE

$$2x^3 - 13x^2 - 22x + 3 = 0 .$$

e_0	q_1	e_1	q_2	e_2	q_3	e_3
	6.5		0		0	
0	8.192308	1.692308	-1.828672	-0.136364	0.136364	0
0	7.814554	-0.377754	-1.440749	0.010169	0.126195	0
0	7.884200	0.069646	-1.511286	-0.000891	0.127086	0
0	7.870850	-0.013350	-1.497861	0.000075	0.127011	0
0	7.873391	0.002541	-1.500408	-0.000006	0.127017	0
0	7.872907	-0.000484	-1.499924	0		
0	7.872999	0.000092	-1.500016			
0	7.872981	-0.000018	-1.499998			
0	7.872984	0.000003	-1.500001			
0	7.872983	-0.000001	-1.500000			

The method can be made to work also for complex and multiple roots. Details can be found, for example, in [5].

It may be observed that multiple roots of algebraic equations will give rise to difficulties with any of the methods we have discussed here. As a matter of fact, all classical methods require a large amount of skill and good judgment for isolation and separation of the roots. It has turned out to be unexpectedly difficult to anticipate all more or less peculiar phenomena which can appear in such special cases. Against this background, Lehmer [6] has constructed a method which seems to be highly insensitive to clustering among the roots. The method is founded on the well-known fact that the integral

$$\frac{1}{2\pi i} \int_c \frac{P'(z)}{P(z)} \, dz \, ,$$

where $P(z)$ is a polynomial and C a closed curve (e.g., the unit circle), gives the number of roots within C. Since we know that the value of the integral is an integer, we can use fairly coarse methods for this evaluation. The method converges geometrically with a quotient $\frac{5}{12}$, and rather long computing times are necessary. On the other hand, this is the price one has to pay to get rid of all complications.

2.7. Equations with several unknowns

We distinguish between linear systems of equations, which will be treated separately in Chapter 4, and nonlinear systems, which may even be transcendental. This latter case can sometimes be treated by a series-expansion technique. We will here limit ourselves to two unknowns. Let the equations be

$$\begin{cases} F(x, y) = 0 \, , \\ G(x, y) = 0 \, . \end{cases}$$

Suppose that (x_0, y_0) is an approximate solution and let h and k be corrections which we shall determine:

$$\begin{cases} F(x_0 + h, y_0 + k) = 0 \, , \\ G(x_0 + h, y_0 + k) = 0 \, . \end{cases}$$

Expanding in Taylor series and truncating after the first-order terms, we get:

$$\begin{cases} F(x_0, y_0) + h \left(\dfrac{\partial F}{\partial x} \right)_0 + k \left(\dfrac{\partial F}{\partial y} \right)_0 = 0 \, , \\ G(x_0, y_0) + h \left(\dfrac{\partial G}{\partial x} \right)_0 + k \left(\dfrac{\partial G}{\partial y} \right)_0 = 0 \, . \end{cases}$$

This linear system in h and k gives the next approximation. In practical work, the derivatives can be replaced with the corresponding difference quotients, and only the values of the functions have to be computed.

Another way which is sometimes practicable is the following. Let the system be $F_1(x_1, x_2, \ldots, x_n) = 0$; $F_2(x_1, x_2, \ldots, x_n) = 0$; $\ldots$; $F_n(x_1, x_2, \ldots, x_n) = 0$. Form the function $f(x_1, x_2, \ldots, x_n) = F_1^2 + F_2^2 + \cdots + F_n^2$ (where F_k is supposed to be real). Then we have $f(x_1, x_2, \ldots, x_n) \geq 0$. A solution of the system obviously corresponds to a minimum of f with the function value $= 0$. Starting from some initial point $(x_1^0, x_2^0, \ldots, x_n^0)$, we compute a number of function values along the straight line $x_2 = x_2^0$; $x_3 = x_3^0$; $\ldots$; $x_n = x_n^0$, that is, only the first variable x_1 is changed. On this line we search for a minimum (at least one minimum is known to exist, since f is bounded downwards). The corresponding value $x_1 = x_1^1$ is fixed, and then we repeat the whole procedure, varying x_2 alone. After a sufficient number of steps, we reach a minimum point, and if the function value is 0, it represents a solution. Alternatively, one can also consider

$$f(x_1, x_2, \ldots, x_n) = |F_1| + |F_2| + \cdots + |F_n| \,.$$

In both cases $f = c$ represents a family of "equipotential" surfaces, and the convergence will be particularly fast if the movements can be performed orthogonally to the surfaces. If this is the case, the procedure is called "method of steepest descent."

REFERENCES

[1] Wilkinson: "The Evaluation of the Zeros of Ill-conditioned Polynomials," *Num. Math.*, **1**, 150–180 (1959).

[2] Zurmühl: Praktische Mathematik (Springer Verlag, Berlin, 1961).

[3] Brodetsky-Smeal: "On Graeffe's Method for Complex Roots," *Proc. Camb. Phil. Soc.*, **22**, 83 (1924).

[4] Muller: "A Method for Solving Algebraic Equations Using an Automatic Computer," *MTAC*, 208–215 (1956).

[5] National Bureau of Standards: *Appl. Math. Series*, No. 49 (Washington, 1956).

[6] D. H. Lehmer: "A Machine Method for Solving Polynomial Equations," *Jour. ACM*, 151–162 (April 1961).

[7] Stiefel: *Einführung in die Numerische Mathematik* (Stuttgart, 1961).

[8] Traub: *Iterative Methods for the Solution of Equations*, (Prentice-Hall, Englewood Cliffs, N.J., 1964).

[9] Wilkinson: *Rounding Errors in Algebraic Processes* (Her Majesty's Stationery Office, London 1963).

EXERCISES

1. The equation $x^6 = x^4 + x^3 + 1$ has one root between 1 and 2. Find this root to six decimals.

2. The equation $x^4 - 5x^3 - 12x^2 + 76x - 79 = 0$ has two roots close to $x = 2$. Find these roots to four decimals.

3. Solve the equation $\log x = \cos x$ to five correct decimals.

4. Find the smallest positive root of the equation $\tan x + \tanh x = 0$ to five correct decimals.

5. Find the abscissa of the inflection point of the curve $y = e^{-x} \log x$ to five correct decimals.

6. What positive value of x makes the function $y = \tan x / x^2$ a minimum?

7. The mean-value theorem states that for a differentiable function,

$$f(x) - f(a) = (x - a) \cdot f'[a + p(x - a)] .$$

Find a positive value x such that $p = \frac{1}{2}$ when $f(x) = \arctan x$ and $a = 0$ (five decimals).

8. When $a > 1$ and $0 < b < \frac{1}{2}$, the equation $1/(e^{x/a} - 1) - a/(e^x - 1) - (a - 1)b = 0$ has one positive root. Calculate this root to four decimals when $a = 5$, $b = \frac{1}{4}$.

9. The function

$$y = \frac{1.00158 - 0.40222x}{1 + 0.636257x} - e^{-x} ,$$

is examined for $0 \leq x \leq 1$. Find the maxima and minima (also including the end points) and sketch the corresponding curve (five decimals).

10. Find the smallest value of a such that $a\sqrt{x} \geq \sin x$ for all positive values of x (five decimals).

11. Determine to five decimals the constant K such that the x-axis is a tangent of the curve $y = Ke^{x/10} - \log x$ $(x > 0)$.

12. Find the smallest value α such that $e^{-\alpha x} \leq 1/(1 + x^2)$ for all $x > 0$ (six decimals).

13. Calculate the area between the two curves $y = \cos x$ and $y = e^{-x}$ $(0 \leq x < \pi/2$; five decimals).

14. One wants to compute the positive root of the equation $x = a - bx^2$ (a and b positive) by using the iterative method $x_{k+1} = a - bx_k^2$. What is the condition for convergence?

15. The iterative method $x_{n+1} = f(x_n)$ for solution of the equation $x = f(x)$ when converging has an error which decreases approximately geometrically. Make use of this fact for constructing a much better approximation from three consecutive values x_{n-1}, x_n, and x_{n+1}.

16. Find the smallest positive root of the equation

$$1 - x + \frac{x^2}{(2!)^2} - \frac{x^3}{(3!)^2} + \frac{x^4}{(4!)^2} - \cdots = 0$$

correct to four decimals.

17. Find the intersection between the curves

$$\begin{cases} y = e^x - 2 , \\ y = \log (x + 2) , \end{cases}$$

to four decimals $(x > 0)$.

18. For values of k slightly less than 1, the equation $\sin x = kx$ has two roots $x = \pm x_0$ close to 0. These roots can be computed by expanding $\sin x$ in a power series and putting $6(1 - k) = s$; $x^2 = y$. Then s is obtained as a power series in y. Next, a new series expansion is attempted: $y = s + \alpha_1 s^2 + \alpha_2 s^3 + \alpha_3 s^4 + \cdots$, with unknown coefficients α_1, $\alpha_2, \alpha_3, \ldots$ Find the first three coefficients, and use the result for solving the equation $\sin x = 0.95x$ (six decimals).

19. The k-fold zero ξ of the equation $f(x) = 0$ shall be determined. If $k = 1$, Newton-Raphson's method gives a quadratic convergence, while $k > 1$ makes the convergence rate geometrical. Show that the modified formula

$$x_{n+1} = x_n - k \cdot \frac{f(x_n)}{f'(x_n)}$$

will always make the convergence quadratic in the neighborhood of ξ. Use the formula for finding a double root close to zero of the equation $x^5 - 7x^4 + 10x^3 + 10x^2 - 7x + 1 = 0$.

20. Solve the equation $x^3 - 5x^2 - 17x + 20 = 0$ by using Graeffe's method (squaring three times).

21. Find a constant c such that the curves $y = 2 \sin x$ and $y = \log x - c$ touch each other in the neighborhood of $x = 8$. Also calculate the coordinates of the contact point to five places.

22. Determine the largest value α_0 of α so that $x^{1/\alpha} \geq \log x$ for all positive values of x. If $\alpha > \alpha_0$, the inequality is not satisfied in a certain interval of x. Find the value of α for which the length of the interval is 100 (two places).

23. If one attempts to solve the equation $x = 1.4 \cos x$ by using the formula $x_{n+1} = 1.4 \cos x_n$, then in the limit, x_n will oscillate between two well-defined values a and b. Find these and the correct solution to four decimal places.

24. The equation $1 - x + x^2/2! - x^3/3! + \cdots - x^n/n! = 0$ (n an odd integer) has one real solution ξ which is positive and, of course, depends upon n. Evaluate $\lim_{n \to \infty} \xi/n$ by using Stirling's asymptotic formula $(n - 1)! \sim \sqrt{2\pi} e^{-n} n^{n-1/2}$.

25. The curves $3x^2 - 2xy + 5y^2 - 7x + 6y - 10 = 0$ and $2x^2 + 3xy - y^2 - 4 = 0$ have one intersection point in the first quadrant. Find its coordinates to four places.

26. The equation $\tan x = x$ has an infinite number of solutions drawing closer and closer to $a = (n + \frac{1}{2})\pi$, $n = 1, 2, 3, \ldots$ Put $x = a - z$, where z is supposed to be a small number, expand $\tan z$ in powers of z, and put $z = A/a + B/a^3 + C/a^5 + \cdots$ Determine the constants A, B, and C, and with this approximation compute the root corresponding to $n = 6$ to six decimals.

27. The equation $\sin x = 1/x$ has infinitely many solutions in the neighborhood of the points $x = n\pi$. For sufficiently large even values of n, the roots ξ_n can be written as $n\pi + \alpha + A\alpha^3 + B\alpha^5 + C\alpha^7 + \cdots$, where $A, B, C, \ldots$ are constants and $\alpha = 1/n\pi$. Find A and B.

28. Solve the equation $x^4 - 8x^3 + 39x^2 - 62x + 50 = 0$, using Bairstow's method. Start with $p = q = 0$.

29. The equation $\sin(xy) = y - x$ defines y as a function of x. When x is close to 1, the function has a maximum. Find its coordinates to three places.

30. The equation $f(x) = 0$ is given. One starts from $x_0 = \xi + \varepsilon$, where ξ is the unknown root. Further, $y_0 = f(x_0)$, $x_1 = x_0 - f(x_0)/f'(x_0)$, and $y_1 = f(x_1)$ are computed. Last a straight line is drawn through the points

$$(x_1, y_1) \quad \text{and} \quad \left(\frac{x_0 + x_1}{2}, \frac{y_0}{2} \right).$$

If ε is sufficiently small, the intersection between the line and the x-axis gives a good approximation of ξ. To what power of ε is the error term proportional?

31. Let $\{x_i\}$ be a sequence defined through $z_i = x_i - (\Delta x_i)^2/\Delta^2 x_i$, where $\Delta x_i = x_{i+1} - x_i$, $\Delta^2 x_i = x_{i+2} - 2x_{i+1} + x_i$. Compute $\lim_{i \to \infty} (z_i - \alpha)/(x_i - \alpha)$ when $x_{i+1} - \alpha = (K + \sigma_i)(x_i - \alpha)$ with $|K| < 1$ and $\sigma_i \to 0$.

Chapter 3

Matrices

3.0. Definitions

Matrix calculus is a most important tool in pure mathematics as well as in a series of applications representing fields as far apart from one another as theoretical physics, elasticity, and economics. Matrices are usually introduced in connection with the theory of coordinate transformations, but here we will, in essence, treat them independently. The determinant conception is supposed to be known although the fundamental rules will be discussed briefly.

Unless specified otherwise, the numbers involved are supposed to be complex. A *matrix* is now defined as an ordered rectangular array of numbers. We denote the number of rows by m and the number of columns by n. A matrix of this kind will be characterized by the type symbol (m, n). The element in the ith row and the kth column of the matrix A is denoted by $(A)_{ik} = a_{ik}$. The whole matrix is usually written

$$A = \begin{pmatrix} a_{11} & a_{12} & \cdots & a_{1n} \\ a_{21} & a_{22} & \cdots & a_{2n} \\ \vdots & & & \\ a_{m1} & a_{m2} & \cdots & a_{mn} \end{pmatrix},$$

or sometimes in the more compact form

$$A = (a_{ik}).$$

In the latter case, however, it must be agreed upon what type of matrix it is. If $m = n$, the matrix is *quadratic;* when $m = 1$, we have a *row vector*, and for $n = 1$, a *column vector*. If $m = n = 1$, the matrix is usually identified with the number represented by this single element. In this and following chapters small letters printed in boldface indicate vectors, while capital letters in boldface indicate matrices.

3.1. Matrix operations

Addition and subtraction are defined only for matrices of the same type, that is, for matrices with the same number of rows and the same number of columns.

47

If A and B are matrices of type (m, n), then $C = A \pm B$ is defined by $c_{ik} = a_{ik} \pm b_{ik}$. Further, if λ is an ordinary number, λA is the matrix with the general element equal to λa_{ik}. If all $a_{ik} = 0$, the matrix is said to be a null matrix.

The definition of matrix multiplication is closely connected with the theory of coordinate transformations. We start with the linear transformation

$$z_i = \sum_{r=1}^{n} a_{ir} y_r \qquad (i = 1, 2, \ldots, m) .$$

If the quantities $y_1, y_2, \ldots, y_n$ can be expressed by means of the quantities $x_1, x_2, \ldots, x_p$ through a linear transformation

$$y_r = \sum_{k=1}^{p} b_{rk} x_k \qquad (r = 1, 2, \ldots, n) ,$$

we find by substitution

$$z_i = \sum_{r=1}^{n} a_{ir} \sum_{k=1}^{p} b_{rk} x_k = \sum_{k=1}^{p} \left(\sum_{r=1}^{n} a_{ir} b_{rk} \right) x_k .$$

In this way we obtain z_i, expressed directly by means of x_k through a composed transformation, and we can write

$$z_i = \sum_{k=1}^{p} c_{ik} x_k \qquad \text{with} \qquad c_{ik} = \sum_{r=1}^{n} a_{ir} b_{rk} .$$

In view of this, the matrix multiplication should be defined as follows. Let A be of type (m, n) and B of type (n, p). Then $C = AB$ if

$$c_{ik} = \sum_{r=1}^{n} a_{ir} b_{rk} , \tag{3.1.1}$$

where C will be of type (m, p).

It should be observed that the commutative law of multiplication is not satisfied. Even if BA is defined (which is the case if $m = p$), and even if BA is of the same type as AB (which is the case if $m = n = p$), we have in general

$$BA \neq AB .$$

Only for some rather special matrices is the relation $AB = BA$ valid.

Transposition means exchange of rows and columns. We will denote the transposed matrix of A by A^T. Thus $(A^T)_{ik} = (A)_{ki}$. *Conjugation* of a matrix means that all elements are conjugated; the usual notation is A^*. If a matrix A is transposed and conjugated at the same time, the following symbol is often used:

$$(A^*)^T = (A^T)^* = A^H .$$

Here we will treat only quadratic matrices (n, n), row vectors $(1, n)$, and column vectors $(n, 1)$. First of all, we define the unit matrix or the identity matrix I:

$$(I)_{ik} = \delta_{ik} ,$$

where δ_{ik} is the Kronecker symbol

$$\delta_{ik} = \begin{cases} 0 & \text{if } i \neq k, \\ 1 & \text{if } i = k. \end{cases}$$

Thus we have

$$I = \begin{pmatrix} 1 & 0 & 0 \cdots 0 \\ 0 & 1 & 0 \cdots 0 \\ \vdots & & \\ 0 & 0 & 0 \cdots 1 \end{pmatrix}.$$

In a quadratic matrix, the elements a_{ii} form a diagonal, the so-called *main diagonal*. Hence the unit matrix is characterized by ones in the main diagonal, zeros otherwise. Now form $AI = B$:

$$b_{ik} = \sum_{r=1}^{n} a_{ir}\delta_{rk} = a_{ik},$$

since $\delta_{rk} = 1$ only when $r = k$.

Analogously, we form $IA = C$ with

$$c_{ik} = \sum_{r=1}^{n} \delta_{ir}a_{rk} = a_{ik}.$$

Hence

$$AI = IA = A.$$

The sum of the diagonal elements is called the *trace* of the matrix:

$$\text{Tr } A = \sum_{i=1}^{n} a_{ii}.$$

A real matrix A is said to be *positive definite* if $x^H A x > 0$ for all vectors $x \neq 0$. If, instead, $x^H A x \geq 0$, A is said to be *positive semidefinite*.

The determinant of a matrix A is a pure number, denoted by $\det A$, which can be computed from A by use of certain rules. Generally, it is the sum of all possible products of elements where exactly one element has been taken from every row and every column, with a plus or a minus sign appended according as the permutation of indices is even or odd.

$$\det A = \sum \pm a_{1,i_1} a_{2,i_2} \cdots a_{n,i_n}.$$

The formula can be rewritten in a more convenient way if we introduce the concept of *algebraic complement*. It is defined in the following way. The algebraic complement A_{ik} of an element a_{ik} is

$$A_{ik} = (-1)^{i+k} \det \alpha_{ik},$$

where α_{ik} is the matrix which is obtained when the ith row and the kth column are deleted from the original matrix. Thus we have

$$\det A = \sum_{i=1}^{n} a_{ik} A_{ik} = \sum_{k=1}^{n} a_{ik} A_{ik}.$$

Further,

$$\sum_{k=1}^{n} a_{ik}A_{jk} = 0 \qquad \text{if} \quad i \neq j\,,$$

and

$$\sum_{i=1}^{n} a_{ij}A_{ik} = 0 \qquad \text{if} \quad j \neq k\,.$$

The general rule is impossible to use in practical work because of the large number of terms $(n!)$; already for $n = 10$, we get more than 3 million terms. Instead, the following rules, which are completely sufficient, are used:

1. A determinant remains unchanged if one row (column) multiplied by a number λ is added to another row (column).

2. If in the ith row or kth column only the element $a_{ik} \neq 0$, then

$$\det A = (-1)^{i+k} a_{ik} \det \boldsymbol{\alpha}_{ik} = a_{ik}A_{ik}\,.$$

3. If

$$A = \begin{pmatrix} a & b \\ c & d \end{pmatrix},$$

then

$$\det A = \begin{vmatrix} a & b \\ c & d \end{vmatrix} = ad - bc\,.$$

From these rules it follows, for example, that $\det A = 0$ if two rows or two columns are proportional or, more generally, if one row (column) can be written as a linear combination of some of the other rows (columns).

If $\det A = 0$, A is said to be *singular;* if $\det A \neq 0$, A is *regular.* Further, if $AB = C$, we have $\det A \det B = \det C$.

Now let A be given with $\det A \neq 0$. We will try to find a matrix X such that $AX = I$. When formulated algebraically, the problem leads to n systems of linear equations, all with the same coefficient matrix but with different right-hand sides. Since A is regular, every system has exactly one solution (a column vector), and these n column vectors together form a uniquely determined matrix X. In a similar way, we see that the system $YA = I$ also has a unique solution Y. Premultiplication of $AX = I$ with Y and postmultiplication of $YA = I$ with X give the result

$$YAX = YI = Y\,; \qquad YAX = IX = X\,,$$

and hence $X = Y$. In this way we have proved that if A is regular, a right inverse and a left inverse exist and are equal. The inverse is denoted by A^{-1}:

$$AA^{-1} = A^{-1}A = I\,.$$

Last, we form a new matrix C with the elements $c_{ik} = A_{ki}$. The matrix C is called the *adjoint* matrix of A and is sometimes denoted by $\tilde{A}$. From the expansion theorem for determinants, we get

$$\tilde{A}A = A\tilde{A} = (\det A)I,$$

or

$$A^{-1} = (\det A)^{-1}\tilde{A}. \tag{3.1.2}$$

3.2. Fundamental computing laws for matrices

As is understood directly, the associative and commutative laws are valid for addition:

$$(A + B) + C = A + (B + C),$$
$$A + B = B + A.$$

Further, the associative and distributive laws for multiplication are fulfilled:

$$(AB)C = A(BC),$$
$$A(B + C) = AB + AC,$$
$$(A + B)C = AC + BC.$$

For transposing, we have the laws

$$(A + B)^T = A^T + B^T,$$
$$(AB)^T = B^T A^T,$$

and for inversion,

$$(AB)^{-1} = B^{-1}A^{-1}.$$

First, we prove $(AB)^T = B^T A^T$. Put $A^T = C$, $B^T = D$, and $AB = G$. Then $c_{ik} = a_{ki}$, $d_{ik} = b_{ki}$, and further

$$(G^T)_{ik} = g_{ki} = \sum a_{kr}b_{ri} = \sum c_{rk}d_{ir} = \sum d_{ir}c_{rk} = (DC)_{ik}.$$

Hence $G^T = DC$ or $(AB)^T = B^T A^T$. It follows directly from the definition that $(AB)^{-1} = B^{-1}A^{-1}$, since

$$(AB)(B^{-1}A^{-1}) = (B^{-1}A^{-1})(AB) = I.$$

From the definition of the inverse, we have $AA^{-1} = I$ and $(A^T)^{-1}A^T = I$. Transposing the first of these equations, we get $(A^{-1})^T A^T = I$, and comparing, we find $(A^T)^{-1} = (A^{-1})^T$.

If in a determinant, $n - p$ rows and $n - p$ columns are removed, we obtain a minor or subdeterminant of order p. If the matrix A has the property that all minors of order $r + 1$ are zero, while there is at least one minor of order r which is not zero, the matrix A is said to be of *rank* r.

Suppose that we have n vectors $u_1, u_2, \ldots, u_n$ (row vectors or column vectors), each with n components. If the relation

$$c_1 u_1 + c_2 u_2 + \cdots + c_n u_n = 0$$

has the only solution $c_1 = c_2 = \cdots = c_n = 0$, the vectors are said to be *linearly independent*. In this case they form a *basis*, and every n-dimensional vector can be written as a linear combination of the base vectors.

If a matrix is of rank r, an arbitrary row vector or column vector can be written as a linear combination of at most r vectors of the remaining ones, since otherwise we would have $r + 1$ linearly independent vectors and the rank would be at least $r + 1$. Hence we can also define the rank as the number of linearly independent row vectors or column vectors.

A system of n vectors $\boldsymbol{u}_i$ in n dimensions such that $\boldsymbol{u}_i^H \boldsymbol{u}_k = \delta_{ik}$ is said to be *orthonormal*. An orthonormal system can always be constructed from n linearly independent vectors through suitable linear combinations.

3.3. Eigenvalues and eigenvectors

If for a given matrix A one can find a number λ and a vector $\boldsymbol{u}$ such that the equation $A\boldsymbol{u} = \lambda\boldsymbol{u}$ is satisfied, λ is said to be an *eigenvalue* or *characteristic value* and $\boldsymbol{u}$ an *eigenvector* or *characteristic vector* of the matrix A. The equation can also be written

$$(A - \lambda I)\boldsymbol{u} = 0 \,.$$

This is a linear homogeneous system of equations. As is well known, non-trivial solutions exist only if $\det(A - \lambda I) = 0$. Thus we obtain an algebraic equation of degree n in λ. Explicitly, it has the following form:

$$\begin{vmatrix} a_{11} - \lambda & a_{12} & a_{13} \cdots & a_{1n} \\ a_{21} & a_{22} - \lambda & a_{23} \cdots & a_{2n} \\ \vdots & & & \\ a_{n1} & a_{n2} & a_{n3} \cdots a_{nn} - \lambda \end{vmatrix} = 0 \,. \tag{3.3.1}$$

We notice at once that one term in this equation is

$$(a_{11} - \lambda)(a_{22} - \lambda) \cdots (a_{nn} - \lambda)$$

and that all other terms contain λ to at most the $(n - 2)$-power. Hence the equation can be written:

$$\lambda^n - (a_{11} + a_{22} + \cdots + a_{nn})\lambda^{n-1} + \cdots + (-1)^n \det A = 0 \,.$$

Thus

$$\sum_{i=1}^{n} \lambda_i = \operatorname{Tr} A \,; \qquad \prod_{i=1}^{n} \lambda_i = \det A \,. \tag{3.3.2}$$

Equation (3.3.1) is called the *characteristic equation;* the roots are the n eigenvalues of the matrix A. Thus a square matrix of order n has always n eigenvalues; in special cases some or all of them may coincide.

If the equation $Au = \lambda u$ is premultiplied by A, we get $A^2 u = \lambda Au = \lambda^2 u$ and, analogously, $A^m u = \lambda^m u$. From this we see that if A has the eigenvalues $\lambda_1, \lambda_2, \ldots, \lambda_n$, then A^m has the eigenvalues $\lambda_1^m, \lambda_2^m, \ldots, \lambda_n^m$, and the eigenvectors are the same. Putting $S_r = \sum_{i=1}^{n} \lambda_i^r$, we have directly $\operatorname{Tr} A^r = S_r$.

Now we suppose that the characteristic equation of A has the form

$$f(x) = x^n + c_1 x^{n-1} + \cdots + c_n = 0 .$$

Then we also have

$$f(x) \equiv (x - \lambda_1)(x - \lambda_2) \cdots (x - \lambda_n)$$

and

$$f'(x) = \frac{f(x)}{x - \lambda_1} + \frac{f(x)}{x - \lambda_2} + \cdots + \frac{f(x)}{x - \lambda_n} .$$

The quotient $f(x)/(x - \lambda)$ is easily obtained:

$$\frac{f(x)}{x - \lambda} = x^{n-1} + (\lambda + c_1)x^{n-2} + (\lambda^2 + c_1\lambda + c_2)x^{n-3} + \cdots ,$$

and summing these equations for $\lambda = \lambda_1, \lambda_2, \ldots, \lambda_n$, we find

$$f'(x) = nx^{n-1} + (S_1 + nc_1)x^{n-2} + (S_2 + c_1 S_1 + nc_2)x^{n-3} + \cdots .$$

Identifying with $f'(x) = nx^{n-1} + (n - 1)c_1 x^{n-2} + \cdots$, we get

$$S_1 + c_1 = 0 ,$$
$$S_2 + c_1 S_1 + 2c_2 = 0 ,$$
$$\vdots$$
$$S_{n-1} + c_1 S_{n-2} + \cdots + c_{n-2}S_1 + (n - 1)c_{n-1} = 0 .$$

Last, we form $f(\lambda_1) + f(\lambda_2) + \cdots + f(\lambda_n) = 0$, and obtain

$$S_n + c_1 S_{n-1} + \cdots + c_{n-1}S_1 + nc_n = 0 .$$

These are the well-known *Newton identities*. The coefficients of the characteristic equation can be computed by solving recursively a linear system of equations containing $\operatorname{Tr} A, \operatorname{Tr} A^2, \ldots, \operatorname{Tr} A^n$. This method has been given by Leverrier, who, in about 1840 (together with Adams), computed the position of Neptune from observed irregularities in the orbit of Uranus.

If a matrix B is formed from a given matrix A by $B = S^{-1}AS$, where S is a regular matrix, A is said to undergo a *similarity transformation*. It is easy to prove that the characteristic equation is invariant during such a transformation, since

$$\det(B - \lambda I) = \det[S^{-1}(A - \lambda I)S]$$
$$= \det S^{-1} \det(A - \lambda I) \det S = \det(A - \lambda I) .$$

(We have, of course, $\det S^{-1} \det S = \det(S^{-1}S) = 1$.)

Hence A and B have the same eigenvalues. If u is an eigenvector of A, and v is an eigenvector of B, both with the same eigenvalue λ, we have

$$A u = \lambda u ; \qquad B v = \lambda v .$$

But $B = S^{-1}AS$, and consequently, $S^{-1}ASv = \lambda v$ or $A(Sv) = \lambda(Sv)$. Hence Sv is an eigenvector of A corresponding to the eigenvalue λ, and we have

$$u = Sv .$$

Because the determinant remains constant when rows and columns are interchanged, the eigenvalues of A and A^T are the same, while the eigenvectors, as a rule, are different.

We will now suppose that all eigenvalues of the matrix A are different, and further, that the eigenvectors of A and A^T are $u_1, u_2, \ldots, u_n,$ and $v_1, v_2, \ldots, v_n,$ respectively. Thus we have

$$\begin{cases} A u_i = \lambda_i u_i , \\ A^T v_k = \lambda_k v_k . \end{cases}$$

The first equation is transposed and multiplied from the right by v_k.

$$u_i^T A^T v_k = \lambda_i u_i^T v_k .$$

The second equation is multiplied from the left by u_i^T.

$$u_i^T A^T v_k = \lambda_k u_i^T v_k .$$

Subtracting, we find $(\lambda_i - \lambda_k) u_i^T v_k = 0$ or $u_i^T v_k = 0$, since $\lambda_i \neq \lambda_k$. The two groups of vectors u_i and v_i are said to be *biorthogonal*. Using the fact that an arbitrary vector w can be repesented as $w = \sum c_i v_i$ we see that $u_i^T v_i \neq 0$, because $u_i^T v_i = 0$ would imply $u_i^T w = 0$ and $u_i = 0$. Since the eigenvectors are determined, except for a constant factor, it is obvious that they can be chosen to satisfy the condition $u_i^T v_i = 1$. From the column vectors $u_1, u_2, \ldots, u_n$ and $v_1, v_2, \ldots, v_n$, we form two matrices, U and V, and further from the eigenvalues $\lambda_1, \lambda_2, \ldots, \lambda_n$, a diagonal matrix D. Then we have directly

$$U^T V = V^T U = I ,$$

that is

$$U^T = V^{-1} ; \qquad V^T = U^{-1} .$$

Further, we have $AU = UD$ and $A^T V = VD$ and, consequently,

$$U^{-1}AU = V^{-1}A^T V = D .$$

This is an example of the transformation of a matrix to diagonal form.

In the special case when A is symmetric, we get $U = V$ and $U^T = U^{-1}$. A matrix with this property is said to be *orthogonal*. Usually this concept is obtained by defining a vector transformation

$$x = Ou ; \qquad y = Ov$$

and requiring that the scalar product be invariant. Thus

$$x^T y = (Ou)^T Ov = u^T O^T Ov = u^T v \,,$$

which gives $O^T O = I$. Later on we will treat special matrices and their properties in more detail.

We will now take up the problem of giving estimates for the eigenvalues of a matrix A. Suppose that λ is a characteristic value of A with the corresponding eigenvector x. The components of x are denoted by $x_1, x_2, \ldots, x_n$. Choose M in such a way that $|x_M| = \max_r |x_r|$. Then we have directly

$$\lambda x_M = \sum_{r=1}^{n} a_{Mr} x_r$$

and

$$|\lambda|\, |x_M| = \left| \sum a_{Mr} x_r \right| \leq \sum |a_{Mr}|\, |x_r| \leq |x_M| \sum |a_{Mr}| \,,$$

from which we obtain the estimate

$$|\lambda| \leq \sum_{r=1}^{n} |a_{Mr}| \,.$$

However, in general, M is not known and instead we have to accept the somewhat weaker estimate

$$|\lambda| \leq \max_i \sum_{k=1}^{n} |a_{ik}| \,. \tag{3.3.3}$$

The transposed matrix has the same set of eigenvalues and hence we also have the estimate

$$|\lambda| \leq \max_k \sum_{i=1}^{n} |a_{ik}| \,, \tag{3.3.4}$$

and the better of these estimates is, of course, preferred. Writing

$$(\lambda - a_{MM})x_M = \sum_{r \neq M} a_{Mr} x_r \,,$$

we find in an analogous way

$$|\lambda - a_{MM}| \leq \sum_{r \neq M} |a_{Mr}| \,,$$

$$|\lambda - a_{MM}| \leq \sum_{r \neq M} |a_{rM}| \,.$$

Introducing the notations $P_i = \sum_{k \neq i} |a_{ik}|$ and $Q_i = \sum_{k \neq i} |a_{ki}|$, we can formulate the inequalities as follows. Any characteristic value λ must lie inside at least one of the circles

$$|\lambda - a_{ii}| \leq P_i \tag{3.3.5}$$

and also inside at least one of the circles

$$|\lambda - a_{ii}| \leq Q_i \,. \tag{3.3.6}$$

Suppose that we have a matrix A which can be transformed into diagonal form: $S^{-1}AS = D$. Using the previous inequalities on the matrix D, we find trivially $|\lambda - \lambda_i| \leq 0$, which means that the circles have degenerated to points. Now assume that $S = S_1 S_2 \cdots S_N$, where every S_k induces a "small" change of A. Then we can gradually transform D back to A. The points will then become circles which grow more and more. From the beginning the circles are disjoint and *every* circle contains exactly *one* eigenvalue. If two circles overlap partially, then exactly two eigenvalues are included, and so on. In this way one can often get more precise information on the distribution of the eigenvalues, at least so far as the domains are disjoint.

Combining the obtained inequalities, one can easily understand that all eigenvalues must lie inside the intersection of the union of the first group of circles and the union of the second group of circles.

The estimates discussed above are usually attributed to Gershgorin, but they have been improved by Ostrowski, who found that P_i and Q_i could be replaced by $R_i = P_i^\alpha Q_i^{1-\alpha}$ $(0 \leq \alpha \leq 1)$. When $\alpha = 0$ or $\alpha = 1$, the inequalities of Gershgorin appear again.

Finally, Brauer has shown that the two unions of n circles each can be replaced by two unions each consisting of $n(n-1)/2$ Cassini ovals

$$|z - a_{jj}||z - a_{kk}| \leq P_j P_k \qquad (k > j; \quad j = 1(1)n - 1)$$

and $|z - a_{jj}||z - a_{kk}| \leq Q_j Q_k$, respectively, and all eigenvalues lie within the intersection of these two unions.

Here we will also show how an approximate eigenvalue can be improved if we also know an approximate eigenvector. Let A be a given matrix, λ_0 and x be an approximate eigenvalue and eigenvector, $\lambda = \lambda_0 + \delta\lambda$ and y be the exact eigenvalue and eigenvector. Then

$$\begin{cases} Ax - \lambda_0 x = \varepsilon, \\ Ay - (\lambda_0 + \delta\lambda)y = 0. \end{cases}$$

Hence $x^T A = \lambda_0 x^T + \varepsilon^T$ and $x^T Ay = \lambda_0 x^T y + \varepsilon^T y$. On the other hand, $x^T Ay = (\lambda_0 + \delta\lambda)x^T y$, and comparing we get:

$$\delta\lambda = \frac{\varepsilon^T y}{x^T y} \simeq \frac{\varepsilon^T x}{x^T x}.$$

EXAMPLES

1.

$$A = \begin{pmatrix} 3 & 2 & 7 \\ -1 & 0 & 5 \\ 4 & -3 & 6 \end{pmatrix}.$$

We find that λ must lie inside the union of the circles

$$|\lambda - 3| \leq 9; \qquad |\lambda| \leq 6; \qquad |\lambda - 6| \leq 7$$

and also inside the union of the circles

$$|\lambda - 3| \leq 5 ; \qquad |\lambda| \leq 5 ; \qquad |\lambda - 6| \leq 12 .$$

In the first group, $|\lambda| \leq 6$ is a subset of $|\lambda - 3| \leq 9$. In the second group, the first two circles are subsets of the third. Hence the final domain is the union of $|\lambda - 3| \leq 9$ and $|\lambda - 6| \leq 7$.

Since $\max_i \sum_{k=1}^{n} |a_{ik}| = 13$ and $\max_k \sum_{i=1}^{n} |a_{ik}| = 18$, we also have $|\lambda| \leq 13$, but this condition gives no improvement. The eigenvalues are approximately 9.56 and $-0.28 \pm 3.50i$ and clearly lie inside the domain which has just been defined.

2.

$$A = \begin{pmatrix} 10 & 7 & 8 & 7 \\ 7 & 5 & 6 & 5 \\ 8 & 6 & 10 & 9 \\ 7 & 5 & 9 & 10 \end{pmatrix} ; \qquad \lambda_0 = 30 ; \qquad x = \begin{pmatrix} 0.53 \\ 0.38 \\ 0.55 \\ 0.52 \end{pmatrix} .$$

We find that

$$\varepsilon = \begin{pmatrix} 0.10 \\ 0.11 \\ 0.20 \\ 0.16 \end{pmatrix}$$

and $\delta\lambda = 0.288$. Hence $\lambda \simeq 30.288$ (the exact eigenvalue is $\lambda = 30.288685$).

The treated case concerning localization of characteristic values of a matrix can easily be generalized to arbitrary algebraic equations. Consider the equation

$$f(z) = z^n + a_1 z^{n-1} + \cdots + a_n = 0 .$$

This equation can be understood as the characteristic equation of a certain matrix A:

$$A = \begin{pmatrix} 0 & 0 \cdots 0 & -a_n \\ 1 & 0 \cdots 0 & -a_{n-1} \\ 0 & 1 \cdots 0 & -a_{n-2} \\ \vdots & & \\ 0 & 0 \cdots 1 & -a_1 \end{pmatrix} .$$

This is the so-called companion matrix, with the characteristic equation $\det (A - \lambda I) = 0$ or

$$\begin{vmatrix} -\lambda & 0 \cdots & 0 & -a_n \\ 1 & -\lambda \cdots & 0 & -a_{n-1} \\ \vdots & & & \\ 0 & 0 \cdots -\lambda & -a_2 \\ 0 & 0 \cdots & 1 & -\lambda - a_1 \end{vmatrix} = 0 .$$

We multiply the last line by λ and add to the preceding one, multiply this line by λ, and add to the preceding one, and so on, and in this way we find $f(\lambda) = 0$.

Now we use the estimates with P_i and Q_i and obtain the result: Every root of the equation $f(z)$ lies inside one of the two circles

$$|z| = 1 , \qquad |z + a_1| = \sum_{r=2}^{n} |a_r| .$$

Every root also lies inside one of the two circles

$$|z| = \max \left(\max_{1 < r < n} (1 + |a_r|), |a_n| \right) \qquad \text{and} \qquad |z + a_1| = 1 .$$

The estimate $|\lambda| \leq \max_k \sum_{i=1}^{n} |a_{ik}|$ gives the sufficient condition $\sum_{r=1}^{n} |a_r| \leq 1$ that all roots of the equation $f(z) = 0$ lie inside the unit circle. Naturally, this is easy to prove directly. Supposing that $f(z) = 0$ has a root z_0 whose absolute value is $r > 1$, we find that

$$z_0^n = -(a_1 z_0^{n-1} + \cdots + a_n)$$

and

$$r^n = |a_1 z_0^{n-1} + \cdots + a_n|$$
$$\leq |a_1| r^{n-1} + |a_2| r^{n-2} + \cdots + |a_n| < r^{n-1}(|a_1| + |a_2| + \cdots + |a_n|) < r^{n-1} ,$$

which is absurd.

3.4. Special matrices

In many applications real symmetric matrices play an important role. However, they are a special case of the so-called Hermitian matrices, characterized through the property $H^H = H$. Such matrices are extremely important in modern physics, and, to a large extent, this is due to the fact that all the eigenvalues of any Hermitian matrix are real. Let λ be an eigenvalue and u the corresponding eigenvector:

$$Hu = \lambda u . \tag{3.4.1}$$

We get by transposing and conjugating:

$$u^H H^H = u^H H = \lambda^* u^H . \tag{3.4.2}$$

Premultiplication of (3.4.1) by u^H and postmultiplication of (3.4.2) by u gives

$$u^H H u = \lambda^* u^H u = \lambda u^H u ,$$

and since $u^H u \neq 0$, we get $\lambda^* = \lambda$, that is, λ real. Again we return to (3.4.1). The condition $H^H = H$ can also be written $H^T = H^*$. Conjugating (3.4.1), we get

$$H^* u^* = H^T u^* = \lambda u^* ,$$

which means that u^* is an eigenvector of H^T if u is an eigenvector of H, while the eigenvalue is the same. We then repeat the argument of Section 3.3, where the eigenvectors were composed of matrices U and V with $U^T V = I$. As has just been proved, we have for Hermitian matrices $V = U^*$ and $U^T U^* = I$, or after conjugation,

$$U^H = U^{-1} . \tag{3.4.3}$$

Matrices with this property are said to be *unitary*, and hence we have proved that a Hermitian matrix can be transformed to a diagonal matrix by use of a suitable unitary matrix. Previously we have shown that a real symmetric matrix can be diagonalized by a suitable orthogonal matrix, which is obviously a special case of this more general property. The special case is well known from analytic geometry (transformation of conic sections to normal form).

We have seen that Hermitian matrices can be diagonalized by unitary matrices. It might be of interest to investigate if such transformations are always possible also for larger groups of matrices than Hermitian. We shall then prove that a necessary and sufficient condition that A can be diagonalized with a unitary matrix is that A is *normal*, that is, $A^H A = A A^H$. This condition is trivially satisfied if A is Hermitian.

(1) The condition is necessary, for if $UAU^H = D$, we get $A = U^H D U$ and

$$AA^H = U^H D U U^H D^H U$$
$$= U^H D D^H U = U^H D^H D U = U^H D^H U U^H D U = A^H A .$$

(2) The condition is sufficient, because according to Schur's lemma (which is proved in Section 3.7) an arbitrary matrix A can be factorized in $U^H T U$ where T is, for example, upper triangular. The condition $AA^H = A^H A$ implies $U^H T U U^H T^H U = U^H T^H U U^H T U$, that is, $T T^H = T^H T$. We now compare the $(1, 1)$-element on both sides and find:

$$\sum_{k=1}^{n} |t_{1k}|^2 = |t_{11}|^2 , \quad \text{that is,} \quad t_{1k} = 0 , \quad k = 2, 3, \ldots, n .$$

Next we compare the $(2, 2)$-element:

$$\sum_{k=2}^{n} |t_{2k}|^2 = |t_{22}|^2$$

(since $t_{12} = 0$) and hence $t_{2k} = 0$, $k = 3, 4, \ldots, n$. Consequently T is not only triangular but also diagonal.

Consider $Ux = \lambda x$, where U is a unitary matrix. Transposing and conjugating, we find that $x^H U^H = x^H U^{-1} = \lambda^* x^H$. Postmultiplication with $Ux = \lambda x$ gives us $x^H U^{-1} U x = \lambda^* \lambda x^H x$, and since $x^H x > 0$, we get $|\lambda| = 1$.

Analogous calculations can be performed also for other types of matrices, and the result is presented in the table below.

Type of matrix	Definition	Eigenvalues	Eigenvectors
Hermitian	$A^H = A$	λ real	u complex
Real, symmetric	$A^T = A^* = A$	λ real	u real
Anti-Hermitian	$A^H = -A$	λ purely imaginary	u complex
Antisymmetric	$A^T = -A$		If $\lambda \neq 0$, then $u^T u = 0$
Real, antisymmetric	$A^T = -A;\ A^* = A$	λ purely imaginary (or 0)	u complex (except when $\lambda = 0$)
Orthogonal	$A^T = A^{-1}$		If $\lambda \neq \pm 1$, then $u^T u = 0$
Real, orthogonal	$A^T = A^{-1};\ A^* = A$	$|\lambda| = 1$	u complex
Unitary	$A^H = A^{-1}$	$|\lambda| = 1$	u complex

Finally, we point out that if U_1 and U_2 are unitary, then also $U = U_1 U_2$ is unitary:

$$U^H = U_2^H U_1^H = U_2^{-1} U_1^{-1} = (U_1 U_2)^{-1} = U^{-1}.$$

Similarly, we can show that if O_1 and O_2 are orthogonal, then the same is true for $O = O_1 O_2$. For a real orthogonal matrix, the characteristic equation is real, and hence complex roots are conjugate imaginaries. Since the absolute value is 1, they must be of the form $e^{i\theta}, e^{-i\theta}$; further, a real orthogonal matrix of odd order must have at least one eigenvalue $+1$ or -1.

Wilkinson [10] has obtained rigorous error bounds for computed eigensystems, and we shall here give a brief account of his result for Hermitian matrices defined through the property $H^H = H$. Let $u_1, u_2, \ldots, u_n$ be an orthonormal set of eigenvectors of the Hermitian matrix H corresponding to the real eigenvalues $\lambda_1, \lambda_2, \ldots, \lambda_n$. Further, let λ be an approximate eigenvalue and x the corresponding approximate eigenvector assumed to be normalized. Then

$$x = \sum_{i=1}^{n} \alpha_i u_i,$$

and further

$$x^H x = \sum_{i=1}^{n} |\alpha_i|^2 = 1.$$

Let the residual vector be $r = Hx - \lambda x$, with the Euclidean norm ε (cf. 3.6) such that $r^H r = \varepsilon^2$. Then we shall prove that for at least one eigenvalue we have $|\lambda_i - \lambda| \leq \varepsilon$. Rewriting r we find

$$r = \sum_{1}^{n} (\lambda_i - \lambda)\alpha_i u_i,$$

and therefore

$$\varepsilon^2 = \sum_{1}^{n} |\lambda_i - \lambda|^2 |\alpha_i|^2.$$

Now, if $|\lambda_i - \lambda| > \varepsilon$ for all i, then we would get

$$\sum_1^n |\lambda_i - \lambda|^2 |\alpha_i|^2 > \varepsilon^2 \sum_1^n |\alpha_i|^2 = \varepsilon^2 ,$$

and hence we conclude that $|\lambda_i - \lambda| \leq \varepsilon$ for at least one value of i. So far, we have not used the fact that H is Hermitian.

Now we seek a real value μ which minimizes the Euclidean norm of $Hx - \mu x$.

$$|Hx - \mu x|^2 = (Hx - \mu x)^H (Hx - \mu x) = x^H H^2 x - 2\mu x^H Hx + \mu^2 .$$

Hence we get $\mu = x^H Hx$, which is the so-called *Rayleigh quotient*. [If x is not normalized, we get instead $x^H Hx / x^H x$; cf. (6.1.5)]. Using the notation λ_R for μ, we obtain the residual vector $r = Hx - \lambda_R x$ with the norm ε, where $r^H r = \varepsilon^2$. We suppose that λ_R is close to λ_1 and that $|\lambda_R - \lambda_i| \geq a$ for $i = 2, 3, \ldots, n$. We know, of course, that $|\lambda_R - \lambda_1| \leq \varepsilon$, but it is possible to find a much better result. For, by definition,

$$\lambda_R = x^H Hx = \left(\sum \alpha_i^* u_i^H\right)\left(\sum \alpha_i \lambda_i u_i\right) = \sum \lambda_i |\alpha_i|^2$$

and

$$Hx - \lambda_R x = \sum_1^n \alpha_i (\lambda_i - \lambda_R) u_i .$$

Thus we get

$$|Hx - \lambda_R x|^2 = \sum_1^n |\alpha_i|^2 |\lambda_i - \lambda_R|^2 \leq \varepsilon^2 ,$$

and hence *a fortiori*

$$\varepsilon^2 \geq \sum_2^n |\alpha_i|^2 |\lambda_i - \lambda_R|^2 \geq a^2 \sum_2^n |\alpha_i|^2 .$$

Thus

$$|\alpha_1|^2 = 1 - \sum_2^n |\alpha_i|^2 \geq 1 - \frac{\varepsilon^2}{a^2} .$$

From $\lambda_R = \sum_1^n \lambda_i |\alpha_i|^2$, we further obtain

$$\lambda_R \sum_1^n |\alpha_i|^2 = \sum_1^n \lambda_i |\alpha_i|^2$$

or

$$|\alpha_1|^2 (\lambda_R - \lambda_1) = \sum_2^n (\lambda_i - \lambda_R) |\alpha_i|^2 .$$

Hence

$$|\alpha_1|^2 |\lambda_R - \lambda_1| \leq \sum_2^n |\lambda_i - \lambda_R| |\alpha_i|^2$$

$$= \sum_2^n \frac{1}{|\lambda_i - \lambda_R|} |\lambda_i - \lambda_R|^2 |\alpha_i|^2 \leq \frac{1}{a} \sum_2^n |\lambda_i - \lambda_R|^2 |\alpha_i|^2 \leq \frac{\varepsilon^2}{a}$$

and

$$|\lambda_R - \lambda_1| \leq \frac{\varepsilon^2}{a} \Big/ \left(1 - \frac{\varepsilon^2}{a^2}\right) .$$

In this derivation a few additional computations gave an improved eigenvalue together with an error bound for this new value.

EXAMPLE

$$H = \begin{pmatrix} 10 & 7 & 8 & 7 \\ 7 & 5 & 6 & 5 \\ 8 & 6 & 10 & 9 \\ 7 & 5 & 9 & 10 \end{pmatrix}; \qquad x = \begin{pmatrix} 0.53 \\ 0.38 \\ 0.55 \\ 0.52 \end{pmatrix}.$$

We then find $x^H H x = 30.2340$, $x^H x = 0.9982$, and $\mu = 30.2885$ (the correct eigenvalue is $\lambda = 30.288685$).

3.5. Matrix functions

Let us start with a polynomial $f(z) = z^n + a_1 z^{n-1} + \cdots + a_n$, where z is a variable and $a_1, a_2, \ldots, a_n$ are real or complex constants. Replacing z by A, where A is a square matrix, we get a matrix polynomial

$$f(A) = A^n + a_1 A^{n-1} + \cdots + a_{n-1}A + a_n I.$$

Here $f(A)$ is a square matrix of the same order as A. We will now investigate eigenvalues and eigenvectors of $f(A)$. Suppose that λ is a characteristic value and u the corresponding characteristic vector of A: $Au = \lambda u$. Premultiplication with A gives the result $A^2 u = \lambda A u = \lambda^2 u$, that is λ^2 is an eigenvalue of A^2; the eigenvector is unchanged. In the same way we find that $f(\lambda)$ is an eigenvalue of the matrix $f(A)$.

It is natural to generalize to an infinite power series:

$$f(z) = a_0 + a_1 z + a_2 z^2 + \cdots.$$

We suppose that the series is convergent in the domain $|z| < R$. Then it can be shown (see Section 3.8) that if A is a square matrix with all eigenvalues less than R in absolute value, then the series $a_0 I + a_1 A + a_2 A^2 + \cdots$ is convergent. The sum of the series will be denoted by $f(A)$. Some series of this kind converge for all finite matrices, for example,

$$e^A = I + A + \frac{A^2}{2!} + \frac{A^2}{3!} + \cdots,$$

while, for example, the series $(I - A)^{-1} = I + A + A^2 + \cdots$ converges only if all eigenvalues of A are less than 1 in absolute value.

If a matrix A undergoes a similarity transformation with a matrix S,

$$S^{-1}AS = B, \tag{3.5.1}$$

then the same is true for A^2: $S^{-1}A^2 S = (S^{-1}AS)(S^{-1}AS) = B^2$. Obviously, this can be directly generalized to polynomials: $S^{-1}f(A)S = f(B)$. Generalization to the case when f also contains negative powers is immediate, since

$$S^{-1}A^{-1}S = (S^{-1}AS)^{-1} = B^{-1} \, .$$

We also conclude directly that

$$S^{-1}\left(\frac{f(A)}{g(A)}\right) S = \frac{f(B)}{g(B)} \, ,$$

where f and g are polynomials. The division has a good meaning, since we have full commutativity. Under certain conditions the results obtained can be further generalized, but we do not want to go into more detail right now.

Cayley-Hamilton's theorem: Any square matrix A satisfies its own characteristic equation. Let $f(\lambda)$ be the characteristic polynomial. Thus

$$(-1)^n f(\lambda) = \det(\lambda I - A) = \lambda^n + c_1 \lambda^{n-1} + \cdots + c_n \, .$$

First, we give a proof in the special case when all eigenvalues are different and hence the eigenvectors form a regular matrix S. Let D be a diagonal matrix with $d_{ii} = \lambda_i$. Then $S^{-1}AS = D$, and further

$$S^{-1}f(A)S = f(D) = \begin{pmatrix} f(\lambda_1) & 0 & \cdots & 0 \\ 0 & f(\lambda_2) & \cdots & 0 \\ \vdots & & & \\ 0 & 0 & \cdots & f(\lambda_n) \end{pmatrix} = 0 \, .$$

Premultiplication with S and postmultiplication with S^{-1} gives

$$f(A) = 0 \, . \tag{3.5.2}$$

A general proof can be obtained from this by observing that the coefficients of the characteristic polynomial are continuous functions of the matrix elements. Hence the case when some eigenvalues are equal can be considered as a limiting case, and in fact, with small modifications the given proof is still valid.

However, we will also give another proof due to Perron. For this, we need the concept of *polynomial matrix*, by which we mean a matrix whose elements are polynomials in a variable. Such a matrix can be written as a sum of matrices, each one containing just one power of the variable, or as a polynomial with constant matrices as coefficients.

Now we consider the adjoint matrix C of the matrix $\lambda I - A$. All elements of C are polynomials of λ of degree not more than $n - 1$. Hence C can be written

$$C = C_1 \lambda^{n-1} + C_2 \lambda^{n-2} + \cdots + C_n \, ,$$

where $C_1, C_2, \ldots, C_n$ are certain constant matrices. Thus

$$(\lambda I - A)(C_1 \lambda^{n-1} + C_2 \lambda^{n-2} + \cdots + C_n) \equiv \det(\lambda I - A) \cdot I$$
$$= (\lambda^n + c_1 \lambda^{n-1} + \cdots + c_n) I \, .$$

Identifying on both sides, we get:

$$\begin{cases} C_1 \qquad\quad = I, \\ C_2 - AC_1 = c_1 I, \\ C_3 - AC_2 = c_2 I, \\ \vdots \\ \quad\;\; - AC_n = c_n I, \end{cases}$$

Premultiplication of the first equation by A^n, the second by A^{n-1}, and so on, followed by addition, gives $A^n + c_1 A^{n-1} + \cdots + c_n I = 0$, and hence $f(A) = 0$.

EXAMPLE

$$A = \begin{pmatrix} 4 & 2 \\ 1 & 3 \end{pmatrix}; \qquad \lambda I - A = \begin{pmatrix} \lambda - 4 & -2 \\ -1 & \lambda - 3 \end{pmatrix};$$

$$C = \begin{pmatrix} \lambda - 3 & 2 \\ 1 & \lambda - 4 \end{pmatrix} = \lambda I + \begin{pmatrix} -3 & 2 \\ 1 & -4 \end{pmatrix};$$

$$(\lambda I - A)C = \begin{pmatrix} (\lambda - 4)(\lambda - 3) - 2 & 0 \\ 0 & (\lambda - 4)(\lambda - 3) - 2 \end{pmatrix} = (-1)^2 f(\lambda) I;$$

$$C_1 = I; \qquad C_2 = \begin{pmatrix} -3 & 2 \\ 1 & -4 \end{pmatrix};$$

$$(\lambda I - A)(\lambda I + C_2) = f(\lambda) I = (\lambda^2 - 7\lambda + 10) I;$$

$$[\lambda^2 + (C_2 - A)\lambda - AC_2] I = (\lambda^2 - 7\lambda + 10) I;$$

$$A \begin{vmatrix} C_2 - A = -7I \\ -AC_2 = 10I; \end{vmatrix}$$

$$-A^2 = -7A + 10I \qquad \text{or} \qquad A^2 - 7A + 10I = 0.$$

A shorter and more direct proof is the following. Denote by e_k a column vector with 1 as the kth element and 0 elsewhere. Then we have

$$A^T e_j = \sum_k a_{jk} e_k \qquad \text{or} \qquad \sum_k (a_{jk} - \delta_{jk} A^T) e_k = 0.$$

Let $c_{jk}(\lambda)$ be the algebraic complement of $a_{jk} - \lambda \delta_{jk}$ in the matrix $A - \lambda I$. Hence

$$\sum_j c_{jr}(\lambda)(a_{jk} - \lambda \delta_{jk}) = f(\lambda) \delta_{rk},$$

and this equation is valid identically in λ. Replacing λ by the matrix A^T, we find

$$f(A^T) e_r = \sum_k f(A^T) \delta_{rk} r_k = \sum_j c_{jr}(A^T) \sum_k (a_{jk} - A^T \delta_{jk}) e_k = 0.$$

This relation is valid for all r, and hence we have $f(A^T) = 0$ or $f(A) = 0$.

Cayley-Hamilton's theorem plays an important role in the theory of matrices as well as in many different applications. It should be mentioned here that for every square matrix A, there exists a unique polynomial $m(\lambda)$ of lowest degree with the leading coefficient 1 such that $m(A) = 0$. It is easy to prove that $m(\lambda)$ is a factor of $f(\lambda)$, since we can write $f(\lambda) = m(\lambda)q(\lambda) + r(\lambda)$, where $r(\lambda)$ is of lower degree than $m(\lambda)$. Replacing λ by A, we get

$$f(A) = m(A)q(A) + r(A) \qquad \text{or} \qquad r(A) = 0 ,$$

since $f(A) = 0$ and $m(A) = 0$. However, this is in contradiction to our supposition that $m(\lambda)$ is the minimum polynomial. In most cases we have $m(\lambda) = f(\lambda)$.

Now consider a matrix polynomial in A:

$$F(A) = A^N + a_1 A^{N-1} + \cdots + a_N I \qquad (N \geq n) .$$

Dividing $F(x)$ by $f(x)$, we get

$$F(x) = f(x)Q(x) + R(x) ,$$

where $R(x)$ is of degree $\leq n - 1$. Replacing x by A, we find

$$F(A) = f(A)Q(A) + R(A) = R(A) . \tag{3.5.3}$$

Without proof we mention that this formula is also valid for functions other than polynomials.

EXAMPLE

Consider e^A where A is a two-by-two-matrix. From the preceding discussion, we have

$$e^A = aA + bI .$$

Suppose that the eigenvalues of A are λ_1 and λ_2 ($\lambda_1 \neq \lambda_2$). Then e^A and $aA + bI$ have the eigenvalues e^{λ_1}, e^{λ_2}, and $a\lambda_1 + b$, $a\lambda_2 + b$, respectively. Hence

$$\begin{cases} e^{\lambda_1} = a\lambda_1 + b , \\ e^{\lambda_2} = a\lambda_2 + b . \end{cases}$$

From this system we solve for a and b:

$$a = \frac{e^{\lambda_2} - e^{\lambda_1}}{\lambda_2 - \lambda_1} ; \qquad b = \frac{\lambda_2 e^{\lambda_1} - \lambda_1 e^{\lambda_2}}{\lambda_2 - \lambda_1} ,$$

or written by use of determinants:

$$a = \frac{\begin{vmatrix} 1 & 1 \\ e^{\lambda_1} & e^{\lambda_2} \end{vmatrix}}{\begin{vmatrix} 1 & 1 \\ \lambda_1 & \lambda_2 \end{vmatrix}} ; \qquad b = \frac{\begin{vmatrix} e^{\lambda_1} & e^{\lambda_2} \\ \lambda_1 & \lambda_2 \end{vmatrix}}{\begin{vmatrix} 1 & 1 \\ \lambda_1 & \lambda_2 \end{vmatrix}} .$$

In the general case we try to determine the coefficients a_r in

$$F(A) = a_{n-1}A^{n-1} + \cdots + a_1 A + a_0 I, \tag{3.5.4}$$

and in this case we find

$$a_r = \frac{D_r}{D} \quad \text{with} \quad D = \begin{vmatrix} 1 & 1 & \cdots 1 \\ \lambda_1 & \lambda_2 & \cdots \lambda_n \\ \vdots & & \\ \lambda_1^{n-1} & \lambda_2^{n-1} \cdots \lambda_n^{n-1} \end{vmatrix}; \tag{3.5.5}$$

D_r is obtained by replacing the elements $\lambda_1^r, \ldots, \lambda_n^r$ in D by $F(\lambda_1), \ldots, F(\lambda_n)$.

An alternative expression for $F(A)$ is due to Sylvester. We will first treat a three-by-three-matrix A with the eigenvalues $\lambda_1, \lambda_2,$ and λ_3 (supposed to be different). Then $F(A)$ should be a polynomial of second degree, and we will show that it can be written

$$F(A) = \frac{(A - \lambda_2 I)(A - \lambda_3 I)}{(\lambda_1 - \lambda_2)(\lambda_1 - \lambda_3)} F(\lambda_1)$$

$$+ \frac{(A - \lambda_3 I)(A - \lambda_1 I)}{(\lambda_2 - \lambda_3)(\lambda_2 - \lambda_1)} F(\lambda_2) + \frac{(A - \lambda_1 I)(A - \lambda_2 I)}{(\lambda_3 - \lambda_1)(\lambda_3 - \lambda_2)} F(\lambda_3). \tag{3.5.6}$$

Let S be the matrix formed by the eigenvectors of A. Then

$$S^{-1}AS = D = \begin{pmatrix} \lambda_1 & 0 & 0 \\ 0 & \lambda_2 & 0 \\ 0 & 0 & \lambda_3 \end{pmatrix}$$

and

$$S^{-1}F(A)S = F(D) = \begin{pmatrix} F(\lambda_1) & 0 & 0 \\ 0 & F(\lambda_2) & 0 \\ 0 & 0 & F(\lambda_3) \end{pmatrix}.$$

Premultiplying the right-hand side of (3.5.6) by S^{-1} and postmultiplying by S, we get

$$\frac{(D - \lambda_2 I)(D - \lambda_3 I)}{(\lambda_1 - \lambda_2)(\lambda_1 - \lambda_3)} F(\lambda_1) + \cdots$$

$$= \begin{pmatrix} \lambda_1 - \lambda_2 & 0 & 0 \\ 0 & 0 & 0 \\ 0 & 0 & \lambda_3 - \lambda_2 \end{pmatrix} \begin{pmatrix} \lambda_1 - \lambda_3 & 0 & 0 \\ 0 & \lambda_2 - \lambda_3 & 0 \\ 0 & 0 & 0 \end{pmatrix} \frac{F(\lambda_1)}{(\lambda_1 - \lambda_2)(\lambda_1 - \lambda_3)} + \cdots$$

$$= \begin{pmatrix} F(\lambda_1) & 0 & 0 \\ 0 & 0 & 0 \\ 0 & 0 & 0 \end{pmatrix} + \begin{pmatrix} 0 & 0 & 0 \\ 0 & F(\lambda_2) & 0 \\ 0 & 0 & 0 \end{pmatrix} + \begin{pmatrix} 0 & 0 & 0 \\ 0 & 0 & 0 \\ 0 & 0 & F(\lambda_3) \end{pmatrix} = F(D).$$

Multiplying by S from the left and by S^{-1} from the right, we regain (3.5.6). Generalization to matrices of arbitrary order is trivial.

EXAMPLES

1. Compute A^{-3} when

$$A = \begin{pmatrix} 4 & 2 \\ 1 & 3 \end{pmatrix}.$$

We find $\lambda_1 = 5$ and $\lambda_2 = 2$. Hence

$$A^{-3} = \frac{\begin{pmatrix} 2 & 2 \\ 1 & 1 \end{pmatrix}}{3} \frac{1}{5^3} - \frac{\begin{pmatrix} -1 & 2 \\ 1 & -2 \end{pmatrix}}{3} \frac{1}{2^3} = \begin{pmatrix} 0.047 & -0.078 \\ -0.039 & 0.086 \end{pmatrix}.$$

2. Given

$$A = \begin{pmatrix} 5 & -3 \\ 2 & -2 \end{pmatrix};$$

find $A^{1/2}$.

Since the new characteristic values can be combined with different signs $(\pm\sqrt{\lambda_1}, \pm\sqrt{\lambda_2}, \dots)$ we generally obtain 2^n solutions of the matrix equation $X^2 = A$. If A is real, then X becomes real only if all eigenvalues of A are real and positive. In this example with $\lambda_1 = 4$, $\lambda_2 = -1$ we will restrict ourselves to the special matrix $A^{1/2}$ corresponding to the values 2 and i. Hence

$$A^{1/2} = \begin{pmatrix} 6 & -3 \\ 2 & -1 \end{pmatrix} \frac{2}{5} + \begin{pmatrix} 1 & -3 \\ 2 & -6 \end{pmatrix} \frac{i}{-5} = \frac{1}{5} \begin{pmatrix} 12 - i & -6 + 3i \\ 4 - 2i & -2 + 6i \end{pmatrix}.$$

3. Given

$$A = \begin{pmatrix} 4 & 2 \\ 1 & 3 \end{pmatrix};$$

find $\log A$, that is, the matrix X for which $e^X = A$.

We obtain directly:

$$\log A = \frac{1}{3} \begin{pmatrix} \log 50 & \log 6.25 \\ \log 2.5 & \log 20 \end{pmatrix} = \begin{pmatrix} 1.30401 & 0.61086 \\ 0.30543 & 0.99858 \end{pmatrix}.$$

3.6. Norms of vectors and matrices

The *norm of a vector* x is defined as a pure number $|x|$ fulfilling the following conditions:

1. $|x| > 0$ for $x \neq 0$ and $|0| = 0$.

2. $|cx| = |c|\,|x|$ for an arbitrary complex number c.

3. $|x + y| \leq |x| + |y|$.

The most common norms are:

  (a) The maximum norm $|x| = \max_i |x_i|$,
  (b) The absolute norm $|x| = \sum_i |x_i|$.
  (c) The Euclidean norm $|x| = (\sum_i |x_i|^2)^{1/2}$.

All these norms are special cases of the more general l_p-norm, defined by

$$|x| = \left\{ \sum_i |x_i|^p \right\}^{1/p},$$

for $p = \infty$, $p = 1$, and $p = 2$, respectively. We will mostly use the Euclidean norm. In this case the inequality by Cauchy-Schwarz is fulfilled:

$$|x^H y| \leq |x| \, |y| \, .$$

Analogously, a *matrix norm* is a pure number $||A||$ such that:

1. $||A|| > 0$ if $A \neq 0$ and $||0|| = 0$;
2. $||cA|| = |c| \, ||A||$ for arbitrary complex c ;   (3.6.2)
3. $||A + B|| \leq ||A|| + ||B||$;
4. $||AB|| \leq ||A|| \, ||B||$.

The most common matrix norms are:

(a) The maximum norms

$$||A|| = ||A||_\infty = \max_i \sum_k |a_{ik}|$$

and

$$||A|| = ||A||_1 = \max_k \sum_i |a_{ik}| \, ,$$

respectively.

(b) The Euclidean norm

$$N(A) = [\mathrm{Tr}\,(A^H A)]^{1/2} = [\mathrm{Tr}\,(AA^H)]^{1/2} = \left[\sum_{i,k} |a_{ik}|^2 \right]^{1/2} .$$

(c) The Hilbert or spectral norm $||A||_2 = \sqrt{\lambda_1}$, where λ_1 is the largest eigenvalue of $H = A^H A$.

Here we point out that from an arbitrary matrix norm $||A||$, we can construct another norm $||A_S||$ defined by $||A_S|| = ||S^{-1}AS||$, where S is a regular matrix. As is easily verified, the relations (3.6.2) are valid for $||A||_S$ if they are satisfied for $||A||$.

In most applications matrices appear together with vectors. For this reason the relations (3.6.2) should be augmented with a supplementary condition connecting vector and matrix norms. To a given vector norm, we define a *compatible* matrix norm by

$$||A|| = \sup_{x \neq 0} \frac{|Ax|}{|x|} \, .$$

Hence the condition of compatibility can be written $|Ax| \leq ||A|| \, |x|$. A matrix norm defined in this way is said to be *subordinate* to the vector norm under discussion. We note that with this definition we always have $||I|| = 1$. In particular, denoting the Euclidean matrix norm by $N(A)$, we observe that $N(I) = n^{1/2}$. Hence this norm is not subordinate to any vector norm. Nevertheless, we will use it for certain estimates, since it is so easily computed.

We will now show that for a given vector norm, the compatible matrix norm satisfies (3.6.2). The first two conditions are practically trivial. The third condition follows if we take a vector x with norm 1 for which $|(A + B)x|$ takes its maximum value. Then

$$\|A + B\| = |(A + B)x|$$
$$= |Ax + Bx| \leq |Ax| + |Bx| \leq \|A\|\,|x| + \|B\|\,|x| = \|A\| + \|B\| .$$

Last, the fourth condition follows in a similar way. Again let x be a vector with norm 1 but now such that $|ABx|$ takes its maximum value. Hence

$$\|AB\| = |ABx| \leq \|A\|\,|Bx| \leq \|A\|\,\|B\|\,|x| = \|A\|\,\|B\| .$$

Put $\|A\| = \max_i \sum_k |a_{ik}|$ and choose a vector x such that $|x|_\infty = m$, that is, $\max_i |x_i| = m$. Then we have for the vector $y = Ax$:

$$|y_i| = \left| \sum_{k=1}^{n} a_{ik} x_k \right| \leq m \sum_{k=1}^{n} |a_{ik}|$$

and further

$$|y|_\infty = \max_i |y_i| \leq m \max_i \sum_k |a_{ik}| .$$

We shall then prove that equality can occur. Let p be an index such that

$$\sum_k |a_{pk}| = \max_i \sum_k |a_{ik}|$$

and further suppose $|x_k| = m$ with sign $(x_k) =$ sign (a_{pk}) for $k = 1, 2, \ldots, n$. This implies $y_p = m \sum_k |a_{pk}|$, that is, $|y|_\infty = m \max_i \sum_k |a_{ik}|$. Hence we have

$$\sup_{x \neq 0} \frac{|Ax|_\infty}{|x|_\infty} = \max_i \sum_k |a_{ik}| .$$

The proof that the second maximum norm is compatible with the absolute norm is similar. Choose a vector x such that $|x|_1 = \sum_i |x_i| = m$ and form $y = Ax$. Then we have

$$|y|_1 = |Ax|_1 = \sum_i \left| \sum_k a_{ik} x_k \right| \leq \sum_i \sum_k |a_{ik}|\,|x_k|$$

$$\leq \sum_k |x_k| \sum_{i=1}^{n} |a_{ik}| \leq \left\{ \max_k \sum_{i=1}^{n} |a_{ik}| \right\} \cdot \sum_k |x_k|$$

$$= m \cdot \max_k \sum_{i=1}^{n} |a_{ik}| .$$

Hence we have $|Ax|/|x| \leq \max_k \sum_i |a_{ik}|$. We shall now prove that equality can occur and suppose that $\sum_{i=1}^{n} |a_{ik}|$ takes its greatest value for $k = q$. We then choose the vector x in such a way that $x_q = m$ while all other components are equal to zero. Then follows:

$$|Ax| = \sum_{i=1}^{n} \left| \sum_k a_{ik} x_k \right| = m \sum_{i=1}^{n} |a_{iq}| = m \max_k \sum_{i=1}^{n} |a_{ik}| .$$

Thus we have

$$\sup_{x \neq 0} \frac{|Ax|_1}{|x|_1} = \max_k \sum_i |a_{ik}| .$$

However, in the following we will concentrate on the Euclidean vector norm and the Hilbert matrix norm, and show first that they are compatible. Starting from a vector x with norm 1, we have $|Ax|^2 = (Ax)^H Ax = x^H A^H A x$. The matrix $H = A^H A$ is Hermitian, and we assume that it has the eigenvalues $\lambda_1 \geq \lambda_2 \geq \cdots \geq \lambda_n \geq 0$. The vector x is now written as a linear combination of the eigenvectors $u_1, u_2, \ldots, u_n$: $x = a_1 u_1 + a_2 u_2 + \cdots + a_n u_n$. Since $|x| = 1$, we have $x^H x = |a_1|^2 + |a_2|^2 + \cdots + |a_n|^2 = 1$ and further,

$$x^H H x = \lambda_1 |a_1|^2 + \lambda_2 |a_2|^2 + \cdots + \lambda_n |a_n|^2$$
$$\leq \lambda_1(|a_1|^2 + |a_2|^2 + \cdots + |a_n|^2) = \lambda_1 .$$

But $||A|| = \max_{|x|=1} |Ax|$, and choosing $x = u_1$, we find directly $||A|| = \sqrt{\lambda_1}$. In the following discussion we will use the notation $||A||$ for the Hilbert norm of a matrix.

For the Euclidean matrix norm, we have $N(A + B) \leq N(A) + N(B)$ obtained simply as the usual triangle inequality in n^2 dimensions. Further, let $C = AB$ and denote the column vectors of B and C by $b_1, b_2, \ldots, b_n$ and $c_1, c_2, \ldots, c_n$, respectively. Then

$$Ab_k = c_k; \qquad N(B) = \left(\sum_k b_k^H b_k\right)^{1/2} \qquad \text{and} \qquad N(AB) = \left(\sum_k c_k^H c_k\right)^{1/2} .$$

But

$$||A|| = \sup_{x \neq 0} \frac{|Ax|}{|x|} \geq \frac{|c_k|}{|b^k|} ,$$

since $|c_k|/|b_k|$ corresponds to a special choice of x. Hence $|c_k|^2 \leq ||A||^2 |b_k|^2$. Summing over k and taking the square root we find

$$N(AB) \leq ||A|| \, N(B) . \tag{3.6.3}$$

If $B = I$, we get $N(A) \leq n^{1/2} ||A||$. Further, let a_i^T be the ith row vector of A. Then the vector Ax has the components $a_i^T x$, and hence

$$|Ax| = \left(\sum_i |a_i^T x|^2\right)^{1/2} \leq \left(\sum_i |a_i|^2 |x|^2\right)^{1/2} = N(A) |x| .$$

From this we see that $|Ax|/|x| \leq N(A)$ and also that $||A|| \leq N(A)$. Using (3.6.3), we then obtain

$$N(AB) \leq N(A) \, N(B) . \tag{3.6.4}$$

Summing up, we have the estimates

$$\begin{cases} ||A|| \leq N(A) \leq n^{1/2} ||A|| , \\ n^{-1/2} N(A) \leq ||A|| \leq N(A) . \end{cases} \tag{3.6.5}$$

If the elements of a matrix A are bounded through $|a_{ik}| \leq c$, then we also have

$$|\text{Tr}\,(A)| \leq nc\,; \qquad N(A) \leq nc\,; \qquad ||A|| \leq nc\,.$$

Now we choose two vectors, x and y, both nonzero. Observing that

$$y^H A x = (x^H A^H y)^*$$

is a pure number, we have, using Cauchy-Schwarz' inequality,

$$\frac{|y^H A x|}{|x|\,|y|} \leq \frac{|Ax|}{|x|}\frac{|y|}{|y|} = \frac{|Ax|}{|x|}\,,$$

and putting $y = Ax$, we obtain further

$$\frac{|y^H A x|}{|x|\,|y|} = \frac{|Ax|\,|Ax|}{|Ax|\,|x|} = \frac{|Ax|}{|x|}\,;$$

that is, the equality sign can be valid in the inequality. Hence we also have

$$||A|| = \sup_{x,y\neq 0} \frac{|y^H A x|}{|x|\,|y|}\,. \tag{3.6.6}$$

On the other hand, we have

$$||A^H|| = \sup_{x,y\neq 0} \frac{|x^H A^H y|}{|y|\,|x|} = \sup_{x,y\neq 0} \frac{|y^H A x|}{|x|\,|y|} = ||A||\,.$$

For a moment we consider the matrix $A^H A$ and obtain

$$\sup_{x\neq 0} \frac{|x^H A^H A x|}{|x|^2} \leq \sup_{x,y\neq 0} \frac{|y^H A^H A x|}{|x|\,|y|}\,,$$

since the left-hand side is a special case of the right-hand side. However, the left-hand side is $\sup_{x\neq 0} |Ax|^2/|x|^2 = ||A||^2$, while the right-hand side is $= ||A^H A||$. Hence $||A^H A|| \geq ||A||^2$. On the other hand, the inequality $||BC|| \leq ||B||\,||C||$ gives $||A^H A|| \leq ||A^H||\,||A|| = ||A||^2$. Hence we finally get

$$||A^H A|| = ||A||^2 = ||A^H||^2\,. \tag{3.6.7}$$

From this relation we can compute $||A||$ in an alternative way. If U is unitary, we have

$$|Ux| = (x^H U^H U x)^{1/2} = (x^H x)^{1/2} = |x|\,.$$

Putting $H = A^H A$, we see that H is Hermitian and can be diagonalized with the aid of a unitary matrix U: $U^H H U = C$, where C is diagonal. Then we have

$$||C|| = \sup_{|x|=1} |U^H H U x|$$

$$= \sup_{|x|=1} (x^H U^H H^H U U^H H U x)^{1/2} = \sup_{|y|=1} (y^H H^H H y)^{1/2} = ||H||\,,$$

where $y = Ux$. Thus the norm remains invariant under a unitary transformation. But the norm of C is trivial: $||C|| = \lambda_1$, where λ_1 is the largest eigenvalue of $H = A^H A$. This matrix is positive semidefinite, since for arbitrary x we have

$$x^H H x = x^H A^H A x = (Ax)^H A x \geq 0 .$$

Then all the characteristic roots λ_i of H are ≥ 0. Let λ and u satisfy $A^H A u = \lambda u$. Then $u^H A^H A u = \lambda u^H u$, that is, $\lambda = \{(Au)^H Au\}/u^H u \geq 0$. Hence

$$||A|| = ||A^H|| = \sqrt{\lambda_1} . \tag{3.6.8}$$

If $H = (h_{ik})$ is a Hermitian matrix, then we have

$$N(H) = [\mathrm{Tr}\,(H^H H)]^{1/2} = [\mathrm{Tr}\,(H^2)]^{1/2} = \left(\sum_i \lambda_i^2\right)^{1/2},$$

if λ_i are the eigenvalues of H. For a moment introduce $v = (\sum |h_{ik}|^2)^{1/2}$. Then we get from (3.6.5)

$$\frac{v}{\sqrt{n}} \leq |\lambda_1| \leq v ,$$

where λ_1 is the numerically largest eigenvalue of H. It should be observed that all eigenvalues of the special Hermitian matrix $A^H A$ are positive, while an arbitrary Hermitian matrix may also have negative eigenvalues.

Next we are going to deduce some estimates of norms in connection with inverses of matrices.

If $||A|| < 1$, the inverse of $I - A$ exists, since

$$|(I - A)x| = |x - Ax| \geq |x| - |Ax| \geq |x| - ||A||\,|x| = (1 - ||A||)\,|x| ,$$

and hence $(I - A)x \neq 0$ for an arbitrary $x \neq 0$, which means that $I - A$ is regular.

From $I = (I - A)(I - A)^{-1} = (I - A)^{-1} - A(I - A)^{-1}$, we have

$$||A||\,||(I - A)^{-1}|| \geq ||A(I - A)^{-1}||$$
$$= ||(I - A)^{-1} - I|| \geq ||(I - A)^{-1}|| - 1 ,$$

and consequently $||(I - A)^{-1}|| \leq 1/(1 - ||A||)$.

Analogously, by treating $I = (I + A)(I + A)^{-1}$ in a similar way, we find that $||(I + A)^{-1}|| \geq 1/(1 + ||A||)$. Since $||-A|| = ||A||$, we can write

$$\begin{cases} \dfrac{1}{1 + ||A||} \leq ||(I + A)^{-1}|| \leq \dfrac{1}{1 - ||A||} , \\[2mm] \dfrac{1}{1 + ||A||} \leq ||(I - A)^{-1}|| \leq \dfrac{1}{1 - ||A||} . \end{cases} \tag{3.6.9}$$

Both inequalities are valid under the assumption that $||A|| < 1$.

Last, we will show one more inequality which is of interest in connection with the inversion of matrices. We have

$$\begin{aligned}
||(A + B)^{-1} - A^{-1}|| &= ||A^{-1} - (A + B)^{-1}|| \\
&= ||A^{-1} - (I + A^{-1}B)^{-1}A^{-1}|| \le ||A^{-1}|| \, ||I - (I + A^{-1}B)^{-1}|| \\
&= ||A^{-1}|| \, ||(I + A^{-1}B)^{-1}(I + A^{-1}B - I)|| \\
&\le ||A^{-1}|| \, ||A^{-1}B|| \, ||(I + A^{-1}B)^{-1}|| \le ||A^{-1}|| \frac{\alpha}{1 - \alpha} \, ,
\end{aligned}$$

where $\alpha = ||A^{-1}B||$ is assumed to be < 1. Hence

$$||(A + B)^{-1} - A^{-1}|| \le ||A^{-1}|| \frac{\alpha}{1 - \alpha} \, . \tag{3.6.10}$$

We have previously defined the spectral norm and shown that $||A|| = \sqrt{\lambda_1}$, where λ_1 is the largest eigenvalue of $A^H A$. We now define also the spectral radius $\rho(A) = \max_{1 \le i \le n} |\lambda_i|$ where λ_i are the eigenvalues of A. For each matrix norm corresponding to a vector norm $\rho(A) \le ||A||$, because there exist a number λ and a vector x such that $Ax = \lambda x$ and $|\lambda| = \rho(A)$. Hence

$$|Ax| = \rho(A) |x| \, ,$$

that is, $\rho(A) = |Ax|/|x| \le \sup |Ay|/|y| = ||A||$. If specially A is Hermitian, then we have $\rho(A) = ||A||$.

In the theory of differential equations, a measure $\mu(A)$ is sometimes used. This quantity can also take negative values, and hence it is no matrix norm. It is defined in the following way:

$$\mu(A) = \lim_{\varepsilon \to +0} \frac{||I + \varepsilon A|| - 1}{\varepsilon} \, . \tag{3.6.11}$$

Neglecting quadratic and higher order terms in ε, we have from (3.6.7):

$$||I + \varepsilon A|| = ||I + \varepsilon(A + A^H)||^{1/2}$$

and further from (3.6.8):

$$\mu(A) = \lambda_1 \, , \tag{3.6.12}$$

where λ_1 is the largest eigenvalue of the matrix $H = \frac{1}{2}(A + A^H)$. For further details see, for example, Dahlquist [5].

3.7. Transformation of vectors and matrices

An arbitrary real nonsingular square matrix can be interpreted as a system of n linearly independent row- or column-vectors. We are now going to show how a system of *orthogonal* vectors can be constructed from them. Denoting the initial vectors by $a_1, a_2, \ldots, a_n$, we form

$$
\begin{cases}
w_1 = a_1 , \\
w_2 = a_2 - (a_2^T w_1 / w_1^T w_1) w_1 , \\
\vdots \\
w_i = a_i - \left\{ \sum_{k=1}^{i-1} [a_i^T w_k / w_k^T w_k] \right\} w_k , \\
\vdots \\
w_n = a_n - \left\{ \sum_{k=1}^{n-1} [a_n^T w_k / w_k^T w_k] \right\} w_k .
\end{cases}
$$

It is obvious that $w_i^T w_k = 0$ if $i \neq k$. These equations can be written in matrix form

$$WB = A ,$$

where $W = (w_1, w_2, \ldots, w_n)$, $A = (a_1, a_2, \ldots, a_n)$, and

$$
B = \begin{pmatrix}
1 & b_{12} & b_{13} & \cdots & b_{1n} \\
0 & 1 & b_{23} & \cdots & b_{2n} \\
0 & 0 & 1 & \cdots & b_{3n} \\
\vdots & & & & \vdots \\
0 & 0 & 0 & \cdots & 1
\end{pmatrix}
$$

with $b_{ik} = w_i^T a_k / w_i^T w_i$. This technique is known as Gram-Schmidt's orthogonalization process.

Previously we showed that an arbitrary Hermitian matrix can be brought to diagonal form by a unitary transformation

$$U^H H U = D .$$

We also noticed that if the n eigenvectors of an arbitrary matrix A are linearly independent, they form a regular matrix S such that

$$S^{-1} A S = D .$$

It is easy to show that if all eigenvalues are different, then the eigenvectors are linearly independent. First, it is trivial that the eigenvector $u \neq 0$ cannot belong to two different values λ_1 and λ_2. Now suppose that the eigenvector u belonging to λ is a linear combination of the eigenvectors u_i belonging to λ_i:

$$u = \sum c_i u_i .$$

Operating with A on both sides, we get

$$Au = \lambda u = \lambda \sum c_i u_i = \sum c_i \lambda_i u_i ,$$

that is, we obtain a linear relation between the u_i contrary to the supposition.

If we also have multiple eigenvalues, the situation is considerably more complicated. Here we will restrict ourselves to stating an important theorem by Jordan without proof:

Every matrix A can be transformed with a regular matrix S as follows:

$$S^{-1}AS = J,$$

where

$$J = \begin{pmatrix} J_1 & 0 \cdots 0 \\ 0 & J_2 \cdots 0 \\ & \cdot \cdot \\ & \cdot \cdot \\ 0 & 0 \cdots J_m \end{pmatrix}.$$ (3.7.1)

The J_i are so-called Jordan boxes, that is, submatrices of the form

$$\lambda_i, \begin{pmatrix} \lambda_i & 1 \\ 0 & \lambda_i \end{pmatrix}, \begin{pmatrix} \lambda_i & 1 & 0 \\ 0 & \lambda_i & 1 \\ 0 & 0 & \lambda_i \end{pmatrix}, \begin{pmatrix} \lambda_i & 1 & 0 & 0 \\ 0 & \lambda_i & 1 & 0 \\ 0 & 0 & \lambda_i & 1 \\ 0 & 0 & 0 & \lambda_i \end{pmatrix}, \cdots$$

In one box we have the same eigenvalue λ_i in the main diagonal, but it can appear also in other boxes. The representation (3.7.1) is called *Jordan's normal form*. If, in particular, all λ_i are different, J becomes a diagonal matrix. For proof and closer details, see, for example, Gantmacher [1] or Schreier-Sperner, II [3].

The occurrence of multiple eigenvalues causes degeneration of different kinds. If an eigenvalue is represented in a Jordan-box with at least 2 rows, we obtain a lower number of linearly independent eigenvectors than the order of the matrix. A matrix with this property is said to be *defective*. If one or more eigenvalues are multiple but in such a way that the corresponding Jordan-boxes contain just one element (that is, no ones are present above the main diagonal), we have another degenerate case characterized by the fact that the minimum polynomial is of lower degree than the characteristic polynomial. A matrix with this property is said to be *derogatory*. Both these degenerate cases can appear separately or combined.

For later use we will now prove a lemma due to Schur. Given an arbitrary matrix A, one can find a unitary matrix U such that $T = U^{-1}AU$ is triangular.

Let the eigenvalues of A be $\lambda_1, \lambda_2, \ldots, \lambda_n$ and determine U_1 so that

$$U_1^{-1}AU_1 = A_1,$$

where A_1 has the following form:

$$A_1 = \begin{pmatrix} \lambda_1 & \alpha_{12} & \alpha_{13} & \cdots & \alpha_{1n} \\ 0 & \alpha_{22} & \alpha_{23} & \cdots & \alpha_{2n} \\ 0 & \alpha_{32} & \alpha_{33} & \cdots & \alpha_{3n} \\ \vdots & & & & \\ 0 & \alpha_{n2} & \alpha_{n3} & \cdots & \alpha_{nn} \end{pmatrix}.$$

As is easily found, the first column of U_1 must be a multiple of x_1, where $Ax_1 = \lambda_1 x_1$. Otherwise, the elements of U_1 can be chosen arbitrarily. The choice can be made in such a way that U_1 becomes unitary; essentially this corresponds to the well-known fact that in an n-dimensional space, it is always possible to choose n mutually orthogonal unit vectors. Analogously, we determine a unitary matrix U_2 which transforms the elements $\alpha_{32}, \alpha_{42}, \ldots, \alpha_{n2}$ to zero. This matrix U_2 has all elements in the first row and the first column equal to zero except the corner element, which is 1. The $(n-1)$-dimensional matrix obtained when the first row and the first column of A_1 are removed has, of course, the eigenvalues $\lambda_2, \lambda_3, \ldots, \lambda_n$, and the corresponding eigenvectors $x_2', x_3', \ldots, x_n'$, where x_i' is obtained from x_i by removing the first component. It is obvious that after $n-1$ rotations of this kind, a triangular matrix is formed and further that the total transformation matrix $U = U_1 U_2 \cdots U_{n-1}$ is unitary, since the product of two unitary matrices is also unitary.

Last, we will show that, by use of a similarity transformation, any matrix A can be transformed to a triangular matrix with the off-diagonal elements arbitrarily small. We start by transforming A to triangular form:

$$T = U^{-1}AU.$$

Then we form $B = D^{-1}TD$, where D is a diagonal matrix with the nonzero elements $\varepsilon, \varepsilon^2, \ldots, \varepsilon^n$. Hence $b_{ik} = t_{ik}\varepsilon^{k-i}$ $(k \geq i)$; that is, the diagonal elements are unchanged, and the other elements can be made arbitrarily small by choosing ε sufficiently small. The matrix B has been obtained from A through $B = S^{-1}AS$, with $S = UD$. Here B is triangular with off-diagonal elements arbitrarily small and the eigenvalues in the main diagonal.

We mention here that in many applications almost diagonal matrices (band matrices) as well as almost triangular matrices (Hessenberg matrices) are quite frequent. In the first case elements not equal to zero are present only in the main diagonal and the p closest diagonals on both sides $(p < n - 1)$. The most interesting case corresponds to $p = 1$ (tridiagonal matrtices). In the second case all elements are zero below (over) the main diagonal except in the diagonal closest to the main diagonal (upper and lower Hessenberg matrix).

3.8. Limits of matrices

A matrix A is said to approach the limit 0 if all elements converge to zero. From (3.6.5) we get directly that if $||A|| \to 0$, then $A \to 0$, and vice versa. A simple generalization is the following. A sequence of matrices $A_1, A_2, \ldots$ converges toward the matrix A if every element of A_N converges to the corresponding element of A. A necessary and sufficient condition for convergence is that $||A - A_N|| \to 0$.

The consecutive powers A^N of a matrix A converge to zero if and only if all eigenvalues of A lie inside the unit circle. This follows directly if A can be transformed to diagonal form:

$$S^{-1}AS = D \, .$$

Then we have $A^N = SD^N S^{-1}$, where D is diagonal with the diagonal elements $\lambda_1, \lambda_2, \ldots, \lambda_n$. Hence we get the condition $\lambda_i^N \to 0$ for $i = 1, 2, \ldots, n$.

If A cannot be transformed to diagonal form, we can use Jordan's normal form. We demonstrate the technique on the series

$$f(A) = a_0 I + a_1 A + a_2 A^2 + \cdots ,$$

and consider a Jordan box J:

$$J = \begin{pmatrix} \lambda & 1 & 0 \cdots 0 \\ 0 & \lambda & 1 \cdots 0 \\ \vdots & & & \vdots \\ 0 & 0 & 0 \cdots \lambda \end{pmatrix} \qquad (p \text{ rows and columns}).$$

By induction it is easy to show that

$$J^N = \begin{pmatrix} \lambda^N & \binom{N}{1}\lambda^{N-1} \cdots \binom{N}{p-1}\lambda^{N-p+1} \\ 0 & \lambda^N & \cdots \binom{N}{p-2}\lambda^{N-p+2} \\ \vdots & & \\ 0 & 0 & \cdots \qquad \lambda^N \end{pmatrix}$$

and more generally

$$f(J) = \begin{pmatrix} f(\lambda) & f'(\lambda)/1! \cdots f^{(p-1)}(\lambda)/(p-1)! \\ 0 & f(\lambda) & \cdots f^{(p-2)}(\lambda)/(p-2)! \\ \vdots & & \\ 0 & 0 & \cdots \qquad f(\lambda) \end{pmatrix} .$$

Hence, for convergence it is necessary and sufficient that the series

$$\sum_N a_N \binom{N}{p-1} \lambda^N$$

converges for every eigenvalue; p is the largest order for those Jordan boxes which belong to this eigenvalue λ. Thus it is obvious that the matrix series converges if $|\lambda_1| < R$, where λ_1 is the absolutely largest eigenvalue and R is the convergence radius for the series $\sum_N a_N z^N$.

Alternatively, we can perform the transformation to almost triangular form just described in Section 3.7:

$$B = S^{-1}AS \, .$$

If the series $a_0 I + a_1 A + a_2 A^2 + \cdots + a_N A^N$ is transformed in the same way, we get, in the main diagonal, elements of the form $a_0 + a_1 \lambda + a_2 \lambda^2 + \cdots + a_N \lambda^N$.

From this we see directly that the matrix series converges for such matrices A that $|\lambda_i| < R$ for all i.

Derivatives and integrals of matrices whose elements are functions of real or complex variables are defined by performing the operations on the individual elements. Care must be exercised, however, because the matrices need not commutate. We give the following examples:

$$\frac{d}{dt}(AB) = A\frac{dB}{dt} + \frac{dA}{dt}B.$$

In particular, if $A = B$, we have

$$\frac{d}{dt}(A^2) = A\frac{dA}{dt} + \frac{dA}{dt}A.$$

Differentiating the identity $AA^{-1} = I$, we get

$$A\frac{dA^{-1}}{dt} + \frac{dA}{dt}A^{-1} = 0 \qquad \text{and} \qquad \frac{dA^{-1}}{dt} = -A^{-1}\frac{dA}{dt}A^{-1}.$$

Problems of this kind appear, above all, in the theory of ordinary differential equations (see Chapter 14).

REFERENCES

[1] Gantmacher: *Matrizenrechnung*, I (translated from the Russian) (Berlin, 1958).

[2] Faddeeva: *Computational Methods of Linear Algebra* (translated from the Russian) (Dover, New York, 1959).

[3] Schreier-Sperner: *Analytische Geometrie*, II (Berlin, 1935).

[4] MacDuffee: *The Theory of Matrices* (Chelsea, New York, 1956).

[5] Dahlquist: *Stability and Error Bounds in the Numerical Integration of Ordinary Differential Equations* (Dissertation, Uppsala, 1958).

[6] Zurmühl: *Matrizen* (Berlin, 1961).

[7] Bodewig: *Matrix Calculus* (Amsterdam, 1959).

[8] National Bureau of Standards: "*Basic Theorems in Matrix Theory*" *Appl. Math. Series*, No. 57 (Washington, 1960).

[9] Varga: *Matrix Iterative Analysis* (Prentice-Hall, Englewood Cliffs, N. J., 1962).

[10] Wilkinson: "Rigorous Error Bounds for Computed Eigensystems," *Comp. Jour.*, **4**, (Oct. 1961).

[11] Fox: *An Introduction to Numerical Linear Algebra* (Clarendon Press, Oxford, 1964).

EXERCISES

1. Show that an antisymmetric matrix of odd order is singular.

2. v is a column vector with n elements. From this vector we form a matrix $A = vv^T$. Show (a) that A is symmetric; (b) that $A^2 = cA$, where c is a constant (depending upon v); (c) that if u is a vector such that $u^T v = 0$, then u is an eigenvector of A corresponding to the eigenvalue 0; (d) that v is an eigenvector of A (the eigenvalue to be computed).

3. Find a $(3, 3)$-matrix which has the eigenvalues $\lambda_1 = 6$, $\lambda_2 = 2$, and $\lambda_3 = -1$, given that the corresponding eigenvectors are

$$u_1 = \begin{pmatrix} 2 \\ 3 \\ -2 \end{pmatrix}; \qquad u_2 = \begin{pmatrix} 9 \\ 5 \\ 4 \end{pmatrix}; \qquad u_3 = \begin{pmatrix} 4 \\ 4 \\ -1 \end{pmatrix}.$$

4. A is a matrix of order (n, n), where all $a_{ik} = 0$ for $i \le k$. Show that $(I - A)^{-1} = I + A + A^2 + \cdots + A^{n-1}$, where I is the unit matrix.

5. The matrix

$$A = \begin{pmatrix} 6 & 2 \\ 1 & 5 \end{pmatrix}.$$

is given. Then $\cos A$ can be written in the form $aA + bI$. Find the constants a and b.

6. Show that $B = e^A$ is an orthogonal matrix if A is antisymmetric. Find B if

$$A = \begin{pmatrix} 0 & \alpha \\ -\alpha & 0 \end{pmatrix}.$$

7. The matrix

$$A = \begin{pmatrix} 0.9 & 0.2 \\ 0.3 & 0.4 \end{pmatrix}$$

is given. Find $\lim_{n \to \infty} A^n$, for example, by proving that the elements of successive matrices A^n form partial sums in convergent geometric series.

8. A and B are $n \times n$-matrices. Prove that the relation $AB - BA = I$ cannot be valid.

9. Both x and y are vectors with n rows. Show that $\det(I - xy^T) = 1 - x^T y$.

10. A $(2, 2)$-matrix

$$A = \begin{pmatrix} a & b \\ c & d \end{pmatrix}$$

with complex elements is given. A is transformed with a unitary matrix U:

$$U^{-1}AU = A' = \begin{pmatrix} a' & b' \\ c' & d' \end{pmatrix},$$

where we choose

$$U = \begin{pmatrix} p & -p\alpha^* \\ p\alpha & p \end{pmatrix}$$

with p real. Then p and α are determined from the conditions that U is unitary and that A' is an upper triangular matrix (that is, $c' = 0$). Find p and α expressed in a, b, c, and d, and use the result for computing a' and d' if

$$A = \begin{pmatrix} 9 & 10 \\ -2 & 5 \end{pmatrix}.$$

11. Two quadratic matrices of the same order, A and B, are symmetric with respect to both diagonals. Show that the matrix $C = AB + BA$ has the same property.

12. A quadratic matrix which has exactly one element equal to 1 in every row and every column and all other elements equal to zero is called a permutation matrix. Show that a permutation matrix is orthogonal and that all eigenvalues are 1 in absolute value.

13. A matrix A is said to be normal if $A^H A = A A^H$. If A is normal it can be transformed by a unitary matrix U to a diagonal matrix B: $B = U^{-1}AU$. Find U and B when

$$A = \begin{pmatrix} a & b \\ -b & a \end{pmatrix},$$

a and b real.

14. One wants to construct an (n, n)-matrix P_r such that $P_r a_1 = P_r a_2 = \cdots = P_r a_r = 0$, where $a_1, a_2, \ldots, a_r$ are given vectors with n components $(r < n)$ which are linearly independent. A matrix A_r of type (r, n) is formed by combining the row vectors $a_1^T, a_2^T, \ldots, a_r^T$ in this order. Show that $P_r = I - A_r^T (A_r A_r^T)^{-1} A_r$ has the desired property (P_r is called a projection matrix) and that P_r is singular when $r \geq 1$. Finally, compute P_r for $r = 2$ when $a_1^T = (1, 3, 0, -2)$ and $a_2^T = (0, 4, -1, 3)$.

15. A is a nonquadratic matrix such that one of the matrices AA^T and $A^T A$ is regular. Use this fact for determining one matrix X such that $AXA = A$. Then find at least one matrix X if

$$A = \begin{pmatrix} 2 & 1 & 0 & -1 \\ 0 & 3 & 1 & 1 \\ -1 & -3 & -1 & 0 \end{pmatrix}.$$

(A matrix X with this property is called a pseudo-inverse of the matrix A.)

Chapter 4

Linear systems of equations

> *"Mine is a long and sad tale"* said the
> Mouse, turning to Alice and sighing.
> *"It is a long tail, certainly"* said Alice,
> looking down with wonder at the Mouse's tail,
> *"but why do you call it sad?"*
>
> LEWIS CARROLL.

4.0. Introduction

We shall now examine a linear system of m equations with n unknowns. If
$m > n$, as a rule the equations cannot be satisfied. In a subsequent chapter we
will discuss how "solutions" are explored which agree with the given equations
as well as possible. If $m < n$, the system usually has an infinite number of solutions. In this chapter we will treat only the case $m = n$.

Let A be the coefficient matrix, x the solution vector, and y the known right-
hand-side vector. In the normal case, we have $y \neq 0$ and $\det A \neq 0$. As is well
known, we have then exactly one solution x. If $y = 0$, and $\det A \neq 0$, we have
only the trivial solution $x = 0$. If $\det A = 0$ but $y \neq 0$, then there exists a finite
solution only in exceptional cases. Last, if $\det A = 0$ and $y = 0$, apart from the
trivial solution $x = 0$, we also have an infinity of nontrivial solutions.

Now suppose that $\det A = D \neq 0$ and $y \neq 0$. According to Cramer's rule,
the system $Ax = y$ has the solution

$$x_r = D_r/D \,, \tag{4.0.1}$$

where D_r is the determinant, which is obtained when the rth column in D is replaced by y. Cramer's rule, of course, is identical with the formula $x = A^{-1}y$,
where A^{-1} is taken from Formula (3.1.2).

After the solution (4.0.1) had been obtained, the whole problem was considered to be solved once and forever. This is true from a purely theoretical
but certainly not from a numerical point of view. Of course, Cramer's rule is
satisfactory when $n = 3$ or $n = 4$, but what happens if $n = 50, 100$, or more?
How does one evaluate a determinant of this magnitude? Using the definition
directly, one will find the number of necessary operations to be of the order $n!$,
and, as is well known, this expression increases very rapidly with n. Already
$n = 50$ would give rise to more than 10^{64} operations, and this enormous numerical task is, of course, beyond reach even of the fastest computers. Even if the

determinants are evaluated in the best possible way, the number of operations is proportional to n^4, while the methods we are going to discuss need only computational work of the order n^3.

We will distinguish mainly between *direct* and *iterative* methods. Among the direct methods, we will first mention the elimination method by Gauss, with modifications by Crout and Jordan, and then the triangularization method. Among the iterative methods, we will give greatest emphasis to the Gauss-Seidel method and the successive over-relaxation method by Young.

4.1. The elimination method by Gauss

We are now going to examine the system $Ax = y$, assuming det $A \neq 0$ and $y \neq 0$. Explicitly the system has the following form:

$$\begin{cases} a_{11}x_1 + a_{12}x_2 + \cdots + a_{1n}x_n = y_1, \\ a_{21}x_1 + a_{22}x_2 + \cdots + a_{2n}x_n = y_2, \\ \vdots \\ a_{n1}x_1 + a_{n2}x_2 + \cdots + a_{nn}x_n = y_n. \end{cases} \qquad (4.1.1)$$

We start by dividing the first equation by a_{11} (if $a_{11} = 0$, the equations are permutated in a suitable way), and then we subtract this equation multiplied by a_{21}, $a_{31}, \ldots, a_{n1}$ from the second, third, $\ldots$, nth equation. Next we divide the second equation by the coefficient a'_{22} of the variable x_2 (this element is called the *pivot element*), and then x_2 is eliminated in a similar way from the third, fourth, $\ldots$, nth equation. This procedure is continued as far as possible, and finally $x_n, x_{n-1}, \ldots$, x_1 are obtained by back-substitution.

In normal cases the method is satisfactory, but some difficulties may arise. As a matter of fact, it may occur that the pivot element, even if it is different from zero, is very small and gives rise to large errors. The reason is that the small coefficient usually has been formed as the difference between two almost equal numbers. One tries to get around this difficulty by suitable permutations, but so far no completely satisfactory strategy seems to have been invented.

When the elimination is completed, the system has assumed the following form:

$$\begin{cases} c_{11}x_1 + c_{12}x_2 + \cdots + c_{1n}x_n = z_1, \\ \qquad\quad c_{22}x_2 + \cdots + c_{2n}x_n = z_2, \\ \qquad\qquad\qquad\quad \vdots \\ \qquad\qquad\qquad\qquad c_{nn}x_n = z_n. \end{cases} \qquad (4.1.2)$$

The new coefficient matrix is an *upper triangular* matrix; the diagonal elements c_{ii} are usually equal to 1.

In practical work it is recommended that an extra column be carried, with the sum of all the coefficients in one row. They are treated exactly like the other coefficients, and in this way we get an easy check throughout the computation by comparing the row sums with the numbers in this extra column.

EXAMPLE

$$\begin{cases} 10x - 7y + 3z + 5u = 6, \\ -6x + 8y - z - 4u = 5, \\ 3x + y + 4z + 11u = 2, \\ 5x - 9y - 2z + 4u = 7. \end{cases}$$ (4.1.3)

First, we eliminate x, using the first equation:

$$\begin{cases} x - 0.7y + 0.3z + 0.5u = 0.6, \\ 3.8y + 0.8z - u = 8.6, \\ 3.1y + 3.1z + 9.5u = 0.2, \\ -5.5y - 3.5z + 1.5u = 4. \end{cases}$$ (4.1.4)

Since the numerically largest y-coefficient belongs to the fourth equation, we permutate the second and fourth equations. After that, y is eliminated:

$$\begin{cases} x - 0.7y + 0.3z + 0.5u = 0.6, \\ y + 0.63636z - 0.27273u = -0.72727, \\ -1.61818z + 0.03636u = 11.36364, \\ 1.12727z + 10.34545u = 2.45455. \end{cases}$$ (4.1.5)

Now z is also eliminated, which gives

$$\begin{cases} x - 0.7y + 0.3z + 0.5u = 0.6, \\ y + 0.63636z - 0.27273u = -0.72727, \\ z - 0.02247u = -7.02247, \\ 10.37079u = 10.37079. \end{cases}$$ (4.1.6)

The final solution is now easily obtained: $u = 1$, $z = -7$, $y = 4$, and $x = 5$.

Jordan's modification means that the elimination is performed not only in the equations below but also in the equations above. In this way, we finally obtain a diagonal (even unit) coefficient matrix, and we have the solution without further computation. In the example just treated, the system (4.1.4) is obtained unchanged. In the next two steps, we get

$$\begin{cases} x + 0.74545z + 0.30909u = 0.09091, \\ y + 0.63636z - 0.27273u = -0.72727, \\ -1.61818z + 0.03636u = 11.36364, \\ 1.12727z + 10.34545u = 2.45455. \end{cases}$$ (4.1.7)

$$\begin{cases} x + 0.32584u = 5.32582, \\ y - 0.25843u = 3.74156, \\ z - 0.02447u = -7.02248, \\ 10.37078u = 10.37078. \end{cases}$$ (4.1.8)

The last step in the elimination is trivial and gives the same result as before.

We will now compare the methods of Gauss and Jordan with respect to the number of operations. An elementary calculation gives the following results (note that in Gauss' method the operations during the back-substitution must be included):

Method	Addition and subtraction	Multiplication	Division
Gauss	$\dfrac{n(n-1)(2n+5)}{6}$	$\dfrac{n(n-1)(2n+5)}{6}$	$\dfrac{n(n+1)}{2}$
Jordan	$\dfrac{n(n-1)(n+1)}{2}$	$\dfrac{n(n-1)(n+1)}{2}$	$\dfrac{n(n+1)}{2}$

Thus the number of operations is essentially $n^3/3$ for Gauss' method and $n^3/2$ for Jordan's method, and hence the former method should usually be preferred.

4.2. Triangularization

As we have already noted, the Gaussian elimination leads to an upper triangular matrix where all diagonal elements are 1. We shall now show that the elimination can be interpreted as a multiplication of the original matrix A (augmented by the vector y) by a suitable lower triangular matrix. Hence, in three dimensions, we put

$$\begin{pmatrix} l_{11} & 0 & 0 \\ l_{21} & l_{22} & 0 \\ l_{31} & l_{32} & l_{33} \end{pmatrix} \begin{pmatrix} a_{11} & a_{12} & a_{13} \\ a_{21} & a_{22} & a_{23} \\ a_{31} & a_{32} & a_{33} \end{pmatrix} = \begin{pmatrix} 1 & r_{12} & r_{13} \\ 0 & 1 & r_{23} \\ 0 & 0 & 1 \end{pmatrix}.$$

In this way we get 9 equations with 9 unknowns (6 l-elements and 3 r-elements). A computational scheme is given in Section (4.3). Finally, an upper triangular matrix remains to be inverted, giving a similar matrix as the result.

EXAMPLE

$$\begin{cases} 2x + y + 4z = 12\,, \\ 8x - 3y + 2z = 20\,, \\ 4x + 11y - z = 33\,. \end{cases}$$

We now perform the following operation:

$$\begin{pmatrix} \frac{1}{2} & 0 & 0 \\ \frac{4}{7} & -\frac{1}{7} & 0 \\ \frac{50}{189} & -\frac{1}{21} & -\frac{1}{27} \end{pmatrix} \begin{pmatrix} 2 & 1 & 4 & 12 \\ 8 & -3 & 2 & 20 \\ 4 & 11 & -1 & 33 \end{pmatrix} = \begin{pmatrix} 1 & \frac{1}{2} & 2 & 6 \\ 0 & 1 & 2 & 4 \\ 0 & 0 & 1 & 1 \end{pmatrix},$$

which means that the system has been reduced to:

$$\begin{cases} x + \frac{1}{2}y + 2z = 6\,, \\ \phantom{x + \frac{1}{2}}y + 2z = 4\,, \\ \phantom{x + \frac{1}{2}y + 2}z = 1\,. \end{cases}$$

The inverse of this coefficient matrix is

$$\begin{pmatrix} 1 & -\frac{1}{2} & -1 \\ 0 & 1 & -2 \\ 0 & 0 & 1 \end{pmatrix},$$

which multiplied with the vector

$$\begin{pmatrix} 6 \\ 4 \\ 1 \end{pmatrix}$$

gives the solution

$$\begin{pmatrix} 3 \\ 2 \\ 1 \end{pmatrix}.$$

If the lower and upper triangular matrices are denoted by L and R, respectively, we have $LA = R$ or $A = L^{-1}R$. Since L^{-1} is also a lower triangular matrix, we can find a factorization of A as the product of one L-matrix and one R-matrix. Replacing L^{-1} by L, we have $LR = A$ and obtain n^2 equations in $n(n + 1)$ unknowns. Conveniently, we can choose $l_{ii} = 1$ or $r_{ii} = 1$. In this way the system $Ax = y$ is resolved into two simpler systems:

$$Lz = y; \qquad Rx = z.$$

Both these systems can easily be solved by back-substitution.

EXAMPLE
$$\begin{cases} x_1 + 2x_2 + 3x_3 = 14, \\ 2x_1 + 5x_2 + 2x_3 = 18, \\ 3x_1 + x_2 + 5x_3 = 20. \end{cases}$$

Hence

$$\begin{pmatrix} 1 & 0 & 0 \\ 2 & 1 & 0 \\ 3 & -5 & 1 \end{pmatrix} \begin{pmatrix} z_1 \\ z_2 \\ z_3 \end{pmatrix} = \begin{pmatrix} 14 \\ 18 \\ 20 \end{pmatrix} \quad \text{and} \quad \begin{pmatrix} z_1 \\ z_2 \\ z_3 \end{pmatrix} = \begin{pmatrix} 14 \\ -10 \\ -72 \end{pmatrix},$$

$$\begin{pmatrix} 1 & 2 & 3 \\ 0 & 1 & -4 \\ 0 & 0 & -24 \end{pmatrix} \begin{pmatrix} x_1 \\ x_2 \\ x_3 \end{pmatrix} = \begin{pmatrix} 14 \\ -10 \\ -72 \end{pmatrix} \quad \text{and} \quad \begin{pmatrix} x_1 \\ x_2 \\ x_3 \end{pmatrix} = \begin{pmatrix} 1 \\ 2 \\ 3 \end{pmatrix}.$$

The former system is solved in the order z_1, z_2, z_3, and the latter in the order x_3, x_2, x_1.

If A is symmetric and positive definite, we can produce a particulary simple triangularization in the form $A = LL^T$, where, as before,

$$L = \begin{pmatrix} l_{11} & 0 & \cdots & 0 \\ l_{21} & l_{22} & \cdots & 0 \\ \vdots & & & \\ l_{n1} & l_{n2} & \cdots & l_{nn} \end{pmatrix}.$$

The elements l_{ik} are determined from $l_{11}^2 = a_{11}$; $l_{11}l_{21} = a_{12}$; ..., $l_{11}l_{n1} = a_{1n}$, $l_{21}^2 + l_{22}^2 = a_{22}$; $l_{21}l_{31} + l_{22}l_{32} = a_{23}$; ... Obviously, it may occur that some terms become purely imaginary, but this does not imply any special complications. We can easily obtain the condition that all elements are real. For if L is real, then $A = LL^T$ is positive definite, since

$$x^T A x = x^T L L^T x = z^T z \quad \text{with} \quad z = L^T x .$$

But z is real, and hence $z^T z > 0$. On the other hand, if A is positive definite, then L must also be real.

The system $Ax = y$ can now be solved in two steps:

$$Lz = y ; \quad L^T x = z .$$

This method is known as the square-root method (Banachiewicz and Dwyer) and, as has already been pointed out, it assumes that A is symmetric.

4.3. Crout's method

In practical work the Gaussian elimination is often performed according to a modification suggested by Crout. Starting from the system (4.1.1), we eliminate in the usual way and obtain the following system (it is assumed that no permutations have been necessary):

$$\begin{cases} x_1 + a_{12}'x_2 + a_{13}'x_3 + \cdots + a_{1n}'x_n = z_1 , \\ \quad x_2 + a_{23}'x_3 + \cdots + a_{2n}'x_n = z_2 , \\ \quad \vdots \\ \quad x_n = z_n . \end{cases} \tag{4.3.1}$$

A certain equation (i) in this system is the result of the subtraction from the corresponding equation in (4.1.1) of multiples of the $(i - 1)$ first equations in the latter system. Hence the equation (i) in (4.1.1) can be written as a linear combination of the first i equations in (4.3.1). The first equation of (4.1.1) is obtained by multiplication of the corresponding equation in (4.3.1) by a certain constant a_{11}':

$$a_{11}'x_1 + a_{11}'a_{12}'x_2 + \cdots + a_{11}'a_{1n}'x_n = a_{11}'z_1 . \tag{4.3.2a}$$

The second equation of (4.1.1) is obtained through multiplication of the first and second equation in (4.3.1) by a_{21}' and a_{22}' respectively, and addition of the results:

$$a_{21}'x_1 + (a_{21}'a_{12}' + a_{22}')x_2 + (a_{21}'a_{13}' + a_{22}'a_{23}')x_3 + \cdots$$
$$+ (a_{21}'a_{1n}' + a_{22}'a_{2n}')x_n = a_{21}'z_1 + a_{22}'z_2 . \tag{4.3.2b}$$

The remaining equations are formed in a similar way, and then the coefficients are identified, and all constants a_{ik}' can be determined.

Especially, we get the following equations for the right-hand sides:

$$\begin{cases} a'_{11}z_1 & = y_1, \\ a'_{21}z_1 + a'_{22}z_2 & = y_2, \\ \vdots \\ a'_{n1}z_1 + a'_{n2}z_2 + \cdots + a'_{nn}z_n = y_n. \end{cases} \qquad (4.3.3)$$

For convenience, we introduce the more compact notations P and Q for the coefficient matrices of (4.3.3) and (4.3.1), respectively, not including the diagonal elements in Q. Hence

$$p_{ik} = \begin{cases} a'_{ik} & \text{for} \quad i \geq k, \\ 0 & \text{for} \quad i < k, \end{cases} \qquad q_{ik} = \begin{cases} 0 & \text{for} \quad i \geq k, \\ a'_{ik} & \text{for} \quad i < k, \end{cases}$$

or $P + Q = A'$. The systems (4.1.1), (4.3.1), and (4.3.3) now take the form:

$$\begin{cases} Ax = y, \\ (Q + I)x = z, \\ Pz = y. \end{cases} \qquad (4.3.4)$$

The second equation is premultiplied with P and gives

$$P(Q + I)x = Pz = y = Ax.$$

With different right members we get other solutions x, and from n linearly independent x-vectors we can form a nonsingular matrix X. In this way we get

$$P(Q + I)X = AX$$

and hence $P(Q + I) = A$. This corresponds to a triangularization of A with all diagonal elements in the upper triangular matrix equal to 1. Augmenting the matrix $Q + I$ with the new column z, and the matrix A with the new column y, we can write

$$P(Q + I \mid z)\,(A \mid y). \qquad (4.3.5)$$

Now we have to perform only the matrix multiplication on the left-hand side and observe that the formation of products is usually interrupted before we reach the end: either we meet a zero in the ith row of P (this happens if $i < k$), or we find a zero in the kth column (this happens if $i > k$). If $i = k$, the row and the column are of equal length; this case is conveniently brought together with the case $i > k$. If $i < k$, the summation will run from 1 to i; if $i \geq k$, it will run from 1 to k. In both cases we take the last term in the sum separately, and hence

$$\begin{cases} \displaystyle\sum_{r=1}^{i-1} a'_{ir}a'_{rk} + a'_{ii}a'_{ik} = a_{ik}, & i < k, \\[2ex] \displaystyle\sum_{r=1}^{k-1} a'_{ir}a'_{rk} + \quad a'_{ik} = a_{ik}, & i \geq k, \\[2ex] \displaystyle\sum_{r=1}^{i-1} a'_{ir}z_r + a'_{ii}z_i = y_i. \end{cases} \qquad (4.3.6)$$

These equations are conveniently rewritten in the following way:

$$\begin{cases} a'_{ik} = a_{ik} - \sum_{r=1}^{k-1} a'_{ir} a'_{rk}\,, & i \geq k\,, \\[2mm] a'_{ik} = \dfrac{1}{a'_{ii}}\left(a_{ik} - \sum_{r=1}^{i-1} a'_{ir} a'_{rk}\right), & i < k\,, \\[2mm] z_i = \dfrac{1}{a'_{ii}}\left(y_i - \sum_{r=1}^{i-1} a'_{ir} z_r\right). & \end{cases} \qquad (4.3.7)$$

The computation is performed in the following order: The first column (unchanged), the rest of the first row (including z_1), the rest of the second column, the rest of the second row (including z_2), and so on. Last, the vector x is computed from the equation

$$(Q + I)x = z.$$

The ith row in this system has the form

$$x_i + \sum_{r=i+1}^{n} a'_{ir} x_r = z_i\,, \qquad (4.3.8)$$

and the components of x are computed from the bottom upward through back-substitution.

In the special case when A is symmetric, we have with $Q + I = R$:

$$PR = A\,; \qquad R^T P^T = A^T = A\,.$$

Now we put $P = LD$, where L is a lower triangular matrix with all diagonal elements equal to 1, and D is a diagonal matrix with $d_{ii} = a'_{ii}$. Hence $P^T = DL^T$, and from this we get $LDR = R^T DL^T$. Since the partition is unique, we have $R = L^T = D^{-1}P^T$, and finally $P^T = DR$ or

$$a'_{ki} = a'_{ik} \cdot a'_{ii}\,; \qquad i < k\,. \qquad (4.3.9)$$

EXAMPLE

$$\begin{cases} 2x_1 - 6x_2 + 8x_3 = 24\,, \\ 5x_1 + 4x_2 - 3x_3 = 2\,, \\ 3x_1 + x_2 + 2x_3 = 16\,. \end{cases}$$

$$(A, y) = \begin{pmatrix} 2 & -6 & 8 & 24 \\ 5 & 4 & -3 & 2 \\ 3 & 1 & 2 & 16 \end{pmatrix};$$

$$(A', z) = \begin{pmatrix} 2 & -3 & 4 & 12 \\ 5 & 19 & -\frac{23}{19} & -\frac{58}{19} \\ 3 & 10 & \frac{40}{19} & 5 \end{pmatrix}.$$

Hence $x_3 = 5$; $x_2 - \frac{23}{19} \cdot 5 = -\frac{58}{19}$; $x_2 = 3$; $x_1 - 3 \cdot 3 + 4 \cdot 5 = 12$; $x_1 = 1$. The relation $P(Q + I) = A$ in this case has the following form:

$$\begin{pmatrix} 2 & 0 & 0 \\ 5 & 19 & 0 \\ 3 & 10 & \frac{40}{19} \end{pmatrix} \begin{pmatrix} 1 & -3 & 4 \\ 0 & 1 & -\frac{23}{19} \\ 0 & 0 & 1 \end{pmatrix} = \begin{pmatrix} 2 & -6 & 8 \\ 5 & 4 & -3 \\ 3 & 1 & 2 \end{pmatrix}.$$

4.4. Error estimates

We start from a linear system of equations $Ax = y$ and suppose that the elements in A and y are less than 1 in absolute value; if this is not the case, the equations are scaled in a suitable way. Further, we assume that all computations are done with s digits and that a_{ik} and y_i are given with this precision; these initial values will be considered to be exact. When a result is obtained with more than s digits, round-off must be introduced. The corresponding maximal error is $\varepsilon = \frac{1}{2}N^{-s}$, where N is the base of the number system; usually $N = 2$ or $N = 10$.

The subsequent analysis is due to Wilkinson [1], [2]. The general idea is to search for a perturbed system $(A + \delta A)x = y + \delta y$ whose *exact* solution coincides with the approximate solution we obtain when we solve $Ax = y$ by use of Gaussian elimination. The perturbations are selected to reproduce exactly the recorded multipliers m_{ik}.

In order to simplify the discussion, we will first consider the case $n = 4$. During the elimination we have four different systems of equations $A^{(r)}x = y^{(r)}$, where $r = 1$ corresponds to the initial system $Ax = y$. The last system has its first equation unchanged from the first system, its second equation unchanged from the second system, and its third equation from the third system:

$$\begin{cases} a_{11}^{(1)}x_1 + a_{12}^{(1)}x_2 + a_{13}^{(1)}x_3 + a_{14}^{(1)}x_4 \equiv y_1^{(1)}, \\ \qquad\qquad a_{22}^{(2)}x_2 + a_{23}^{(2)}x_3 + a_{24}^{(2)}x_4 \equiv y_2^{(2)}, \\ \qquad\qquad\qquad\qquad a_{33}^{(3)}x_3 + a_{34}^{(3)}x_4 \equiv y_3^{(3)}, \\ \qquad\qquad\qquad\qquad\qquad\qquad a_{44}^{(4)}x_4 \equiv y_4^{(4)}. \end{cases} \qquad (4.4.1)$$

The sign $\equiv$ means that the equation is satisfied exactly. The first three of these equations are found also in the third system, where the last two equations read

$$\begin{cases} a_{33}^{(3)}x_3 + a_{34}^{(3)}x_4 \equiv y_3^{(3)}, \\ a_{43}^{(3)}x_3 + a_{44}^{(3)}x_4 \equiv y_3^{(3)}. \end{cases} \qquad (4.4.2)$$

Now we assume that the absolute values of all $a_{ik}^{(r)}$ and $y_i^{(r)}$ are less than 1; otherwise, this can be achieved by suitable scaling. Then x_3 is eliminated by muliplying the first equation in (4.4.2) by a constant m_{43}, and adding to the second equation. We find directly:

$$\begin{cases} m_{43} = -a_{43}^{(3)}/a_{33}^{(3)}, \\ a_{44}^{(4)} = a_{44}^{(3)} + m_{43}a_{34}^{(3)}, \\ y_4^{(4)} = y_4^{(3)} + m_{43}y_3^{(3)}. \end{cases} \qquad (4.4.3)$$

These values are, of course, rounded. The exact values are:

$$-m_{43} \equiv (a_{43}^{(3)}/a_{33}^{(3)}) + \eta_{43}; \qquad |\eta_{43}| \leq \tfrac{1}{2}N^{-s} = \varepsilon;$$
$$-m_{43}a_{33}^{(3)} \equiv a_{43}^{(3)} + \varepsilon_{43}^{(3)}; \qquad |\varepsilon_{43}^{(3)}| = |a_{33}^{(3)}\eta_{43}| \leq |\eta_{43}| \leq \varepsilon.$$

Analogously,

$$a_{44}^{(4)} \equiv a_{44}^{(3)} + m_{43}a_{34}^{(3)} + \varepsilon_{44}^{(3)}; \qquad |\varepsilon_{44}^{(3)}| \leq \varepsilon;$$
$$y_{4}^{(4)} \equiv y_{4}^{(3)} + m_{43}y_{3}^{(3)} + \varepsilon_{4}^{(3)}; \qquad |\varepsilon_{4}^{(3)}| \leq \varepsilon.$$

Here $\varepsilon_{43}^{(3)}$, $\varepsilon_{44}^{(3)}$, and $\varepsilon_{4}^{(3)}$ are the perturbations which have to be added to $a_{43}^{(3)}$, $a_{44}^{(3)}$, and $y_{4}^{(3)}$ in order that the third system, partly reproduced by (4.4.2), should become equivalent to the fourth system (4.4.1). Hence

$$(\delta A^{(3)} \mid \delta y^{(3)}) \leq \varepsilon \begin{pmatrix} 0 & 0 & 0 & 0 & | & 0 \\ 0 & 0 & 0 & 0 & | & 0 \\ 0 & 0 & 0 & 0 & | & 0 \\ 0 & 0 & 1 & 1 & | & 1 \end{pmatrix}.$$

Now we consider the transition between the second and the third system. We have, for example, $a_{33}^{(3)} \equiv a_{33}^{(2)} + m_{32}a_{23}^{(2)} + \varepsilon_{33}^{(2)}$; $|\varepsilon_{33}^{(2)}| \leq \varepsilon$. Similar estimates are valid for $\varepsilon_{32}^{(2)}$, $\varepsilon_{34}^{(2)}$, and $\varepsilon_{3}^{(2)}$. On the other hand, the *fourth* equation in the second system must be constructed in such a way that the *perturbed* fourth equation in the third system is reproduced exactly. We certainly have, for example,

$$y_{4}^{(3)} \equiv y_{4}^{(2)} + m_{42}y_{2}^{(2)} + \varepsilon_{4}^{(2)}; \qquad |\varepsilon_{4}^{(2)}| \leq \varepsilon,$$

but what has to be reproduced is the quantity $y_{4}^{(3)} + \varepsilon_{4}^{(3)}$. This means that in the previous perturbation matrix, 1 must be added to all numbers that are not zero, and further that the rectangle of nonzero numbers should be bordered with ones. Thus we get

$$(\delta A^{(2)} \mid \delta y^{(2)}) \leq \varepsilon \begin{pmatrix} 0 & 0 & 0 & 0 & | & 0 \\ 0 & 0 & 0 & 0 & | & 0 \\ 0 & 1 & 1 & 1 & | & 1 \\ 0 & 1 & 2 & 2 & | & 2 \end{pmatrix}.$$

In a similar way, we find

$$(\delta A^{(1)} \mid \delta y^{(1)}) \leq \varepsilon \begin{pmatrix} 0 & 0 & 0 & 0 & | & 0 \\ 1 & 1 & 1 & 1 & | & 1 \\ 1 & 2 & 2 & 2 & | & 2 \\ 1 & 2 & 3 & 3 & | & 3 \end{pmatrix}. \tag{4.4.4}$$

We have now reached the following stage. The system

$$(A^{(1)} + \delta A^{(1)})x = y^{(1)} + \delta y^{(1)} \tag{4.4.5}$$

on repeated elimination gives the following systems:

$$(A^{(2)} + \delta A^{(2)})x = y^{(2)} + \delta y^{(2)} \ ;$$
$$(A^{(3)} + \delta A^{(3)})x = y^{(3)} + \delta y^{(3)} \ ;$$
$$A^{(4)}x = y^{(4)} \ .$$

Further, the factors used during the elimination are exactly identical with those computed during the approximate reduction.

We shall now discuss the solution of the final (triangular) set $A^{(4)}x = y^{(4)}$ written in full in (4.4.1), and suppose that we get a computed solution vector ξ. For example, from the second equation in the system, we get

$$x_2 \equiv \frac{y_2^{(2)} - a_{23}^{(2)}x_3 - a_{24}^{(2)}x_4}{a_{22}^{(2)}} \qquad \text{(If the division were exact!)} \ .$$

However, we actually compute

$$\xi_2 \equiv \frac{y_2^{(2)} - a_{23}^{(2)}\xi_3 - a_{24}^{(2)}\xi_4}{a_{22}^{(2)}} + \eta_2 \ ; \qquad |\eta_2| \le \varepsilon \ ,$$

where η_2 is the round-off error from the division. The numerator is assumed to be computed with a negligible rounding error (for example, in a double-length accumulator). Hence

$$a_{22}^{(2)}\xi_2 + a_{23}^{(2)}\xi_3 + a_{24}^{(2)}\xi_4 \equiv y_2^{(2)} + \delta z_2 \ ,$$

where $|\delta z_2| = |\eta_2 \cdot a_{22}^{(2)}| \le |\eta_2| \le \varepsilon$.

It should be observed that the absolute value of x_2 need *not* be less than 1, and for this reason we must compute x_2 to s fractional places, which might mean more than s significant digits. However, we shall ignore this complication here. Hence it is clear that ξ is the *exact* solution of the modified system

$$A^{(4)}x = y^{(4)} + \delta z \qquad \text{with} \qquad |\delta z_i| < \varepsilon \ .$$

It is now easy to see how the system (4.4.5) should be modified to give the system $A^{(4)}x = y^{(4)} + \delta z$ on elimination, and replacing (4.4.5) by

$$(A^{(1)} + \delta A^{(1)})x = y^{(1)} + \delta y^{(1)} + \delta y^{(0)} \ ,$$

we obtain

$$\delta y^{(0)} = \begin{pmatrix} \delta z_1 \\ -m_{21}\delta z_1 + \delta z_2 \\ -m_{31}\delta z_1 - m_{32}\delta z_2 + \delta z_3 \\ -m_{41}\delta z_1 - m_{42}\delta z_2 - m_{43}\delta z_3 + \delta z_4 \end{pmatrix} \ .$$

Hence

$$\delta y^{(0)} \le \varepsilon \begin{pmatrix} 1 \\ 2 \\ 3 \\ 4 \end{pmatrix} \ .$$

Generalizing to n equations, we get the following result: The solution obtained by Gaussian elimination of the system $Ax = y$ with round-off to s fractional places is an exact solution to another system $(A + \delta A)x = y + \delta y$, where

$$(\delta A \mid \delta y) \leq \varepsilon \begin{pmatrix} 0 & 0 & 0 & 0 \cdots & 0 & 0 & 1 \\ 1 & 1 & 1 & 1 \cdots & 1 & 1 & 3 \\ 1 & 2 & 2 & 2 \cdots & 2 & 2 & 5 \\ \vdots & & & & & & \\ 1 & 2 & 3 & 4 \cdots & (n-2) & (n-2) & 2n-3 \\ 1 & 2 & 3 & 4 \cdots & (n-1) & (n-1) & 2n-1 \end{pmatrix} \qquad (4.4.6)$$

with $\varepsilon = \frac{1}{2}N^{-s}$. If the computations are performed with s significant digits instead of s fractional places, one naturally obtains slightly worse estimates. It should be observed, however, that we have worked throughout with maximal errors. In practical computations, the errors are usually much smaller, as has also been observed in numerical examples. More detailed accounts of these and related questions can be found in the papers by Wilkinson (also consult [3]).

Last, we will also discuss briefly what can be said with regard to the errors in the solution itself. Suppose that x is the exact solution and x' the approximate solution. Hence x and x' are exact solutions of the systems

$$\begin{cases} Ax = y, \\ (A + \delta A)x' = y + \delta y. \end{cases} \qquad (4.4.7)$$

From this we get

$$\begin{aligned} x' - x &= (A + \delta A)^{-1}(y + \delta y) - A^{-1}y \\ &= \{(A + \delta A)^{-1} - A^{-1}\}y + (A + \delta A)^{-1}\delta y. \end{aligned}$$

Using the triangular inequality and equation (3.6.10), we find, with $\alpha = \|A^{-1}\delta A\|$,

$$|x' - x| \leq |\{(A + \delta A)^{-1} - A^{-1}\}y| + |(A + \delta A)^{-1}\delta y|$$

$$\leq \|A^{-1}\| \frac{\alpha}{1 - \alpha} |y| + \|(A + \delta A)^{-1}\| \cdot |\delta y|.$$

But

$$\begin{aligned} \|(A + \delta A)^{-1}\| &= \|(A + \delta A)^{-1} - A^{-1} + A^{-1}\| \\ &\leq \|(A + \delta A)^{-1} - A^{-1}\| + \|A^{-1}\| \\ &\leq \|A^{-1}\| \frac{\alpha}{1 - \alpha} + \|A^{-1}\| \\ &= \frac{\|A^{-1}\|}{1 - \alpha}. \end{aligned}$$

Hence we finally have

$$|x' - x| \leq \frac{||A^{-1}||}{1 - \alpha} (\alpha|y| + |\delta y|) \, . \tag{4.4.8}$$

First we note that the norm of the inverse of A plays a decisive role for the magnitude of the errors in the solution. Also the constant α strongly depends on this norm, since $\alpha = ||A^{-1}\delta A|| \leq ||A^{-1}|| \cdot ||\delta A||$. But we can say practically nothing of $||A^{-1}||$ without computing the inverse, which, of course, is closely connected with the solution of the system.

Systems of equations with the property that small changes in the coefficients give rise to large changes in the solution are said to be *ill-conditioned*. A measure of the numerical difficulties associated with a certain matrix is given by the so-called condition numbers

$$\begin{cases} M = nM(A)M(A^{-1}) \, , \\ N = n^{-1}N(A)N(A^{-1}) \, , \\ P = \lambda/\mu \, , \end{cases} \tag{4.4.10}$$

where $M(A) = \max_{ik} |a_{ik}|$, and further λ is the largest and μ the smallest absolute value of the characteristic roots of A (cf. Section 3.6). In particular, if A is Hermitian, we have $P = ||A|| \cdot ||A^{-1}||$.

In general, the numbers M, N, and P become large at the same time. Taking Wilson's matrix (see below) as an example, we obtain $M = 2720$, $N = 752$, and $P = 2984$. Large condition numbers, as a rule, indicate numerical difficulties, especially in the solution of linear systems of equations and in the inversion of matrices.

A well-known example has been presented by T. S. Wilson:

$$\begin{cases} 10x + 7y + 8z + 7w = 32 \, , \\ 7x + 5y + 6z + 5w = 23 \, , \\ 8x + 6y + 10z + 9w = 33 \, , \\ 7x + 5y + 9z + 10w = 31 \, . \end{cases}$$

Putting $x = 6$, $y = -7.2$, $z = 2.9$, and $w = -0.1$, we obtain the left-hand sides equal to 32.1, 22.9, 32.9, and 31.1, respectively, and we are inclined to believe that we are near the exact solution. However, setting instead $x = 1.50$, $y = 0.18$, $z = 1.19$, and $w = 0.89$, we obtain 32.01, 22.99, 32.99, and 31.01, but in fact we are still far from the correct solution $x = y = z = w = 1$.

4.5. Iterative methods

We again start from the system $Ax = y$ and assume that we have obtained an approximate solution x_0. Putting $x = x_0 + \varepsilon_0$, we get $Ax = Ax_0 + A\varepsilon_0 = y$, and $A\varepsilon_0 = y - Ax_0$. The vector $r_0 = y - Ax_0$ is called the *residual vector*; it is, of course, 0 for the exact soultion. If the computations are done with s decimals, it is in general suitable to scale r_0 with the factor 10^s before we solve the system $A\varepsilon_0 = r_0$, as before.

Now we shall assume that we know an approximate inverse $B \simeq A^{-1}$. The procedure just described can be considered as the first step in an iteration chain. Obviously, we have $r_{n-1} = y - Ax_{n-1}$ and $x_n = x_{n-1} + Br_{n-1}$ for $n = 1$, and now we take these relations as definitions of r_n and x_n. First we shall prove that

$$r_n = (I - AB)^{n+1}y .$$

To begin with, the relation is correct for $n = 0$. Next we suppose that it is correct for $n - 1$: $r_{n-1} = (I - AB)^n y$. The relation $x_n = x_{n-1} + Br_{n-1}$ is premultiplied with A and subtracted from y:

$$y - Ax_n = y - Ax_{n-1} - ABr_{n-1}$$

or

$$r_n = r_{n-1} - ABr_{n-1} = (I - AB)r_{n-1} = (I - AB)^{n+1}y .$$

Hence we also have

$$\begin{cases} x_0 = By , \\ x_1 = x_0 \quad + Br_0 , \\ x_2 = x_1 \quad + Br_1 , \\ \quad \vdots \\ x_n = x_{n-1} + Br_{n-1} , \end{cases}$$

and after addition,

$$x_n = B(y + r_0 + r_1 + \cdots + r_{n-1}) = B(I + E + E^2 + \cdots + E^n)y ,$$

where $E = I - AB$. Now assume that $||E|| = \varepsilon < 1$; then $||E^2|| \leq ||E|| \cdot ||E|| = \varepsilon^2$ and analogously, $||E^n|| \leq \varepsilon^n$. This suffices to secure the convergence of the series $I + E + E^2 + \cdots$ to $(I - E)^{-1}$, and hence

$$\lim_{n \to \infty} x_n = B(I - E)^{-1}y = B(AB)^{-1}y = A^{-1}y .$$

In particular we have for $n = 1$: $x_1 = B(2I - AB)y = (2B - BAB)y$ (compare with the iteration formula $x_{n+1} = 2x_n - ax_n^2$ for $1/a$).

First we shall discuss an iterative method constructed by Jacobi. We write the coefficient matrix A in the form $D + C$ where D contains all diagonal elements of A and zeros elsewhere. Starting with $x_0 = 0$, for example, we form successively $x_1, x_2, \ldots$ by use of the formula

$$x_{n+1} = -D^{-1}Cx_n + D^{-1}y = Bx_n + c .$$

Thus we get

$$x_1 = c ; \quad x_2 = (I + B)c ; \quad x_3 = (I + B + B^2)c ; \quad \ldots$$

Assuming $||\boldsymbol{B}|| < 1$, we obtain

$$\lim_{n \to \infty} \boldsymbol{x}_n = (\boldsymbol{I} - \boldsymbol{B})^{-1}\boldsymbol{c} = (\boldsymbol{I} - \boldsymbol{B})^{-1}\boldsymbol{D}^{-1}\boldsymbol{y} = \{\boldsymbol{D}(\boldsymbol{I} - \boldsymbol{B})\}^{-1}\boldsymbol{y}$$

$$= (\boldsymbol{D} + \boldsymbol{C})^{-1}\boldsymbol{y} = \boldsymbol{A}^{-1}\boldsymbol{y} .$$

In general, without giving precise conditions, we conclude that convergence is obtained if the matrix $\boldsymbol{A}$ has a pronounced diagonal dominance.

We shall now discuss the method of Gauss-Seidel. Starting from the system (4.1.1), we construct a series of approximate solutions in the following way The first vector $\boldsymbol{x}^{(0)}$ is chosen to be $\boldsymbol{0}$. We put $x_2 = x_3 = \cdots = x_n = 0$ in the first equation, and solve for $x_1 = x_1^{(1)}$. In the second equation, we put $x_3 = x_4 = \cdots = x_n = 0$ and further $x_1 = x_1^{(1)}$, and then we solve for $x_2 = x_2^{(1)}$, and so on. In this way we obtain a vector

$$\boldsymbol{x}^{(1)} = \begin{pmatrix} x_1^{(1)} \\ \vdots \\ x_n^{(1)} \end{pmatrix},$$

which ought to be a better approximation than $\boldsymbol{x}^{(0)}$, and the whole procedure can be repeated. Introducing the matrices

$$\boldsymbol{A}_1 = \begin{pmatrix} a_{11} & 0 & 0 & \cdots & 0 \\ a_{21} & a_{22} & 0 & \cdots & 0 \\ \vdots & & & & \\ a_{n1} & a_{n2} & a_{n3} & \cdots & a_{nn} \end{pmatrix}; \quad \boldsymbol{A}_2 = \begin{pmatrix} 0 & a_{12} & a_{13} & \cdots & a_{1n} \\ 0 & 0 & a_{23} & \cdots & a_{2n} \\ \vdots & & & & \\ 0 & 0 & 0 & \cdots & 0 \end{pmatrix};$$

$$\boldsymbol{A}_1 + \boldsymbol{A}_2 = \boldsymbol{A} ,$$

we can formulate the Gauss-Seidel iteration method: $\boldsymbol{A}_1\boldsymbol{x}^{(p+1)} = \boldsymbol{y} - \boldsymbol{A}_2\boldsymbol{x}^{(p)}$.

Choosing $\boldsymbol{x}^{(0)} = \boldsymbol{0}$, we get for $p = 0$, $\boldsymbol{x}^{(1)} = \boldsymbol{A}_1^{-1}\boldsymbol{y}$. Further, for $p = 1$, we have $\boldsymbol{x}^{(2)} = \boldsymbol{A}_1^{-1}(\boldsymbol{y} - \boldsymbol{A}_2\boldsymbol{A}_1^{-1}\boldsymbol{y}) = (\boldsymbol{I} - \boldsymbol{A}_1^{-1}\boldsymbol{A}_2)\boldsymbol{A}_1^{-1}\boldsymbol{y}$ and analogously for $p = 2$, we have $\boldsymbol{x}^{(3)} = (\boldsymbol{I} - \boldsymbol{A}_1^{-1}\boldsymbol{A}_2 + \boldsymbol{A}_1^{-1}\boldsymbol{A}_2\boldsymbol{A}_1^{-1}\boldsymbol{A}_2)\boldsymbol{A}_1^{-1}\boldsymbol{y}$. Putting $\boldsymbol{A}_1^{-1}\boldsymbol{A}_2 = \boldsymbol{E}$, we find by induction that $\boldsymbol{x}^{(p)} = (\boldsymbol{I} - \boldsymbol{E} + \boldsymbol{E}^2 - \cdots + (-1)^{p-1}\boldsymbol{E}^{p-1})\boldsymbol{A}_1^{-1}\boldsymbol{y}$. When $p \to \infty$ the series within parentheses converges if $||\boldsymbol{E}|| < 1$, and the final value is

$$\boldsymbol{x} = (\boldsymbol{I} + \boldsymbol{E})^{-1}\boldsymbol{A}_1^{-1}\boldsymbol{y} = \{\boldsymbol{A}_1(\boldsymbol{I} + \boldsymbol{E})\}^{-1}\boldsymbol{y} = (\boldsymbol{A}_1 + \boldsymbol{A}_2)^{-1}\boldsymbol{y} = \boldsymbol{A}^{-1}\boldsymbol{y} .$$

The condition for convergence of Gauss-Seidel's method is that the absolutely largest eigenvalue of $\boldsymbol{A}_1^{-1}\boldsymbol{A}_2$ must be absolutely less than 1. For practical purposes the following criterion can be used: We have convergence if for $i = 1, 2, \ldots, n$, $|a_{ii}| > S_i$, where $S_i = \sum_{k \neq i} |a_{ik}|$. We also form $\rho_i = S_i/(|a_{ii}| - S_i)$ and put $\rho = \max_i \rho_i$. Then the method converges if all $\rho_i > 0$, and it converges more rapidly the smaller ρ is. An empirical rule is that ρ should be < 2 in order to produce a sufficiently fast convergence. Essentially, it is important that the matrix has a clear diagonal dominance.

On the whole we can say that Gauss-Seidel's method converges twice as fast as Jacobi's; a proof for this is given at the end of this section.

EXAMPLE

$$\begin{cases} 8x - 3y + 2z = 20 , \\ 4x + 11y - z = 33 , \\ 6x + 3y + 12z = 36 . \end{cases}$$

i	x_i	y_i	z_i
1	2.5	2.1	1.2
2	2.988	2.023	1.000
3	3.0086	1.9969	0.9965
4	2.99971	1.99979	1.00020
5	2.999871	2.000065	1.000048

Systems of equations with a large number of unknowns appear essentially when one wants to solve ordinary or partial differential equations by difference technique. The coefficient matrices are often sparse and moreover they usually possess a special characteristic which we shall call property A. This property is defined in the following way. Let n be the order of the matrix and W the set $\{1, 2, 3, \ldots, n\}$. Then there must exist two disjoint subsets S and T such that $S \cup T = W$ and further $a_{ik} \neq 0$ implies $i = k$ or $i \in S, k \in T$, or $i \in T, k \in S$. This means that the matrix after suitable row-permutations, each followed by the corresponding column permutation, can be written in the form

$$\begin{pmatrix} D_1 & E \\ F & D_2 \end{pmatrix} ,$$

where D_1 and D_2 are diagonal matrices. For example, the matrix

$$\begin{pmatrix} 2 & -1 & 0 & 0 \\ -1 & 2 & -1 & 0 \\ 0 & -1 & 2 & -1 \\ 0 & 0 & -1 & 2 \end{pmatrix}$$

has property A, and we can choose $S = \{1, 3\}, T = \{2, 4\}$.

For systems of equations with coefficient matrices of this kind, D. M. Young has developed a special technique, *Successive Over-Relaxation, SOR*. We shall further assume A symmetric and $a_{ii} > 0$; $a_{ii} > \sum_{k \neq i} |a_{ik}|$ for all values of i. As before we split A into two parts $A = D + C$, D containing the diagonal elements. The equation $(D + C)x = y$ is now rewritten in the form $(I + D^{-1}C)x = D^{-1}y$ or $x = Bx + c$, where $B = -D^{-1}C$ and $c = D^{-1}y$. Thus all diagonal elements of B are equal to zero, and we can split B into one upper and one lower triangular matrix: $B = R + L$. Then the SOR-method is defined through

$$x^{(n+1)} = (1 - \omega)x^{(n)} + \omega\{Lx^{(n+1)} + Rx^{(n)} + c\} ,$$

where ω is the so-called relaxation factor. When $0 < \omega < 1$ we have *under-*

relaxation; when $\omega > 1$ we have *over-relaxation*. If $\omega = 1$, we are back with Gauss-Seidel's method. Solving for $x^{(n+1)}$, we get

$$x^{(n+1)} - (I - \omega L)^{-1}\{(1 - \omega)I + \omega R\}x^{(n)} + (I - \omega L)^{-1}\omega c .$$

It is then clear that Jacobi's, Gauss-Seidel's, and Young's methods can be represented in the form $x^{(n+1)} = Mx^{(n)} + d$ according to the following:

Jacobi	$M = -D^{-1}C = B$	$d = D^{-1}y = c$
Gauss-Seidel	$M = (I - L)^{-1}R$	$d = (I - L)^{-1}$
SOR	$M = (I - \omega L)^{-1}\{(1 - \omega)I + \omega R\}$	$d = (I - \omega L)^{-1}\omega c .$

The convergence speed will essentially depend upon the properties of the matrix M, convergence being obtained only if $\rho(M) < 1$, where $\rho(M)$ is the spectral radius of M. For the SOR-method we encounter the important question of how ω should be chosen to make $\rho(M)$ as small as possible.

It is easy to see that the eigenvalues of A do not change during the permutations which are performed for creating B. We notice, for example, that $\mu I - B$ is of the following type:

$$\begin{pmatrix} \mu & 0 & 0 & 0 & 0 & * & * & * \\ 0 & \mu & 0 & 0 & 0 & * & * & * \\ 0 & 0 & \mu & 0 & 0 & * & * & * \\ 0 & 0 & 0 & \mu & 0 & * & * & * \\ 0 & 0 & 0 & 0 & \mu & * & * & * \\ * & * & * & * & * & \mu & 0 & 0 \\ * & * & * & * & * & 0 & \mu & 0 \\ * & * & * & * & * & 0 & 0 & \mu \end{pmatrix} = \begin{pmatrix} \mu I_1 & E \\ F & \mu I_2 \end{pmatrix} .$$

From well-known determinantal rules it is clear that the same number of elements from E and from F must enter an arbitrary term in the determinant. Two conclusions can be drawn from this. First: the characteristic equation in the example above must be $\mu^8 + a_1\mu^6 + a_2\mu^4 + a_3\mu^2 = 0$ and in general $\mu^r P_s(\mu^2) = 0$, where $s = (n - r)/2$. Now B need not be symmetrical, but

$$B' = D^{1/2}BD^{-1/2} = -D^{1/2}(D^{-1}C)D^{-1/2} = -D^{-1/2}CD^{-1/2}$$

is symmetric. In fact, putting for a moment $S = D^{-1/2}$, we have

$$(SCS)^T = S^T C^T S^T = SCS ,$$

since S is diagonal and C symmetric (because A was supposed to be symmetric). Hence, B and B' have the same eigenvalues which must be real: $0, 0, \ldots$, $\pm \mu_1, \pm \mu_2, \ldots$ The second conclusion which can be made is the following. If all elements of E are multiplied by a factor $k \neq 0$ and all elements of F are divided by the same factor, the value of the determinant is unchanged.

We now derive an important relationship between an eigenvalue λ of the matrix $M = (I - \omega L)^{-1}\{(1 - \omega)I + \omega R\}$ and an eigenvalue μ of B. The equation $\det(M - \lambda I) = 0$ can be written

$$\det\{(I - \omega L)^{-1}[\omega R - (\omega - 1)I - (I - \omega L)\lambda]\} = 0 ,$$

that is,

$$\det\left([R + \lambda L] - \frac{\lambda + \omega - 1}{\omega} I\right) = 0 ,$$

As has just been discussed we move a factor $\lambda^{1/2}$ from L to R which does not change the value of the determinant, and then we divide all elements by $\lambda^{1/2}$:

$$\det\left([R + L] - \frac{\lambda + \omega - 1}{\omega \lambda^{1/2}} I\right) = 0 .$$

But $R + L = B$ and if μ is an eigenvalue of B, we must have

$$\mu = \frac{\lambda + \omega - 1}{\omega \lambda^{1/2}} .$$

Since $b_{ii} = 0$ and $\sum_{k \neq i}|b_{ik}| < 1$ (note that $a_{ii} > \sum_{k \neq i}|a_{ik}|$), we have according to Gershgorin $\mu^2 < 1$. We can now express λ in terms of μ; conveniently we put $z = \lambda^{1/2}$ and obtain

$$z = \tfrac{1}{2}\omega\mu \pm \sqrt{\tfrac{1}{4}\omega^2\mu^2 - \omega + 1} .$$

For certain values of ω we get real solutions; for other values, complex solutions. The limit between these two possibilities is determined by the equation $\tfrac{1}{4}\mu^2\omega^2 - \omega + 1 = 0$ with the two roots:

$$\omega_1 = 2(1 - \sqrt{1 - \mu^2})/\mu^2 ; \qquad \omega_2 = 2(1 + \sqrt{1 - \mu^2})/\mu^2 .$$

Real solutions z are obtained if $\omega \leq \omega_1$ or $\omega \geq \omega_2$, and complex solutions z if $\omega_1 < \omega < \omega_2$. In the real case only the *greater* solution

$$z_1 = \tfrac{1}{2}\omega\mu + \sqrt{\tfrac{1}{4}\omega^2\mu^2 - \omega + 1}$$

is of interest. A simple calculation shows that $dz_1/d\omega < 0$ for $\omega < \omega_1$, while $dz_1/d\omega > 0$ for $\omega > \omega_2$. When $\omega \to \omega_1$ from the left, the derivative goes toward $-\infty$, and when $\omega \to \omega_2$ from the right, the derivative will approach $+\infty$. When $\omega_1 < \omega < \omega_2$, we have

$$z = \tfrac{1}{2}\omega\mu \pm i\sqrt{\omega - 1 - \tfrac{1}{4}\omega^2\mu^2} ,$$

that is, $|z|^2 = \omega - 1$. If $|\lambda| = |z|^2$ is represented as a function of ω, we get the result as shown in Fig. 4.5. Using this figure, we can draw all the conclusions we need. First it is obvious that the optimal value of the relaxation-factor is $\omega_b = \omega_1$, that is,

$$\omega_b = \frac{2(1 - \sqrt{1 - \mu^2})}{\mu^2} ,$$

where μ is the spectral radius of B. It is also easy to understand that a value ω which is a little too big is far less disastrous than a value which is a little too small since the left derivative is infinite. For convergence we must have $|\lambda| = 1$, and hence $0 < \omega < 2$, but we have SOR only when $1 < \omega < 2$.

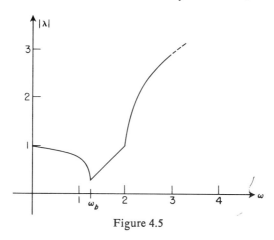

Figure 4.5

In more complicated cases when μ is not known one must resort to guesses or experiments. A correct choice of ω may imply considerable savings in computation effort, and there are realistic examples which indicate that the work may be reduced by a factor of 10–100 compared to Gauss-Seidel's method.

We shall now illustrate the method on a simple example.

$$
\begin{array}{rcl}
4x_1 \quad\quad + x_3 + x_4 &=& 1 \\
4x_2 \quad\quad + x_4 &=& 2 \\
x_1 \quad + 4x_3 \quad\quad &=& 3 \\
x_1 + x_2 \quad + 4x_4 &=& 4
\end{array}
$$

Exact solution:
$$
\begin{aligned}
\xi_1 &= -41/209 = -0.196172 \\
\xi_2 &= 53/209 = 0.253589 \\
\xi_3 &= 167/209 = 0.799043 \\
\xi_4 &= 206/209 = 0.985646 .
\end{aligned}
$$

The coefficient matrix obviously fulfills all conditions posed before. Now μ is determined from the equation

$$
\begin{vmatrix}
\mu & 0 & -\frac{1}{4} & -\frac{1}{4} \\
0 & \mu & 0 & -\frac{1}{4} \\
-\frac{1}{4} & 0 & \mu & 0 \\
-\frac{1}{4} & -\frac{1}{4} & 0 & \mu
\end{vmatrix} = 0 ,
$$

that is, $16\mu^4 - 3\mu^2 + \frac{1}{16} = 0$. The largest positive root is $\mu = (\sqrt{5} + 1)/8$ giving $\omega = 2(1 - \sqrt{1 - \mu^2})/\mu^2 = 1.04464$ and $\lambda = \omega - 1 = 0.04464$. In the table below we compare the fifth approximation in different methods. It turns out that lower approximations become best for larger values of ω, and the error, defined as $e = \sum |x_i - \xi_i|$ in the fifth approximation, is actually smallest for $\omega = 1.050$ when we get an accuracy of exactly 6 decimals.

	x_1	x_2	x_3	x_4	e
Jacobi	-0.184570	0.260742	0.798828	0.985352	0.019265
Gauss-Seidel	-0.195862	0.253780	0.798965	0.985520	0.000705
SOR ($\omega = 1.04464$)	-0.196163	0.253594	0.799042	0.985644	0.000018
Exact	-0.196172	0.253589	0.799043	0.985646	—

The convergence speed for the iteration method $x^{(n+1)} = Mx^{(n)} + c$ is defined through $R = -\log \rho(M)$, where $\rho(M)$ is the spectral radius for M. In the example above we find:

$$R_1 = -\log \frac{\sqrt{5} + 1}{8} = 0.9 \quad \text{for Jacobi's method,}$$

$$R_2 = -2 \log \frac{\sqrt{5} + 1}{8} = 1.8 \quad \text{for Gauss-Seidel's method,}$$

$$R_3 = -\log 0.04464 = 3.1 \quad \text{for SOR.}$$

The relationship $\mu = (\lambda + \omega - 1)/(\lambda^{1/2}\omega)$ used for $\omega = 1$ gives $\lambda = \mu^2$, where λ is the spectral radius for Gauss-Seidel's matrix, and μ the spectral radius for B, that is, the matrix of Jacobi's method. Hence we have the general result that the convergence for Gauss-Seidel's method is twice as fast as for Jacobi's method.

4.6. Complex systems of equations

Suppose that A and B are real, nonsingular matrices of type (n, n) and that x, y, u, and v are real column vectors with n rows. Then the system

$$(A + iB)(x + iy) = u + iv \tag{4.6.1}$$

is equivalent to

$$\begin{cases} Ax - By = u, \\ Ay + Bx = v. \end{cases}$$

Eliminating, we find

$$\begin{cases} (B^{-1}A + A^{-1}B)x = B^{-1}u + A^{-1}v, \\ (B^{-1}A + A^{-1}B)y = B^{-1}v - A^{-1}u. \end{cases} \tag{4.6.2}$$

Both these systems can be solved simultaneously as *one* system with two different right-hand sides, since the coefficient matrix is the same. If A and B are inverted by Jordan's method, and the system is solved by using Gaussian elimination, the total number of multiplications will be approximately $7n^3/3$. Alternatively, we can also write

$$\begin{pmatrix} A & -B \\ B & A \end{pmatrix} \begin{pmatrix} x \\ y \end{pmatrix} = \begin{pmatrix} u \\ v \end{pmatrix},$$

which can be solved by $(2n)^3/3 = 8n^3/3$ multiplications.

REFERENCES

[1] Wilkinson: "Rounding Errors in Algebraic Processes," *Proc. Int. Conf. Inf. Proc.,* UNESCO, 44–53 (1959).

[2] *Id.:* "Error Analysis of Direct Methods of Matrix Inversion," *Jour. ACM,* **8,** 281–330 (1961).

[3] National Physical Laboratory: *Modern Computing Methods* (London, 1961).

[4] Young: "Iterative Methods for Solving Partial Difference Equations of Elliptic Type," *Trans. Am. Math. Soc.,* **76,** 92–111 (1954).

[5] National Bureau of Standards: Appl. Math. Series, No. 29, *Simultaneous Linear Equations and the Determination of Eigenvalues* (Washington, 1953).

[6] *Id.:* Appl. Math. Series, No. 39, *Contributions to the Solution of Systems of Linear Equations and the Determination of Eigenvalues* (1954).

[7] Bargmann, Montgomery, von Neumann: *Solution of Linear Systems of High Order,* (Institute for Advanced Study, Princeton, 1946).

[8] Wilkinson: *Rounding Errors in Algebraic Processes* (Her Majesty's Stationery Office, London, 1963).

EXERCISES

1. Using Crout's method, solve the system

$$\begin{cases} x + 2y - 12z + 8v = 27\,, \\ 5x + 4y + 7z - 2v = 4\,, \\ -3x + 7y + 9z + 5v = 11\,, \\ 6x - 12y - 8z + 3v = 49\,. \end{cases}$$

2. Using Crout's method, solve the system

$$\begin{cases} x + 2y + 3z + 4v = 20\,, \\ 3x - 2y + 8z + 4v = 26\,, \\ 2x + y - 4z + 7v = 10\,, \\ 4x + 2y - 8z - 4v = 2\,. \end{cases}$$

3. Using Crout's method, solve the system

$$\begin{cases} 2x + 10y - 6z + 4u + 8v = 8\,, \\ -3x - 12y - 9z + 6u + 3v = 3\,, \\ -x + y - 34z + 15u + 18v = 29\,, \\ 4x + 18y + 4u + 14v = -2\,, \\ 5x + 26y - 19z + 25u + 36v = 23\,. \end{cases}$$

4. Using an iterative method, solve the following system (three correct decimals):

$$\begin{cases} x_1 + 10x_2 + \quad x_3 & = 10 \text{ ,} \\ 2x_1 \qquad + 20x_3 + \quad x_4 & = 10 \text{ ,} \\ \quad 3x_2 \qquad\qquad + 30x_5 + 3x_6 & = 0 \text{ ,} \\ 10x_1 + \quad x_2 \qquad\qquad\qquad - \quad x_6 & = 5 \text{ ,} \\ \qquad\qquad\qquad 2x_4 - \quad 2x_5 + 20x_6 & = 5 \text{ ,} \\ \qquad\qquad x_3 + 10x_4 - \quad x_5 & = 0 \text{ .} \end{cases}$$

5. Certain linear systems of equations can be treated conveniently by Gauss-Seidel's iterative technique. After a simple generalization, the method can be used also on some nonlinear systems. Find in this way a solution of the system

$$\begin{cases} x - 0.1y^2 + 0.05z^2 = 0.7 \text{ ,} \\ y + 0.3x^2 - 0.1xz = 0.5 \text{ ,} \\ z + 0.4y^2 + 0.1xy = 1.2 \text{ .} \end{cases}$$

(Accuracy desired: three decimals.)

6. The system

$$\begin{cases} ax + by + c = 0 \text{ ,} \\ dx + ey + f = 0 \text{ ,} \end{cases}$$

can be solved by minimizing the function $z = (ax + by + c)^2 + (dx + ey + f)^2$ We start from an approximate solution (x_k, y_k) and construct a better one by first keeping $y = y_k$ constant and varying x; the minimum point is called x_{k+1}. Next we keep $x = x_{k+1}$ constant and vary y; the minimum point is called y_{k+1}. The procedure is then repeated. Use this method to solve the system

$$\begin{cases} 5x + 2y - 11 = 0 \text{ ,} \\ x - 3y - 9 = 0 \text{ ,} \end{cases}$$

with $x_0 = y_0 = 0$ (six iterations; six decimals).

7. Show that the method in the previous example, when applied on the system $Ax = y$, is equivalent to forming the symmetric system $A^T A x = A^T y$.

8. Find the optimal relaxation factor for the system

$$\begin{cases} 2x - y & = 7 \text{ ,} \\ -x + 2y - z = 1 \text{ ,} \\ -y + 2z = 1 \text{ .} \end{cases}$$

Then perform 4 iterations with ω rounded to 1 decimal.

Chapter 5

Matrix inversion

*Great fleas have little fleas upon their back
to bite 'em,
and little fleas have lesser fleas, and so ad
infinitum.
The great fleas themselves in turn have
greater fleas to go on,
while these again have greater still, and
greater still, and so on.*

5.0. Introduction

We suppose that A is a square nonsingular matrix of type (n, n). Our problem is to determine the matrix X which satisfies the relation $AX = I$. We denote by $x_1, x_2, \ldots, x_n$ the vectors formed by the first, second, $\ldots, n$th column of X, and analogously we define the unit vectors $I_1, I_2, \ldots, I_n$. Hence the equation $AX = I$ can be replaced by n linear systems of equations, all with the same coefficient matrix: $Ax_r = I_r$ $(r = 1, 2, \ldots, n)$. All these systems have unique solutions, since $\det A \neq 0$. The solution can be performed by Gauss' elimination method, but, as a rule, Jordan's modification is preferred. This is due to the fact that although Gauss' method is faster, Jordan's modification demands less memory space for storing intermediate results.

5.1. Jordan's method

Since the coefficient matrix is the same for all systems, we start by writing A and I side by side. Then x_1 is eliminated from all equations but the first one, x_2 from all equations but the second one, and so on. As is easily understood, only n^2 nontrivial elements need to be stored during the whole process. Every step in the elimination gives one column in the unit matrix to the left, and at the end we have I to the left and A^{-1} to the right. After p steps we have the scheme shown at the top of p. 104. The following formula is used for the pth reduction of the pth row:

$$a_{pk}^p = \frac{a_{pk}^{p-1}}{a_{pp}^{p-1}} . \tag{5.1.1}$$

For the other rows we have

$$a_{jk}^p = a_{jk}^{p-1} - a_{jp}^{p-1} \frac{a_{pk}^{p-1}}{a_{pp}^{p-1}} . \tag{5.1.2}$$

103

	(1)	(2)···(p)	(p+1) ··· (n)	(1)	(2) ··· (p)	(p+1)···(n)
(1)	1	0 ··· 0	$a^p_{1,p+1}$ ··· $a^p_{1,n}$	a^p_{11} a^p_{12} ··· a^p_{1p}		0 ··· 0
(2)	0	1 ··· 0	$a^p_{2,p+1}$ ··· a^p_{2n}	a^p_{21} a^p_{22} ··· a^p_{2p}		0 ··· 0
⋮						
(p)	0	0 ··· 1	$a^p_{p,p+1}$ ··· $a^p_{p,n}$	$a^p_{p,1}$ $a^p_{p,2}$ ··· a^p_{pp}		0 ··· 0
(p+1)	0	0 ··· 0	$a^p_{p+1,p+1}$ ··· $a^p_{p+1,n}$	$a^p_{p+1,1}$ $a^p_{p+1,2}$ ··· $a^p_{p+1,p}$		1 ··· 0
⋮						
(n)	0	0 ··· 0	$a^p_{n,p+1}$ ··· a^p_{nn}	a^p_{n1} a^p_{n2} ··· a^p_{np}		0 ··· 1

After n steps we are through and have obtained the inverse A^{-1}; only the n^2 elements within the frame need to be stored during the computation. The element $a^p_{p+1,p+1}$, printed in boldface type, is the pivot element for the next step.

EXAMPLE

$$A = \begin{pmatrix} 50 & 107 & 36 \\ 25 & 54 & 20 \\ 31 & 66 & 21 \end{pmatrix}.$$

50	107	36	1	0	0	1	2.14	0.72	0.02	0	0
25	54	20	0	1	0 → 0	0.5	2	−0.5	1	0
31	66	21	0	0	1	0	−0.34	−1.32	−0.62	0	1

		1	0	−7.84	2.16	−4.28	0
	→ 0	1	4	−1	2	0
		0	0	0.04	−0.96	0.68	1

		1	0	0	−186	129	196
	→ 0	1	0	95	−66	−100
		0	0	1	−24	17	25

5.2. Triangularization

As in Section 4.2, the matrix A is written as the product of two triangular matrices denoted by L and R. Hence $A = LR$, and after inversion we get $A^{-1} = R^{-1}L^{-1}$. Inversion of a triangular matrix can be performed directly by use of the definition; the inverse is of the same form as the original matrix. Put, for example,

$$\begin{pmatrix} l_{11} & 0 & 0 \\ l_{21} & l_{22} & 0 \\ l_{31} & l_{32} & l_{33} \end{pmatrix} \begin{pmatrix} x_{11} & 0 & 0 \\ x_{21} & x_{22} & 0 \\ x_{31} & x_{32} & x_{33} \end{pmatrix} = \begin{pmatrix} 1 & 0 & 0 \\ 0 & 1 & 0 \\ 0 & 0 & 1 \end{pmatrix}.$$

Then the unknowns can be determined successively from the following equations:

$$\begin{cases} l_{11}x_{11} & = 1 , \\ l_{21}x_{11} + l_{22}x_{21} & = 0 , \\ l_{31}x_{11} + l_{32}x_{21} + l_{33}x_{31} & = 0 , \end{cases} \quad \begin{cases} l_{22}x_{22} & = 1 , \\ l_{32}x_{22} + l_{33}x_{32} = 0 , \end{cases} \quad l_{33}x_{33} = 1 .$$

Analogously,

$$\begin{pmatrix} r_{11} & r_{12} & r_{13} \\ 0 & r_{22} & r_{23} \\ 0 & 0 & r_{33} \end{pmatrix} \begin{pmatrix} y_{11} & y_{12} & y_{13} \\ 0 & y_{22} & y_{23} \\ 0 & 0 & y_{33} \end{pmatrix} = \begin{pmatrix} 1 & 0 & 0 \\ 0 & 1 & 0 \\ 0 & 0 & 1 \end{pmatrix} ,$$

gives the following system:

$$\begin{cases} r_{33}y_{33} = 1 , \\ r_{22}y_{23} + r_{23}y_{33} = 0 , \\ r_{11}y_{13} + r_{12}y_{23} + r_{13}y_{33} = 0 , \end{cases} \quad \begin{cases} r_{22}y_{22} = 1 , \\ r_{11}y_{12} + r_{12}y_{22} = 0 , \end{cases} \quad r_{11}y_{11} = 1 .$$

The method described here is sometimes called the unsymmetrical Choleski method.

EXAMPLE

$$A = \begin{pmatrix} 50 & 107 & 36 \\ 25 & 54 & 20 \\ 31 & 66 & 21 \end{pmatrix} , \quad L = \begin{pmatrix} 5 & 0 & 0 \\ 2.5 & 1 & 0 \\ 3.1 & -0.68 & 0.2 \end{pmatrix} ,$$

$$R = \begin{pmatrix} 10 & 21.4 & 7.2 \\ 0 & 0.5 & 2 \\ 0 & 0 & 0.2 \end{pmatrix} .$$

As mentioned before, the partition is not unique; the diagonal elements in one of the matrices are at our disposal. We easily find

$$R^{-1} = \begin{pmatrix} 0.1 & -4.28 & 39.2 \\ 0 & 2 & -20 \\ 0 & 0 & 5 \end{pmatrix} , \quad L^{-1} = \begin{pmatrix} 0.2 & 0 & 0 \\ -0.5 & 1 & 0 \\ -4.8 & 3.4 & 5 \end{pmatrix} ,$$

and

$$R^{-1}L^{-1} = \begin{pmatrix} -186 & 129 & 196 \\ 95 & -66 & -100 \\ -24 & 17 & 25 \end{pmatrix} ,$$

in accordance with earlier results. If, instead, we choose the diagonal elements of L equal to 1, we obtain

$$L = \begin{pmatrix} 1 & 0 & 0 \\ 0.5 & 1 & 0 \\ 0.62 & -0.68 & 1 \end{pmatrix} .$$

The remaining matrix can be split into a product of a diagonal matrix D and an upper triangular matrix R, with all diagonal elements equal to 1. Hence

$$D = \begin{pmatrix} 50 & 0 & 0 \\ 0 & 0.5 & 0 \\ 0 & 0 & 0.04 \end{pmatrix}$$

and

$$R = \begin{pmatrix} 1 & 2.14 & 0.72 \\ 0 & 1 & 4 \\ 0 & 0 & 1 \end{pmatrix}.$$

As is easily understood, the partition $A = LDR$ is now unique; the inverse becomes $A^{-1} = R^{-1}D^{-1}L^{-1}$. In our special example we get

$$A^{-1} = \begin{pmatrix} 1 & -2.14 & -7.84 \\ 0 & 1 & -4 \\ 0 & 0 & 1 \end{pmatrix} \begin{pmatrix} 0.02 & 0 & 0 \\ 0 & 2 & 0 \\ 0 & 0 & 25 \end{pmatrix} \begin{pmatrix} 1 & 0 & 0 \\ -0.5 & 1 & 0 \\ -0.96 & 0.68 & 1 \end{pmatrix}$$

$$= \begin{pmatrix} -186 & 129 & 196 \\ 95 & -66 & -100 \\ -24 & 17 & 25 \end{pmatrix}.$$

Since det $L =$ det $R = 1$, we also get det $A =$ det $D = d_1 d_2 \cdots d_n$.

5.3. Choleski's method

Now we assume that the matrix A is real symmetric and positive definite. Then we can write $A = LL^T$, where L is a lower triangular matrix.* As pointed out before, some elements of L might become imaginary, but this will cause no extra trouble. The inverse is obtained from

$$A^{-1} = (L^T)^{-1}L^{-1} = (L^{-1})^T L^{-1}, \tag{5.3.1}$$

and we have to perform only one inversion of a triangular matrix and one matrix multiplication.

EXAMPLE

$$A = \begin{pmatrix} 1 & 2 & 6 \\ 2 & 5 & 15 \\ 6 & 15 & 46 \end{pmatrix}, \qquad L = \begin{pmatrix} 1 & 0 & 0 \\ 2 & 1 & 0 \\ 6 & 3 & 1 \end{pmatrix};$$

$$L^{-1} = \begin{pmatrix} 1 & 0 & 0 \\ -2 & 1 & 0 \\ 0 & -3 & 1 \end{pmatrix}, \qquad (L^{-1})^T L^{-1} = \begin{pmatrix} 5 & -2 & 0 \\ -2 & 10 & -3 \\ 0 & -3 & 1 \end{pmatrix}.$$

* In some special cases such a partition does not exist.

When the computations are done by hand, Choleski's method, which is also called the square-root method, seems to be quite advantageous.

5.4. The escalator method

This method makes use of the fact that if the inverse of a matrix A_n of type (n, n) is known, then the inverse of the matrix A_{n+1}, where one extra row has been added at the bottom and one extra column to the right, can easily be obtained. We put

$$A = \left(\frac{A_1 \mid A_2}{A_3^T \mid a} \right) \quad \text{and} \quad A^{-1} = \left(\frac{X_1 \mid X_2}{X_3^T \mid x} \right),$$

where A_2 and X_2 are column vectors, A_3^T and X_3^T row vectors, and a and x ordinary numbers. Further, A_1^{-1} is assumed to be known. The following equations are obtained:

$$\begin{cases} A_1 X_1 + A_2 X_3^T = I, & (5.4.1.a) \\ A_1 X_2 + A_2 x = 0, & (5.4.1.b) \\ A_3^T X_1 + a X_3^T = 0, & (5.4.1.c) \\ A_3^T X_2 + ax = 1. & (5.4.1.d) \end{cases}$$

From (b) we get $X_2 = -A_1^{-1} A_2 x$, which, inserted in (d), gives $(a - A_3^T A_1^{-1} A_2) x = 1$. Hence x and then also X_2 can be determined. Next we get from (a):

$$X_1 = A_1^{-1}(I - A_2 X_3^T),$$

which, inserted in (c), gives

$$(a - A_3^T A_1^{-1} A_2) X_3^T = -A_3^T A_1^{-1},$$

and hence X_3^T is known. Finally, we obtain X_1 from (a), and the inverse A^{-1} has been computed. In this way we can increase the dimension number successively.

EXAMPLE

$$A = \begin{pmatrix} 13 & 14 & 6 & 4 \\ 8 & -1 & 13 & 9 \\ 6 & 7 & 3 & 2 \\ \hline 9 & 5 & 16 & 11 \end{pmatrix}, \qquad A_1 = \begin{pmatrix} 13 & 14 & 6 \\ 8 & -1 & 13 \\ 6 & 7 & 3 \end{pmatrix};$$

$$A_1^{-1} = \tfrac{1}{94} \begin{pmatrix} 94 & 0 & -188 \\ -54 & -3 & 121 \\ -62 & 7 & 125 \end{pmatrix}, \qquad A_2 = \begin{pmatrix} 4 \\ 9 \\ 2 \end{pmatrix};$$

$$A_3^T = (9, 5, 16); \qquad\qquad a = 11.$$

Using the scheme just described, we find

$$x = -94 ; \qquad X_2 = \begin{pmatrix} 0 \\ -1 \\ 65 \end{pmatrix} , \qquad X_3^T = (-416, 97, 913) ;$$

$$X_1 = \begin{pmatrix} 1 & 0 & -2 \\ -5 & 1 & 11 \\ 287 & -67 & -630 \end{pmatrix}$$

and finally

$$A^{-1} = \begin{pmatrix} 1 & 0 & -2 & 0 \\ -5 & 1 & 11 & -1 \\ 287 & -67 & -630 & 65 \\ -416 & 97 & 913 & -94 \end{pmatrix} .$$

The escalator method can be interpreted as a special case of a more general partition method which can be used, for example, for inverting matrices of such a large order that all elements cannot be accommodated in the memory of a computer at one time. Put

$$\begin{pmatrix} A & B \\ C & D \end{pmatrix} \begin{pmatrix} X & Y \\ Z & V \end{pmatrix} = \begin{pmatrix} I & 0 \\ 0 & I \end{pmatrix} ,$$

where A and D are quadratic matrices, not necessarily of the same order, and we obtain

$$\begin{cases} AX + BZ = I , \\ AY + BV = 0 , \\ CX + DZ = 0 , \\ CY + DV = I , \end{cases} \text{and} \quad \begin{cases} X = (A - BD^{-1}C)^{-1} , \\ Y = -A^{-1}B(D - CA^{-1}B)^{-1} , \\ Z = -D^{-1}C(A - BD^{-1}C)^{-1} , \\ V = (D - CA^{-1}B)^{-1} . \end{cases}$$

5.5. Complex matrices

It is easy to compute the inverse of the matrix $A + iB$ if at least one of the matrices A and B is nonsingular. We put

$$(A + iB)(X + iY) = I$$

and obtain, if A^{-1} exists:

$$X = (A + BA^{-1}B)^{-1} ; \qquad Y = -A^{-1}B(A + BA^{-1}B)^{-1} ;$$

and if B^{-1} exists:

$$X = B^{-1}A(AB^{-1}A + B)^{-1} ; \qquad Y = -(AB^{-1}A + B)^{-1} .$$

If A and B are both regular, the two expressions for X and Y are, of course, identical; we have, for example,

$$B^{-1}A(AB^{-1}A + B)^{-1} = (A^{-1}B)^{-1} \cdot (AB^{-1}A + B)^{-1}$$
$$= \{(AB^{-1}A + B)(A^{-1}B)\}^{-1} = (A + BA^{-1}B)^{-1} .$$

Somewhat more complicated is the case when A and B are both singular but $A + iB$ is regular. We can then proceed as follows. Consider the matrices $F = A + rB$ and $G = B - rA$, where r is a real number. Then there must exist a number r such that, for example, F becomes regular. In order to prove this, put det $F = \det(A + rB) = f(r)$. Obviously, $f(r)$ is a polynomial of degree n, and if $f(r) = 0$ for all r, we would also get $f(i) = 0$ against our assumption that $A + iB$ is regular.

From $F + iG = (1 - ir)(A + iB)$, we get $(A + iB)^{-1} = (1 - ir)(F + iG)^{-1}$. Here $(F + iG)^{-1}$ can be computed as before, since F is regular.

5.6. Iterative methods

A matrix A with $||A|| < 1$ is given, and we want to compute $(I - A)^{-1}$. Put $C_0 = I$ and $C_{n+1} = I + AC_n$; it is clear that $C_n = I + A + A^2 + \cdots + A^n$. As we have shown previously, C_n converges to $(I - A)^{-1}$ when $n \to \infty$.

Now suppose instead that we want to compute A^{-1} and that we know an approximate inverse B. Forming $AB - I = E$, we get $A^{-1} = B(I + E)^{-1} = B(I - E + E^2 - \cdots)$ if the series converges. The condition for convergence is $||E|| < 1$. In practical computation a strong improvement is usually obtained already for $n = 1$. In this case, we can write $A^{-1} \cong B(I - E) = B(2I - AB)$, or putting $B_0 = B$, we obtain $B_{n+1} = B_n(2I - AB_n)$ in complete analogy to Formula (2.2.6). With $AB_n = I + E_n$, we find $AB_{n+1} = I - E_n^2$, and hence convergence is quadratic. Moreover, by use of induction, we can easily prove that $B_n = B_0(I - E + E^2 - \cdots + (-1)^{2n-1}E^{2n-1})$, where $E = E_0 = AB - I$, as before.

EXAMPLE

$$A = \begin{pmatrix} 5 & 2 \\ 3 & -1 \end{pmatrix}, \qquad B = \begin{pmatrix} 0.1 & 0.2 \\ 0.3 & -0.4 \end{pmatrix},$$

$$E = \begin{pmatrix} 0.1 & 0.2 \\ 0 & 0 \end{pmatrix}, \qquad E^2 = \begin{pmatrix} 0.01 & 0.02 \\ 0 & 0 \end{pmatrix};$$

$$B(I - E + E^2) = \begin{pmatrix} 0.1 & 0.2 \\ 0.3 & -0.4 \end{pmatrix} \begin{pmatrix} 0.91 & -0.18 \\ 0 & 1 \end{pmatrix} = \begin{pmatrix} 0.091 & 0.182 \\ 0.273 & -0.454 \end{pmatrix}$$

while

$$A^{-1} = \begin{pmatrix} \frac{1}{11} & \frac{2}{11} \\ \frac{3}{11} & -\frac{5}{11} \end{pmatrix} .$$

EXERCISES

1. Find the inverse of the matrix

$$A = \begin{pmatrix} 1 & \frac{1}{3} & \frac{1}{5} \\ \frac{1}{3} & \frac{1}{5} & \frac{1}{7} \\ \frac{1}{5} & \frac{1}{7} & \frac{1}{9} \end{pmatrix},$$

by using Jordan's method.

2. Find the inverse of the matrix

$$A = \begin{pmatrix} 3 & 7 & 8 & 15 \\ 2 & 5 & 6 & 11 \\ 2 & 6 & 10 & 19 \\ 4 & 11 & 19 & 38 \end{pmatrix}.$$

3. Find the inverse of the matrix

$$A = \begin{pmatrix} 1 & 7 & 10 & 3 \\ 2 & 19 & 27 & 8 \\ 0 & 12 & 17 & 4 \\ 5 & 2 & 6 & 3 \end{pmatrix}.$$

4. Find the inverse of Pascal's matrix

$$A = \begin{pmatrix} 1 & 1 & 1 & 1 & 1 \\ 1 & 2 & 3 & 4 & 5 \\ 1 & 3 & 6 & 10 & 15 \\ 1 & 4 & 10 & 20 & 35 \\ 1 & 5 & 15 & 35 & 70 \end{pmatrix}.$$

5. Using Choleski's method, find the inverse of the matrix

$$A = \begin{pmatrix} 1 & 2 & 0.5 & 1 \\ 2 & 5 & 0 & -2 \\ 0.5 & 0 & 2.25 & 7.5 \\ 1 & -2 & 7.5 & 27 \end{pmatrix}.$$

6. Factorize the matrix

$$A = \begin{pmatrix} -2 & 4 & 8 \\ -4 & 18 & -16 \\ -6 & 2 & -20 \end{pmatrix}$$

in the form LDR, and use the result for computation of $A^{-1} = R^{-1}D^{-1}L^{-1}$.

7. A is an (n, n)-matrix with all elements equal to 1. Find a number α such that $I + \alpha A$ becomes the inverse of $I - A$.

8. Compute the condition numbers M and N for the matrix

$$A = \begin{pmatrix} 1 & 2 & 3 & 4 \\ 2 & 6 & 12 & 20 \\ 6 & 24 & 60 & 120 \\ 24 & 120 & 360 & 840 \end{pmatrix}.$$

9. Compute the inverse of

$$A = \begin{pmatrix} 5+i & 4+2i \\ 10+3i & 8+6i \end{pmatrix}.$$

10. Find the inverse of

$$A_4 = \begin{pmatrix} 3 & -1 & 10 & 2 \\ 5 & 1 & 20 & 3 \\ 9 & 7 & 39 & 4 \\ 1 & -2 & 2 & 1 \end{pmatrix}$$

by repeated use of the escalator method.

11. The inverse of a matrix A is known: $A^{-1} = B$. E_{ik} is a matrix with all elements zero except the element in the ith row and kth column which is 1. Find the inverse of $A + \delta E_{ik}$, where δ is a small number (δ^2 and higher terms can be neglected). Then use the result on the matrices

$$A = \begin{pmatrix} 1 & 0 & -2 & 0 \\ -5 & 1 & 11 & -1 \\ 287 & -67 & -630 & 65 \\ -416 & 97 & 913 & -94 \end{pmatrix} \quad \text{and} \quad B = \begin{pmatrix} 13 & 14 & 6 & 4 \\ 8 & -1 & 13 & 9 \\ 6 & 7 & 3 & 2 \\ 9 & 5 & 16 & 11 \end{pmatrix},$$

where the element 913 is changed to 913.01.

12. A matrix with the property $a_{ik} = 0$ for $|i - k| > \alpha$, where α is an integer ≥ 1, is called a band matrix. Consider a band matrix A with diagonal elements $a_1, a_2, \ldots, a_n$ and the elements next to the main diagonal equal to $1(\alpha = 1)$. Show that we can write $A = LDL^T$, where

$$L = \begin{pmatrix} 1 & 0 & 0 \cdots 0 \\ c_2 & 1 & 0 \cdots 0 \\ 0 & c_3 & 1 \cdots 0 \\ \vdots & & \\ 0 & 0 & 0 \cdots c_n\ 1 \end{pmatrix} \quad \text{and} \quad D = \begin{pmatrix} d_1 & 0 & 0 \cdots 0 \\ 0 & d_2 & 0 \cdots 0 \\ 0 & 0 & d_3 \cdots 0 \\ \vdots & & \\ 0 & 0 & 0 \cdots d_n \end{pmatrix}$$

and $d_1 = a_1$; $c_{k+1} = 1/d_k$; $d_{k+1} = a_{k+1} - c_{k+1}$ (we suppose that $d_k \neq 0$). Use this method on the matrix

$$A = \begin{pmatrix} 1 & 1 & 0 & 0 & 0 \\ 1 & 2 & 1 & 0 & 0 \\ 0 & 1 & 3 & 1 & 0 \\ 0 & 0 & 1 & 4 & 1 \\ 0 & 0 & 0 & 1 & 5 \end{pmatrix}.$$

13. A and B are given matrices of type (n, n). We have the estimates $|a_{ik}| < \varepsilon_1$; $|b_{ik}| < \varepsilon_2$. Form $C = A + B$ and $D = AB$ and show that $|c_{ik}| < \varepsilon_1 + \varepsilon_2$; $|d_{ik}| < n\varepsilon_1\varepsilon_2$.

14. A is a given matrix and B an approximate inverse (0th approximation). Using $E = AB - I$, we construct a series of improved approximations. Find an error estimate for the pth approximation when $|e_{ik}| < \varepsilon$ and $|b_{ik}| < \alpha$.

15. Let

$$A = \begin{pmatrix} 18 & -12 & 23 & -42 \\ -17 & 8 & -72 & -22 \\ -31 & 42 & 54 & 45 \\ 77 & -33 & -16 & -63 \end{pmatrix}$$

and

$$B = 10^{-3} \begin{pmatrix} -19 & -2 & 12 & 23 \\ -3 & 21 & 34 & 19 \\ 12 & -8 & 2 & -4 \\ -25 & -11 & -3 & 2 \end{pmatrix},$$

where B is an approximate inverse of A. Compute the second approximation of A^{-1} and estimate the error according to Exercise 14.

16. A is an $n \times n$ matrix with elements $a_{ik} = 2\delta_{ik} - \delta_{i-1,k} - \delta_{i+1,k}$, where

$$\delta_{rs} = \begin{cases} 1 & \text{if } r = s, \\ 0 & \text{if } r \neq s. \end{cases}$$

Show that $A^{-1} = B$, where

$$b_{ik} = \frac{\min{(i, k)}[n + 1 - \max{(i, k)}]}{n + 1}.$$

17. The $n \times n$ nonsingular matrix A is given. The rows are denoted by $A_1^T, A_2^T, \ldots,$ A_n^T, while the columns of A^{-1} are denoted by $C_1, C_2, \ldots, C_n$. The matrix B is obtained from A by replacing A_r^T by a^T. Show that B is nonsingular if $\lambda = a^T C_r$ is not equal to zero and compute the columns $D_1, D_2, \ldots, D_n$ in the inverse B^{-1} by trying $D_i = C_i + \alpha_i C_r$ $(i \neq r)$ and $D_r = \alpha_r C_r$.

Chapter 6

Algebraic eigenvalue problems

Das also war des Pudels Kern!   GOETHE.

6.0. Introduction

Determination of eigenvalues and eigenvectors of matrices is one of the most important problems of numerical analysis. Theoretically, the problem has been reduced to finding the roots of an algebraic equation and to solving n linear homogeneous systems of equations. In practical computation, as a rule, this method is unsuitable, and better methods must be applied.

When there is a choice between different methods, the following questions should be answered:

(a) Are both eigenvalues and eigenvectors asked for, or are eigenvalues alone sufficient?

(b) Are only the absolutely largest eigenvalue(s) of interest?

(c) Does the matrix have special properties (real symmetric, Hermitian, and so on)?

If the eigenvectors are not needed less memory space is necessary, and further, if only the largest eigenvalue is wanted, a particularly simple technique can be used. Except for a few special cases a direct method for computation of the eigenvalues from the equation det $(\lambda I - A) = 0$ is never used. Further it turns out that practically all methods depend on transforming the initial matrix one way or other without affecting the eigenvalues. The table on p. 114 presents a survey of the most important methods giving initial matrix, type of transformation, and transformation matrix. As a rule, the transformation matrix is built up successively, but the resulting matrix need not have any simple properties, and if so, this is indicated by a horizontal line. It is obvious that such a compact table can give only a superficial picture; moreover, in some cases the computation is performed in two steps. Thus a complex matrix can be transformed to a normal matrix following Eberlein, while a normal matrix can be diagonalized following Goldstine-Horwitz. Incidentally, both these procedures can be performed simultaneously giving a unified method as a result. Further, in some cases we have recursive techniques which differ somewhat in principle from the other methods.

It is not possible to give here a complete description of all these methods because of the great number of special cases which often give rise to difficulties. However, methods which are important in principle will be treated carefully

Initial matrix	Technique	Transformation matrix	Method, originator
Real	Iteration and deflation	—	Power method
Real, symmetric	Diagonalization	Orthogonal	Jacobi
Hermitian	Diagonalization	Unitary	Generalized Jacobi
Normal	Diagonalization	Unitary	Goldstine-Horwitz
Real, symmetric	Tridiagonalization	Orthogonal (rotations)	Givens
Real, symmetric	Tridiagonalization	Orthogonal (reflections)	Householder
Real	Tridiagonalization	—	Lanczos
Real	Tridiagonalization	—	La Budde
Real	Triangularization	—	Ruthishauser (LR)
Hessenberg	Triangularization	Unitary	Francis (QR)
Complex	Triangularization	Unitary	Lotkin, Greenstadt
Complex	Reduction to Hessenberg form	Unitary (rotations)	Givens
Complex	Reduction to Hessenberg form	Unitary (reflections)	Householder
Complex	Reduction to normal matrix	—	Patricia Eberlein
Tridiagonal	Sturm sequence and interpolation		Givens, Wilkinson, Wielandt
Hessenberg	Recursive evaluation and interpolation		Hyman, Wielandt

and in other cases at least the main features will be discussed. On the whole we can distinguish four principal groups with respect to the kind of transformation used initially:

1. Diagonalization,
2. Almost diagonalization (tridiagonalization),
3. Triangularization,
4. Almost triangularization (reduction to Hessenberg form).

The determination of the eigenvectors is trivial in the first case and almost trivial in the third case. In the other two cases a recursive technique is easily established which will work without difficulties in nondegenerate cases. To a certain amount we shall discuss the determination of eigenvectors, for example, Wilkinson's technique which tries to avoid a dangerous error accumulation. Also Wielandt's method, aiming at an improved determination of approximate eigenvectors, will be treated.

6.1. The power method

We assume that the eigenvalues of A are $\lambda_1, \lambda_2, \ldots, \lambda_n$, where $|\lambda_1| > |\lambda_2| \geq \cdots \geq |\lambda_n|$. Now we let A operate repeatedly on a vector v, which we express as a linear combination of the eigenvectors

$$v = c_1 v_1 + c_2 v_2 + \cdots + c_n v_n \,. \tag{6.1.1}$$

Then we have

$$Av = c_1 A v_1 + c_2 A v_2 + \cdots + c_n A v_n = \lambda_1 \left(c_1 v_1 + c_2 \frac{\lambda_2}{\lambda_1} v_2 + \cdots + c_n \frac{\lambda_n}{\lambda_1} v_n \right)$$

and through iteration we obtain

$$A^p v = \lambda_1^p \left\{ c_1 v_1 + c_2 \left(\frac{\lambda_2}{\lambda_1} \right)^p v_2 + \cdots + c_n \left(\frac{\lambda_n}{\lambda_1} \right)^p v_n \right\}. \tag{6.1.2}$$

For large values of p, the vector

$$c_1 v_1 + c_2 \left(\frac{\lambda_2}{\lambda_1} \right)^p v_2 + \cdots + c_n \left(\frac{\lambda_n}{\lambda_1} \right)^p v_n$$

will converge toward $c_1 v_1$, that is, the eigenvector of λ_1. The eigenvalue is obtained as

$$\lambda_1 = \lim_{p \to \infty} \frac{(A^{p+1} v)_r}{(A^p v)_r} \,, \qquad r = 1, 2, \ldots, n \,, \tag{6.1.3}$$

where the index r signifies the rth component in the corresponding vector. The rate of convergence is determined by the quotient λ_2/λ_1; convergence is faster the smaller $|\lambda_2/\lambda_1|$ is. For numerical purposes the algorithm just described can be formulated in the following way. Given a vector y_k, we form two other vectors, y_{k+1} and z_{k+1}:

$$\begin{cases} z_{k+1} = Ay_k\,, \\ y_{k+1} = z_{k+1}/\alpha_{k+1}\,, \\ \alpha_{k+1} = \max_r |(z_{k+1})_r|\,. \end{cases} \tag{6.1.4}$$

The initial vector y_0 should be chosen in a convenient way, often one tries a vector with all components equal to 1.

EXAMPLE

$$A = \begin{pmatrix} 1 & -3 & 2 \\ 4 & 4 & -1 \\ 6 & 3 & 5 \end{pmatrix}\,.$$

Starting from

$$y_0 = \begin{pmatrix} 1 \\ 1 \\ 1 \end{pmatrix}\,,$$

we find that

$$y_5 = \begin{pmatrix} 0.3276 \\ 0.0597 \\ 1 \end{pmatrix}\,, \qquad y_{10} = \begin{pmatrix} 0.3007 \\ 0.0661 \\ 1 \end{pmatrix}\,,$$

$$y_{15} = \begin{pmatrix} 0.3000 \\ 0.0667 \\ 1 \end{pmatrix}\,, \qquad \text{and} \qquad z_{16} = \begin{pmatrix} 2.0999 \\ 0.4668 \\ 7.0001 \end{pmatrix}\,.$$

After round-off, we get

$$\lambda_1 = 7 \qquad \text{and} \qquad v_1 = \begin{pmatrix} 9 \\ 2 \\ 30 \end{pmatrix}\,.$$

If the matrix A is Hermitian and all eigenvalues are different, the eigenvectors, as shown before, are orthogonal. Let x be the vector obtained after p iterations:

$$x = c_1 v_1 + c_2 \left(\frac{\lambda_2}{\lambda_1}\right)^p v_2 + \cdots + c_n \left(\frac{\lambda_n}{\lambda_1}\right)^p v_n = c_1 v_1 + \varepsilon_2 v_2 + \cdots + \varepsilon_n v_n\,.$$

We suppose that all v_i are normalized:

$$v_i^H v_k = \delta_{ik}\,.$$

Then we have

$$Ax = c_1\lambda_1 v_1 + \varepsilon_2\lambda_2 v_2 + \cdots + \varepsilon_n\lambda_n v_n$$

and

$$x^H Ax = \lambda_1|c_1|^2 + \lambda_2|\varepsilon_2|^2 + \cdots + \lambda_n|\varepsilon_n|^2 .$$

Further, $x^H x = |c_1|^2 + |\varepsilon_2|^2 + \cdots + |\varepsilon_n|^2$. When p increases, all ε_i tend to zero, and with $x = \text{const } A^p(\sum c_i v_i)$, we get Rayleigh's quotient

$$\lambda_1 = \lim_{p\to\infty} \frac{x^H Ax}{x^H x} . \tag{6.1.5}$$

EXAMPLE

With

$$A = \begin{pmatrix} 10 & 7 & 8 & 7 \\ 7 & 5 & 6 & 5 \\ 8 & 6 & 10 & 9 \\ 7 & 5 & 9 & 10 \end{pmatrix} \quad \text{and} \quad x_0 = \begin{pmatrix} 1 \\ 1 \\ 1 \\ 1 \end{pmatrix},$$

we obtain for $p = 1, 2$, and 3, $\lambda_1 = 29.75, 30.287$, and 30.288662, respectively, compared with the correct value 30.28868. The corresponding eigenvector is

$$\begin{pmatrix} 0.95761 \\ 0.68892 \\ 1 \\ 0.94379 \end{pmatrix} .$$

The quotients of the individual vector components give much slower convergence; for example, $(x_3)_1/(x_2)_1 = 30.25987$.

The power method can easily be modified in such a way that certain other eigenvalues can also be computed. If, for example, A has an eigenvalue λ, then $A - qI$ has an eigenvalue $\lambda - q$. Using this principle, we can produce the two outermost eigenvalues. Further, we know that λ^{-1} is an eigenvalue of A^{-1} and analogously that $(\lambda - q)^{-1}$ is an eigenvalue of $(A - qI)^{-1}$. If we know that an eigenvalue is close to q, we can concentrate on that, since $(\lambda - q)^{-1}$ becomes large as soon as λ is close to q.

We will now discuss how the absolutely next largest eigenvalue can be calculated if we know the largest eigenvalue λ_1 and the corresponding eigenvector x_1. Let a^T be the first row vector of A and form

$$A_1 = A - x_1 a^T . \tag{6.1.6}$$

Here x_1 is supposed to be normalized in such a way that the first component is 1. Hence the first row of A_1 is zero. Now let λ_2 and x_2 be an eigenvalue and the corresponding eigenvector with the first component of x_2 equal to 1. Then we have

$$A_1(x_1 - x_2) = A(x_1 - x_2) - x_1 a^T(x_1 - x_2) = \lambda_1 x_1 - \lambda_2 x_2 - (\lambda_1 - \lambda_2)x_1$$
$$= \lambda_2(x_1 - x_2),$$

since $a^T x_1 = \lambda_1$ and $a^T x_2 = \lambda_2$ (note that the first component of x_1 as well as of x_2 is 1).

Thus λ_2 is an eigenvalue and $x_1 - x_2$ is an eigenvector of A_1. Since $x_1 - x_2$ has the first component equal to 0, the first column of A_1 is irrelevant, and in fact we need consider only the $(n - 1, n - 1)$-matrix, which is obtained when the first row and first column of A are removed. We determine an eigenvector of this matrix, and by adding a zero as first component, we get a vector z. Then we obtain x_2 from the relation

$$x_2 = x_1 + cz.$$

Multiplying with a^T we find $a^T x_2 = a^T x_1 + ca^T z$, and hence $c = (\lambda_2 - \lambda_1)/a^T z$. When λ_2 and x_2 have been determined, the process, which is called *deflation*, can be repeated.

EXAMPLE

The matrix
$$A = \begin{pmatrix} -306 & -198 & 426 \\ 104 & 67 & -147 \\ -176 & -114 & 244 \end{pmatrix},$$

has an eigenvalue $\lambda_1 = 6$ and the corresponding eigenvector

$$\begin{pmatrix} 2 \\ -1 \\ 1 \end{pmatrix},$$

or normalized,

$$x_1 = \begin{pmatrix} 1 \\ -\frac{1}{2} \\ \frac{1}{2} \end{pmatrix}.$$

Without difficulty we find

$$A_1 = \begin{pmatrix} 0 & 0 & 0 \\ -49 & -32 & 66 \\ -23 & -15 & 31 \end{pmatrix}.$$

Now we need consider only

$$B_1 = \begin{pmatrix} -32 & 66 \\ -15 & 31 \end{pmatrix},$$

and we find the eigenvalues $\lambda_2 = -2$ and $\lambda_3 = 1$, which are also eigenvalues of the original matrix A. The two-dimensional eigenvector belonging to $\lambda_2 = -2$ is

$$\binom{11}{5},$$

and hence

$$x_2 = x_1 + cz = \begin{pmatrix} 1 \\ -\frac{1}{2} \\ \frac{1}{2} \end{pmatrix} + c \begin{pmatrix} 0 \\ 11 \\ 5 \end{pmatrix}.$$

Since $a^T z = -48$, we get $c = \frac{1}{6}$ and

$$x_2 = \begin{pmatrix} 1 \\ \frac{4}{3} \\ \frac{4}{3} \end{pmatrix} \quad \text{or} \quad \begin{pmatrix} 3 \\ 4 \\ 4 \end{pmatrix}.$$

With $\lambda_3 = 1$, we find

$$x_3 = \begin{pmatrix} 1 \\ -\frac{1}{2} \\ \frac{1}{2} \end{pmatrix} + c \begin{pmatrix} 0 \\ 2 \\ 1 \end{pmatrix}$$

and $a^T z = 30$. Hence $c = -\frac{1}{6}$ and

$$x_3 = \begin{pmatrix} 1 \\ -\frac{5}{6} \\ \frac{1}{3} \end{pmatrix} \quad \text{or} \quad \begin{pmatrix} 6 \\ -5 \\ 2 \end{pmatrix},$$

and all eigenvalues and eigenvectors are known.

If A is Hermitian, we have $x_1^H x_2 = 0$ when $\lambda_1 \neq \lambda_2$. Now suppose that $x_1^H x_1 = 1$, and form

$$A_1 = A - \lambda_1 x_1 x_1^H. \tag{6.1.7}$$

It is easily understood that the matrix A_1 has the same eigenvalues and eigenvectors as A except λ_1, which has been replaced by zero. In fact, we have $A_1 x_1 = A x_1 - \lambda_1 x_1 x_1^H x_1 = \lambda_1 x_1 - \lambda_1 x_1 = 0$ and $A_1 x_2 = A x_2 - \lambda_1 x_1 x_1^H x_2 = \lambda_2 x_2$, and so on. Then we can again use the power method on the matrix A_1.

EXAMPLE

$$A = \begin{pmatrix} 10 & 7 & 8 & 7 \\ 7 & 5 & 6 & 5 \\ 8 & 6 & 10 & 9 \\ 7 & 5 & 9 & 10 \end{pmatrix}; \quad \lambda_1 = 30.288686; \quad x_1 = \begin{pmatrix} 0.528561 \\ 0.380255 \\ 0.551959 \\ 0.520933 \end{pmatrix};$$

$$A_1 = \begin{pmatrix} 1.53804 & 0.91234 & -0.83654 & -1.33984 \\ 0.91234 & 0.62044 & -0.35714 & -0.99979 \\ -0.83654 & -0.35714 & 0.77228 & 0.29097 \\ -1.33984 & -0.99979 & 0.29097 & 1.78053 \end{pmatrix}.$$

With the starting vector

$$y_0 = \begin{pmatrix} 1 \\ 1 \\ -1 \\ -1 \end{pmatrix},$$

we find the following values for Rayleigh's quotient: $\lambda_2 = 3.546, 3.8516$, and 3.85774 compared with the correct value 3.858057.

If the numerically largest eigenvalue of a real matrix A is complex, $\lambda \cdot e^{i\varphi}$, then $\lambda \cdot e^{-i\varphi}$ must also be an eigenvalue. It is also clear that if x_1 is the eigenvector belonging to $\lambda e^{i\varphi}$, then x_1^* is the eigenvector belonging to $\lambda e^{-i\varphi}$.

Now suppose that we use the power method with a real starting vector x: $x = c_1 x_1 + c_1^* x_1^* + \cdots$ Then we form $A^m x$, with m so large that the contributions from all the other eigenvectors can be neglected. Further, a certain component of $A^m x$ is denoted by p_m. Then $p_m \simeq c\lambda^m \xi(e^{mi\varphi + i\theta + i\psi} + e^{-(mi\varphi + i\theta + i\psi)})$, where $c_1 = ce^{i\theta}$ and the initial component of x_1 corresponding to p is $\xi e^{i\varphi}$. Hence

$$p_m \simeq 2c\lambda^m \cos(m\varphi + \alpha),$$

where we have put $\theta + \psi = \alpha$. Now we form

$$p_m p_{m+2} - p_{m+1}^2 \simeq 4c^2\xi^2\lambda^{2m+2}[\cos(m\varphi + \alpha)\cos((m+2)\varphi + \alpha) - \cos^2((m+1)\varphi + \alpha)] = -4c^2\xi^2\lambda^{2m+2}\sin^2\varphi.$$

Hence

$$\lim_{m\to\infty} \frac{p_m p_{m+2} - p_{m+1}^2}{p_{m-1}p_{m+1} - p_m^2} = \lambda^2. \tag{6.1.8}$$

Then we easily find

$$\lim_{m\to\infty} \frac{\lambda^2 p_m + p_{m+2}}{2\lambda p_{m+1}} = \cos\varphi. \tag{6.1.9}$$

In particular, if $\varphi = \pi$, that is, if the numerically largest eigenvalues are of the form $\pm\lambda$ with real λ, then we have the simpler formula

$$\lim_{m\to\infty} \frac{p_{m+2}}{p_m} = \lambda^2. \tag{6.1.10}$$

6.2. Jacobi's method

In many applications we meet the problem of diagonalizing real, symmetric matrices. This problem is particularly important in quantum mechanics.

In Chapter 3 we proved that for a real symmetric matrix A, all eigenvalues are real, and that there exists a real orthogonal matrix O such that $O^{-1}AO$ is diagonal. We shall now try to produce the desired orthogonal matrix as a product of very special orthogonal matrices. Among the off-diagonal elements we choose the numerically largest element: $|a_{ik}| = \max$. The elements a_{ii}, a_{ik}, $a_{ki}(=a_{ik})$, and a_{kk} form a $(2,2)$-submatrix which can easily be transformed to diagonal form. We put

$$O = \begin{pmatrix} \cos\varphi & -\sin\varphi \\ \sin\varphi & \cos\varphi \end{pmatrix},$$

and get

$$D = O^{-1}AO = \begin{pmatrix} \cos\varphi & \sin\varphi \\ -\sin\varphi & \cos\varphi \end{pmatrix} \begin{pmatrix} a_{ii} & a_{ik} \\ a_{ik} & a_{kk} \end{pmatrix} \begin{pmatrix} \cos\varphi & -\sin\varphi \\ \sin\varphi & \cos\varphi \end{pmatrix}, \quad (6.2.1)$$

$$d_{ii} = a_{ii}\cos^2\varphi + 2a_{ik}\sin\varphi\cos\varphi + a_{kk}\sin^2\varphi ,$$

$$d_{ik} = d_{ki} = -(a_{ii} - a_{kk})\sin\varphi\cos\varphi + a_{ik}(\cos^2\varphi - \sin^2\varphi) ,$$

$$d_{kk} = a_{ii}\sin^2\varphi - 2a_{ik}\sin\varphi\cos\varphi + a_{kk}\cos^2\varphi .$$

Now choose the angle φ such that $d_{ik} = d_{ki} = 0$, that is, $\tan 2\varphi = 2a_{ik}/(a_{ii} - a_{kk})$. This equation gives 4 different values of φ, and in order to get as small rotations as possible we claim $-\pi/4 \leq \varphi \leq \pi/4$. Putting

$$R = \sqrt{(a_{ii} - a_{kk})^2 + 4a_{ik}^2} \quad \text{and} \quad \sigma = \begin{cases} 1 & \text{if} \quad a_{ii} \geq a_{kk} , \\ -1 & \text{if} \quad a_{ii} < a_{kk} , \end{cases}$$

we obtain:

$$\begin{cases} \sin 2\varphi = 2\sigma a_{ik}/R , \\ \cos 2\varphi = \sigma(a_{ii} - a_{kk})/R , \end{cases}$$

since the angle 2φ must belong to the first quadrant if $\tan 2\varphi > 0$ and to the fourth quadrant if $\tan 2\varphi < 0$. Hence we have for the angle φ:

$$\varphi = \tfrac{1}{2}\arctan\left(2a_{ik}/(a_{ii} - a_{kk})\right) \quad \text{if} \quad a_{ii} \neq a_{kk} ,$$

$$\varphi = \begin{cases} \pi/4 & \text{when} \quad a_{ik} > 0 \\ -\pi/4 & \text{when} \quad a_{ik} < 0 \end{cases} \quad \text{if} \quad a_{ii} = a_{kk} ,$$

where the value of the arctan-function is chosen between $-\pi/2$ and $\pi/2$. After a few simple calculations we get finally:

$$\begin{cases} d_{ii} = \tfrac{1}{2}(a_{ii} + a_{kk} + \sigma R) , \\ d_{kk} = \tfrac{1}{2}(a_{ii} + a_{kk} - \sigma R) , \\ d_{ik} = d_{ki} = 0 . \end{cases} \quad (6.2.2)$$

(Note that $d_{ii} + d_{kk} = a_{ii} + a_{kk}$ and $d_{ii}d_{kk} = a_{ii}a_{kk} - a_{ik}^2$.)

We perform a series of such two-dimensional rotations; the transformation matrices have the form given above in the elements (i,i), (i,k), (k,i), and (k,k) and are identical with the unit matrix elsewhere. Each time we choose such values i and k that $|a_{ik}| = \max$. We shall show that with the notation $P_r = O_1 O_2 \cdots O_r$, the matrix $B_r = P_r^{-1}AP_r$ for increasing r will approach a diagonal matrix D with the eigenvalues of A along the main diagonal. Then it is obvious that we get the eigenvectors as the corresponding columns of $O = \lim_{r \to \infty} P_r$ since we have $O^{-1}AO = D$, that is, $AO = OD$. Let x_k be the kth column vector of O and λ_k the kth diagonal element of D. Then we have

$$Ax_k = \lambda_k x_k .$$

If $\sum_{k \neq i} |a_{ik}|$ is denoted by ε_i, we know from Gershgorin's theorem that $|a_{ii} - \lambda| < \varepsilon_i$ for some value of i, and if the process has been brought sufficiently far, *every* circle defined in this way contains exactly one eigenvalue. Thus it is easy to see when sufficient accuracy has been attained and the procedure can be discontinued.

The convergence of the method has been examined by von Neumann and Goldstine in the following way. We put $\tau^2(A) = \sum_i \sum_{k \neq i} a_{ik}^2 = N^2(A) - \sum_i a_{ii}^2$ and, as before, $B = O^{-1}AO$. The orthogonal transformation affects only the ith row and column and the kth row and column. Taking only off-diagonal elements into account, we find for $r \neq i$ and $r \neq k$ relations of the form

$$\begin{cases} a'_{ir} = a_{ir} \cos \varphi + a_{kr} \sin \varphi , \\ a'_{kr} = -a_{ir} \sin \varphi + a_{kr} \cos \varphi , \end{cases}$$

and hence $a'^2_{ir} + a'^2_{kr} = a^2_{ir} + a^2_{kr}$. Thus $\tau^2(A)$ will be changed only through the cancellation of the elements a_{ik} and a_{ki}, that is,

$$\tau^2(A') = \tau^2(A) - 2a_{ik}^2 .$$

Since a_{ik} was the absolutely largest of all $n(n - 1)$ off-diagonal elements, we have

$$a_{ik}^2 \geq \frac{\tau^2(A)}{n(n - 1)} ,$$

and

$$\tau^2(A') \leq \tau^2(A) \cdot \left(1 - \frac{2}{n(n - 1)}\right) < \tau^2(A) \cdot \exp\left(-\frac{2}{n(n - 1)}\right) .$$

Hence we get the final estimate,

$$\tau(A') < \tau(A) \cdot \exp\left(-\frac{1}{n(n - 1)}\right) . \tag{6.2.3}$$

After N iterations, $\tau(A)$ has decreased with at least the factor $\exp\left(-N/n(n - 1)\right)$, and for a sufficiently large N we come arbitrarily close to the diagonal matrix containing the eigenvalues.

In a slightly different modification, we go through the matrix row by row, performing a rotation as soon as $|a_{ik}| > \varepsilon$. Here ε is a prescribed tolerance which, of course, has to be changed each time the whole matrix has been passed. This modification seems to be more powerful than the preceding one.

The method was first suggested by Jacobi. It has proved very efficient for diagonalization of real symmetric matrices on automatic computers.

EXAMPLE

$$A = \begin{pmatrix} 10 & 7 & 8 & 7 \\ 7 & 5 & 6 & 5 \\ 8 & 6 & 10 & 9 \\ 7 & 5 & 9 & 10 \end{pmatrix}.$$

Choosing $i = 3$, $k = 4$, we obtain, $\tan 2\varphi = 18/(10 - 10) = \infty$ and $\varphi = 45°$. After the first rotation, we have

$$A_1 = \begin{pmatrix} 10 & 7 & 15/\sqrt{2} & -1/\sqrt{2} \\ 7 & 5 & 11/\sqrt{2} & -1/\sqrt{2} \\ 15/\sqrt{2} & 11/\sqrt{2} & 19 & 0 \\ -1/\sqrt{2} & -1/\sqrt{2} & 0 & 1 \end{pmatrix}.$$

Here we take $i = 1$, $k = 3$, and obtain $\tan 2\varphi = 15\sqrt{2}/(10 - 19)$ and $\varphi = -33°. 5051$. After the second rotation we have

$$A_2 = \begin{pmatrix} 2.978281 & 1.543214 & 0 & -0.589611 \\ 1.543214 & 5 & 10.349806 & -0.707107 \\ 0. & 10.349806 & 26.021719 & -0.390331 \\ -0.589611 & -0.707107 & -0.390331 & 1 \end{pmatrix}$$

and after 10 rotations we have

$$A_{10} = \begin{pmatrix} 3.858056 & 0 & -0.000656 & -0.001723 \\ 0 & 0.010150 & 0.000396 & 0.000026 \\ -0.000656 & 0.000396 & 30.288685 & 0.001570 \\ -0.001723 & 0.000026 & 0.001570 & 0.843108 \end{pmatrix}.$$

After 17 rotations the diagonal elements are 3.85805745, 0.01015005, 30.28868533, and 0.84310715, while the remaining elements are equal to 0 to 8 decimals accuracy. The sum of the diagonal elements is 35.99999999 and the product 1.00000015 in good agreement with the exact characteristic equation:

$$\lambda^4 - 35\lambda^3 + 146\lambda^2 - 100\lambda + 1 = 0.$$

Generalization to Hermitian matrices, which are very important in modern physics, is quite natural. As has been proved before, to a given Hermitian matrix H we can find a unitary matrix U such that $U^{-1}HU$ becomes a diagonal matrix. Apart from trivial factors, a two-dimensional unitary matrix has the form

$$U = \begin{pmatrix} \cos\varphi & -\sin\varphi \cdot e^{-i\theta} \\ \sin\varphi \cdot e^{i\theta} & \cos\varphi \end{pmatrix}.$$

A two-dimensional Hermitian matrix

$$H = \begin{pmatrix} a & b - ic \\ b + ic & d \end{pmatrix}$$

is transformed to diagonal form by $U^{-1}HU = D$, where

$$\begin{cases} d_{11} = a \cos^2 \varphi + d \sin^2 \varphi + 2(b \cos \theta + c \sin \theta) \sin \varphi \cos \varphi \,, \\ d_{22} = a \sin^2 \varphi + d \cos^2 \varphi - 2(b \cos \theta + c \sin \theta) \sin \varphi \cos \varphi \,, \\ d_{12} = d_{21}^* = (d - a) \sin \varphi \cos \varphi e^{-i\theta} - (b + ic) \sin^2 \varphi e^{-2i\theta} + (b - ic) \cos^2 \varphi \,. \end{cases}$$

Putting $d_{12} = 0$, we separate the real and imaginary parts and then multiply the resulting equations, first by $\cos \theta$ and $\sin \theta$, then by $-\sin \theta$ and $\cos \theta$, and finally add them together. Using well-known trigonometric formulas, we get

$$\begin{cases} b \sin \theta - c \cos \theta = 0 \,, \\ (a - d) \sin 2\varphi - 2(b \cos \theta + c \sin \theta) \cos 2\varphi = 0 \,. \end{cases} \tag{6.2.4}$$

In principle we obtain θ from the first equation and then φ can be solved from the second. Rather arbitrarily we demand $-\pi/2 \le \theta \le \pi/2$ and hence

$$\begin{cases} \sin \theta = c\sigma/r \,, \\ \cos \theta = b\sigma/r \,, \end{cases}$$

where

$$\sigma = \begin{cases} 1 & \text{when} \quad b \ge 0 \,, \\ -1 & \text{when} \quad b < 0 \,, \end{cases} \quad \text{and} \quad r = \sqrt{b^2 + c^2} \,.$$

Since $b \cos \theta + c \sin \theta = \sigma r$ the remaining equation has the solution

$$\begin{cases} \sin 2\varphi = 2\sigma\tau r/R \,, \\ \cos 2\varphi = \tau(a - d)/R \,, \end{cases}$$

with $\tau = \pm 1$ and $R = \sqrt{(a - d)^2 + 4r^2} = \sqrt{(a - d)^2 + 4(b^2 + c^2)}$. Now we want to choose φ according to $-\pi/4 \le \varphi \le \pi/4$ in order to get as small a rotation as possible which implies

$$\tau = \begin{cases} 1 & \text{for} \quad a \ge d \,, \\ -1 & \text{for} \quad a < d \,. \end{cases}$$

The following explicit solution is now obtained (note that b and c cannot both be equal to 0 because then H would already be diagonal):

$$\begin{cases} b \ne 0: & \theta = \arctan (c/b) \,, \\ b = 0: & \theta = \begin{cases} \pi/2 & \text{if} \quad c > 0 \,, \\ -\pi/2 & \text{if} \quad c < 0 \,, \end{cases} \\ a - d \ne 0: & \varphi = \tfrac{1}{2} \arctan \left(2\sigma r/(a - d) \right) \,, \\ a - d = 0: & \varphi = \begin{cases} \pi/4 & \text{if} \quad b \ge 0 \,, \\ -\pi/4 & \text{if} \quad b < 0 \,. \end{cases} \end{cases} \tag{6.2.5}$$

As usual the value of the arctan-function must be chosen between $-\pi/2$ and

$\pi/2$. The element d_{11} can now be written

$$d_{11} = \tfrac{1}{2}(a + d) + \tfrac{1}{2}(a - d)\cos 2\varphi + \sigma r \sin 2\varphi$$

and consequently:

$$\begin{cases} d_{11} = \tfrac{1}{2}(a + d + \tau R)\,, \\ d_{22} = \tfrac{1}{2}(a + d - \tau R)\,. \end{cases} \tag{6.2.6}$$

If $c = 0$ we get $\theta = 0$ and recover the result in Jacobi's method.

This procedure can be used repeatedly on larger Hermitian matrices, where the unitary matrices differ from the unit matrix only in four places. In the places (i, i), (i, k), (k, i), and (k, k), we introduce the elements of our two-dimensional matrix. The product of the special matrices $U_1, U_2, \ldots, U_k$ is a new unitary matrix approaching U when k is increased.

Finally we mention that a normal matrix (defined through $A^H A = A A^H$) can always be diagonalized with a unitary matrix. The process can be performed following a technique suggested by Goldstine and Horwitz [8] which is similar to the method just described for Hermitian matrices. The reduction of an arbitrary complex matrix to normal form can be accomplished through a method given by Patricia Eberlein [10]. In practice, both these processes are performed simultaneously.

6.3. Givens' method

Again we assume that the matrix A is real and symmetric. In Givens' method we can distinguish among three different phases. The first phase is concerned with $\tfrac{1}{2}(n - 1)(n - 2)$ orthogonal transformations, giving as result a band matrix with unchanged characteristic equation. In the second phase a sequence of functions $f_i(x)$ is generated, and it is shown that it forms a Sturm sequence, the last member of which is the characteristic polynomial. With the aid of the sign changes in this sequence, we can directly state how many roots *larger* than the inserted value x the characteristic equation has. By testing for a number of suitable values x, we can obtain all the roots. During the third phase, the eigenvectors are computed. The orthogonal transformations are performed in the following order. The elements a_{22}, a_{23}, a_{32}, and a_{33} define a two-dimensional subspace, and we start by performing a rotation in this subspace. This rotation affects all elements in the second and third rows and in the second and third columns. However, the quantity φ defining the orthogonal matrix

$$O = \begin{pmatrix} \cos \varphi & -\sin \varphi \\ \sin \varphi & \cos \varphi \end{pmatrix}$$

is now determined from the condition $a'_{13} = a'_{31} = 0$ and not, as in Jacobi's method, by $a'_{23} = a'_{32} = 0$. We have $a'_{13} = -a_{12}\sin\varphi + a_{13}\cos\varphi = 0$ and $\tan\varphi = a_{13}/a_{12}$. The next rotation is performed in the $(2, 4)$-plane with the new

φ determined from $a'_{14} = a'_{41} = 0$, that is, $\tan \varphi = a_{14}/a'_{12}$. [Note that the $(1, 2)$-element was changed during the preceding rotation.] Now all elements in the second and fourth rows and in the second and fourth columns are changed, and it should be particularly observed that the element $a'_{13} = 0$ is not affected. In the same way, we make the elements $a_{15}, \ldots, a_{1n}$ equal to zero by rotations in the $(2, 5)$-, $\ldots$, $(2, n)$-planes.

Now we pass to the elements $a_{24}, a_{25}, \ldots, a_{2n}$, and they are all set to zero by rotations in the planes $(3, 4), (3, 5), \ldots, (3, n)$. During the first of these rotations, the elements in the third and fourth rows and in the third and fourth columns are changed, and we must examine what happens to the elements a'_{13} and a'_{14} which were made equal to zero earlier. We find

$$a''_{13} = \quad a'_{13} \cos \varphi + a'_{14} \sin \varphi = 0 \, ,$$
$$a''_{14} = -a'_{13} \sin \varphi + a'_{14} \cos \varphi = 0 \, .$$

Further, we get $a''_{24} = -a'_{23} \sin \varphi + a'_{24} \cos \varphi = 0$ and $\tan \varphi = a'_{24}/a'_{23}$. By now the procedure should be clear, and it is easily understood that we finally obtain a band matrix, that is, such a matrix that $a_{ik} = 0$ if $|i - k| > p$. In this special case we have $p = 1$. Now we put

$$
B = \begin{pmatrix}
\alpha_1 & \beta_1 & 0 & \cdots & & 0 \\
\beta_1 & \alpha_2 & \beta_2 & 0 & & 0 \\
0 & \beta_2 & \alpha_3 & \ddots & & \vdots \\
\vdots & \vdots & & \ddots & & \beta_{n-1} \\
0 & 0 & \cdots & & \beta_{n-1} & \alpha_n
\end{pmatrix} ;
\qquad (6.3.1)
$$

B has been obtained from A by a series of orthogonal transformations,

$$A_1 = O_1^{-1}AO_1 \, ,$$
$$A_2 = O_2^{-1}A_1O_2 = (O_1O_2)^{-1}A(O_1O_2) \, ,$$
$$\vdots$$
$$B = A_s = O_s^{-1}A_{s-1}O_s = (O_1O_2 \cdots O_s)^{-1}A(O_1O_2 \cdots O_s) = O^{-1}AO \, ,$$

with $s = \frac{1}{2}(n - 1)(n - 2)$. In Chapter 3 it was proved that A and B have the same eigenvalues and further that, if u is an eigenvector of A and v an eigenvector of B (both with the same eigenvalue), then we have $u = Ov$. Thus the problem has been reduced to the computation of eigenvalues and eigenvectors of the band matrix B.

We can suppose that all $\beta_j \neq 0$ [otherwise $\det (B - \lambda I)$ could be split into two determinants of lower order]. Now we form the following sequence of functions:

$$f_i(\lambda) = (\lambda - \alpha_i)f_{i-1}(\lambda) - \beta_{i-1}^2 f_{i-2}(\lambda) \, , \qquad (6.3.2)$$

with $f_0(\lambda) = 1$ and $\beta_0 = 0$. We find at once that $f_1(\lambda) = \lambda - \alpha_1$, which can be interpreted as the determinant of the $(1, 1)$-element in the matrix $\lambda I - B$.

Analogously, we have $f_2(\lambda) = (\lambda - \alpha_1)(\lambda - \alpha_2) - \beta_1^2$, which is the $(1, 2)$-minor of $\lambda I - B$. By induction, it is an easy matter to prove that $f_n(\lambda)$ is the characteristic polynomial.

Next we shall examine the roots of the equation $f_i(\lambda) = 0$, $i = 1, 2, \ldots, n$. For $i = 1$ we have the only root $\lambda = \alpha_1$. For $i = 2$ we observe that $f_2(-\infty) > 0$; $f_2(\alpha_1) < 0$; $f_2(+\infty) > 0$. Hence we have two real roots σ_1 and σ_2 with, for example, $-\infty < \sigma_1 < \alpha_1 < \sigma_2 < +\infty$. For $i = 3$ we will use a method which can easily be generalized to an induction proof. Then we write $f_2(\lambda) = (\lambda - \sigma_1)(\lambda - \sigma_2)$ and obtain from (6.3.2):

$$f_3(\lambda) = (\lambda - \alpha_3)(\lambda - \sigma_1)(\lambda - \sigma_2) - \beta_2^2(\lambda - \alpha_1) .$$

Now it suffices to examine the sign of $f_3(\lambda)$ in a few suitable points:

λ	$-\infty$	σ_1	σ_2	$+\infty$
sign $[f_3(\lambda)]$	$-$	$+$	$-$	$+$

We see at once that the equation $f_3(\lambda) = 0$ has three real roots ρ_1, ρ_2, and ρ_3 such that $-\infty < \rho_1 < \sigma_1 < \rho_2 < \sigma_2 < \rho_3 < +\infty$. In general, if $f_{i-2}(\lambda)$ has the roots $\sigma_1, \sigma_2, \ldots, \sigma_{i-2}$ and $f_{i-1}(\lambda) = 0$ the roots $\rho_1, \rho_2, \ldots, \rho_{i-1}$, then

$$f_i(\lambda) = (\lambda - \alpha_i)(\lambda - \rho_1)(\lambda - \rho_2) \cdots (\lambda - \rho_{i-1})$$
$$- \beta_{i-1}^2(\lambda - \sigma_1)(\lambda - \sigma_2) \cdots (\lambda - \sigma_{i-2}) ,$$

where

$$-\infty < \rho_1 < \sigma_1 < \rho_2 < \sigma_2 < \cdots < \rho_{i-2} < \sigma_{i-2} < \rho_{i-1} < +\infty .$$

By successively putting $\lambda = -\infty$, $\rho_1, \rho_2, \ldots, \rho_{i-1}$, and $+\infty$, we find that $f_i(\lambda)$ has different signs in two arbitrary consecutive points. Hence $f_i(\lambda) = 0$ has i real roots, separated by the roots of $f_{i-1}(\lambda) = 0$.

We are now going to study the number of sign changes $V(\rho)$ in the sequence $f_0(\rho), f_1(\rho), \ldots, f_n(\rho)$. It is evident that $V(-\infty) = n$ and $V(\infty) = 0$. Suppose that a and b are two such real numbers that $f_i(\lambda) \neq 0$ in the closed interval $a \leq \lambda \leq b$. Then obviously $V(a) = V(b)$. First we examine what happens if the equation $f_i(\lambda) = 0$, $1 < i < n$, has a root ρ in the interval. From $f_{i+1}(\lambda) = (\lambda - \alpha_{i+1})f_i(\lambda) - \beta_i^2 f_{i-1}(\lambda)$ it follows for $\lambda = \rho$ that $f_{i+1}(\rho) = -\beta_i^2 f_{i-1}(\rho)$. Hence $f_{i-1}(\rho)$ and $f_{i+1}(\rho)$ have different signs, and clearly this is also true in an interval $\rho - \varepsilon < \lambda < \rho + \varepsilon$. Suppose, for example, that $f_{i-1}(\rho) < 0$; then we may have the following combination of signs:

λ	f_{i-1}	f_i	f_{i+1}
$\rho - \varepsilon$	$-$	$-$	$+$
$\rho + \varepsilon$	$-$	$+$	$+$.

Hence, the number of sign changes does not change when we pass through a root of $f_i(\lambda) = 0$ if $i < n$. When $i = n$, however, the situation is different.

Suppose, for example, that n is odd. Denoting the roots of $f_n(\lambda) = 0$ by ρ_1, $\rho_2, \ldots, \rho_n$ and the roots of $f_{n-1}(\lambda) = 0$ by $\sigma_1, \sigma_2, \ldots, \sigma_{n-1}$, we have

λ	f_0	f_1	$\cdots f_{n-2}$	f_{n-1}	f_n
$\rho_1 - \varepsilon$	$+$	$-$	$\cdots \quad -$	$+$	$-$
$\rho_1 + \varepsilon$	$+$	$-$	$\cdots \quad -$	$+$	$+$.

Then we see that $V(\rho_1 - \varepsilon) - V(\rho_1 + \varepsilon) = 1$. Now we let λ increase until it reaches the neighborhood of σ_1, where we find the following scheme:

λ	$\cdots f_{n-2}$	f_{n-1}	f_n
$\sigma_1 - \varepsilon$	$\cdots \quad -$	$-$	$+$
$\sigma_1 + \varepsilon$	$\cdots \quad -$	$+$	$+$.

Hence $V(\sigma_1 - \varepsilon) - V(\sigma_1 + \varepsilon) = 0$. Then we let λ increase again (now a sign change of f_{n-2} may appear, but, as shown before, this does not affect V) until we reach the neighborhood of ρ_2, where we have

λ	$\cdots f_{n-1}$	f_n
$\rho_2 - \varepsilon$	$\cdots \quad -$	$+$
$\rho_2 + \varepsilon$	$\cdots \quad -$	$-$

and hence $V(\rho_2 - \varepsilon) - V(\rho_2 + \varepsilon) = 1$. Proceeding in the same way through all the roots ρ_i, we infer that the number of sign changes decreases by one unit each time a root ρ_i is passed. Hence we have proved that if $\varphi(\lambda)$ is the number of eigenvalues of the matrix B which are larger than λ, then

$$V(\lambda) = \varphi(\lambda) . \tag{6.3.3}$$

The sequence $f_i(\lambda)$ is called a *Sturm sequence*. The described technique makes it possible to compute all eigenvalues in a given interval ("telescope method").

For the third phase, computation of the eigenvectors, we shall follow J. H. Wilkinson in [2]. Let λ_1 be an exact eigenvalue of B. Thus we search for a vector x such that $Bx = \lambda_1 x$. Since this is a homogeneous system in n variables, and since det $(B - \lambda_1 I) = 0$, we can obtain a nontrivial solution by choosing $n - 1$ equations and determine the components of x (apart from a constant factor); the remaining equation must then be automatically satisfied. In practical work it turns out, even for quite well-behaved matrices, that the result to a large extent depends on *which* equation was excluded from the beginning. Essentially, we can say that the serious errors which appear on an unsuitable choice of equation to be excluded depend on numerical compensations; thus round-off errors achieve a dominant influence.

Let us assume that the ith equation is excluded, while the others are solved by elimination. The solution (supposed to be exact) satisfies the $n - 1$ equations used for elimination but gives an error δ when inserted into the ith equation.

Actually, we have solved the system

$$\begin{cases} \beta_{j-1}x_{j-1} + (\alpha_j - \lambda)x_j + \beta_j x_{j+1} = 0 \,, & j \neq i \,, \\ \beta_{i-1}x_{i-1} + (\alpha_i - \lambda)x_i + \beta_i x_{i+1} = \delta \,, & \delta \neq 0 \,. \end{cases} \qquad (6.3.4)$$

(We had to use an approximation λ instead of the exact eigenvalue λ_1.) Since constant factors may be omitted, this system can be written in a simpler way:

$$(B - \lambda I)x = e_i \,, \qquad (6.3.5)$$

where e_i is a column vector with the ith component equal to 1 and the others equal to 0. If the eigenvectors of B are $v_1, v_2, \ldots, v_n$, this vector e_i can be expressed as a linear combination, that is,

$$e_i = \sum_{j=1}^{n} c_{ij}v_j \,, \qquad (6.3.6)$$

and from (6.3.5) we get

$$x = \sum_{j=1}^{n} c_{ij}(B - \lambda I)^{-1}v_j = \sum_{j=1}^{n} c_{ij} \frac{1}{\lambda_j - \lambda} v_j \,. \qquad (6.3.7)$$

Now let $\lambda = \lambda_1 + \varepsilon$, and we obtain

$$x = -\frac{c_{i1}}{\varepsilon} v_1 + \sum_{j=2}^{n} c_{ij} \frac{1}{\lambda_j - \lambda_1 - \varepsilon} v_j \,. \qquad (6.3.8)$$

Under the assumption that $c_{i1} \neq 0$, our solution x approaches v_1 as $\varepsilon \to 0$ (apart from trivial factors). However, it may well happen that c_{i1} is of the same order of magnitude as ε (that is, the vector e_i is almost orthogonal to v_1), and under such circumstances it is clear that the vector x in (6.3.8) cannot be a good approximation of v_1. Wilkinson suggests that (6.3.5) be replaced by

$$(B - \lambda I)x = b \,, \qquad (6.3.9)$$

where we have the vector b at our disposal. This system is solved by Gaussian elimination, where it should be observed that the equations are permutated properly to make the pivot element as large as possible. The resulting system is written:

$$\begin{cases} p_{11}x_1 + p_{12}x_2 + p_{13}x_3 & = c_1 \,, \\ \quad\quad p_{22}x_2 + p_{23}x_3 + p_{24}x_4 & = c_2 \,, \\ \quad\quad\quad\quad\quad \vdots & \\ \quad\quad\quad\quad\quad\quad p_{n,n}x_n & = c_n \,. \end{cases} \qquad (6.3.10)$$

As a rule, most of the coefficients $p_{13}, p_{24}, \ldots$ are zero. Since the c_i have been obtained from the b_i which we had at our disposal, we could as well choose the constants c_i deliberately. It seems to be a reasonable choice to take all c_i

equal to 1; no eigenvector should then be disregarded. Thus we choose

$$c = \sum_{r=1}^{n} e_r .$$ (6.3.11)

The system is solved, as usual, by back-substitution, and last, the vector x is normalized. Even on rather pathological matrices, good results have been obtained by Givens' method.

6.4. Householder's method

This method, also, has been designed for real, symmetric matrices. We shall essentially follow the presentation given by Wilkinson [4]. The first step consists of reducing the given matrix A to a band matrix. This is done by orthogonal transformations representing *reflections*. The orthogonal matrices, will be denoted by P_r with the general structure

$$P = I - 2ww^T .$$ (6.4.1)

Here w is a column vector such that

$$w^T w = 1 .$$ (6.4.2)

It is evident that P is symmetric. Further, we have

$$P^T P = (I - 2ww^T)(I - 2ww^T) = I - 4ww^T + 4ww^T ww^T = I ;$$

that is, P is also orthogonal.

The matrix P acting as an operator can be given a simple geometric interpretation. Let P operate on a vector x from the left:

$$Px = (I - 2ww^T)x = x - 2(w^T x)w .$$

In Fig. 6.4 the line L is perpendicular to the unit vector w in a plane defined by w and x. The distance from the endpoint of x to L is $|x| \cos (x, w) = w^T x$, and the mapping P means a reflection in a plane perpendicular to w.

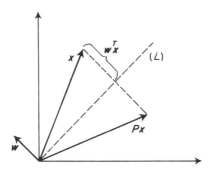

Figure 6.4

Those vectors w which will be used are constructed with the first $(r - 1)$ components zero, or

$$w_r^T = (0, 0, \ldots, 0, x_r, x_{r+1}, \ldots, x_n) .$$

With this choice we form $P_r = I - 2w_r w_r^T$. Further, by (6.4.2) we have

$$x_r^2 + x_{r+1}^2 + \cdots + x_n^2 = 1 .$$

Now put $A = A_1$ and form successively

$$A_r = P_r A_{r-1} P_r , \tag{6.4.3}$$

$r = 2, 3, \ldots, n - 1$. At the first transformation, we get zeros in the positions $(1, 3), (1, 4), \ldots, (1, n)$ and in the corresponding places in the first column. The final result will become a band matrix as in Givens' method. The matrix A_{r-1} contains $n - r$ elements in the row $(r - 1)$, which must be reduced to zero by transformation with P_r; this gives $n - r$ equations for the $n - r + 1$ elements $x_r, x_{r+1}, \ldots, x_n$, and further we have the condition that the sum of the squares must be 1.

We carry through one step in the computation in an example:

$$A = \begin{pmatrix} a_1 & b_1 & c_1 & d_1 \\ b_1 & b_2 & c_2 & d_2 \\ c_1 & c_2 & c_3 & d_3 \\ d_1 & d_2 & d_3 & d_4 \end{pmatrix},$$

$$w_2^T = (0, x_2, x_3, x_4) ; \qquad x_2^2 + x_3^2 + x_4^2 = 1 .$$

The transformation $P_2 A P_2$ must now produce zeros instead of c_1 and d_1. Obviously, the matrix P_2 has the following form:

$$P_2 = \begin{pmatrix} 1 & 0 & 0 & 0 \\ 0 & 1 - 2x_2^2 & -2x_2 x_3 & -2x_2 x_4 \\ 0 & -2x_2 x_3 & 1 - 2x_3^2 & -2x_3 x_4 \\ 0 & -2x_2 x_4 & -2x_3 x_4 & 1 - 2x_4^2 \end{pmatrix}.$$

Since in the first row of P_2 only the first element is not zero, for example, the $(1, 3)$-element of $P_2 A P_2$ can become zero only if the corresponding element is zero already in $A P_2$. Putting $p_1 = b_1 x_2 + c_1 x_3 + d_1 x_4$, we find that the first row of $A P_2$ has the following elements:

$$a_1 , \qquad b_1 - 2p_1 x_2 , \qquad c_1 - 2p_1 x_3 , \qquad d_1 - 2p_1 x_4 .$$

Now we claim that

$$\begin{cases} c_1 - 2p_1 x_3 = 0 , \\ d_1 - 2p_1 x_4 = 0 . \end{cases} \tag{6.4.4}$$

Since we are performing an orthogonal transformation, the sum of the squares of the elements in a row is invariant, and hence

$$a_1^2 + (b_1 - 2p_1x_2)^2 = a_1^2 + b_1^2 + c_1^2 + d_1^2 .$$

Putting $S = (b_1^2 + c_1^2 + d_1^2)^{1/2}$, we obtain

$$b_1 - 2p_1x_2 = \pm S .$$ (6.4.5)

Multiplying (6.4.5) by x_2 and (6.4.4) by x_3 and x_4, we get

$$b_1x_2 + c_1x_3 + d_1x_4 - 2p_1(x_2^2 + x_3^2 + x_4^2) = \pm Sx_2 .$$

The sum of the first three terms is p_1 and further $x_2^2 + x_3^2 + x_4^2 = 1$. Hence

$$p_1 = \mp Sx_2 .$$ (6.4.6)

Inserting this into (6.4.5), we find that $x_2^2 = \frac{1}{2}(1 \mp b_1/S)$, and from (6.4.4), $x_3 = \mp c_1/2Sx_2$ and $x_4 = \mp d_1/2Sx_2$.

In the general case, two square roots have to be evaluated, one for S and one for x_2. Since we have x_2 in the denominator, we obtain the best accuracy if x_2 is large. This is accomplished by choosing a suitable sign for the square-root extraction for S. Thus the quantities ought to be defined as follows:

$$x_2^2 = \frac{1}{2}\left(1 + \frac{b_1 \cdot \text{sign } b_1}{S}\right) .$$ (6.4.7)

The sign for this square root is irrelevant and we choose plus. Hence we obtain for x_3 and x_4:

$$x_3 = \frac{c_1 \text{ sign } b_1}{2Sx_2} ; \qquad x_4 = \frac{d_1 \text{ sign } b_1}{2Sx_2} .$$ (6.4.8)

The end result is a band matrix whose eigenvalues and eigenvectors are computed exactly as in Givens' method. In order to get an eigenvector v of A, an eigenvector x of the band matrix has to be multiplied by the matrix $P_2, P_3, \ldots,$ P_{n-1}; this should be done by iteration:

$$\begin{cases} x_{n-1} = P_{n-1}x , \\ x_{n-2} = P_{n-2}x_{n-1} , \\ \quad \vdots \\ v \equiv x_2 = P_2x_3 . \end{cases}$$ (6.4.9)

6.5. Lanczos' method

The reduction of real symmetric matrices to tridiagonal form can be accomplished through methods devised by Givens and Householder. For arbitrary matrices a similar reduction can be performed by a technique suggested by Lanczos. In this method two systems of vectors are constructed, $x_1, x_2, \ldots, x_n$ and $y_1, y_2, \ldots, y_n$, which are *biorthogonal*; that is, for $j \neq k$, we have $x_j^H y_k = 0$.

The initial vectors x_1 and y_1 can be chosen arbitrarily though in such a way that $x_1^H y_1 \neq 0$. The new vectors are formed according to the rules

$$
\begin{cases}
x_{k+1} = Ax_k - \sum_{j=1}^{k} a_{jk} x_j \,, \\
y_{k+1} = A^H y_k - \sum_{j=1}^{k} b_{jk} y_j \,.
\end{cases}
$$

The coefficients are determined from the biorthogonality condition, and for $j = 1, 2, \ldots, k$, we form:

$$
0 = y_j^H x_{k+1} = y_j^H A x_k - \sum_{r=1}^{k} a_{rk} y_j^H x_r = y_j^H A x_k - a_{jk} y_j^H x_j \,.
$$

If $y_j^H x_j \neq 0$, we get

$$
a_{jk} = \frac{y_j^H A x_k}{y_j^H x_j} \,.
$$

Analogously

$$
b_{jk} = \frac{x_j^H A^H y_k}{x_j^H y_j} \,.
$$

Let us now consider the numerator in the expression for a_{jk} when $j \leq k - 2$:

$$
y_j^H A x_k = (x_k^H A^H y_j)^* = \left\{ x_k^H \left(y_{j+1} + \sum_{r=1}^{j} b_{rj} y_r \right) \right\}^* = 0 \,,
$$

because of the biorthogonality. Hence we have $a_{jk} = 0$ for $j \leq k - 2$, and similarly we also have $b_{jk} = 0$ under the same condition. In this way the following simpler formulas are obtained:

$$
\begin{cases}
x_{k+1} = Ax_k - (a_{k-1,k} x_{k-1} + a_{kk} x_k) \,, \\
y_{k+1} = A^H y_k - (b_{k-1,k} y_{k-1} + b_{kk} y_k) \,.
\end{cases}
$$

If the vectors $x_1, x_2, \ldots, x_n$ are considered as columns in a matrix X and if further a tridiagonal matrix J is formed from the coefficients $a_{k-1,k}$ and a_{kk} with one's in the remaining diagonal:

$$
J = \begin{pmatrix}
a_{11} & a_{12} & 0 & 0 & \cdots & 0 & 0 \\
1 & a_{22} & a_{23} & 0 & \cdots & 0 & 0 \\
0 & 1 & a_{33} & a_{34} & \cdots & 0 & 0 \\
\vdots & & & & & & \vdots \\
0 & 0 & 0 & \cdots & 1 & a_{n-1,n-1} & a_{n-1,n} \\
0 & 0 & 0 & \cdots & 0 & 1 & a_{nn}
\end{pmatrix} ;
$$

then we can simply write $AX = XJ$, and provided the vectors $x_1, x_2, \ldots, x_n$ are linearly independent

$$
J = X^{-1} A X \,.
$$

If similar matrices are formed from the vectors $y_1, y_2, \ldots, y_n$ and from the coefficients $b_{k-1,k}, b_{kk}$, we get

$$K = Y^{-1}A^H Y.$$

Certain complications may arise, for example, that some x- or y-vector may become zero, but it can also happen that $x_j^H y_j = 0$ even if $x_j \neq 0$ and $y_j \neq 0$. The simplest way out is to choose other initial vectors even if it is sometimes possible to get around the difficulties by modifying the formulas themselves.

Obviously, Lanczos' method can be used also with real symmetric or Hermitian matrices. Then one chooses just *one* sequence of vectors which must form an orthogonal system. For closer details, particularly concerning the determination of the eigenvectors, Lanczos' paper [1] should be consulted; a detailed discussion of the degenerate cases is given by Causey and Gregory [9].

Here we also mention still one method for tridiagonalization of arbitrary real matrices, first given by La Budde. Space limitations prevent us from a closer discussion, and instead we refer to the original paper [11].

6.6. Other methods

Among other interesting methods we mention the LR-method. Starting from a matrix $A = A_1$, we split it into two triangular matrices $A_1 = L_1 R_1$ with $l_{ii} = 1$, and then we form $A_2 = R_1 L_1$. Since $A_2 = R_1 A_1 R_1^{-1}$, the new matrix A_2 has the same eigenvalues as A_1. Then we treat A_2 in the same way as A_1, and so on, obtaining a sequence of matrices $A_1, A_2, A_3, \ldots$, which in general converges toward an upper triangular matrix. If the eigenvalues are real, they will appear in the main diagonal. Even the case in which complex eigenvalues are present can be treated without serious complications. Closer details are given in [5], where the method is described by its inventor, H. Rutishauser.

Here we shall also examine the more general eigenvalue problem,

$$\det (A - \lambda B) = 0,$$

where A and B are symmetric and, further, B is positive definite. Then we can split B according to $B = LL^T$, where L is a lower triangular matrix. Hence

$$A - \lambda B = A - \lambda LL^T = L(L^{-1}A(L^T)^{-1} - \lambda I)L^T,$$

and $\det (A - \lambda B) = (\det L)^2 \cdot \det (C - \lambda I)$, where $C = L^{-1}A(L^{-1})^T$. Since $C^T = C$, the problem has been reduced to the usual type treated before.

6.7. Complex matrices

For computing eigenvalues and eigenvectors of arbitrary complex matrices (also, real nonsymmetric matrices fall naturally into this group), we shall first discuss a triangularization method suggested by Lotkin [6] and Greenstadt [7].

The method depends on the lemma by Schur stating that for each square matrix A there exists a unitary matrix U such that $U^{-1}AU = T$, where T is a (lower or upper) triangular matrix (see Section 3.7). In practical computation one tries to find U as a product of essentially two-dimensional unitary matrices, using a procedure similar to that described for Hermitian matrices in Section 6.2. It is possible to give examples for which the method does not converge (the sum of the squares of the absolute values of the subdiagonal elements is not monotonically decreasing, cf. [13]), but in practice convergence is obtained in many cases.

We start by examining the two-dimensional case and put

$$A = \begin{pmatrix} a & b \\ c & d \end{pmatrix}; \qquad U = \begin{pmatrix} p & -q^* \\ q & p \end{pmatrix} \qquad (p \text{ real}). \qquad (6.7.1)$$

From $U^H U = I$, we get $p^2 + |q|^2 = 1$. Further, we suppose that $A' = U^{-1}AU$, where

$$A' = \begin{pmatrix} a' & b' \\ c' & d' \end{pmatrix},$$

and obtain

$$\begin{cases} a' = d + (a - d)p^2 + p(bq + cq^*) , \\ b' = bp^2 - cq^{*2} + (d - a)pq^* , \\ c' = cp^2 + (d - a)pq - bq^2 , \\ d' = a + (d - a)p^2 - p(bq + cq^*) . \end{cases} \qquad (6.7.2)$$

Clearly we have $a' + d' = a + d$. Claiming $c' = 0$, we find with $q = \alpha p$,

$$\alpha^2 - \frac{d - a}{b} \alpha = \frac{c}{b} ,$$

and

$$\alpha = \frac{1}{2b} \left(d - a \pm \sqrt{(d - a)^2 + 4bc} \right) . \qquad (6.7.3)$$

Here we conveniently choose the sign that makes $|\alpha|$ as small as possible; with $p = \cos \theta$ and $q = \sin \theta \cdot e^{i\varphi}$, we get $|\alpha| = \tan \theta$. Hence α is obtained directly from the elements a, b, c, and d. Normally, we must take the square root of a complex number, and this can be done by the formula

$$\sqrt{A + iB} = \pm (1/\sqrt{2})(\sqrt{C + A} + i \text{ sign } B\sqrt{C - A}) ,$$

where $C = \sqrt{A^2 + B^2}$. When α has been determined, we get p and q from

$$\begin{cases} p = (1 + |\alpha|^2)^{-1/2} , \\ q = \alpha p . \end{cases} \qquad (6.7.4)$$

Now we pass to the main problem and assume that A is an arbitrary complex matrix (n, n). We choose that element *below* the main diagonal which is largest

in absolute value, and perform an essentially two-dimensional unitary transformation which makes this element zero. This procedure is repeated until
$\sum_{i>k} |a_{ik}|^2$ is less than a given tolerance. Denoting the triangular matrix by T,
we have $U^{-1}AU = T$ with $U = U_1 U_2 \cdots U_N$, where $U_1, U_2, \ldots, U_N$ are the individual essentially two-dimensional unitary matrices. Clearly, the eigenvalues
are t_{rr}, $r = 1, 2, \ldots, n$.

In order to compute the eigenvectors, we start with the triangular matrix,
and we shall restrict ourselves to the case when all eigenvalues are different.
We see directly that the vector whose first component is 1 and whose other
components are 0 is an eigenvector belonging to the first eigenvalue t_{11}. Next
we see that we can determine such a value y_{12} that the vector with first component y_{12} and second component 1 becomes an eigenvector. The condition is
$t_{11}y_{12} + t_{12} = t_{22}y_{12}$, from which we can determine y_{12} (we suppose that $t_{11} \neq t_{22}$).
We proceed in the same way and collect all eigenvectors to a triangular matrix
Y, and further we form a diagonal matrix Λ with the diagonal elements t_{rr},
$r = 1, 2, \ldots, n$. Then we have

$$T = \begin{vmatrix} t_{11} & t_{12} & t_{13} \cdots t_{1n} \\ 0 & t_{22} & t_{23} \cdots t_{2n} \\ 0 & 0 & t_{33} \cdots t_{3n} \\ \vdots & & & \\ 0 & 0 & 0 \cdots t_{nn} \end{vmatrix} ; \qquad Y = \begin{vmatrix} 1 & y_{12} & y_{13} \cdots y_{1n} \\ 0 & 1 & y_{23} \cdots y_{2n} \\ 0 & 0 & 1 \cdots y_{3n} \\ \vdots & \vdots & \vdots \\ 0 & 0 & 0 \cdots 1 \end{vmatrix} ; \qquad (6.7.5)$$

$$\Lambda = \begin{vmatrix} t_{11} & 0 \cdots 0 \\ 0 & t_{22} \cdots 0 \\ \vdots & \\ 0 & 0 \cdots t_{nn} \end{vmatrix} ,$$

and obviously $TY = Y\Lambda$. The quantities y_{ik}, $k > i$, are computed recursively
from the relation

$$y_{ik} = \sum_{r=i+1}^{k} \frac{t_{ir} y_{rk}}{t_{kk} - t_{ii}} , \qquad (6.7.6)$$

valid for $i = 1, 2, \ldots, (k - 1)$. First we put $y_{kk} = 1$ and then we use (6.7.6)
for $i = k - 1, k - 2, \ldots, 1$, and in this way the eigenvector y_k belonging to
the eigenvalue t_{kk} can be determined. Last, we obtain the eigenvectors x_k of
the original matrix A from

$$x_k = U y_k . \qquad (6.7.7)$$

When the method is used on a computer, we must reserve two memory places
for each element (even if A is real). Only in special cases are all results real.
The method described here depends on annihilation of a subdiagonal element,
which, however, does *not* guarantee that $\tau^2 = \sum_{i>k} |a_{ik}|^2$ decreases. An alter-
native technique can be constructed aiming at minimization of τ^2 by choosing
θ and φ conveniently. The equations become fairly complicated but can be
solved numerically and, as a rule, the minimum point need not be established

to a very high degree of accuracy. The value of the method is still difficult to estimate as it has not yet been tried sufficiently in practice.

6.8. Hyman's method

A matrix is said to be of upper Hessenberg form if $a_{ik} = 0$ for $i - k > 1$ and of lower Hessenberg form if $a_{ik} = 0$ for $k - i > 1$. In the following we shall choose to work with the upper Hessenberg form. An arbitrary complex matrix can quite easily be reduced to Hessenberg form which will now be demonstrated. Essentially the reduction goes as in Givens' method demanding $(n - 1)(n - 2)/2$ steps. In the general case the transformations are unitary, but for real matrices we can use real (orthogonal) transformations which is a great advantage.

Starting from an arbitrary complex matrix $C = (c_{ik})$ we perform a two-dimensional rotation in the (i, k)-plane under the condition $c'_{k,i-1} = 0$ $(i = 2, 3, \ldots, n - 1; k = i + 1, i + 2, \ldots, n)$. The first rotation occurs in the $(2, 3)$-plane with the condition $c'_{31} = 0$, next in the $(2, 4)$-plane with $c'_{41} = 0$, and so on. In this way all elements in the first column except c'_{11} and c'_{21} are annihilated. After that we rotate in the $(3, 4)$-plane with $c'_{42} = 0$, and so on. Introduce the notations

$$\begin{cases} c_{i,i-1} = a + bi \,, \\ c_{k,i-1} = c + di \,, \end{cases} \quad \text{and} \quad U = \begin{pmatrix} p & -q^* \\ q & p \end{pmatrix},$$

with $p = \cos \theta$, $q = \sin \theta e^{i\varphi}$ making U unitary. Then we get

$$c'_{k,i-1} = -q c_{i,i-1} + p c_{k,i-1} = 0$$

and splitting into real and imaginary parts:

$$\begin{cases} c \cos \theta = a \sin \theta \cos \varphi - b \sin \theta \sin \varphi \,, \\ d \cos \theta = a \sin \theta \sin \varphi - b \sin \theta \cos \varphi \,. \end{cases}$$

Squaring and adding we get $(c^2 + d^2) \cos^2 \theta = (a^2 + b^2) \sin^2 \theta$, and a trivial elimination also gives the angle φ:

$$\begin{cases} \tan \theta = \sqrt{(c^2 + d^2)/(a^2 + b^2)} \,, \\ \tan \varphi = (ad - bc)/(ac + bd) \,. \end{cases}$$

In the real case we have $b = d = 0$ giving $\varphi = 0$ and $\tan \theta = c/a$.

We shall also briefly show how the reduction can be made by use of reflections following Householder (also cf. [13]). Putting $A = A_0$, we shall describe one step in the reduction leading from A_{r-1} to A_r, where

$$A_{r-1} = \left(\begin{array}{c|c} H_{r-1} & C_{r-1} \\ \hline 0 & b_{r-1} \mid B_{r-1} \end{array} \right) \begin{array}{l} \}r \\ \}n - r \end{array}$$
$$\underbrace{\phantom{H_{r-1}}}_{r-1} \quad \underbrace{\phantom{b_{r-1} B_{r-1}}}_{n-r}$$

and

$$H_{r-1} = \begin{pmatrix} * & * & * & * \cdots * & * \\ * & * & * & * \cdots * & * \\ & * & * & * \cdots * & * \\ & & * & * \cdots * & * \\ & \mathbf{0} & & & \\ & & & * & * \end{pmatrix}$$

is of upper Hessenberg form. The matrix A_r will then be produced through $A_r = P_r A_{r-1} P_r$ leaving H_{r-1} as well as the null-matrix of dimension $(n-r) \times (r-1)$ unchanged while the vector b_{r-1} must be annihilated except for the first component. In this way a new Hessenberg matrix H_r of dimension $(r+1) \times (r+1)$ is formed by moving one new row and one new column to H_{r-1}. Now we choose

$$P_r = \left(\begin{array}{c|c} I & 0 \\ \hline 0 & \underbrace{Q_r} \end{array} \right) \}n - r$$
$$\underbrace{}_{n-r}$$

with $Q_r = I - 2w_r w_r^H$ and $w_r^H w_r = 1$, w_r being a column vector with $(n - r)$ elements. A simple computation gives

$$P_r A_{r-1} P_r = \left(\begin{array}{c|c|c} H_{r-1} & C_{r-1} Q_r \\ \hline 0 & a_{r-1} & Q_r B_{r-1} Q_r \end{array} \right) \quad \text{with} \quad a_{r-1} = \begin{pmatrix} \alpha_r \\ 0 \\ 0 \\ \vdots \\ 0 \end{pmatrix}.$$

Hence $Q_r b_{r-1} = a_{r-1}$ and $a_{r-1}^H a_{r-1} = b_{r-1}^H Q_r^H Q_r b_{r-1} = b_{r-1}^H b_{r-1}$, that is, $|\alpha_r| = |b_{r-1}|$ (we suppose Euclidean vector norm). Further $b_{r-1} = Q_r^H a_{r-1}$ and $e_1^T b_{r-1} = e_1^T Q_r^H a_{r-1} = (1 - |w_1|^2)\alpha_r$, and since $1 - |w_1|^2$ is real, $\arg \alpha_r = \arg(e_1^T b_{r-1})$. Here e_1 is a vector with the first component $= 1$ and all other components $= 0$. Thus the argument of α_r is equal to the argument of the top element of the vector b_{r-1}. Finally, since

$$Q_r b_{r-1} = (I - 2w_r w_r^H) b_{r-1} = b_{r-1} - (2w_r^H b_{r-1}) w_r = a_{r-1},$$

we get

$$w_r = \frac{b_{r-1} - a_{r-1}}{|b_{r-1} - a_{r-1}|},$$

and A_r is completely determined. If this procedure is repeated we finally reach the matrix A_{n-2} which is of upper Hessenberg form.

After having produced a Hessenberg matrix $A = U^{-1} C U$ with the same eigenvalues as C we now turn to the computation of these. Let x be a vector with unknown components and λ an arbitrary (complex) number. Then form the following linear system of equations:

$$\begin{cases} a_{11}x_1 + a_{12}x_2 + \cdots + a_{1n}x_n = \lambda x_1 , \\ a_{21}x_1 + a_{22}x_2 + \cdots + a_{2n}x_n = \lambda x_2 , \\ \qquad\qquad\qquad\qquad\vdots \\ a_{n-1,n-2}x_{n-2} + a_{n-1,n-1}x_{n-1} + a_{n-1,n}x_n = \lambda x_{n-1} , \\ \qquad\qquad\qquad a_{n,n-1}x_{n-1} + a_{nn}x_n = \lambda x_n . \end{cases} \qquad (6.8.2)$$

We suppose that all elements $a_{i,i-1} \neq 0$, $i = 2, 3, \ldots, n$, and as the system is homogeneous we can choose, for example, $x_n = 1$. Then we can solve x_{n-1} as a first-degree polynomial from the last equation. From the next to last equation, x_{n-2} is obtained as a second-degree polynomial in λ, and so on; and finally x_1 is determined from the second equation as a polynomial of degree $n - 1$ in λ. If all these values of $x_1, x_2, \ldots, x_n$ are inserted into the expression $(\lambda - a_{11})x_1 - a_{12}x_2 - \cdots - a_{1n}x_n$, we get as result the characteristic polynomial $f(\lambda) = \det(\lambda I - A)$ apart from a constant factor. It is now possible to compute the eigenvalues from these results by interpolation, first linear and then quadratic.

The method described here is such that the values obtained are *exact* eigenvalues of another matrix differing only slightly from A ("reverse error computation," cf. Section 4.4. and Wilkinson, [18], p. 147). It is quite possible that the errors in the eigenvalues may become large; this means that the eigenvalue problem *per se* is ill-conditioned and we can expect numerical difficulties irrespective of the method chosen.

The interpolation can be improved by computing the first and possibly also the second derivative of the characteristic polynomial by iteration. The following formulas are used:

$$x_{i-1} = \frac{(\lambda - a_{ii})x_i - \sum_{k=i+1}^n a_{ik}x_k}{a_{i,i-1}} ; \qquad x_n = 1 , \qquad a_{1,0} = 1 ;$$

$$y_{i-1} = \frac{(\lambda - a_{ii})y_i + x_i - \sum_{k=i+1}^n a_{ik}y_k}{a_{i,i-1}} ; \qquad y_n = 0 ;$$

$$z_{i-1} = \frac{(\lambda - a_{ii})z_i + 2y_i - \sum_{k=i+1}^n a_{ik}z_k}{a_{i,i-1}} ; \qquad z_n = 0 .$$

Denoting the characteristic polynomial by $f(\lambda)$ $(= \lambda^n + \cdots)$ and putting

$$p_{00} = \left\{ \prod_{i=2}^n a_{i,i-1} \right\}^{-1} ,$$

we have

$$f(\lambda) = \frac{x_0}{p_{00}} ; \qquad f'(\lambda) = \frac{y_0}{p_{00}} ; \qquad f''(\lambda) = \frac{z_0}{p_{00}} .$$

Then Newton-Raphson's method can be used (possibly a variant independent of multiplicity):

$$\lambda_{n+1} = \lambda_n - \frac{x_0}{y_0} \quad \text{or} \quad \lambda_{n+1} = \lambda_n - \frac{x_0 y_0}{y_0^2 - x_0 z_0} .$$

Example

$$A = \begin{pmatrix} 6 & 3 & -4 & 2 \\ 2 & 1 & 5 & -3 \\ 0 & 3 & 7 & 1 \\ 0 & 0 & 2 & 5 \end{pmatrix} .$$

Starting with $\lambda_0 = 5$, we compute x_i and y_i:

i	x_i	y_i
4	1	0
3	0	$\frac{1}{2}$
2	$-\frac{1}{3}$	$-\frac{1}{3}$
1	$\frac{5}{6}$	$-\frac{25}{12}$
0	$-\frac{11}{6}$	$\frac{71}{12}$

Hence $\lambda_1 = 5 + \frac{22}{71} \simeq 5.3$ (correct value 5.374). Let us now once again write down expressions as polynomials in λ for the set $x_1, x_2, \ldots, x_n$ which is also extended by an extra variable associated with the first equation of the system:

$$\begin{cases} x_0 &= p_{00}\lambda^n \phantom{^{-1}} + p_{01}\lambda^{n-1} + \cdots + p_{0n} , \\ x_1 &= p_{11}\lambda^{n-1} + p_{12}\lambda^{n-2} + \cdots + p_{1n} , \\ \vdots \\ x_i &= p_{ii}\lambda^{n-i} + p_{i,i+1}\lambda^{n-i-1} + \cdots + p_{in} , \\ \vdots \\ x_{n-1} &= p_{n-1,n-1}\lambda + p_{n-1,n} , \\ x_n &= p_{nn} \, (=1) . \end{cases}$$

Using $Ax = \lambda x$, we get

$$\begin{aligned} x_{i-1} &= \frac{(\lambda - a_{ii})x_i - a_{i,i+1}x_{i+1} - \cdots - a_{in}x_n}{a_{i,i-1}} \\ &= \frac{(\lambda - a_{ii}) \sum_{k=i}^{n} p_{ik}\lambda^{n-k} - a_{i,i+1} \sum_{k=i+1}^{n} p_{i+1,k}\lambda^{n-k} - \cdots - a_{in}p_{nn}}{a_{i,i-1}} . \end{aligned}$$

Comparing powers λ^{n-k} on both sides, we find

$$a_{i,i-1}p_{i-1,k} = p_{i,k+1} - \sum_{r=i}^{k} a_{ir}p_{rk} \tag{6.8.4}$$

$$(i = 1, 2, \ldots, n; \quad k = i-1, i, \ldots, n; \quad a_{1,0} = 1, \quad p_{i,n+1} = 0).$$

This relation can be written in matrix form; for $n = 4$ we get

$$
\begin{pmatrix}
1 & a_{11} & a_{12} & a_{13} & a_{14} \\
0 & a_{21} & a_{22} & a_{23} & a_{24} \\
0 & 0 & a_{32} & a_{33} & a_{34} \\
0 & 0 & 0 & a_{43} & a_{44} \\
0 & 0 & 0 & 0 & 1
\end{pmatrix}
\begin{pmatrix}
p_{00} & p_{01} & p_{02} & p_{03} & p_{04} \\
0 & p_{11} & p_{12} & p_{13} & p_{14} \\
0 & 0 & p_{22} & p_{23} & p_{24} \\
0 & 0 & 0 & p_{33} & p_{34} \\
0 & 0 & 0 & 0 & p_{44}
\end{pmatrix}
$$

$$
=
\begin{pmatrix}
p_{11} & p_{12} & p_{13} & p_{14} & 0 \\
0 & p_{22} & p_{23} & p_{24} & 0 \\
0 & 0 & p_{33} & p_{34} & 0 \\
0 & 0 & 0 & p_{44} & 0 \\
0 & 0 & 0 & 0 & 1
\end{pmatrix}.
\qquad (6.8.5)
$$

Taking determinants on both sides we find in particular $p_{00} = (a_{21}a_{32}a_{43})^{-1}$, which can easily be generalized to arbitrary n. The characteristic equation is now

$$
p_{00}\lambda^n + p_{01}\lambda^{n-1} + \cdots + p_{0n} = 0 .
$$

This method of generating the characteristic equation and then the eigenvalues by some standard technique, as a rule is unsuitable for stability reasons except possibly if the computation of coefficients can be performed within the rational field.

Computation of the eigenvectors from (6.8.2) cannot be recommended since instability will occur. However, it is possible to use an idea by Wielandt, in principle inverse interpolation. Let λ_0 be an approximate eigenvalue ($\lambda_0 \simeq \lambda_1$, where λ_1 is an exact eigenvalue) and u_0 an approximate eigenvector. Further we suppose that the exact eigenvectors are $v_1, v_2, \ldots, v_n$. Successive iterations are now computed from

$$
(A - \lambda_0 I)u_{i+1} = s_i u_i ,
$$

where s_i is a suitable scale factor compensating for the fact that $A - \lambda_0 I$ is almost singular which will cause no trouble otherwise. Now suppose, for example, $u_0 = \sum \alpha_r v_r$ which leads to

$$
(A - \lambda_0 I)u_1 = s_0 \sum \alpha_r v_r ,
$$

that is, $u_1 = s_0 \sum \alpha_r (A - \lambda_0 I)^{-1} v_r = s_0 \sum \alpha_r (\lambda_r - \lambda_0)^{-1} v_r \simeq s_0 \alpha_1 (\lambda_1 - \lambda_0)^{-1} v_1$ and we have the same effect as when the power method is applied. The solution of the system $(A - \lambda_0 I)u_{i+1} = s_i u_i$ is performed by standard means, for example, following Gauss or Crout.

If the eigenvector problem is ill-conditioned in itself, this technique, of course, will not help. Consider, for example,

$$
A = \begin{pmatrix} 1 & \varepsilon_1 \\ \varepsilon_2 & 1 \end{pmatrix}
$$

with the well-conditioned eigenvalues $1 \pm \sqrt{\varepsilon_1 \varepsilon_2}$. The eigenvectors

$$\begin{pmatrix} 1 \\ \sqrt{\varepsilon_2/\varepsilon_1} \end{pmatrix} \quad \text{and} \quad \begin{pmatrix} 1 \\ -\sqrt{\varepsilon_2/\varepsilon_1} \end{pmatrix}$$

on the other hand are ill-conditioned since small errors in ε_1 and ε_2 can give rise to very large errors in the eigenvectors.

Thus, Hyman's method will produce eigenvalues and eigenvectors to general complex matrices provided they are first transformed to Hessenberg form. The method has the advantage that the computations can be performed in the real or complex domain depending on the special circumstances.

6.9. The QR-method by Francis

The main principle of the QR-method is to produce a (for example upper) triangular matrix by the aid of unitary transformations. The technique is completely general but becomes too complicated for arbitrary matrices. For this reason the matrix is usually first transformed to Hessenberg form as described in the preceding section.

The following theorem is essential in this connection: Every regular square matrix A can be written as $A = QR$ where Q is unitary and R upper triangular, and both are uniquely determined. The existence of such a partition can be proved by a method resembling Gram-Schmidt's orthogonalization. Suppose that A is regular and put

$$QR = A \ .$$

The column vectors of Q and A are denoted by q_i and a_i. If we now multiply all the rows in Q by the first column of R we get (since only the first element in the R-column is not equal to 0) $r_{11} q_1 = a_1$. But since Q is unitary, we also have $q_i^H q_k = \delta_{ik}$; hence $r_{11} = |a_1|$ (apart from a trivial factor $e^{i\theta_1}$); then also q_1 is uniquely determined. Now we assume that the vectors $q_1, q_2, \ldots, q_{k-1}$ are known. Multiplying all rows of Q by the kth column of R, we find

$$\sum_{i=1}^{k} r_{ik} q_i = a_k \ .$$

Next, for $j = 1, 2, \ldots, k - 1$, we multiply from the left with q_j^H and obtain $r_{jk} = q_j^H a_k$. Further we also have

$$r_{kk} q_k = a_k - \sum_{i=1}^{k-1} r_{ik} q_i \ .$$

The right-hand side is certainly not equal to 0, because otherwise we would have linear dependence between the vectors a_i. Hence we find

$$r_{kk} = \left| a_k - \sum_{i=1}^{k-1} r_{ik} q_i \right| ,$$

and the vector q_k is also determined. Obviously, uniqueness is secured apart from factors $e^{i\theta_k}$ in the diagonal elements of R, and hence the condition $r_{kk} > 0$ will determine Q and R completely.

Now we start from $A = A_1$ and form sequences of matrices A_s, Q_s, and R_s by use of the following algorithm:

$$A_s = Q_s R_s \,; \qquad A_{s+1} = Q_s^H A_s Q_s = Q_s^H Q_s R_s Q_s = R_s Q_s \,.$$

This means that first A_s is partitioned in a product of a unitary and an upper triangular matrix, and then A_{s+1} is computed as $R_s Q_s$. It also means that A_{s+1} is formed from A_s through a similarity transformation with a unitary matrix. Next, we put

$$P_s = Q_1 Q_2 \cdots Q_s \qquad \text{and} \qquad R_s R_{s-1} \cdots R_1 = U_s \,,$$

and so we obtain

$$A_{s+1} = Q_s^{-1} A_s Q_s = Q_s^{-1}(Q_{s-1}^{-1} A_{s-1} Q_{s-1}) Q_s = \cdots = P_s^{-1} A_1 P_s \,.$$

Then we form

$$P_s U_s = Q_1 Q_2 \cdots Q_s R_s R_{s-1} \cdots R_1 = Q_1 Q_2 \cdots Q_{s-1} A_s R_{s-1} \cdots R_1$$
$$= P_{s-1} A_s U_{s-1} \,.$$

But $A_s = P_{s-1}^{-1} A_1 P_{s-1}$ and consequently $P_{s-1} A_s = A_1 P_{s-1}$ which gives

$$P_s U_s = A_1 P_{s-1} U_{s-1} = \cdots = A_1^{s-1} P_1 U_1 = A_1^{s-1} Q_1 R_1 = A_1^s \,.$$

Here P_s is unitary and U_s upper triangular, and in principle they could be computed from A_1^s by partition in a product of a unitary and an upper triangular matrix. In this way we would also obtain A_{s+1} through a similarity transformation:

$$A_{s+1} = P_s^{-1} A_1 P_s \,.$$

We now assert that the matrix A_{s+1} for increasing s more and more will approach an upper triangular matrix. We do not give a complete proof but restrict ourselves to the main points; further details can be found in [12]. The following steps are needed for the proof:

1. $A = A_1$ is written $A_1 = XDX^{-1} = XDY$ which is always possible if all eigenvalues are different (this restriction can be removed afterward). Further we assume $d_{ii} = \lambda_i$ with $|\lambda_1| > |\lambda_2| > \cdots$

2. Then we also have $P_s U_s = A_1^s = XD^s Y$.

3. X is partitioned as $X = Q_X R_X$ and Y as $Y = L_Y R_Y$ where Q_X is unitary, R_X and R_Y upper triangular, and L_Y lower triangular with ones in the main diagonal (both partitions are unique). For the latter partition a permutation might be necessary.

4. Then we get $P_s U_s = Q_X R_X D^s L_Y D^{-s} D^s R_Y$. The decisive point is that $D^s L_Y D^{-s}$, as is easily shown, in subdiagonal elements will contain quotients

$(\lambda_i/\lambda_1)^s$ so that this matrix actually will approach the identity matrix. If so, we are left with $P_s U_s \simeq Q_X R_X D^s R_Y$.

5. But P_s and Q_X are unitary while U_s and $R_X D^s R_Y$ are upper triangular. Since the partition is unique, we can draw the conclusion that

$$\lim_{s \to \infty} P_s = Q_X \qquad \text{and} \qquad U_s \simeq R_X D^s R_Y \,.$$

6. $A_{s+1} = P_s^H A_1 P_s = P_s^H X D X^{-1} P_s \to Q_X^H X D X^{-1} Q_X \to R_X D R_X^{-1}$ (since $X = Q_X R_X$ and $R_X = Q_X^H X$). The matrix $R_X D R_X^{-1}$ is an upper triangular matrix with the same eigenvalues as A.

As already mentioned, the QR-method will become too laborious for arbitrary matrices, and instead it is used on special matrices, preferably Hessenberg or symmetric band-matrices. The method has good stability properties and seems to be one of the most promising at present. Several algorithms in ALGOL treating this method have already been published [15, 16].

REFERENCES

[1] Lanczos: "An Iteration Method for the Solution of the Eigenvalue Problem of Linear Differential and Integral Operators," *J. of Res.*, **45**, 255 (1950).

[2] Wilkinson: "The Calculation of the Eigenvectors of Codiagonal Matrices," *The Computer Journal*, **1**, 90–96.

[3] Givens: *Numerical Computation of the Characteristic Values of a Real Symmetric Matrix*, Oak Ridge National Laboratory, ORNL–1574 (1954).

[4] Wilkinson: "Householder's Method for the Solution of the Algebraic Eigenproblem," *The Computer Journal*, 23–27 (April 1960).

[5] National Bureau of Standards: *Appl. Math. Series*, No. 49 (Washington, 1958).

[6] Lotkin: "Characteristic Values for Arbitrary Matrices," *Quart. Appl. Math.*, **14**, 267–275 (1956).

[7] Greenstadt: *A Method for Finding Roots of Arbitrary Matrices*, MTAC, **9**, 47–52 (1955).

[8] Goldstine, Horwitz: *A Procedure for the Diagonalization of Normal Matrices;* Jour. Assoc. Comp. Mach. 6, 1959 pp. 176–195.

[9] Causey, Gregory: *On Lanczos' Algorithm for Tridiagonalizing Matrices;* SIAM Review Vol. 3, No. 4, Oct. 1961.

[10] Eberlein: *A Jacobi-like Method for the Automatic Computation of Eigenvalues and Eigenvectors of an Arbitrary Matrix;* J. SIAM Vol. 10, No. 1, March 1962.

[11] La Budde: *The Reduction of an Arbitrary Real Square Matrix to Tridiagonal Form Using Similarity Transformations;* Math. Comp. 17 (1963) p. 433.

[12] Francis: *The QR Transformation—A Unitary Analogue to the LR Transformation;* Comp. Jour. 4 (1961) pp. 265–271 and 332–345.

[13] Fröberg: *On Triangularization of Complex Matrices by Two-Dimensional Unitary Transformations;* BIT 5 : 4 (1965) p. 230.

[14] Householder, Bauer: *On Certain Methods for Expanding the Characteristic Polynomial;* Num. Math 1 (1959) p. 29.

[15] Businger: *Eigenvalues of a real symmetric matrix by the QR method;* Algorithm 253, Comm. ACM 8 : 4, p. 217.

[16] Ruhe: *Eigenvalues of a complex matrix by the QR method;* BIT 6 : 4 (1966), p. 350.

[17] Hyman: *Eigenvalues and Eigenvectors of General Matrices;* Ass. Comp. Mach. (June 1957).

[18] Wilkinson: *The Algebraic Eigenvalue Problem;* Oxford Univ. Press 1965.

[19] Wilkinson: *Convergence of the LR, QR and Related Algorithms;* Comp. Jour. 8 (1966), p. 77.

EXERCISES

1. Find the largest eigenvalue of the matrix

$$A = \begin{pmatrix} 25 & -41 & 10 & -6 \\ -41 & 68 & -17 & 10 \\ 10 & -17 & 5 & -3 \\ -6 & 10 & -3 & 2 \end{pmatrix},$$

correct to three places.

2. Find the absolutely smallest eigenvalue and the corresponding eigenvector of the matrix

$$A = \begin{pmatrix} 1 & 2 & -2 & 4 \\ 2 & 12 & 3 & 5 \\ 3 & 13 & 0 & 7 \\ 2 & 11 & 2 & 2 \end{pmatrix},$$

using the fact that λ^{-1} is an eigenvalue of A^{-1} if λ is an eigenvalue of A.

3. Find the largest eigenvalue and the corresponding eigenvector of the Hermitian matrix

$$H = \begin{pmatrix} 8 & -5i & 3 - 2i \\ 5i & 3 & 0 \\ 3 + 2i & 0 & 2 \end{pmatrix}.$$

4. Using the fact that $\lambda - a$ is an eigenvalue of $A - aI$, if λ is an eigenvalue of A, find the highest and the lowest eigenvalue of the matrix

$$A = \begin{pmatrix} 9 & 10 & 8 \\ 10 & 5 & -1 \\ 8 & -1 & 3 \end{pmatrix}.$$

Choose $a = 12$ and the starting vectors

$$\begin{pmatrix} 1 \\ 1 \\ 1 \end{pmatrix} \quad \text{and} \quad \begin{pmatrix} -1 \\ 1 \\ 1 \end{pmatrix},$$

respectively. (Desired accuracy: two decimal places.)

5. The matrix

$$A = \begin{pmatrix} 14 & 7 & 6 & 9 \\ 7 & 9 & 4 & 6 \\ 6 & 4 & 9 & 7 \\ 9 & 6 & 7 & 15 \end{pmatrix}$$

has an eigenvalue close to 4. Compute this eigenvalue to six places, using the matrix $B = (A - 4I)^{-1}$.

6. A is a matrix with one eigenvalue λ_0 and another $-\lambda_0$ (λ_0 real, >0); all remaining eigenvalues are such that $|\lambda| < \lambda_0$. Generalize the power method so that it can be used in this case. Use the result for computing λ_0 for the matrix

$$A = \begin{pmatrix} 2.24 & -2.15 & -7.37 \\ -2.15 & 0.75 & -0.87 \\ -7.37 & -0.87 & -1.99 \end{pmatrix} \quad \text{(two decimals)}.$$

7. Find the largest eigenvalue of the modified eigenvalue problem $Ax = \lambda Bx$ when

$$A = \begin{pmatrix} 1 & 6 & 6 & 4 \\ 6 & 37 & 43 & 16 \\ 6 & 43 & 86 & -27 \\ 4 & 16 & -27 & 106 \end{pmatrix} \quad \text{and} \quad B = \begin{pmatrix} 1 & 2 & -1 & 4 \\ 2 & 5 & 1 & 6 \\ -1 & 1 & 11 & -11 \\ 4 & 6 & -11 & 22 \end{pmatrix}.$$

8. Show that

$$A = \begin{pmatrix} a_0 & a_1 & a_2 & a_3 \\ a_3 & a_0 & a_1 & a_2 \\ a_2 & a_3 & a_0 & a_1 \\ a_1 & a_2 & a_3 & a_0 \end{pmatrix}$$

can be written $a_0 I + a_1 C + a_2 C^2 + a_3 C^3$, where C is a constant matrix. Also find the eigenvalues and the eigenvectors of A. (A is called *circulant*.)

9. In the matrix A of type (n, n) all diagonal elements are a, while all the others are b. Find such numbers p and q that $A^2 - pA + qI = 0$. Use this relation for finding eigenvalues [one is simple, and one is $(n - 1)$-fold] and the eigenvectors of A.

10. Using the LR-method, find A_1, A_2, and A_3 when

$$A = \begin{pmatrix} 7 & 6 \\ 3 & 4 \end{pmatrix}.$$

11. A is a matrix with eigenvalues $\lambda_1, \lambda_2, \ldots, \lambda_n$ (all different). Put

$$B_k = \frac{(A - \lambda_1 I)(A - \lambda_2 I)\cdots(A - \lambda_{k-1}I)(A - \lambda_{k+1}I)\cdots(A - \lambda_n I)}{(\lambda_k - \lambda_1)(\lambda_k - \lambda_2)\cdots(\lambda_k - \lambda_{k-1})(\lambda_k - \lambda_{k+1})\cdots(\lambda_k - \lambda_n)}.$$

Show that $B_k^2 = B_k$.

12. A real, symmetric matrix A has the largest eigenvalue equal to λ_1 and the next largest equal to λ_2, and both are greater than zero. All the other eigenvalues are considerably less in absolute value than these two. By use of the power method one has obtained a vector of the form $u + \varepsilon v$, where u and v are eigenvectors corresponding to the eigenvalues λ_1 and λ_2. From these values a series of consecutive Rayleigh quotients is constructed. Show how λ_1 can be accurately determined from three such quotients R_1, R_2, and R_3.

Chapter 7

Linear operators

*"I could have done it in a much more
complicated way"* said the red Queen,
immensely proud.   LEWIS CARROLL.

7.0. General properties

When one wishes to construct formulas for interpolation, numerical differentiation, quadrature, and summation, the operator technique proves to be a most useful tool. One of the greatest advantages is that one can sketch the type of formula desired in advance and then proceed directly toward the goal. Usually the deduction is also considerably simplified. It must be understood, however, that complete formulas, including remainder terms, are in general not obtained in this way.

The operators are supposed to operate on functions belonging to a linear function space. Such a space is defined as a set F of functions, f, g, $\ldots$, having the properties that $f \in F$, $g \in F$ implies $\alpha f + \beta f \in F$, where α and β are arbitrary constants. A linear operator is then a mapping of F on a linear function space F^*; usually one chooses $F^* = F$. An operator P being linear means that $P(\alpha f + \beta g) = \alpha P f + \beta P g$ for all f and $g \in F$ with α and β arbitrary constants. The operators which we are going to discuss fulfill the associative and distributive laws, but in general they do not commutate. If we let $P_1, P_2, P_3, \ldots$ denote operators, then the following laws are supposed to be valid:

$$\begin{cases} P_1 + (P_2 + P_3) = (P_1 + P_2) + P_3 \, , \\ \quad P_1(P_2 P_3) = (P_1 P_2)P_3 \\ P_1(P_2 + P_3) = P_1 P_2 + P_1 P_3 \, , \end{cases} \qquad (7.0.1)$$

Here we define the sum P of two operators P_1 and P_2 as the operator which transforms the function f into $P_1 f + P_2 f$; the product $P_1 P_2$ is defined as the operator which transforms f into $P_1(P_2 f)$.

Two operators P_1 and P_2 are said to be equal if $P_1 f = P_2 f$ for all functions $f \in F$. We also define the inverse P^{-1} of an operator P. If Q is such an operator that for all f we have $Qg = f$ if $Pf = g$, then Q is said to be an inverse of P. Here we must be careful, however, since it can occur that P^{-1} is not uniquely determined. In order to demonstrate this we suppose that $w(x)$ is a function which is annihilated by P, that is, $Pw(x) = 0$, and further that $f(x)$ represents a possible result of $P^{-1}g(x)$. Then $f(x) + w(x)$ is another possible result. In fact, we have $P(f(x) + w(x)) = Pf(x) = g(x)$. Hence we can write $P^{-1}Pf = f + w$, where w is an arbitrary function annihilated by P.

147

Thus P^{-1} is a right inverse but not a left inverse. If the equation $Pw(x) = 0$ has only the trivial solution $w(x) = 0$, then P^{-1} also becomes a left inverse:

$$PP^{-1} = P^{-1}P = 1 .$$

Then we say briefly that P^{-1} is the inverse of P.

7.1. Special operators

Now we introduce the following 13 operators, including their defining equations:

$$
\left\{
\begin{array}{ll}
Ef(x) = f(x + h) & \text{The shifting operator} \\[4pt]
\varDelta f(x) = f(x + h) - f(x) & \text{The forward-difference operator} \\[4pt]
\nabla f(x) = f(x) - f(x - h) & \text{The backward-difference operator} \\[4pt]
\delta f(x) = f\left(x + \dfrac{h}{2}\right) - f\left(x - \dfrac{h}{2}\right) & \text{The central-difference operator} \\[8pt]
\mu f(x) = \dfrac{1}{2}\left[f\left(x + \dfrac{h}{2}\right) + f\left(x - \dfrac{h}{2}\right)\right] & \text{The mean-value operator} \\[8pt]
\left.
\begin{array}{l}
Sf(x + h) = Sf(x) + f(x) \\[2pt]
Tf(x) = Tf(x - h) + f(x) \\[2pt]
\sigma f\left(x + \dfrac{h}{2}\right) = \sigma f\left(x - \dfrac{h}{2}\right) + f(x)
\end{array}
\right\} & \text{Summation operators} \\[14pt]
Df(x) = f'(x) & \text{The differentiation operator} \\[4pt]
Jf(x) = \displaystyle\int_a^x f(t)\,dt & \begin{array}{l}\text{The indefinite integration operator} \\ (a\ \text{arbitrary constant})\end{array} \\[10pt]
\left.
\begin{array}{l}
J_1 f(x) = \displaystyle\int_x^{x+h} f(t)\,dt \\[6pt]
J_{-1} f(x) = \displaystyle\int_{x-h}^{x} f(t)\,dt \\[6pt]
J_0 f(x) = \displaystyle\int_{x-(h/2)}^{x+(h/2)} f(t)\,dt
\end{array}
\right\} & \text{Definite integration operators}
\end{array}
\right.
$$

(7.1.1)

We observe that in all cases, except for D and J, the interval length h enters the formula in an important way. Actually, this ought to have been marked in some way, for example, by using the notation E_h instead of E. However, no misunderstandings need to be feared, and if one wants to use other intervals, this can easily be denoted, for example,

$$\delta_{2h} f(x) = f(x + h) - f(x - h) .$$

Further, we note that the summation operators are distinguished from the others by the fact that they are defined recursively, and in general this will give rise to an indeterminacy. This will be discussed in some detail later.

In the following we will need relations between different operators. The first eight depend on the interval length h and have a discrete character, while the last five, whether they contain h or not, are connected with operations associated with a passage to the limit. The first group is easily attainable in numerical computation, the last group in analytical. In the rest of this chapter, our main task will be to express the operators of one group in terms of the operators of the other.

First, we take some simple relations and formulas. To begin with we observe that the operator E^{-1} ought to be interpreted by $E^{-1}f(x) = f(x - h)$. Hence $EE^{-1} = E^{-1}E = 1$. The difference operator Δ has the same relation to the summation operator S as the differentiation operator D to the integration operator J. We have clearly

$$DJf(x) = D \int_a^x f(t)\, dt = f(x)\,,$$

but

$$JDf(x) = \int_a^x f'(t)\, dt = f(x) - f(a)\,.$$

In the same way as the integration introduces an indefinite constant, this must also be the case for the summation operator since the summation has to start at some initial value. We can build up a difference scheme and continue to the right in a unique way but not to the left. It is easy to see that if one first forms the sums starting from a certain initial value and then computes the differences, we regain the old values. However, if the differences are formed first and the summation is performed afterward, an indeterminacy must appear. Hence we have:

$$\Delta Sf(x) = f(x)\,,$$
$$S\Delta f(x) = f(x) + c\,.$$

Similar relations hold for the operators T and σ in relation to the difference operators ∇ and δ. Summing up, we can say that the operators J, S, T, and σ are *right inverses* but not *left inverses* to the operators D, Δ, ∇, and δ. This fact is closely related to a property of these latter operators. If one of them, for example D, operates on a polynomial, we get as result a polynomial of lower degree. This can be expressed by saying that the operators are *degrading*, and in such cases the reverse operation must give an indeterminacy. This is also reflected by the fact that the equations $Df(x) = 0$ and $\Delta f(x) = 0$ possess nontrivial solutions ($\neq 0$).

In order to link the discrete operators to the others, we make use of Taylor's formula,

$$f(x + h) = f(x) + hf'(x) + \frac{h^2}{2!}f''(x) + \cdots\,.$$

By means of operator notations, we can write

$$Ef(x) = \left(1 + hD + \frac{h^2D^2}{2!} + \cdots\right)f(x) = e^{hD}f(x)\,.$$

Shift and difference operators vs. differentiation operators

	E	Δ	∇	δ	U
E	E	$\Delta+1$	$(1-\nabla)^{-1}$	$1+\tfrac{1}{2}\delta^2+\delta\sqrt{1+\dfrac{\delta^2}{4}}$	e^{U}
Δ	$E-1$	Δ	$(1-\nabla)^{-1}-1$	$\tfrac{1}{2}\delta^2+\delta\sqrt{1+\dfrac{\delta^2}{4}}$	$e^{U}-1$
∇	$1-E^{-1}$	$1-(1+\Delta)^{-1}$	∇	$-\tfrac{1}{2}\delta^2+\delta\sqrt{1+\dfrac{\delta^2}{4}}$	$1-e^{-U}$
δ	$E^{1/2}-E^{-1/2}$	$\Delta(1+\Delta)^{-1/2}$	$\nabla(1-\nabla)^{-1/2}$	δ	$2\sinh\dfrac{U}{2}$
μ	$\tfrac{1}{2}(E^{1/2}+E^{-1/2})$	$\left(1+\dfrac{\Delta}{2}\right)(1+\Delta)^{-1/2}$	$\left(1-\dfrac{\nabla}{2}\right)(1-\nabla)^{-1/2}$	$\sqrt{1+\dfrac{\delta^2}{4}}$	$\cosh\dfrac{U}{2}$
U	$\log E$	$\log(1+\Delta)$	$\log\dfrac{1}{1-\nabla}$	$2\sinh^{-1}\dfrac{\delta}{2}$	U

This formula can be used only for analytical functions. Further, it should be observed that e^{hD} is defined by its power series expansion; in the following this technique will be used frequently. Then it is clear that Taylor's formula can be written in the compact form

$$E = e^{hD} . \tag{7.1.2}$$

In the table on p. 150 we present the relations between shift and difference operators and the differentiation operator, for which we now prefer the form $U = hD$. In particular we observe that $\delta = 2 \sinh U/2$ and $\mu = \sqrt{1 + \delta^2/4}$; both of these formulas will be used frequently.

For the integration operators, we find the following formulas:

$$\begin{cases} DJ_1 = J_1 D = \Delta , & \Delta J = J_1 , \\ DJ_{-1} = J_{-1} D = \nabla , & \nabla J = J_{-1} , \\ DJ_0 = J_0 D = \delta , & \delta J = J_0 , \\ DJ = 1 . \end{cases} \tag{7.1.3}$$

Further, we also have

$$\begin{cases} J_1 = \dfrac{h\Delta}{\log (1 + \Delta)} = \dfrac{h\Delta}{U} , \\[2ex] J_{-1} = -\dfrac{h\nabla}{\log (1 - \nabla)} = \dfrac{h\nabla}{U} , \\[2ex] J_0 = \dfrac{h\delta}{2 \sinh^{-1} (\delta/2)} = \dfrac{h\delta}{U} . \end{cases} \tag{7.1.4}$$

7.2. Power series expansions with respect to δ

The quantities which are available in numerical work are, apart from the functions themselves, differences of all orders. To obtain the most symmetrical formulas, one ought to use the central difference operator δ. For this reason it is natural that we investigate power series expansions in terms of δ a little more in detail. It should be observed that these expansions, as a rule, ought to contain only *even* powers of δ, since odd powers refer to points between the mesh points, and such function values are usually not known. If, in spite of all precautions, odd powers are obtained, we can multiply with

$$\mu \left(1 + \frac{\delta^2}{4} \right)^{-1/2} \equiv 1 . \tag{7.2.1}$$

The resulting expression can then be written $\mu\delta \cdot P(\delta)$, where P is an even power series. As to the first factor, we have $\mu\delta = \frac{1}{2}(E - E^{-1})$, and hence odd powers have been eliminated.

Our main problem is now expressing $F(U)$ as a power series in δ by use of the relation $\delta = 2\sinh{(U/2)}$. Denoting the inverse function of sinh by $\sinh^{-1}$, we have $U = 2\sinh^{-1}(\delta/2)$, and in principle we merely have to insert this into $F(U)$. We shall now discuss how this is done in practice.

The quantities δ, F, and U will be denoted by x, y, and z, respectively, and we have the relations

$$\begin{cases} y = y(z) \, . \\ x = 2\sinh{(z/2)} \, . \end{cases} \tag{7.2.2}$$

Differentiating, we get

$$dx = \cosh\left(\frac{z}{2}\right) dz = \left(1 + \frac{x^2}{4}\right)^{1/2} dz \quad \text{or} \quad \frac{dz}{dx} = \left(1 + \frac{x^2}{4}\right)^{-1/2} .$$

The general idea is to differentiate y with respect to x and then compute $(d^n y/dx^n)_{x=0}$; from this we easily get the coefficients of the expansion. We consider the following ten functions:

1. $y = z$
2. $y = z/\cosh{(z/2)}$
3. $y = z^2$
4. $y = \cosh pz$
5. $y = \sinh pz/\sinh z$
6. $y = \cosh pz/\cosh{(z/2)}$
7. $y = 1/\cosh{(z/2)}$
8. $y = \sinh z/z$
9. $y = 2\tanh{(z/2)}/z$
10. $y = 2(\cosh z - 1)/z^2$.

Then we find the following expansions:

1.

$$\frac{dy}{dx} = \frac{dz}{dx} = \frac{1}{\cosh{(z/2)}} = \left(1 + \frac{x^2}{4}\right)^{-1/2}$$

$$= \sum_n \binom{-\frac{1}{2}}{n} \frac{x^{2n}}{2^{2n}} = \sum_n \frac{(-\frac{1}{2})(-\frac{3}{2}) \cdots (\frac{1}{2} - n)}{n! \cdot 2^{2n}} x^{2n}$$

$$= \sum_n \frac{(-1)^n \cdot 1 \cdot 3 \cdot 5 \cdots (2n - 1)}{2^{3n} \cdot n!} x^{2n} = \sum_n \frac{(-1)^n \cdot (2n)!}{2^{4n} \cdot (n!)^2} x^{2n} \, .$$

Integrating this relation, we get:

$$y = z = \sum_{n=0}^{\infty} \frac{(-1)^n \cdot (2n)!}{2^{4n} \cdot (n!)^2(2n + 1)} x^{2n+1}$$

$$= x - \frac{x^3}{24} + \frac{3x^5}{640} - \frac{5x^7}{7168} + \frac{35x^9}{294912} - \cdots .$$

2.

$$\begin{cases} y = \dfrac{z}{\cosh{(z/2)}} \, ; \\[2mm] \dfrac{dy}{dx} = \dfrac{\cosh{(z/2)} - (z/2)\sinh{(z/2)}}{\cosh^2{(z/2)}} \dfrac{dz}{dx} = \dfrac{1 - xy/4}{1 + x^2/4} \, . \end{cases}$$

Hence $(x^2 + 4)y' + xy = 4$ (y' means dy/dx). For $x = 0$ we get $y = 0$, and from the equation we get $y' = 1$.

Next we differentiate m times and obtain

$$\begin{cases} (x^2 + 4)y'' & + & 3xy' & + & y & = 0 , \\ (x^2 + 4)y''' & + & 5xy'' & + & 4y' & = 0 , \\ (x^2 + 4)y^{\mathrm{IV}} & + & 7xy''' & + & 9y'' & = 0 , \\ \quad \vdots & & & & \\ (x^2 + 4)y^{(m+1)} & + & (2m + 1)xy^{(m)} & + & m^2 y^{(m-1)} & = 0 . \end{cases}$$

Putting $x = 0$ and $m = 2n$, we obtain $y^{(2n+1)}(0) = -n^2 y^{(2n-1)}(0)$, and hence $y^{(2n+1)}(0) = (-1)^n (n!)^2$. Thus we get

$$y = \sum_{n=0}^{\infty} (-1)^n \frac{(n!)^2}{(2n + 1)!} x^{2n+1}$$

$$= x - \frac{x^3}{6} + \frac{x^5}{30} - \frac{x^7}{140} + \frac{x^9}{630} - \frac{x^{11}}{2772} + \frac{x^{13}}{12012} - \cdots .$$

3.

$$\begin{cases} y = z^2 ; \\ \dfrac{dy}{dx} = 2z\dfrac{dz}{dx} = 2\dfrac{z}{\cosh (z/2)} = 2 \cdot \displaystyle\sum_{n=0}^{\infty} (-1)^n \frac{(n!)^2}{(2n + 1)!} x^{2n+1} , \end{cases}$$

and after integration,

$$y = 2 \cdot \sum_{n=1}^{\infty} (-1)^{n-1} \frac{[(n - 1)!]^2}{(2n)!} x^{2n}$$

$$= x^2 - \frac{x^4}{12} + \frac{x^6}{90} - \frac{x^8}{560} + \frac{x^{10}}{3150} - \frac{x^{12}}{16632} + \cdots .$$

4.

$$\begin{cases} y = \cosh pz ; \\ y' = \dfrac{dy}{dx} = \dfrac{p \sinh pz}{\cosh (z/2)} ; \\ y'' = \dfrac{p^2 \cosh pz}{\cosh^2 (z/2)} - \dfrac{p \sinh pz}{\cosh (z/2)} \dfrac{(1/2) \sinh (z/2)}{\cosh^2 (z/2)} = \dfrac{p^2 y - xy'/4}{1 + x^2/4} . \end{cases}$$

Hence $(x^2 + 4)y'' + xy' - 4p^2 y = 0$ with the initial conditions $x = 0$, $y = 1$, and $y' = 0$. Using the same technique as above, we obtain without difficulty

$$y^{(2n+2)}(0) = (p^2 - n^2)y^{(2n)}(0) ,$$

and hence

$$y = 1 + \frac{p^2 x^2}{2!} + \frac{p^2(p^2 - 1)x^4}{4!} + \frac{p^2(p^2 - 1)(p^2 - 4)x^6}{6!} + \cdots .$$

5. In the preceding formula, we have computed $y = \cosh pz$. Differentiating with respect to x, we find

$$p \sinh pz \frac{1}{\cosh (z/2)} = p^2 x + \frac{p^2(p^2 - 1)x^3}{3!} + \frac{p^2(p^2 - 1)(p^2 - 4)x^5}{5!} + \cdots ,$$

that is,

$$\frac{\sinh pz}{\sinh z} = p \left[1 + \frac{(p^2 - 1)x^2}{3!} + \frac{(p^2 - 1)(p^2 - 4)x^4}{5!} + \cdots \right].$$

6. $y = \cosh pz/\cosh (z/2)$. The following equation is easily obtained:

$$(x^2 + 4)y'' + 3xy' + (1 - 4p^2)y = 0 .$$

For $x = 0$ we have $y = 1$ and $y' = 0$; further, we find

$$(x^2 + 4)y^{(n+1)} + (2n + 1)xy^{(n)} + (n^2 - 4p^2)y^{(n-1)} = 0 .$$

Hence

$$\begin{cases} y''(0) = p^2 - \frac{1}{4} , \\ y^{IV}(0) = (p^2 - \frac{1}{4})(p^2 - \frac{9}{4}) , \\ y^{VI}(0) = (p^2 - \frac{1}{4})(p^2 - \frac{9}{4})(p^2 - \frac{25}{4}) , \\ \vdots \end{cases}$$

and

$$y = 1 + \binom{p + \frac{1}{2}}{2} x^2 + \binom{p + \frac{3}{2}}{4} x^4 + \binom{p + \frac{5}{2}}{6} x^6 + \cdots .$$

7. As was shown in (1), we have

$$\frac{1}{\cosh (z/2)} = \left(1 + \frac{x^2}{4} \right)^{-1/2} = \sum_{n=0}^{\infty} \frac{(-1)^n(2n)!}{2^{4n}(n!)^2} x^{2n}$$

$$= 1 - \frac{x^2}{8} + \frac{3x^4}{128} - \frac{5x^6}{1024} + \frac{35x^8}{32768} - \cdots .$$

8. We start from

$$y = \frac{\sinh z}{z} = \frac{2 \sinh (z/2) \cosh (z/2)}{z} = \frac{2 \sinh (z/2)}{z/\cosh (z/2)} = \frac{x}{z/\cosh (z/2)} .$$

Since y is an even function, we try with $y = a_0 + a_1 x^2 + a_2 x^4 + \cdots$ and obtain, using the expansion of (2),

$$(a_0 + a_1 x^2 + a_2 x^4 + \cdots) \left(x - \frac{x^3}{6} + \frac{x^5}{30} - \frac{x^7}{140} + \cdots \right) \equiv x .$$

Identifying the coefficients, we get

$$y = 1 + \frac{x^2}{6} - \frac{x^4}{180} + \frac{x^6}{1512} - \frac{23x^8}{226800} + \frac{263x^{10}}{14968800} - \cdots .$$

9. Putting $v = z \cdot \cosh(z/2)$, we easily find the equation

$$(x^2 + 4)v' - xv = x^2 + 4 ,$$

and

$$v^{(2n+1)}(0) = -n(n-1)v^{(2n-1)}(0) \quad (n \geq 2) ; \qquad v'(0) = 1 , \quad v'''(0) = \tfrac{1}{2} .$$

Hence

$$v = x + \frac{x^3}{12} - \frac{x^5}{120} + \frac{x^7}{840} - \frac{x^9}{5040} + \frac{x^{11}}{27720} - \cdots .$$

Putting

$$y = \frac{\tanh(z/2)}{(z/2)} = 1 + a_1 x^2 + a_2 x^4 + \cdots$$

and using $vy = 2\sinh(z/2) = x$, we find:

$$y = 1 - \frac{x^2}{12} + \frac{11x^4}{720} - \frac{191x^6}{60480} + \frac{2497x^8}{3628800} - \frac{14797x^{10}}{95800320} + \cdots .$$

10. We obtain directly

$$y = \frac{2(\cosh z - 1)}{z^2} = \frac{4\sinh^2(z/2)}{z^2}$$

$$= \frac{\big(2\sinh(z/2) \cdot \cosh(z/2)\big)\big(2\sinh(z/2)/\cosh(z/2)\big)}{z \cdot z}$$

$$= \frac{\sinh z}{z} \frac{2\tanh(z/2)}{z} ,$$

and hence get $y = x^2/z^2$ as the product of the expansions (8) and (9):

$$y = 1 + \frac{x^2}{12} - \frac{x^4}{240} + \frac{31x^6}{60480} - \frac{289x^8}{3628800} + \frac{317x^{10}}{22809600} - \cdots .$$

Summing up, we have the following results:

$$U = \partial - \frac{\partial^3}{24} + \frac{3\partial^5}{640} - \frac{5\partial^7}{7168} + \frac{35\partial^9}{294912} - \frac{63\partial^{11}}{2883584} + \cdots , \qquad (7.2.3)$$

$$U = \mu\partial \left(1 - \frac{\partial^2}{6} + \frac{\partial^4}{30} - \frac{\partial^6}{140} + \frac{\partial^8}{630} - \frac{\partial^{10}}{2772} + \frac{\partial^{12}}{12012} - \cdots \right),$$

$$\qquad (7.2.4)$$

$$U^2 = \partial^2 - \frac{\partial^4}{12} + \frac{\partial^6}{90} - \frac{\partial^8}{560} + \frac{\partial^{10}}{3150} - \frac{\partial^{12}}{16632} + \cdots , \qquad (7.2.5)$$

$$\cosh pU = 1 + \frac{p^2\partial^2}{2!} + \frac{p^2(p^2-1)\partial^4}{4!} + \frac{p^2(p^2-1)(p^2-4)\partial^6}{6!} + \cdots , \qquad (7.2.6)$$

$$\frac{\sinh pU}{\sinh U} = p + \binom{p+1}{3}\delta^2 + \binom{p+2}{5}\delta^4 + \binom{p+3}{7}\delta^6 + \cdots, \qquad (7.2.7)$$

$$\frac{\cosh pU}{\cosh(U/2)} = 1 + \binom{p+\frac{1}{2}}{2}\delta^2 + \binom{p+\frac{3}{2}}{4}\delta^4 + \binom{p+\frac{5}{2}}{6}\delta^6 + \cdots, \qquad (7.2.8)$$

$$\mu^{-1} = 1 - \frac{\delta^2}{8} + \frac{3\delta^4}{128}$$
$$- \frac{5\delta^6}{1024} + \frac{35\delta^8}{32768} - \frac{63\delta^{10}}{262144} + \frac{231\delta^{12}}{4194304} - \cdots, \qquad (7.2.9)$$

$$\frac{\mu\delta}{U} = 1 + \frac{\delta^2}{6} - \frac{\delta^4}{180} + \frac{\delta^6}{1512} - \frac{23\delta^8}{226800} + \frac{263\delta^{10}}{14968800} - \cdots, \quad (7.2.10)$$

$$\frac{\delta}{\mu U} = 1 - \frac{\delta^2}{12} + \frac{11\delta^4}{720} - \frac{191\delta^6}{60480} + \frac{2497\delta^8}{3628800} - \frac{14797\delta^{10}}{95800320} + \cdots,$$
$$(7.2.11)$$

$$\frac{\delta^2}{U^2} = 1 + \frac{\delta^2}{12} - \frac{\delta^4}{240} + \frac{31\delta^6}{60480} - \frac{289\delta^8}{3628800} + \frac{317\delta^{10}}{22809600} - \cdots.$$
$$(7.2.12)$$

Here we obtained formula (7.2.4) by multiplying the expansions for

$$U/\cosh(U/2) \qquad \text{and} \qquad \mu = \cosh(U/2).$$

The formulas (7.2.3) through (7.2.12) may essentially be regarded as relations between U, on one hand, and δ, on the other, and all of them ultimately depend upon Taylor's formula $E = e^U$. From this point of view, it seems appropriate to discuss the meaning and the validity of this formula a little more in detail. We assume that $f(x)$ is an analytic function, and write the formula as follows:

$$Ef(x) = e^U f(x) \qquad \text{or} \qquad f(x+h) = e^U f(x).$$

We observe that the formula is correct if $f(x) = x^n$. The right-hand side reduces to

$$\left(1 + hD + \frac{h^2 D^2}{2!} + \cdots + \frac{h^n D^n}{n!}\right) x^n,$$

since the function is annihilated by higher powers of D. But for $r \le n$, we have

$$\frac{h^r D^r x^n}{r!} = h^r n(n-1) \cdots (n-r+1) \frac{x^{n-r}}{r!} = \binom{n}{r} h^r x^{n-r},$$

and hence the whole expression is

$$\sum_{r=0}^{n} \frac{h^r D^r}{r!} x^n = \sum_{r=0}^{n} \binom{n}{r} h^r x^{n-r} = (x+h)^n.$$

Clearly, the formula is also valid for polynomials, and a reasonable interpretation of the formula $E = e^{U}$ is the following: The operators E and $1 + hD + h^2D^2/2! + \cdots + h^nD^n/n!$ are equivalent when used on a polynomial $P_n(x)$ of degree n, where n is an arbitrary positive integer. As second example we take the formula $E^p = (1 + \Delta)^p$. If p is an integer, $(1 + \Delta)^p$ is a polynomial, and we get an identity. But if the formula is used on polynomials, it is also valid for all real p. For $f(x) = x^3$ we have

$$\Delta x^3 = (x + h)^3 - x^3 = 3x^2h + 3xh^2 + h^3 ,$$

$$\Delta^2 x^3 = 3h[(x + h)^2 - x^2] + 3h^2[(x + h) - x] = 6h^2(x + h) ,$$

$$\Delta^3 x^3 = 6h^3 ,$$

$$(1 + \Delta)^p x^3 = \sum_{r=0}^{3} \binom{p}{r} \Delta^r x^3 = x^3 + p(3x^2h + 3xh^2 + h^3)$$

$$+ \frac{p(p - 1)}{2} 6h^2(x + h) + \frac{p(p - 1)(p - 2)}{6} 6h^3$$

$$= x^3 + 3x^2ph + 3xp^2h^2 + p^3h^3 = (x + ph)^3 .$$

Taking instead $f(x) = e^x$, we get

$$\Delta e^x = e^{x+h} - e^x = \alpha e^x ,$$

where $\alpha = e^h - 1$. Hence

$$(1 + \Delta)^p e^x = \sum \binom{p}{r} \Delta^r e^x$$

$$= \sum \binom{p}{r} \alpha^r e^x = (1 + \alpha)^p e^x ,$$

provided that the series

$$\sum \binom{p}{r} \alpha^r$$

converges. As is well known, this series is absolutely convergent if $|\alpha| < 1$, that is, if $h < \log 2$. In this case we get

$$(1 + \Delta)^p e^x = (1 + \alpha)^p e^x = e^{ph}e^x = e^{x+ph} ,$$

and hence the formula $E^p = (1 + \Delta)^p$ is valid.

Summing up, we note that the operator formulas need not be universally valid, but we see that they can be used on special functions, in particular on polynomials. In practice they are often used on considerably more general functions, and then, as a rule, the series expansions have to be truncated after a few terms. A rigorous treatment, however, falls outside the scope of this book. Infinite power series in degrading operators in general should be understood as asymptotic series. If such a series is truncated after a certain term, the remainder term will behave asymptotically as the first neglected term as $h \to 0$.

7.3. Factorials

The factorial concept actually falls outside the frame of this chapter, but with regard to later applications we will nevertheless treat it here. The *factorial polynomial* of degree n (n positive integer) is defined by the formula

$$p^{(n)} = p(p-1)(p-2) \cdots (p-n+1) = \frac{p!}{(p-n)!} . \qquad (7.3.1)$$

Among the important properties of such expressions, we note the following:

$$(p+1)^{(n)} - p^{(n)} = (p+1)p(p-1) \cdots (p-n+2)$$
$$- p(p-1) \cdots (p-n+1)$$
$$= p(p-1)(p-2) \cdots (p-n+2)[p+1-(p-n+1)]$$
$$= np(p-1) \cdots (p-n+2)$$

or

$$\Delta p^{(n)} = np^{(n-1)} . \qquad (7.3.2)$$

Here the Δ-symbol operates on the variable p with $h = 1$.

We generalize directly to

$$\Delta^2 p^{(n)} = n(n-1)p^{(n-2)} = n^{(2)} p^{(n-2)}$$

and

$$\Delta^k p^{(n)} = n(n-1)(n-2) \cdots (n-k+1)p^{(n-k)} = n^{(k)} p^{(n-k)} . \qquad (7.3.3)$$

If $k = n$, we get $\Delta^n p^{(n)} = n!$

Until now we have defined factorials only for positive values of n. When $n > 1$, we have $p^{(n)} = (p-n+1)p^{(n-1)}$, and requiring that this formula also hold for $n = 1$ and $n = 0$, we get $p^{(0)} = 1$; $p^{(-1)} = 1/(p+1)$. Using the formula repeatedly for $n = -1, -2, \ldots$, we obtain:

$$p^{(-n)} = \frac{1}{(p+1)(p+2) \cdots (p+n)} = \frac{1}{(p+n)^{(n)}} . \qquad (7.3.4)$$

With this definition, the formula $p^{(n)} = (p-n+1)p^{(n-1)}$, as well as (7.3.2), holds also for negative values of n.

We shall now derive formulas for expressing a factorial as a sum of powers, and a power as a sum of factorials. Putting

$$z^{(n)} = \sum_{k=1}^{n} \alpha_k^{(n)} z^k , \qquad (7.3.5)$$

and using the identity $z^{(n+1)} = (z-n)z^{(n)}$, we obtain, on comparing the coefficients of z^k,

$$\alpha_k^{(n+1)} = \alpha_{k-1}^{(n)} - n\alpha_k^{(n)} , \qquad (7.3.6)$$

with $\alpha_1^{(1)} = 1$ and $\alpha_0^{(n)} = 0$. These numbers are usually called *Stirling's numbers of the first kind;* they are presented in the table below.

n \ k	1	2	3	4	5	6	7	8	9	10
1	1									
2	−1	1								
3	2	−3	1							
4	−6	11	−6	1						
5	24	−50	35	−10	1					
6	−120	274	−225	85	−15	1				
7	720	−1764	1624	−735	175	−21	1			
8	−5040	13068	−13132	6769	−1960	322	−28	1		
9	40320	−109584	118124	−67284	22449	−4536	546	−36	1	
10	−362880	1026576	−1172700	723680	−269325	63273	−9450	870	−45	1

For example, $z^{(5)} = 24z - 50z^2 + 35z^3 - 10z^4 + z^5$.

To obtain a power in terms of factorials, we observe that

$$z \cdot z^{(k)} = (z - k + k)z^{(k)}$$
$$= z^{(k+1)} + kz^{(k)} .$$

Putting

$$z^n = \sum_{k=1}^{n} \beta_k^{(n)} z^{(k)} , \qquad (7.3.7)$$

we obtain

$$z^n = z \cdot z^{n-1} = z \sum_{k=1}^{n-1} \beta_k^{(n-1)} z^{(k)}$$
$$= \sum_{k=1}^{n-1} \beta_k^{(n-1)} (z^{(k+1)} + kz^{(k)}) .$$

Since no constant term is present on the right-hand side of (7.3.7), we have $\beta_0^{(n)} = 0$, and we can let the summation run from $k = 0$ instead of $k = 1$:

$$z^n = \sum_{k=0}^{n-1} \beta_k^{(n-1)} (z^{(k+1)} + kz^{(k)})$$
$$= \sum_{k=1}^{n} \beta_{k-1}^{(n-1)} z^{(k)} + \sum_{k=1}^{n-1} k\beta_k^{(n-1)} z^{(k)}$$
$$= \sum_{k=1}^{n} \beta_k^{(n)} z^{(k)} .$$

Identifying both sides, we get

$$\begin{cases} \beta_k^{(n)} = \beta_{k-1}^{(n-1)} + k\beta_k^{(n-1)} ; & k = 1, 2, \ldots, n - 1 \\ \beta_n^{(n)} = \beta_{n-1}^{(n-1)} = 1 . \end{cases} \qquad (7.3.8)$$

The numbers β are called *Stirling's numbers of the second kind;* they are displayed in the table below:

n\k	1	2	3	4	5	6	7	8	9	10
1	1									
2	1	1								
3	1	3	1							
4	1	7	6	1						
5	1	15	25	10	1					
6	1	31	90	65	15	1				
7	1	63	301	350	140	21	1			
8	1	127	966	1701	1050	266	28	1		
9	1	255	3025	7770	6951	2646	462	36	1	
10	1	511	9330	34105	42525	22827	5880	750	45	1

For example, we have

$$z^5 = z^{(1)} + 15z^{(2)} + 25z^{(3)} + 10z^{(4)} + z^{(5)} \, .$$

Both these tables, counted with n rows and n columns, form triangular matrices which we denote by A and B, respectively. Further, let u and v be the following column vectors:

$$u = \begin{pmatrix} z \\ z^2 \\ \vdots \\ z^n \end{pmatrix} ; \quad \text{and} \quad v = \begin{pmatrix} z^{(1)} \\ z^{(2)} \\ \vdots \\ z^{(n)} \end{pmatrix} .$$

Then we have

$$v = Au ; \qquad u = Bv ; \qquad AB = BA = I \, . \tag{7.3.9}$$

The Stirling numbers $\alpha_k^{(n)}$ and $\beta_k^{(n)}$ are special cases of the so-called Bernoulli numbers $B_\nu^{(n)}$ of order n; they are defined by the expansion

$$\frac{t^n}{(e^t - 1)^n} = \sum_{\nu=0}^{\infty} \frac{t^\nu}{\nu!} B_\nu^{(n)} \, . \tag{7.3.10}$$

Later we shall deal with the case $n = 1$, but otherwise these numbers are beyond the scope of this book. Closer details are given in [1], p. 127. We only mention the following relations:

$$\begin{cases} \alpha_k^{(n)} = \dbinom{n-1}{k-1} B_{n-k}^{(n)} \, , \\[2mm] \beta_k^{(n)} = \dbinom{n}{k} B_{n-k}^{(-n)} \, . \end{cases} \tag{7.3.11}$$

From Formulas (7.3.2) and (7.3.3), we see that the factorials have very attractive properties with respect to the operator $\varDelta$. This ensures that sums of factorials are easily computed. In order to demonstrate this, we regard

$$\sum_{i=m}^{n} f(x_0 + ih) = \sum_{i=m}^{n} f_i \,,$$

where f is a function such that $\Delta F = f$, and obtain

$$\sum_{i=m}^{n} f_i = (F_{m+1} - F_m) + (F_{m+2} - F_{m+1}) + \cdots + (F_{n+1} - F_n)$$
$$= F_{n+1} - F_m \,.$$

Since $\Delta\big(p^{(n+1)}/(n+1)\big) = p^{(n)}$, we get

$$\sum_{p=P}^{Q} p^{(n)} = \frac{(Q+1)^{(n+1)} - P^{(n+1)}}{n+1} \,. \tag{7.3.12}$$

Here P and Q need not be integers; on the other hand, the interval is supposed to be 1.

If, instead, the factorial is defined by

$$p^{(n)} = p(p - h) \cdots (p - nh + h) \,,$$

the formula is slightly modified:

$$\sum_{p=P}^{Q} p^{(n)} = \frac{(Q+h)^{(n+1)} - P^{(n+1)}}{(n+1)h} \,. \tag{7.3.13}$$

We shall return to these formulas in Chapter 11.

In passing, we also mention that one can define central factorials:

$$p^{\{n\}} = \left(p + \frac{n-1}{2}\right)^{(n)} \tag{7.3.14}$$

and central mean factorials:

$$p^{[n]} = \mu p^{\{n\}} = \frac{1}{2}\left[\left(p + \frac{1}{2}\right)^{\{n\}} + \left(p - \frac{1}{2}\right)^{\{n\}}\right]$$
$$= p\left(p + \frac{n}{2} - 1\right)^{(n-1)} \,. \tag{7.3.15}$$

In both cases we have analogous difference formulas:

$$\delta p^{\{n\}} = n p^{\{n-1\}} \,; \qquad \delta p^{[n]} = n p^{[n-1]} \,. \tag{7.3.16}$$

Naturally, one can construct expansions of factorials in powers, and conversely, also in these cases. However, we refrain from doing this, and instead we refer to [2], pp. 54 and 568.

REFERENCES

[1] Milne-Thomson: *The Calculus of Finite Differences* (Mac Millan, London, 1933).

[2] Buckingham: *Numerical Methods* (Pitman, London, 1957).

[3] Milne: *Numerical Calculus* (Princeton University Press, Princeton, 1949).

EXERCISES

1. Show that $\nabla - \varDelta = \varDelta\nabla$ and that $\varDelta + \nabla = \varDelta/\nabla - \nabla/\varDelta$.

2. Prove the relations.

(a) $\mu(f_k g_k) = \mu f_k \mu g_k + \frac{1}{4}\delta f_k \delta g_k$.

(b) $\mu\left(\dfrac{f_k}{g_k}\right) = \dfrac{\mu f_k \mu g_k - \frac{1}{4}\delta f_k \delta g_k}{g_{k-1/2} g_{k+1/2}}$.

(c) $\delta(f_k g_k) = \mu f_k \delta g_k + \mu g_k \delta f_k$.

(d) $\delta\left(\dfrac{f_k}{g_k}\right) = \dfrac{\mu g_k \delta f_k - \mu f_k \delta g_k}{g_{k-1/2} g_{k+1/2}}$.

3. Prove that:

(a) $\displaystyle\sum_{k=0}^{n-1} \varDelta^2 f_k = \varDelta f_n - \varDelta f_0$.

(b) $\displaystyle\sum_{k=0}^{n-1} \delta^2 f_{2k+1} = \tanh(U/2)(f_{2n} - f_0)$.

4. Prove that:

(a) $\varDelta\sqrt{f_k} = \dfrac{\varDelta f_k}{\sqrt{f_k} + \sqrt{f_{k+1}}}$.

(b) $p^{(2n)} = 2^{2n} \left(\dfrac{p}{2}\right)^{(n)} \left(\dfrac{p-1}{2}\right)^{(n)}$ $\quad$ (n integer, >0).

5. A Fibonacci series is characterized by $a_n = a_{n-1} + a_{n-2}$, where a_0 and a_1 are given numbers. Show that if y_n are terms in a Fibonacci series, the same holds for $\varDelta^N y_n, \varDelta^{N-1} y_n, \varDelta^{N-2} y_n, \ldots$

6. Find $\delta^2(x^2), \delta^2(x^3)$, and $\delta^2(x^4)$ when $h = 1$. Use the result for determining a particular solution of the equation $\mu f(x) = 2x^4 + 4x^3 + 3x^2 + 3x + \frac{1}{8}$.

7. A function $f(x)$ is given in equidistant points $x_0, x_1, \ldots, x_n$, where $x_k = x_0 + kh$, and the corresponding differences of different orders are denoted by $\varDelta f, \varDelta^2 f, \varDelta^3 f, \ldots$ By also taking the point midway between the first points, we obtain $x_0, x_{1/2}, x_1, x_{3/2}, \ldots$, and the corresponding differences are denoted by $\varDelta_1 f, \varDelta_1^2 f, \varDelta_1^3 f, \ldots$ Show that

$$\varDelta_1^r = 2^{-r}[\varDelta^r - a\varDelta^{r+1} + b\varDelta^{r+2} - \cdots],$$

and compute the constants a and b.

8. The sequence y_n is formed according to the rule

$$y_n = (-1)^{n-1}\left\{\binom{\frac{1}{2}}{n+1} + \binom{-\frac{1}{2}}{n+1}\right\}.$$

Prove that $\varDelta^n y_n = 0$.

9. Find $\cos pz$ as a power series of x when $x = 2\sin(z/2)$.

10. The expression δy_0 cannot usually be computed directly from a difference scheme. Find its value expressed in known central differences.

11. (a) Show that

$$\varDelta\binom{n}{i+1} = \binom{n}{i}.$$

where $\varDelta$ operates on n, and hence that

$$\sum_{n=1}^{N}\binom{n}{i} = \binom{N+1}{i+1} - \binom{1}{i+1}.$$

(b) Find the coefficients c_i in the expansion

$$n^5 = \sum_{i=0}^{5} c_i \binom{n}{i}$$

and compute the smallest value N such that $\sum_{n=1}^{N} n^5 > 10^{10}$.

12. Express

$$J_0 y_0 = \int_{x_0 - h/2}^{x_0 + h/2} y(t)\, dt$$

in central differences of y.

13. We define

$$\delta^2 f = f(x - h) - 2f(x) + f(x + h)$$

and

$$\delta'^2 f = f(x - nh) - 2f(x) + f(x + nh).$$

Obtain δ'^2 as a power series in δ^2 (the three first terms).

14. A Fibonacci series a_n, $n = 0, 1, 2, \ldots$, is given. Show that

$$a_{r+2n} + (-1)^n a_r = k_n a_{r+n},$$

where k_n are integers forming a new Fibonacci series, and state the first terms in this series.

Chapter 8

Interpolation

Interpolation—that is the art of reading between the lines in a table.

8.0. Introduction

The kind of problem we are going to treat in this chapter can be briefly described in the following way. For a function $f(x, y, z, \ldots)$, certain points $P_i(x_i, y_i, z_i, \ldots)$, $i = 1, 2, \ldots, n$ are known. Find the value of the function at the point $P(\xi, \eta, \zeta, \ldots)$. It is obvious that this problem is strongly underdetermined, since one can prescribe an arbitrary value to this point. In practical cases, however, f is usually a "smooth" function, that is, it varies in a rather regular manner. This is, of course, a somewhat vague condition which ought to be put in a more precise form. First, we shall restrict ourselves to functions of one variable. In the following we shall almost exclusively treat this case.

From now on we consider a function $y = f(x)$ with known values $y_i = f(x_i)$, $i = 1, 2, \ldots, n$. We suppose that the function can be approximated by a certain type of function, and usually we shall be concerned with *polynomial approximation*. In special cases, other kinds of functions (trigonometric, exponential) may occur.

When we have to choose a suitable interpolation method, we should first answer the following questions. Are the given points equidistant? Has a difference scheme been constructed? Is a table of interpolation coefficients available? Should interpolation be performed at the beginning or at the end of a table? Is extremely high accuracy desired? We are now going to describe a number of methods; in each case some kind of advice is given as to the circumstances under which the method should be applied.

8.1. Lagrange's interpolation formula

We assume that for a function $f(x)$ which is continuously differentiable n times, n points (x_1, y_1), (x_2, y_2), $\ldots$, (x_n, y_n) are known. Our problem is then to find the function value y corresponding to a given value x. Clearly, this problem has an infinite number of solutions. Usually $f(x)$ is replaced by a polynomial $P(x)$ of degree $n - 1$, taking the values $y_1, y_2, \ldots, y_n$ for $x = x_1, x_2, \ldots, x_n$. Putting

$$y = a_0 + a_1 x + \cdots + a_{n-1} x^{n-1},$$

we get a linear system of equations in the coefficients $a_0, a_1, \ldots, a_{n-1}$. The determinant (Vandermonde's determinant) is $\prod_{i>k} (x_i - x_k)$ and hence is not equal to 0. Thus there is a unique solution and it is obvious that the desired polynomial can be written

$$
\begin{cases}
P(x) = y = \sum_{k=1}^{n} L_k(x) y_k, & \text{where} \\
L_k(x) = \dfrac{(x - x_1)(x - x_2) \cdots (x - x_{k-1})(x - x_{k+1}) \cdots (x - x_n)}{(x_k - x_1)(x_k - x_2) \cdots (x_k - x_{k-1})(x_k - x_{k+1}) \cdots (x_k - x_n)}.
\end{cases}
\tag{8.1.1}
$$

For $x = x_r$ $(1 \leq r \leq n)$, all terms in the sum vanish except the rth, which takes the value $y = y_r$.

We shall now examine the difference between the given function $f(x)$ and the polynomial $P(x)$ for an arbitrary value x_0 of x. Then it is convenient to use the following functions:

$$
F(x) = \prod_{r=1}^{n} (x - x_r); \qquad F_k(x) = \prod_{r \neq k} (x - x_r).
\tag{8.1.2}
$$

Obviously, we have $F(x) = (x - x_k) F_k(x)$ and $F'(x) = (x - x_k) F'_k(x) + F_k(x)$, and hence $F'(x_k) = F_k(x_k)$. Thus we can also write

$$
P(x) = \sum_{k=1}^{n} \frac{F_k(x)}{F_k(x_k)} y_k = \sum_{k=1}^{n} \frac{F(x)}{(x - x_k) F'(x_k)} y_k.
\tag{8.1.3}
$$

We suppose that the point x_0 lies in the closed interval I bounded by the extreme points of $(x_1, x_2, \ldots, x_n)$ and further that $x_0 \neq x_k$, $k = 1, 2, \ldots, n$. We define the function $G(x) = f(x) - P(x) - R F(x)$, where R is a constant which is determined so that $G(x_0) = 0$. Obviously, we have $G(x) = 0$ for $x = x_0, x_1, x_2, \ldots, x_n$, and by using Rolle's theorem repeatedly, we conclude that $G^{(n)}(\xi) = 0$, where $\xi \in I$. But $G^{(n)}(\xi) = f^{(n)}(\xi) - R \cdot n!$, since $P(x)$ is of degree $n - 1$, and $R = f^{(n)}(\xi)/n!$. Hence, replacing x_0 by x, we obtain (note that ξ is a function of x)

$$
f(x) = P(x) + \frac{f^{(n)}(\xi)}{n!} F(x),
$$

or written explicitly,

$$
\begin{aligned}
f(x) = {} & \frac{(x - x_2)(x - x_3) \cdots (x - x_n)}{(x_1 - x_2)(x_1 - x_3) \cdots (x_1 - x_n)} y_1 + \frac{(x - x_1)(x - x_3) \cdots (x - x_n)}{(x_2 - x_1)(x_2 - x_3) \cdots (x_2 - x_n)} y_2 \\
& + \cdots + \frac{(x - x_1)(x - x_2) \cdots (x - x_{n-1})}{(x_n - x_1)(x_n - x_2) \cdots (x_n - x_{n-1})} y_n \\
& + \frac{f^{(n)}(\xi)}{n!} (x - x_1)(x - x_2) \cdots (x - x_n).
\end{aligned}
\tag{8.1.4}
$$

This is Lagrange's interpolation formula. It can, of course, be differentiated if one wants to compute the derivative in an arbitrary point; it must be remembered, however, that the factor $f^{(n)}(\xi)$ is also a function of x. For the derivative in one of the given points (x_r, y_r), it is possible to obtain a comparatively simple expression. We have

$$P(x) = \sum_{k=1}^{n} \frac{F(x)}{(x - x_k)F'(x_k)} y_k \, ,$$

$$\frac{dP}{dx} = \sum \frac{(x - x_k)F'(x) - F(x)}{(x - x_k)^2 F'(x_k)} y_k \, ,$$

and for $x = x_r$, we find

$$\left(\frac{dP}{dx}\right)_{x=x_r} = \sum_{k \neq r} \frac{F'(x_r)}{(x_r - x_k)F'(x_k)} y_k + \lim_{x \to x_r} \frac{(x - x_r)F'(x) - F(x)}{(x - x_r)^2 F'(x_r)} y_r \, .$$

The last term can be computed, for example, by l'Hospital's rule and becomes $(F''(x_r)/2F'(x_r))y_r$. From $F(x) = \prod_{k=1}^{n}(x - x_k)$, we get, by logarithmical differentiation

$$F'(x) = F(x) \sum_{k} \frac{1}{x - x_k}$$

and

$$F''(x) = F'(x) \sum_{k} \frac{1}{x - x_k} - F(x) \sum_{k} \frac{1}{(x - x_k)^2}$$

$$= F(x) \left[\left(\sum \frac{1}{x - x_k}\right)^2 - \sum \frac{1}{(x - x_k)^2} \right] = 2F(x) \sum_{\substack{i,k \\ i<k}} \frac{1}{(x - x_i)(x - x_k)} \, .$$

If we put $x = x_r$, we get $F(x) = 0$, and we can obtain nonzero contributions only when we have the factor $(x - x_r)$ in the denominator. Hence

$$F''(x_r) = \lim_{x \to x_r} \frac{2F(x)}{x - x_r} \sum_{k \neq r} \frac{1}{x_r - x_k} = 2F'(x_r) \sum_{k \neq r} \frac{1}{x_r - x_k} \, ,$$

and

$$\left(\frac{dP}{dx}\right)_{x=x_r} = \sum_{k \neq r} \frac{1}{x_r - x_k} \left[y_r + \frac{F'(x_r)}{F'(x_k)} y_k \right] . \tag{8.1.5}$$

We now pass to the case when the x-coordinates are equidistant, that is, $x_{k+1} = x_k + h$, $k = 1, 2, \ldots, n - 1$, and distinguish between two cases, namely n even and n odd. First we treat the case when n is even, and start by renumbering the points. This is done because we try to arrange them in such a way that the point x lies in the middle of the interval. The index k now takes the values

$$-\frac{n}{2} + 1, \ -\frac{n}{2} + 2, \ \ldots, \ -1, \ 0, \ 1, \ 2, \ \ldots, \ \frac{n}{2} \, .$$

We put $x_k = x_0 + kh$ and $x = x_0 + ph$, where p is a fraction which preferably should lie between 0 and 1. Our original x_1 now becomes $x_0 + (-n/2 + 1)h$, our original x_2 becomes $x_0 + (-n/2 + 2)h$, and our original x_n becomes $x_0 + (n/2)h$. Hence

$$
\begin{cases}
x_k - x_{1-n/2} = \left(k + \dfrac{n}{2} - 1\right)h \\[2mm]
x_k - x_{2-n/2} = \left(k + \dfrac{n}{2} - 2\right)h \\[2mm]
\vdots \\[1mm]
x_k - x_{k-1} \ = h \\[1mm]
x_k - x_{k+1} \ = -h \\[1mm]
\vdots \\[1mm]
x_k - x_{n/2} \ = -\left(\dfrac{n}{2} - k\right)h .
\end{cases}
\qquad
\begin{cases}
x - x_{1-n/2} = \left(p + \dfrac{n}{2} - 1\right)h \\[2mm]
x - x_{2-n/2} = \left(p + \dfrac{n}{2} - 2\right)h \\[2mm]
\vdots \\[1mm]
x - x_{n/2} \ = \left(p - \dfrac{n}{2}\right)h .
\end{cases}
$$

Let $A_k^n(p)$ be the coefficient of y_k in Lagrange's interpolation formula:

$$
A_k^n(p) = \frac{(-1)^{n/2-k}}{(n/2 + k - 1)! \, (n/2 - k)! \, (p - k)} \prod_{t=1}^{n}\left(p + \frac{n}{2} - t\right). \qquad (8.1.6)
$$

In a similar way, we get for odd values of n:

$$
A_k^n(p) = \frac{(-1)^{(n-1)/2-k}}{\big((n-1)/2 + k\big)! \, \big((n-1)/2 - k\big)! \, (p-k)}
$$

$$
\times \prod_{t=1}^{n}\left(p + \frac{n+1}{2} - t\right). \qquad (8.1.7)
$$

Interpolating the function $f(x) = 1$, we see that $\sum_k A_k^{(n)} p = 1$. Further, we have the remainder term

$$
R_n = \left(p + \frac{n}{2} - 1\right)\left(p + \frac{n}{2} - 2\right)\cdots\left(p - \frac{n}{2}\right)\frac{h^n f^{(n)}(\xi)}{n!}
$$

$$
= \binom{p + n/2 - 1}{n} h^n f^{(n)}(\xi) .
$$

The coefficients $A_k^n(p)$ have been tabulated for $n = 3, 4, \ldots, 11$ and for different values of p [1].

EXAMPLE

For a function $y = y(x)$, 5 points are given according to the table below. Find y for $x = 1.0242$, using Lagrangian five-point interpolation. We have $p = 0.420$, and from the table we obtain the coefficients $A_{-2}, \ldots, A_2$. (Note that A_{-1} and A_2 are negative; the sign is indicated in the heading of the table [1]. The reader should check that $A_{-2} + A_{-1} + A_0 + A_1 + A_2 = 1$.)

x	y	A
1.00	4.6415888	0.02277254
1.01	4.6570095	-0.15523816
1.02	4.6723287	0.78727924
1.03	4.6875481	0.38006584
1.04	4.7026694	-0.03487946

$\sum A_i y_i = 4.6787329$.

Hartree and others have emphasized that the Lagrangian technique must be used with discrimination. Suppose, for example, that we start from the points $(0, 0)$, $(1, 1)$, $(2, 8)$, $(3, 27)$, and $(4, 64)$ on the curve $y = x^3$ and that we want to compute $\sqrt[3]{20}$ by inverse Lagrange interpolation. We find directly that

$$x = y\left[\frac{(y-8)(y-27)(y-64)}{1\cdot(-7)(-26)(-63)}\cdot 1 + \frac{(y-1)(y-27)(y-64)}{8\cdot 7\cdot(-19)(-56)}\cdot 2\right.$$
$$\left.+ \frac{(y-1)(y-8)(y-64)}{27\cdot 26\cdot 19\cdot(-37)}\cdot 3 + \frac{(y-1)(y-8)(y-27)}{64\cdot 63\cdot 56\cdot 37}\cdot 4\right].$$

With $y = 20$, we get $x = -1.3139$ instead of the correct value 2.7144. Linear interpolation gives 2.63, that is, a deviation of only 3%. This example shows clearly that a higher-order formula does not necessarily give a better result than a lower-order formula.

In its general form Lagrange's formula is used only on rare occasions in practical computation. On the other hand, it is extremely valuable in theoretical work within different branches of numerical analysis.

8.2. Hermite's interpolation formula

The Hermitian interpolation is rather similar to the Lagrangian. The difference is that we now seek a polynomial $P(x)$ of degree $2n - 1$ such that in the points $x_1, x_2, \ldots, x_n$, $P(x)$ and $f(x)$, as well as $P'(x)$ and $f'(x)$, coincide. Thus we form

$$P(x) = \sum_{k=1}^{n} U_k(x) f(x_k) + \sum_{k=1}^{n} V_k(x) f'(x_k) . \qquad (8.2.1)$$

Here $U_k(x)$ and $V_k(x)$ are supposed to be polynomials of degree $2n - 1$. Our requirements are fulfilled if we claim that

$$\begin{cases} U_k(x_i) = \delta_{ik} , & V_k(x_i) = 0 , \\ U_k'(x_i) = 0 , & V_k'(x_i) = \delta_{ik} . \end{cases} \qquad (8.2.2)$$

As is easily inferred, we may choose

$$\begin{cases} U_k(x) = W_k(x) L_k(x)^2 , \\ V_k(x) = Z_k(x) L_k(x)^2 , \end{cases} \qquad (8.2.3)$$

where $L_k(x)$ is defined in (8.1.1); clearly, we have $L_k(x_i) = \delta_{ik}$. Now $L_k(x)$ is of degree $n - 1$, and hence W_k and Z_k must be linear functions. By using the conditions (8.2.2), we get

$$\begin{cases} W_k(x_k) = 1, & Z_k(x_k) = 0, \\ W_k'(x_k) = -2L_k'(x_k), & Z_k'(x_k) = 1. \end{cases} \tag{8.2.4}$$

Thus Hermite's interpolation formula takes the form

$$P(x) = \sum_{k=1}^{n} \{1 - 2L_k'(x_k)(x - x_k)\} L_k(x)^2 f(x_k)$$

$$+ \sum_{k=1}^{n} (x - x_k) L_k(x)^2 f'(x_k). \tag{8.2.5}$$

Interpolation of this kind is sometimes called osculating interpolation.

We shall now estimate the magnitude of the error and construct the following function:

$$G(x) = f(x) - P(x) - S \cdot F(x)^2. \tag{8.2.6}$$

Hence $G(x_k) = G'(x_k) = 0$, $k = 1, 2, \ldots, n$. We determine a constant S such that $G(x)$ vanishes in an additional point x_0, that is,

$$S = \frac{f(x_0) - P(x_0)}{F(x_0)^2}. \tag{8.2.7}$$

As in the Lagrangian case, let I be the closed interval bounded by the extreme points in $(x_0, x_1, x_2, \ldots, x_n)$. Since $G(x)$ vanishes in $n + 1$ different points, $x_0, x_1, \ldots, x_n$, we know that $G'(x)$ vanishes in n intermediate points in the interval I. Also, $G'(x)$ vanishes in the points $x_1, x_2, \ldots, x_n$, that is, in $2n$ points in all. Hence $G''(x)$ vanishes $2n - 1$ times in I, $G'''(x)$ vanishes $2n - 2$ times in I, and so on, and finally $G^{(2n)}(x)$ vanishes in at least one point $\xi \in I$, where we must assume that $f(x)$ is continuously differentiable at least $2n$ times. Thus we have

$$G^{(2n)}(\xi) = f^{(2n)}(\xi) - S \cdot (2n)! = 0. \tag{8.2.8}$$

From (8.2.7) and (8.2.8) it follows that

$$f(x_0) = P(x_0) + \frac{f^{(2n)}(\xi)}{(2n)!} F(x_0)^2.$$

This relation is trivially correct also if $x_0 = x_k$, $k = 1, 2, \ldots, n$. Hence x_0 can be replaced by x, and we find the following expression for the complete interpolation formula:

$$f(x) = P(x) + \frac{f^{(2n)}(\xi)}{(2n)!} [F(x)]^2. \tag{8.2.9}$$

Here we also mention a technique to determine a curve through given points in such a way that the resulting curve becomes as "smooth" as possible. In practical applications (shipbuilding, aircraft industry) this is done by a flexible ruler, *spline*. Numerically spline interpolation is usually performed by computing a series of third-degree polynomials, one for each interval, and then one demands continuity for the function and its first and second derivatives in all the given points (nodes).

8.3. Divided differences

If one wants to interpolate by use of function values which are given for non-equidistant points, the Lagrangian scheme is impractical and requires much labor. In this respect the divided differences offer better possibilities.

Let $x_0, x_1, \ldots, x_n$ be $n + 1$ given points. Then we define the *first divided difference* of $f(x)$ between x_0 and x_1:

$$f(x_0, x_1) = \frac{f(x_1) - f(x_0)}{x_1 - x_0} = f(x_1, x_0) . \tag{8.3.1}$$

Analogously, the second divided difference is defined by

$$f(x_0, x_1, x_2) = \frac{f(x_1, x_2) - f(x_0, x_1)}{x_2 - x_0} , \tag{8.3.2}$$

and in a similar way the nth divided difference:

$$f(x_0, x_1, \ldots, x_n) = \frac{f(x_1, x_2, \ldots, x_n) - f(x_0, x_1, \ldots, x_{n-1})}{x_n - x_0} . \tag{8.3.3}$$

By induction it is easy to prove that

$$f(x_0, x_1, \ldots, x_k)$$
$$= \sum_{p=0}^{k} \frac{f(x_p)}{(x_p - x_0)(x_p - x_1) \cdots (x_p - x_{p-1})(x_p - x_{p+1}) \cdots (x_p - x_k)} . \tag{8.3.4}$$

For equidistant arguments we have

$$f(x_0, x_1, \ldots, x_n) = \frac{1}{h^n \cdot n!} \Delta^n f_0 ,$$

where $h = x_{k+1} - x_k$. From this we see that $f(x_0, x_1, \ldots, x_k)$ is a symmetric function of the arguments $x_0, x_1, \ldots, x_k$. If two arguments are equal, we can still attribute a meaning to the difference; we have, for example,

$$f(x_0, x_0) = \lim_{x \to x_0} \frac{f(x) - f(x_0)}{x - x_0} = f'(x_0)$$

and analogously

$$f(\underbrace{x_0, x_0, \ldots, x_0}_{r+1 \text{ arguments}}) = \frac{f^{(r)}(x_0)}{r!} .$$

Finally, we also get

$$f(x, x, x_0, x_1, \ldots, x_n) = \frac{d}{dx} f(x, x_0, x_1, \ldots, x_n) \,.$$

From the defining equations we have

$$\begin{cases} f(x) & = f(x_0) + (x - x_0) \cdot f(x, x_0) \,, \\ f(x, x_0) & = f(x_0, x_1) + (x - x_1) \cdot f(x, x_0, x_1) \,, \\ f(x, x_0, x_1) & = f(x_0, x_1, x_2) + (x - x_2) \cdot f(x, x_0, x_1, x_2) \,, \\ \quad \vdots \\ f(x, x_0, \ldots, x_{n-1}) = f(x_0, x_1, \ldots, x_n) + (x - x_n) \cdot f(x, x_0, x_1, \ldots, x_n) \,. \end{cases}$$

Multiplying the second equation by $(x - x_0)$, the third by $(x - x_0)(x - x_1)$, and so on, and finally the last equation by $(x - x_0)(x - x_1) \cdots (x - x_{n-1})$, and adding we find

$$f(x) = f(x_0) + (x - x_0) \cdot f(x_0, x_1) + (x - x_0)(x - x_1) \cdot f(x_0, x_1, x_2) + \cdots$$
$$+ (x - x_0)(x - x_1) \cdots (x - x_{n-1}) \cdot f(x_0, x_1, \ldots, x_n) + R \,, \qquad (8.3.5)$$

where $R = f(x, x_0, x_1, \ldots, x_n) \prod_{i=0}^{n} (x - x_i)$. This is Newton's interpolation formula with divided differences.

For a moment we put $f(x) = P(x) + R$. Since $P(x)$ is a polynomial of degree n and, further, R vanishes for $x = x_0, x_1, \ldots, x_n$, we have $f(x_k) = P(x_k)$ for $k = 0, 1, 2, \ldots, n$, and clearly $P(x)$ must be identical with the Lagrangian interpolation polynomial. Hence

$$R = \frac{f^{(n+1)}(\xi)}{(n+1)!} \prod_{i=0}^{n} (x - x_i) \,. \qquad (8.3.6)$$

We also find that

$$f(x, x_0, x_1, \ldots, x_n) = \frac{f^{(n+1)}(\xi)}{(n+1)!} \,.$$

EXAMPLE

Find a polynomial satisfied by $(-4, 1245)$, $(-1, 33)$, $(0, 5)$, $(2, 9)$, and $(5, 1335)$.

x	y				
-4	1245				
		-404			
-1	33		94		
		-28		-14	
0	5		10		3
		2		13	
2	9		88		
		442			
5	1335				

(Note that this scheme contains divided differences.)

$$f(x) = 1245 - 404(x + 4) + 94(x + 4)(x + 1)$$
$$- 14(x + 4)(x + 1)x + 3(x + 4)(x + 1)x(x - 2)$$
$$= 3x^4 - 5x^3 + 6x^2 - 14x + 5 .$$

The practical computation is best done by a technique developed by Aitken. The different interpolation polynomials are denoted by $I(x)$, and first we form the linear expression

$$I_{0,1}(x) = \frac{y_0(x_1 - x) - y_1(x_0 - x)}{x_1 - x_0} = \frac{1}{x_1 - x_0}\begin{vmatrix} y_0 & x_0 - x \\ y_1 & x_1 - x \end{vmatrix}.$$

Obviously, $I_{0,1}(x_0) = y_0$ and $I_{0,1}(x_1) = y_1$. Next we form

$$I_{0,1,2}(x) = \frac{I_{0,1}(x)(x_2 - x) - I_{0,2}(x)(x_1 - x)}{x_2 - x_1} = \frac{1}{x_2 - x_1}\begin{vmatrix} I_{0,1}(x) & x_1 - x \\ I_{0,2}(x) & x_2 - x \end{vmatrix}$$

and observe that

$$I_{0,1,2}(x_0) = \frac{y_0(x_2 - x_0) - y_0(x_1 - x_0)}{x_2 - x_1} = y_0 ,$$

$I_{0,1,2}(x_1) = y_1$, and $I_{0,1,2}(x_2) = y_2$. In general, it is easy to prove that if

$$I_{0,1,2,\ldots,n}(x) = \frac{I_{0,1,2\ldots,n-2,n-1}(x_n - x) - I_{0,1,2,\ldots,n-2,n}(x_{n-1} - x)}{x_n - x_{n-1}} ,$$

we have $I_{0,1,2,\ldots,n}(x_k) = y_k$; $k = 0, 1, 2, \ldots, n$. Hence nth degree interpolation can be performed by $n(n + 1)/2$ linear interpolations. Conveniently, this is done by aid of the scheme below.

x_0	y_0				$x_0 - x$
x_1	y_1	$I_{0,1}(x)$			$x_1 - x$
x_2	y_2	$I_{0,2}(x)$	$I_{0,1,2}(x)$		$x_2 - x$
x_3	y_3	$I_{0,3}(x)$	$I_{0,1,3}(x)$	$I_{0,1,2,3}(x)$	$x_3 - x$

EXAMPLE

$$K(x) = \int_0^1 \frac{dt}{\sqrt{(1 - x^2t^2)(1 - t^2)}}$$

is to be computed for $x = 0.4142$. From a table the following values are obtained:

x	$y = K(x)$				
0.30	1.608049				-1142
0.35	1.622528	1.641119			-642
0.40	1.640000	1.644537	1.645508		-142
0.45	1.660886	1.648276	1.645714	1.645567	358
0.50	1.685750	1.652416	1.645954	1.645571	1.645563	858

This interpolation is identical with the Lagrangian, but it has two essential advantages. On one hand, it is much simpler computationally, on the other, it gives a good idea of the accuracy obtained.

8.4. Difference schemes

A difference scheme is constructed as shown in the following example.

y	$\varDelta$	$\varDelta^2$	$\varDelta^3$	$\varDelta^4$
0				
	1			
1		14		
	15		36	
16		50		24
	65		60	
81		110		24
	175		84	
256		194		24
	369		108	
625		302		
	671			
1296				

In general, we have the following picture:

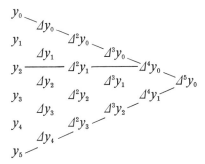

We see directly that the quantities $\varDelta^k y_0$ lie on a straight line sloping down to the right. On the other hand, since $\varDelta = E\nabla$, we have, for example, $\varDelta y_4 = \nabla y_5$; $\varDelta^2 y_3 = \nabla^2 y_5$; $\varDelta^3 y_2 = \nabla^3 y_5$; and so on; and we infer that the quantities $\nabla^k y_n$ lie on a straight line sloping upward to the right. Finally, we also have $\varDelta = E^{1/2}\delta$ and hence, for example, $\varDelta^2 y_1 = E\delta^2 y_1 = \delta^2 y_2$; $\varDelta^4 y_0 = \delta^4 y_2$; and so on. In this way we find that the quantities $\delta^{2k} y_n$ lie on a *horizontal* line. Note that the difference scheme is exactly the same and that it is only a question of notations what the differences are called. For example, we have

$$\varDelta^3 y_1 = \nabla^3 y_4 = \delta^3 y_{5/2} \,.$$

When working with difference schemes, we observe a very characteristic kind of error propagation which we shall now illustrate. Consider a function which is zero in all grid points except one, where it is ε. We obtain the following difference scheme.

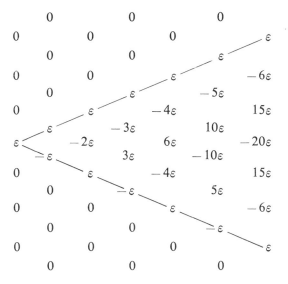

The error propagates in a triangular pattern and grows quickly; apart from the sign we recognize the binomial coefficients in the different columns. In higher-order differences the usual round-off errors appear as tangible irregular fluctuations. Gross errors are easily revealed by use of a difference scheme, and in such cases the picture above should be kept in mind.

The propagation of round-off errors is clearly demonstrated in the scheme below, where we have assumed as large variations as possible between two consecutive values.

$$
\begin{array}{ccccc}
\varepsilon & & -4\varepsilon & & 16\varepsilon \\
& -2\varepsilon & & 8\varepsilon & \\
-\varepsilon & & 4\varepsilon & & -16\varepsilon \\
& 2\varepsilon & & -8\varepsilon & \\
\varepsilon & & -4\varepsilon & & 16\varepsilon \\
& -2\varepsilon & & 8\varepsilon & \\
-\varepsilon & & 4\varepsilon & & -16\varepsilon \\
& 2\varepsilon & & -8\varepsilon & \\
\varepsilon & & -4\varepsilon & & 16\varepsilon \\
& -2\varepsilon & & 8\varepsilon & \\
-\varepsilon & & 4\varepsilon & & -16\varepsilon
\end{array}
$$

Hence, in the worst possible case, we can obtain a doubling of the error for every new difference introduced.

8.5. Interpolation formulas by use of differences

We suppose that we know the values $\ldots, y_{-2}, y_{-1}, y_0, y_1, y_2, \ldots$ of a function for $x = \ldots, x_0 - 2h, x_0 - h, x_0, x_0 + h, x_0 + 2h, \ldots$ and want to compute the function value y_p for $x = x_0 + ph$, where in general $-1 < p < 1$. In the sequel we shall often use the notation $q = 1 - p$. Symbolically, we have

$$y_p = E^p y_0 = (1 + \varDelta)^p y_0 = y_0 + \binom{p}{1} \varDelta y_0 + \binom{p}{2} \varDelta^2 y_0 + \cdots . \qquad (8.5.1)$$

This is *Newton's forward-difference formula*; its validity has, to some extent, been discussed in Section 7.2.

An alternative derivation can be made by use of factorials. Putting

$$y = a_0 + a_1 p^{(1)} + a_2 p^{(2)} + \cdots ,$$

and operating with $\varDelta^k$ on both sides, we get for $p = 0$:

$$\left(\varDelta^k y_p\right)_{p=0} = \varDelta^k y_0 = a_k \cdot k!$$

and hence we again attain formula (8.5.1).

For a moment we put

$$\varphi(p) = y_0 + p \varDelta y_0 + \frac{p(p-1)}{1 \cdot 2} \varDelta^2 y_0 + \cdots + \binom{p}{n} \varDelta^n y_0 ,$$

and we see directly that

$$\begin{cases} \varphi(0) = y_0 , \\ \varphi(1) = y_1 , \\ \quad\vdots \\ \varphi(n) = y_n , \end{cases} \qquad \left(\text{since } \varphi(k) = (1 + \varDelta)^k y_0 = E^k y_0 = y_k\right) .$$

Consequently, $\varphi(p)$ is identical with the Lagrangian interpolation polynomial and the remainder term is the same as in this case:

$$y_p = y_0 + \binom{p}{1} \varDelta y_0 + \binom{p}{2} \varDelta^2 y_0 + \cdots + \binom{p}{n} \varDelta^n y_0 + R_{n+1} ,$$

$$R_{n+1} = \frac{y^{(n+1)}(\xi)}{(n+1)!} h^{n+1} p(p-1) \cdots (p-n) = \binom{p}{n+1} h^{n+1} y^{(n+1)}(\xi) .$$

An analogous formula in ∇ can easily be obtained:

$$y_p = (E^{-1})^{-p} y_0 = (1 - \nabla)^{-p} y_0$$

$$= y_0 + p \nabla y_0 + \frac{p(p+1)}{2!} \nabla^2 y_0 + \cdots . \qquad (8.5.2)$$

This is Newton's backward-difference formula.

The two formulas by Newton are used only occasionally and almost exclusively at the beginning or at the end of a table. More important are formulas which make use of central differences, and a whole series of such formulas with slightly different properties can be constructed.

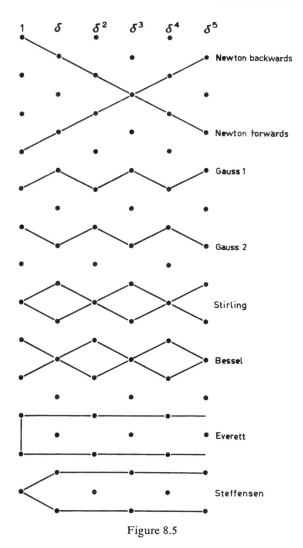

Figure 8.5

Let φ and ψ (with or without indices) denote *even* and *odd* functions of δ, respectively. The best-known formulas have the following structure:

$$
\begin{cases}
1.\ \ y_p = \varphi(\delta)y_0 + \psi(\delta)y_{-1/2} & \text{(Gauss)} \\
2.\ \ y_p = \varphi(\delta)y_0 + \psi(\delta)y_{1/2} & \text{(Gauss)} \\
3.\ \ y_p = \varphi(\delta)y_0 + \mu\psi(\delta)y_0 & \text{(Stirling)} \\
4.\ \ y_p = \mu\varphi(\delta)y_{1/2} + \psi(\delta)y_{1/2} & \text{(Bessel)} \\
5.\ \ y_p = \varphi_0(\delta)y_0 + \varphi_1(\delta)y_1 & \text{(Everett)} \\
6.\ \ y_p = y_0 + \psi_0(\delta)y_{-1/2} + \psi_1(\delta)y_{1/2} & \text{(Steffensen)}
\end{cases}
$$

Naturally, φ and ϕ stand for different functions in different cases. All functions can be obtained by essentially the same technique. We shall use

$$y_p = E^p y_0 = e^{pU} y_0 \quad \text{and} \quad \delta = 2 \sinh \frac{U}{2}.$$

If, in order to use the parity properties of φ and ϕ, we formally change the sign of δ, then we also have to change the sign of U. Hence in case 1 we get

$$\begin{cases} e^{pU} = \varphi + \phi \cdot e^{-U/2}, \\ e^{-pU} = \varphi - \phi \cdot e^{U/2}, \end{cases}$$

whence

$$\varphi = \frac{\cosh (p + 1/2)U}{\cosh (U/2)}$$

$$= 1 + \binom{p+1}{2} \delta^2 + \binom{p+2}{4} \delta^4 + \cdots$$

[cf. (7.2.8)], and

$$\phi = \frac{\sinh pU}{\cosh (U/2)} = 2 \sinh \frac{U}{2} \frac{\sinh pU}{\sinh U}$$

$$= \delta \left[p + \binom{p+1}{3} \delta^2 + \binom{p+2}{5} \delta^4 + \cdots \right]$$

[cf. (7.2.7)]. An analogous formula is obtained in the second case. Hence, the Gaussian interpolation formulas can be written in the following form:

$$y_p = y_0 + \binom{p}{1} \delta y_{-1/2} + \binom{p+1}{2} \delta^2 y_0$$

$$+ \binom{p+1}{3} \delta^3 y_{-1/2} + \binom{p+2}{4} \delta^4 y_0 + \cdots, \qquad (8.5.3)$$

$$y_p = y_0 + \binom{p}{1} \delta y_{1/2} + \binom{p}{2} \delta^2 y_0$$

$$+ \binom{p+1}{3} \delta^3 y_{1/2} + \binom{p+1}{4} \delta^4 y_0 + \cdots. \qquad (8.5.4)$$

In the third case we get

$$\begin{cases} e^{pU} = \varphi + \phi \cosh \dfrac{U}{2}, \\[2mm] e^{-pU} = \varphi - \phi \cosh \dfrac{U}{2}, \end{cases}$$

whence $\varphi = \cosh pU$ and $\phi = \sinh pU / \cosh (U/2)$. Thus we obtain *Stirling's interpolation formula;*

$$y_p = y_0 + p\mu\delta y_0 + \frac{p^2}{2!}\delta^2 y_0 + \binom{p+1}{3}\mu\delta^3 y_0 + \frac{p^2(p^2-1)}{4!}\delta^4 y_0$$

$$+ \binom{p+2}{5}\mu\delta^5 y_0 + \frac{p^2(p^2-1)(p^2-4)}{6!}\delta^6 y_0 + \cdots. \qquad (8.5.5)$$

Here we have used (7.2.6) and (7.2.7) multiplied by $\delta = 2\sinh(U/2)$. More-over, Stirling's formula can also be obtained by adding (8.5.3) and (8.5.4). In the fourth case we obtain the following system:

$$\begin{cases} e^{(p-1/2)U} = \varphi\cosh\dfrac{U}{2} + \psi, \\[2mm] e^{-(p-1/2)U} = \varphi\cosh\dfrac{U}{2} - \psi, \end{cases}$$

whence $\varphi = \cosh(p-\frac{1}{2})U/\cosh(U/2)$ and $\psi = \sinh(p-\frac{1}{2})U$.

Here we can use formula (7.2.8), with p replaced by $p - \frac{1}{2}$, in order to find φ. On the other hand, we cannot use (7.2.7) multiplied by $\sinh U = \mu\delta = \frac{1}{2}(E - E^{-1})$ to obtain ψ, since we would get an even series in δ operating on $y_{3/2}$ and $y_{-1/2}$. Instead, we multiply (7.2.8) by p and integrate in δ:

$$\int \frac{p\cosh pU}{\cosh(U/2)}\,d\delta = \int p\cosh pU\,\frac{dU}{d\delta}\,d\delta = \sinh pU.$$

Also integrating the right-hand side of (7.2.8), we finally get

$$\sinh pU = p\delta + \binom{p+\frac{1}{2}}{2}\frac{p\delta^3}{3} + \binom{p+\frac{3}{2}}{4}\frac{p\delta^5}{5} + \cdots.$$

If we change p to $p - \frac{1}{2}$, we obtain *Bessel's formula:*

$$y_p = \mu y_{1/2} + \left(p - \frac{1}{2}\right)\delta y_{1/2} + \binom{p}{2}\mu\delta^2 y_{1/2} + \binom{p}{2}\frac{p-\frac{1}{2}}{3}\delta^3 y_{1/2}$$

$$+ \binom{p+1}{4}\mu\delta^4 y_{1/2} + \binom{p+1}{4}\frac{p-\frac{1}{2}}{5}\delta^5 y_{1/2} + \cdots. \qquad (8.5.6)$$

In the fifth case we find

$$\begin{cases} e^{pU} = \varphi_0 + \varphi_1 e^{U}, \\ e^{-pU} = \varphi_0 + \varphi_1 e^{-U}, \end{cases}$$

whence $\varphi_0 = \sinh qU/\sinh U$ and $\varphi_1 = \sinh pU/\sinh U$, where $q = 1 - p$. This is the important *Everett formula:*

$$y_p = qy_0 + \binom{q+1}{3}\delta^2 y_0 + \binom{q+2}{5}\delta^4 y_0 + \cdots$$

$$+ py_1 + \binom{p+1}{3}\delta^2 y_1 + \binom{p+2}{5}\delta^4 y_1 + \cdots. \qquad (8.5.7)$$

Finally we have the sixth case:

$$\begin{cases} e^{pU} = 1 + \phi_0 e^{-U/2} + \phi_1 e^{U/2}, \\ e^{-pU} = 1 - \phi_0 e^{U/2} - \phi_1 e^{-U/2}, \end{cases}$$

whence

$$\phi_0 = \frac{\cosh(U/2) - \cosh(p - \frac{1}{2})U}{\sinh U}$$

and

$$\phi_1 = \frac{\cosh(p + \frac{1}{2})U - \cosh(U/2)}{\sinh U}.$$

Observing that $\sinh U = 2 \sinh(U/2)\cosh(U/2) = \delta \cdot \cosh(U/2)$ and using (7.2.8), we obtain *Steffensen's formula*:

$$y_p = y_0 + \binom{1+p}{2}\delta y_{1/2} + \binom{2+p}{4}\delta^3 y_{1/2} + \binom{3+p}{6}\delta^5 y_{1/2} + \cdots$$
$$- \binom{1-p}{2}\delta y_{-1/2} - \binom{2-p}{4}\delta^3 y_{-1/2} - \binom{3-p}{6}\delta^5 y_{-1/2} - \cdots . \quad (8.5.8)$$

The structure of all these formulas can easily be demonstrated by sketching a difference scheme, where the different quantities are represented by points. The column to the left stands for the function values, then we have the first differences, and so on. The two Newton formulas are also included (Fig. 8.5, p. 176).

The two Gaussian interpolation formulas are of interest almost exclusively from a theoretical standpoint. Stirling's formula is suitable for small values of p, for example, $-\frac{1}{4} \leq p \leq \frac{1}{4}$, and Bessel's formula is suitable for values of p not too far from $\frac{1}{2}$, for example, $\frac{1}{4} \leq p \leq \frac{3}{4}$. Everett's formula is perhaps the one which is most generally useful, not the least because the coefficients have been tabulated (this is true also for Bessel's formula), and further because even differences often are tabulated together with the function values. Steffensen's formula might compete if the corresponding conditions were fulfilled.

In many cases one wants just a simple and fast formula, taking into account only first- and second-order differences. Such a formula is easily obtained, for example, from Everett's formula, by putting $\delta^2 y_0 \simeq \delta^2 y_1 \simeq \delta^2 y$ and neglecting higher-order terms. Hence

$$y_p \simeq q y_0 + p y_1 - \frac{pq}{2}\delta^2 y. \quad (8.5.9)$$

Last, we shall also consider the remainder term. When doing this, we shall restrict ourselves to Everett's formula; analogous conditions prevail in the other cases. Putting

$$\varphi_{n-1}(p) = \sum_{i=0}^{n-1} \left\{ \binom{p+i}{2i+1}\delta^{2i} y_1 + \binom{q+i}{2i+1}\delta^{2i} y_0 \right\},$$

we have for $n = 1$, $\varphi_0(p) = p y_1 + q y_0$, and hence $\varphi_0(0) = y_0$, $\varphi_0(1) = y_1$. As a basis for induction, we suppose that $\varphi_{n-1}(i) = y_i$, $i = -n + 1, \ldots, 0, 1, \ldots, n$; as has just been shown, this is fulfilled for $n = 1$. We shall now prove that $\varphi_n(i) = y_i$, $i = -n, \ldots, 0, 1, \ldots, n + 1$. First we observe that

$$\varphi_n(p) = \varphi_{n-1}(p) + \binom{p+n}{2n+1}\delta^{2n}y_1 + \binom{q+n}{2n+1}\delta^{2n}y_0 .$$

Further, $\varphi_{n-1}(p)$ is a polynomial in p of degree $2n - 1$ which takes the values $y_{-n+1}, \ldots, y_0, y_1, \ldots, y_n$ in the $2n$ points $-n + 1, \ldots, 0, 1, \ldots, n$. Hence it must be equal to the Lagrangian interpolation polynomial

$$P_{n-1}(p) = \sum_{k=-n+1}^{n} \frac{(p+n-1)(p+n-2)\cdots(p-n)}{(k+n-1)!\,(n-k)!} \frac{(-1)^{n-k}}{p-k} y_k .$$

By use of direct insertion, we find without difficulty that $\varphi_n(i) = y_i$ for $i = -n + 1, \ldots, 0, 1, \ldots, n$, and hence we need consider only the values $i = -n$ and $i = n + 1$. For symmetry reasons it is sufficient to examine, for example, the case $i = n + 1$. Then we have

$$\varphi_{n-1}(n + 1) = \sum_{k=-n+1}^{n} \frac{2n(2n-1)\cdots 1}{(k+n-1)!\,(n-k)!\,(-1)^{n-k}(n-k+1)} y_k$$

$$= \sum_{k=-n+1}^{n} \frac{(-1)^{n-k}(2n)!}{(n+k-1)!\,(n-k+1)!} y_k$$

$$= \sum_{r=0}^{2n-1} \frac{(-1)^{r-1}(2n)!}{r!\,(2n-r)!} y_{r-n+1} .$$

Of the two remaining terms in $\varphi_n(p)$, the second one is zero, and the first one has the value

$$\binom{2n+1}{2n+1}\delta^{2n}y_1 = \delta^{2n}y_1 = E^{-n}(E-1)^{2n}Ey_0$$

$$= E^{-n+1}\cdot\sum_{r=0}^{2n}(-1)^r\binom{2n}{r}E^r y_0 = \sum_{r=0}^{2n}\frac{(-1)^r\cdot(2n)!}{r!\,(2n-r)!} y_{r-n+1} .$$

The two sums are of different signs and cancel except for the term $r = 2n$ from the latter sum. Hence we get

$$\varphi_n(n + 1) = (-1)^{2n}\frac{(2n)!}{(2n)!\,0!} y_{2n-n+1} = y_{n+1} ,$$

which concludes the proof.

Thus the truncated series is identical with the Lagrangian interpolation polynomial, and the remainder term is also the same. Hence we obtain the complete Everett formula:

$$\left\{ \begin{array}{l} y_p = \displaystyle\sum_{r=0}^{n-1}\binom{q+r}{2r+1}\delta^{2r}y_0 + \sum_{r=0}^{n-1}\binom{p+r}{2r+1}\delta^{2r}y_1 + \binom{p+n-1}{2n}h^{2n}y^{(2n)}(\xi) , \\[2mm] x_0 - (n-1)h \le \xi \le x_0 + nh . \end{array} \right.$$

$$(8.5.10)$$

8.6. Throwback

By throwback we mean a general technique of comprising higher-order differences into lower-order differences, thereby reducing the computational work on interpolation. Differences modified in this way are frequently used in modern tables. Here we shall consider only Everett's formula, and we also restrict ourselves to throwing back the fourth difference on the second. The relevant terms in the formula are of the type

$$py + \binom{p+1}{3}\delta^2 y + \binom{p+2}{5}\delta^4 y + \cdots$$

$$= py + \frac{p(p^2-1)}{6}\left(\delta^2 y - \frac{4-p^2}{20}\delta^4 y\right) + \cdots .$$

When p varies from 0 to 1, the factor $(4-p^2)/20$ varies only slightly, namely, between 0.15 and 0.20, and we will replace this factor by a constant, which still remains to be chosen.

We now introduce so-called *modified differences* $\delta_m^2 y$:

$$\delta_m^2 y_0 = \delta^2 y_0 - C\,\delta^4 y_0 ; \qquad \delta_m^2 y_1 = \delta^2 y_1 - C\,\delta^4 y_1 .$$

The modified interpolation formula then takes the form

$$y_p = py_1 + qy_0 + \binom{p+1}{3}\delta_m^2 y_1 + \binom{q+1}{3}\delta_m^2 y_0 .$$

If the error is denoted by ε, we find

$$\varepsilon(p) = \left[\binom{q+2}{5} + C\binom{q+1}{3}\right]\delta^4 y_0$$

$$+ \left[\binom{p+2}{5} + C\binom{p+1}{3}\right]\delta^4 y_1 . \qquad (8.6.1)$$

Here terms of sixth and higher order have been neglected. Supposing that $\delta^4 y$ varies so slowly that $\delta^4 y_0$ and $\delta^4 y_1$ do not differ appreciably, we find, taking into account that $q = 1 - p$:

$$\varepsilon(p) = \left\{\frac{p(1-p)(2-p)}{6}\left[\frac{4-(1-p)^2}{20} - C\right]\right.$$

$$\left. + \frac{p(1-p)(1+p)}{6}\left[\frac{4-p^2}{20} - C\right]\right\}\delta^4 y$$

or after simplification:

$$\varepsilon(p) = \frac{p(1-p)}{24}(2 - 12C + p - p^2)\,\delta^4 y .$$

Putting $\alpha = 12C - 2$, we examine the function

$$\varphi(p) = p(1-p)(p - p^2 - \alpha) = (p^2 - p)^2 + \alpha(p^2 - p) .$$

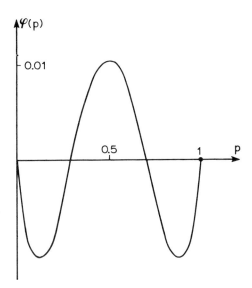

Figure 8.6

For sufficiently small values of α, this function has one maximum and two minima in the interval $0 < p < 1$. We now want to choose α in such a way that the maximum deviation is minimized, and, as is easily inferred, this occurs when all maximum deviations are equal apart from the sign. A simple calculation gives the result $\alpha = (\sqrt{2} - 1)/2$ and $C = (3 + \sqrt{2})/24 = 0.1839$. The error curve then gets the shape shown in Fig. 8.6, and the modified second differences are defined by the formula

$$\delta_m^2 y = \delta^2 y - 0.1839\, \delta^4 y . \qquad (8.6.2)$$

The error due to this approximation is obtained by use of $(8.6.1)$, and a straightforward calculation shows that it is less in absolute value than $0.00122M$, where $M = \max\,(|\delta^4 y_0|, |\delta^4 y_1|)$. If the fourth difference is absolutely less than 400, the error will be less than half a unit in the last place. If, as was done when C was determined, we can suppose that the fourth differences are equal, larger values can be accepted, since

$$|\varepsilon|_{\max} = (3 - 2\sqrt{2})\frac{|\delta^4 y|}{384}$$

and

$$|\varepsilon|_{\max} < \frac{1}{2} \qquad \text{if} \quad |\delta^4 y| < 1119 .$$

The throwback technique was introduced by Comrie. At first he worked with Bessel's formula, which, incidentally, gives the same value for C. He also examined how the method works for higher differences. More details on this matter can be found, for example, in [2].

8.7. Subtabulation

In many cases a function is tabulated with intervals h, although we need values at closer intervals αh, that is, we want to *subtabulate* the function. Let E_1 be the shift operator associated with the interval αh, and $\varDelta_1$ the corresponding difference operator. Then we have $E_1 = E^\alpha$ and

$$\varDelta_1 = (1 + \varDelta)^\alpha - 1 = \alpha\varDelta + \frac{\alpha(\alpha - 1)}{2}\varDelta^2 + \frac{\alpha(\alpha - 1)(\alpha - 2)}{6}\varDelta^3 + \cdots .$$

In order to perform the interpolation, we need different powers of $\varDelta_1$:

$$\varDelta_1^r = \alpha^r\varDelta^r \left\{1 + \frac{r(\alpha - 1)}{2}\varDelta + \frac{r(\alpha - 1)}{24}[4(\alpha - 2) + 3(r - 1)(\alpha - 1)]\varDelta^2 \right.$$

$$+ \frac{r(\alpha - 1)}{48}[2(\alpha - 2)(\alpha - 3) + 4(r - 1)(\alpha - 1)(\alpha - 2)$$

$$\left. + (r - 1)(r - 2)(\alpha - 1)^2]\varDelta^3 + \cdots \right\}. \tag{8.7.1}$$

Usually we have $\alpha = \frac{1}{2}, \frac{1}{5}$, or $\frac{1}{10}$; in the last case, that is, for $\alpha = 0.1$, we find

$$\varDelta_1^r = 10^{-r}\varDelta^r$$

$$\times \left[1 - \frac{9r}{20}\varDelta + \frac{3r(27r + 49)}{800}\varDelta^2 - \frac{3r(81r^2 + 441r + 580)}{16000}\varDelta^3 + \cdots \right].$$

It is now easy to calculate $\varDelta_1, \varDelta_1^2, \varDelta_1^3, \ldots$, and with these values we construct a new difference scheme from which the desired function values can be obtained without difficulty.

In the past, subtabulation was mostly used for computation of tables by use of a desk calculator. Then, a very accurate formula was often employed for obtaining a small number of key values, which formed the basis for subtabulation in one or several stages. These computations were easy to check, since the key values necessarily had to be reproduced. Nowadays the method is mainly of theoretical interest.

8.8. Special and inverse interpolation

Certain functions vary too rapidly for interpolation along the usual lines to be possible. Consider, for example, the function $y = (e^x - e^{-x})/\sin^2 x$. It is quite clear that y is not interpolable for small values of x. On the other hand, the function $z = xy$ is very well behaved and offers no difficulties for interpolation; afterward one has only to divide by x. Another example is presented by the function

$$y = \int_x^\infty \frac{e^{-t}}{t}\,dt ,$$

which becomes infinite for $x = 0$. As is easily inferred, the function $z = y + \log x$ can be interpolated without any complications for positive values of x. Even if one has a function whose analytical character is unknown, a similar technique can often be used. In such cases the regularity of the differences of different orders is used as a test that the method is fairly realistic.

Sometimes the special properties of a function can be used for simplifying the interpolation. A few examples will be presented.

(a) $y = e^x$. Hence

$$e^{x+h} = e^x \cdot e^h = e^x \left(1 + h + \frac{h^2}{2!} + \cdots \right).$$

In larger tables the interval is usually 10^{-4}, and if the series is truncated after the h^2-term, we will obtain about 13 correct figures. As a matter of fact, we can choose h such that $-5 \cdot 10^{-5} < h \leq 5 \cdot 10^{-5}$, and hence $|h^3/3!| < 2.1 \cdot 10^{-14}$.

(b) $y = \sin x$ and $y = \cos x$ are usually tabulated together. We have, for example,

$$\sin (x + h) = \sin x \cos h + \cos x \sin h$$

$$= \sin x \cdot \left(1 - \frac{h^2}{2!} + \cdots \right) + \cos x \cdot \left(h - \frac{h^3}{3!} + \cdots \right).$$

If the interval is 10^{-4}, the formula $\sin (x + h) = \sin x + h \cos x$ will give a maximum error of one unit in the ninth place.

(c) $y = \log x$. Hence

$$\log (x + h) = \log \left[x \left(1 + \frac{h}{x} \right) \right] = \log x + \log \left(1 + \frac{h}{x} \right)$$

$$= \log x + \frac{h}{x} - \frac{h^2}{2x^2} + \cdots .$$

The procedure for inverse interpolation is best demonstrated by an example. Let a function $y = f(x)$ be tabulated together with $\delta^2 y$ and $\delta^4 y$. We are now trying to find such a value x between x_0 and $x_1 = x_0 + h$ that $f(x) = y$, where y is a given value. As before, we denote $(x - x_0)/h$ by p, and using Everett's formula, we obtain the equation

$$y = p y_1 + \binom{p + 1}{3} \delta^2 y_1 + \binom{p + 2}{5} \delta^4 y_1 + (1 - p) y_0$$

$$+ \binom{2 - p}{3} \delta^2 y_0 + \binom{3 - p}{5} \delta^4 y_0 .$$

This is a fifth-degree equation in p, and we will solve it by an iteration technique. First we determine a value p_1 from the equation $p_1 y_1 + (1 - p_1) y_0 = y$. This value is inserted into the $\delta^2 y$-terms, while the $\delta^4 y$-terms are neglected, and in this way we obtain a value p_2 from the equation

$$p_2 y_1 + (1 - p_2) y_0 = y - \left(\frac{p_1 + 1}{3}\right) \delta^2 y_1 - \left(\frac{2 - p_1}{3}\right) \delta^2 y_0 \, .$$

Next we obtain p_3 from

$$p_3 y_1 + (1 - p_3) y_0 = y - \left(\frac{p_2 + 1}{3}\right) \delta^2 y_1$$

$$- \left(\frac{2 - p_2}{3}\right) \delta^2 y_0 - \left(\frac{p_2 + 2}{5}\right) \delta^4 y_1 - \left(\frac{3 - p_2}{5}\right) \delta^4 y_0 \, .$$

If necessary, the procedure is repeated until the values do not change any more. The procedure described here can obviously be carried over to other interpolation formulas.

8.9. Extrapolation

In general it is obvious that extrapolation is a far more delicate process than interpolation. For example, one could point to the simple fact that a function might possess a very weak singularity which is hardly apparent in the given values but nevertheless might have a fatal effect in the extrapolation point. There are, however, a few cases when a fairly safe extrapolation can be performed implying considerable gains in computing time and accuracy.

First we shall discuss *Aitken-extrapolation*. Suppose that we have a sequence $y_0, y_1, y_2, \ldots$ converging toward a value y. We assume the convergence to be geometric, that is,

$$y - y_n = ah^n + \varepsilon_n h^n \, ,$$

where $\varepsilon_n \to 0$ when $h \to 0$. This is often expressed through

$$y - y_n = ah^n + o(h^n) \, .$$

From this we obtain

$$\frac{y - y_{n+1}}{y - y_n} = h + o(h)$$

and

$$\frac{y - y_n}{y - y_{n-1}} = h + o(h) \, .$$

Subtracting we get

$$\frac{y - y_{n+1}}{y - y_n} - \frac{y - y_n}{y - y_{n-1}} = o(h)$$

and hence

$$y_{n-1} y_{n+1} - y_n^2 - y(y_{n-1} - 2y_n + y_{n+1}) = o(h^{2n}) \, .$$

Since $\quad\quad\quad y_{n-1} - 2y_n + y_{n+1} \simeq -ah^{n-1}(1-h)^2 = O(h^{n-1})\,,$

we get

$$y = \frac{y_{n-1}y_{n+1} - y_n^2}{y_{n-1} - 2y_n + y_{n+1}} + o(h^{n+1})$$

or

$$y = y_n^* = y_{n+1} - \frac{(\Delta y_n)^2}{\Delta^2 y_{n-1}} + o(h^{n+1})\,. \tag{8.9.1}$$

The procedure can then be repeated with the new series y_n^*.

We shall also very briefly discuss *Richardson extrapolation*. Usually one wants to compute in a finite process a certain quantity, for example, an integral, by the aid of approximations obtained with different interval lengths. Assume that we have

$$F(h) = F(0) + a_1 h^2 + a_2 h^4 + a_3 h^6 + \cdots . \tag{8.9.2}$$

Using the expression for $F(ph)$ we can eliminate the quadratic term; as a rule we then take $p = 2$. Thus we form

$$G(h) = F(h) - \frac{F(2h) - F(h)}{2^2 - 1} = F(0) - 4a_2 h^4 - 20a_3 h^6 - \cdots . \tag{8.9.3}$$

The procedure can then be repeated:

$$H(h) = G(h) - \frac{G(2h) - G(h)}{2^4 - 1} = F(0) + 64a_3 h^6 + \cdots . \tag{8.9.4}$$

Richardson's technique is used, for example, in connection with numerical quadrature (Romberg-integration) and for solving differential equations.

REFERENCES

[1] National Bureau of Standards: *Tables of Lagrangian Interpolation Coefficients* (Washington, 1944).

[2] *Interpolation and Allied Tables* (London, 1956).

[3] Jordan: *Calculus of Finite Differences* (Chelsea, New York, 1947).

[4] Steffensen: *Interpolation* (Chelsea, New York, 1950).

EXERCISES

1. The table below contains an error which should be located and corrected.

x	$f(x)$	x	$f(x)$
3.60	0.112046	3.65	0.152702
3.61	0.120204	3.66	0.160788
3.62	0.128350	3.67	0.168857
3.63	0.136462	3.68	0.176908
3.64	0.144600		

2. The function $y = f(x)$ is given in the points $(7, 3), (8, 1), (9, 1)$, and $(10, 9)$. Find the value of y for $x = 9.5$, using Lagrange's interpolation formula.

3. From the beginning of a table, the following values are reproduced:

x	$f(x)$
0.001	0.54483 33726
0.002	0.55438 29800
0.003	0.56394 21418
0.004	0.57351 08675
0.005	0.58308 91667

Find the function value for $x = 0.00180$ as accurately as possible.

4. A function $f(x)$ is known in three points, x_1, x_2, and x_3, in the vicinity of an extreme point x_0. Show that

$$x_0 \simeq \frac{x_1 + 2x_2 + x_3}{4} - \frac{f(x_1, x_2) + f(x_2, x_3)}{4f(x_1, x_2, x_3)}.$$

Use this formula to find x_0 when the following values are known:

x	3.0	3.6	3.8
f	0.13515	0.83059	0.26253

5. The function $y = x!$ has a minimum between 0 and 1. Find the abscissa from the data below.

x	$\dfrac{d}{dx} \log (x!)$	δ^2	δ^4
0.46	$-0.00158\ 05620$	-888096	-396
0.47	$+0.00806\ 64890$	-872716	-383

6. The function $y = \exp x$ is tabulated for $x = 0(0.01)1$. Find the maximal error on linear interpolation.

7. The equation $x^3 - 15x + 4 = 0$ has a root close to 0.3. Obtain this root with 6 decimal places, using inverse interpolation (for example, with Bessel's interpolation formula).

8. $f(x_n + ph)$ is denoted by y_{n+p}. Show that

$$y_{n+p} = y_n + p\,\Delta y_{n-1} + \binom{p+1}{2}\Delta^2 y_{n-2} + \binom{p+2}{3}\Delta^3 y_{n-3} + \cdots,$$

assuming that the series converges.

9. Find the constants A, B, C, and D in the interpolation formula

$$f(x_0 + ph) = Af_{-1} + Bf_1 + Chf'_{-1} + Dhf'_1 + R,$$

as well as the order of the error term R. Use the formula to obtain $f(2.74)$ when

$f(2.6) = 0.0218502$ and $f(2.8) = 0.0168553$. The function f is defined by

$$f(x) = \int_x^\infty (e^{-t}/t)\, dt \ .$$

10. Find an interpolation formula which makes use of $y_0, y_0'', y_0^{IV}, \ldots$ and $y_1, y_1'', y_1^{IV}, \ldots$ (up to sixth-order terms).

11. Determine the constants a, b, c, and d in such a way that the formula

$$y_p = ay_0 + by_1 + h^2(cy_0'' + dy_1'') \ ,$$

becomes correct to the highest possible order. Use the formula to compute $Ai(1.1)$ when $Ai(1.0) = 0.135292$ and $Ai(1.2) = 0.106126$. It is known that the function $y = Ai(x)$ satisfies the differential equation $y'' = xy$.

12. The following data are given for a certain function:

x	y	y'
0.4	1.554284	0.243031
0.5	1.561136	-0.089618

In the interval $0.4 < x < 0.5$, y has a maximum. Find its coordinates as accurately as possible.

13. For a function $y = f(x)$ the following data are known: $f(0) = 1.1378$; $f'(0.1) = -3.1950$; $f'(0.2) = -3.1706$; $f(0.3) = 0.1785$. The function has an inflection point in the interval $0 < x < 0.3$. Find its coordinates to three places.

14. A function $f(x, y)$ takes the following values in nine adjacent mesh points:

(6, 4) 8.82948	(7, 4) 11.33222	(8, 4) 14.17946
(6, 5) 9.31982	(7, 5) 11.97257	(8, 5) 14.98981
(6, 6) 9.81019	(7, 6) 12.61294	(8, 6) 15.80018

Find the function value for $x = 7.2$, $y = 5.6$.

15. The function y is given in the table below.

x	y
0.01	98.4342
0.02	48.4392
0.03	31.7775
0.04	23.4492
0.05	18.4542

Find y for $x = 0.0341$.

16. The function

$$y = \int_x^\infty \frac{e^{-t}}{t}\, dt$$

is given in the table below. Find y for $x = 0.0378$.

x	0.00	0.01	0.02	0.03	0.04	0.05	0.06
y	∞	4.0379	3.3547	2.9591	2.6813	2.4679	2.2953

17. The function $y = f(x)$ is supposed to be differentiable three times. Prove the relation

$$f(x) = -\frac{(x - x_1)(x - 2x_0 + x_1)}{(x_1 - x_0)^2} f(x_0) + \frac{(x - x_0)(x - x_1)}{x_0 - x_1} f'(x_0)$$

$$+ \frac{(x - x_0)^2}{(x_1 - x_0)^2} f(x_1) + R(x) ,$$

where $R(x) = \frac{1}{6}(x - x_0)^2(x - x_1)f'''(\xi)$; $x_0 < x, \xi < x_1$.

18. A function $\varphi(x)$ is defined by $f(x) \cdot g(x)$. According to Steffensen, the following formula is valid:

$$\varphi(x_0, x_1, \ldots, x_n) = \sum_{r=0}^{n} f(x_0, x_1, \ldots, x_r) \cdot g(x_r, \ldots, x_n) .$$

Prove this formula in the case $n = 2$.

19. In a table the function values have been rounded to a certain number of decimals. In order to facilitate linear interpolation one also wants to give the first differences. Determine which is best: to use the differences of the rounded values, or to give rounded values of the exact first differences.

20. One wants to compute an approximate value of π by considering regular n-sided polygons with corners on the unit circle, for such values of n that the perimeter P_n can easily be computed. Determine P_n for $n = 3, 4$, and 6 and extrapolate to ∞ taking into account that the error is proportional to $1/n^2$.

21. The function $f(x) = x + x^2 + x^4 + x^8 + x^{16} + \cdots$ is defined for $-1 < x < 1$. Determine

$$c = -\lim_{x \to 1} \left[\frac{\log (1 - x)}{\log 2} + f(x) \right] ,$$

by explicit computations for $x = 0.95, 0.98$, and 0.99 and extrapolation to $x = 1$.

Chapter 9

Numerical differentiation

The function $f(x)$ is supposed to be analytic and tabulated in equidistant points. Our problem is then to compute the derivative, either in a grid point or in an interior point. The first case can be considered as a pure differentiation problem, the second as a combined interpolation and differentiation problem.

From the relation $\varDelta = e^{hD} - 1$, we obtain formally

$$hD = \log(1 + \varDelta) = \varDelta - \tfrac{1}{2}\varDelta^2 + \tfrac{1}{3}\varDelta^3 - \cdots,$$

or

$$f'(x) = \frac{1}{h}\left(\varDelta f(x) - \frac{1}{2}\varDelta^2 f(x) + \frac{1}{3}\varDelta^3 f(x) - \cdots\right). \tag{9.1}$$

In general we have $D^r = h^{-r}[\log(1 + \varDelta)]^r$, where the following expansion can be used:

$$\frac{[\log(1 + x)]^k}{k!} = \sum_{n=k}^{\infty} \frac{\alpha_k^{(n)}}{n!} x^n. \tag{9.2}$$

The coefficients $\alpha_k^{(n)}$ again are Stirling's numbers of the first kind. The relation follows from

$$[\log(1 + x)]^k = \left[\frac{d^k}{dt^k}(1 + x)^t\right]_{t=0}$$

$$= \left[\frac{d^k}{dt^k}\sum_{n=0}^{\infty}\frac{t^{(n)}}{n!}x^n\right]_{t=0}$$

$$= \left[\sum_{n=0}^{\infty}\frac{x^n}{n!}\frac{d^k}{dt^k}\sum_{p=1}^{n}\alpha_p^{(n)}t^p\right]_{t=0}$$

$$= \sum_{n=k}^{\infty}\frac{x^n}{n!}\alpha_k^{(n)} \cdot k!$$

---

* Famous Swedish religious leader of the nineteenth century.

(since we get no contribution if $n < k$). Hence we obtain

$$D^r = h^{-r}\left(\alpha_r^{(r)} + \frac{\alpha_r^{(r+1)}}{r + 1}\, \varDelta + \frac{\alpha_r^{(r+2)}}{(r + 1)(r + 2)}\, \varDelta^2 + \cdots\right) \varDelta^r,$$

or

$$f^{(r)}(x) = h^{-r}\left(\alpha_r^{(r)} \varDelta^r f(x) + \frac{\alpha_r^{(r+1)}}{r + 1}\, \varDelta^{r+1} f(x)\right.$$

$$\left. + \frac{\alpha_r^{(r+2)}}{(r + 1)(r + 2)}\, \varDelta^{r+2} f(x) + \cdots\right). \qquad (9.3)$$

Analogously, we find

$$f^{(r)}(x) = h^{-r}\left(\alpha_r^{(r)} \nabla^r f(x) - \frac{\alpha_r^{(r+1)}}{r + 1}\, \nabla^{r+1} f(x)\right.$$

$$\left. + \frac{\alpha_r^{(r+2)}}{(r + 1)(r + 2)}\, \nabla^{r+2} f(x) - \cdots\right). \qquad (9.4)$$

As before, we prefer expressions containing central differences. Then we can apply (7.2.3) or (7.2.4); the first formula gives

$$y' = Dy = \frac{1}{h} \sum_{n=0}^{\infty} \frac{(-1)^n (2n)!}{2^{4n} (n!)^2 (2n + 1)}\, \delta^{2n+1} y$$

$$= \frac{1}{h}\left(\delta y - \frac{1}{24}\, \delta^3 y + \frac{3}{640}\, \delta^5 y - \frac{5}{7168}\, \delta^7 y + \cdots\right). \qquad (9.5)$$

This formula, however, contains only odd powers of δ and is useless except for the computation of the derivative in the middle between two grid points. Formula (7.2.4) gives instead

$$y' = Dy = \frac{\mu}{h} \sum_{n=0}^{\infty} (-1)^n \frac{(n!)^2}{(2n + 1)!}\, \delta^{2n+1} y$$

$$= \frac{1}{h}\left[\mu \delta y - \frac{1}{6}\, \mu \delta^3 y + \frac{1}{30}\, \mu \delta^5 y - \frac{1}{140}\, \mu \delta^7 y + \frac{1}{630}\, \mu \delta^9 y - \cdots\right]$$

$$= \frac{1}{2h}\left[y_1 - y_{-1} - \frac{1}{6}(\delta^2 y_1 - \delta^2 y_{-1}) + \frac{1}{30}(\delta^4 y_1 - \delta^4 y_{-1}) - \cdots\right]. \qquad (9.6)$$

The second derivative can be obtained by squaring (9.5) or directly from (7.2.5):

$$y'' = \frac{1}{h^2} \sum_{n=0}^{\infty} 2(-1)^n \frac{(n!)^2}{(2n + 2)!}\, \delta^{2n+2} y$$

$$= \frac{1}{h^2}\left[\delta^2 y - \frac{1}{12}\, \delta^4 y + \frac{1}{90}\, \delta^6 y - \frac{1}{560}\, \delta^8 y + \frac{1}{3150}\, \delta^{10} y - \cdots\right]. \qquad (9.7)$$

Combining (9.6) and (9.7), we can obtain desired expressions for higher derivatives. For example, we find

$$hy' = \sum_{n=0}^{k-1} (-1)^n \frac{(n!)^2}{(2n+1)!} \mu\delta^{2n+1}y + (-1)^k \frac{(k!)^2}{(2k+1)!} h^{2k+1}y^{(2k+1)}(\xi) ;$$

$$x - kh < \xi < x + kh .$$

We shall now also seek the derivative in an interior point. We start from Everett's formula, which is differentiated with respect to $p(dx = hdp; q = 1 - p)$:

$$y'_p = \frac{1}{h} \left[y_1 - y_0 + \frac{d}{dp}\left(\frac{p(p^2-1)}{3!}\right)\delta^2 y_1 - \frac{d}{dq}\left(\frac{q(q^2-1)}{3!}\right)\delta^2 y_0 \right.$$

$$\left. + \frac{d}{dp}\left(\frac{p(p^2-1)(p^2-4)}{5!}\right)\delta^4 y_1 - \frac{d}{dq}\left(\frac{q(q^2-1)(q^2-4)}{5!}\right)\delta^4 y_0 + \cdots \right].$$

$$(9.8)$$

For higher derivatives we can continue differentiating with respect to p or q. In particular we have

$$\frac{d}{dp}\left(\frac{p(p^2-1)\cdots(p^2-n^2)}{(2n+1)!}\right)_{p=0} = \frac{(-1)^n \cdot (n!)^2}{(2n+1)!} ,$$

$$\frac{d}{dq}\left(\frac{q(q^2-1)\cdots(q^2-n^2)}{(2n+1)!}\right)_{q=1} = \frac{2(1-4)(1-9)\cdots(1-n^2)}{(2n+1)!}$$

$$= \frac{(-1)^{n-1}\cdot(n-1)!\,(n+1)!}{(2n+1)!} .$$

From this we get still another formula for the first derivative:

$$y'_0 = \frac{1}{h}\left[y_1 - y_0 + \sum_{n=1}^{\infty} \frac{(-1)^n \cdot (n!)^2}{(2n+1)!}\left(\delta^{2n}y_1 + \frac{n+1}{n}\delta^{2n}y_0\right)\right]$$

$$= \frac{1}{h}\left[y_1 - y_0 - \frac{1}{6}\delta^2 y_1 - \frac{1}{3}\delta^2 y_0 \right.$$

$$\left. + \frac{1}{30}\delta^4 y_1 + \frac{1}{20}\delta^4 y_0 - \frac{1}{140}\delta^6 y_1 - \frac{1}{105}\delta^6 y_0 + \cdots \right].$$

$$(9.9)$$

It is easy to understand that, as a rule, significant figures are lost in numerical differentiation. The main term in the computation of $f'(x)$ is, of course,

$$f'(x) \cong \frac{1}{2h}[f(x+h) - f(x-h)] ,$$

and the two numbers within brackets must be of the same order of magnitude. Hence, numerical differentiation, contrary to analytic differentiation, should be considered as a "difficult" operation. Later we shall see that the reverse is true for integration.

EXAMPLE

The function $y = \sin x$ is tabulated in the scheme below. Find the derivative in the point $x = 1$.

x	y	δ	δ^2	δ^3	δ^4
0.7	0.644218				
		73138			
0.8	0.717356		−7167		
		65971		−660	
0.9	0.783327		−7827		79
		58144		−581	
1.0	0.841471		−8408		85
		49736		−496	
1.1	0.891207		−8904		87
		40832		−409	
1.2	0.932039		−9313		
		31519			
1.3	0.963558				

$$y'(1) = \frac{1}{0.2}\left[0.891207 - 0.783327 + \frac{1}{6}(0.008904 - 0.007827)\right.$$
$$\left. + \frac{1}{30}(0.000087 - 0.000079) + \cdots \right]$$
$$= 0.54030 .$$

Tabulated value: $\cos 1 = 0.540302$.

REFERENCES

[1] Milne-Thomson: *The Calculus of Finite Differences* (MacMillan, London, 1933).

EXERCISES

1. A function is given according to the table below. Find the derivative for $x = 0.5$.

2. The function $y = f(x)$ has a minimum in the interval $0.2 < x < 1.4$. Find the x-coordinate of the minimum point.

Table for Ex. 1		Table for Ex. 2	
x	y	x	y
0.35	1.521525	0.2	2.10022
0.40	1.505942	0.4	1.98730
0.45	1.487968	0.6	1.90940
0.50	1.467462	0.8	1.86672
0.55	1.444243	1.0	1.85937
0.60	1.418083	1.2	1.88737
0.65	1.388686	1.4	1.95063

3. A function $y = y(x)$ is given according to the table below.

x	y
0.00	-1.000000
0.05	-0.973836
0.10	-0.896384
0.15	-0.770736
0.20	-0.601987
0.25	-0.397132
0.30	-0.164936
0.35	$+0.084255$
0.40	$+0.338744$
0.45	$+0.585777$
0.50	$+0.811898$

The function satisfies the differential equation

$$(1 - x^2)y'' - 2xy' - \frac{a}{1 - x^2} y + by = 0 .$$

Find the parameters a and b which are both known to be integers.

4. A function $y = y(x)$ is given in the table below. Find the second derivative for $x = 3$.

5. y is a function of x satisfying the differential equation $xy'' + ay' + (x - b)y = 0$, where a and b are known to be integers. Find the constants a and b from the table below.

Table for Ex. 4		Table for Ex. 5	
x	y	x	y
2.94	0.18256 20761	0.8	1.73036
2.96	0.18110 60149	1.0	1.95532
2.98	0.17967 59168	1.2	2.19756
3.00	0.17827 10306	1.4	2.45693
3.02	0.17689 06327	1.6	2.73309
3.04	0.17553 40257	1.8	3.02549
3.06	0.17420 05379	2.0	3.33334
		2.2	3.65563

6. A function $y = y(x)$ is given in the table below. The function is a solution of the equation $x^2y'' + xy' + (x^2 - n^2)y = 0$, where n is a positive integer. Find n.

x	y
85.00	0.03538 78892
85.01	0.03461 98696
85.02	0.03384 90002
85.03	0.03307 53467
85.04	0.03229 89750

Chapter 10

Numerical quadrature

10.0. Introduction

When a definite integral has to be computed by numerical methods, it is essential that the integrand have no singularities in the domain of interest. The only exception is when the integrand contains a so-called weight function as factor; in this case a reasonable singularity can be allowed for the weight function. We shall here give a few examples of the treatment of such "improper" integrals, also taking into account the case when the integration interval is infinite.

First we consider $I = \int_0^1 (\cos x / \sqrt{x}) \, dx$. The integrand is singular in the origin, but the singularity is of the form x^{-p} with $p < 1$, and hence the integral has a finite value. There are several possible ways to compute this value.

(a) Direct series expansion gives

$$I = \int_0^1 dx \left\{ \frac{1}{\sqrt{x}} - \frac{x^{3/2}}{2!} + \frac{x^{7/2}}{4!} - \frac{x^{11/2}}{6!} + \cdots \right\}$$

$$= \left[2x^{1/2} - \frac{x^{5/2}}{2!\,5/2} + \frac{x^{9/2}}{4!\,9/2} - \frac{x^{13/2}}{6!\,13/2} + \cdots \right]_0^1$$

$$= 2 - \frac{1}{5} + \frac{1}{108} - \frac{1}{4680} + \frac{1}{342720} - \cdots = 1.8090484759 \, .$$

(b) The singularity is subtracted and the remaining regular part is computed by aid of a suitable numerical method. In this case we can write

$$I = \int_0^1 \frac{dx}{\sqrt{x}} - \int_0^1 \frac{1 - \cos x}{\sqrt{x}} \, dx = 2 - \int_0^1 \frac{1 - \cos x}{\sqrt{x}} \, dx \, .$$

In the last integral the integrand is regular in the whole interval [for $x = 0$ we get $\lim_{x \to 0} \left((1 - \cos x)/\sqrt{x} \right) = 0$], and at first glance we would not expect any numerical difficulties. However, it turns out that we obtain a very poor accuracy unless we take very small intervals close to the origin. The reason is that higher derivatives are still singular in the origin, and hence we ought

195

to employ a more refined subtraction procedure. For example, we can write instead

$$I = \int_0^1 \frac{1 - x^2/2}{\sqrt{x}} \, dx + \int_0^1 \frac{\cos x - 1 + x^2/2}{\sqrt{x}} \, dx$$

$$= 1.8 + \int_0^1 \frac{\cos x - 1 + x^2/2}{\sqrt{x}} \, dx \, .$$

We gain two advantages in this way: (1) the integrand becomes much smaller; (2) the numerical accuracy is improved considerably.

(c) The substitution $x = t^2$ gives $I = 2 \int_0^1 \cos t^2 \, dt$. The resulting integral has no singularity left in the integrand, and even all derivatives are regular. Hence, direct computation by use of a suitable numerical method will give no difficulties. We can also employ series expansion:

$$I = 2 \int_0^1 \left[1 - \frac{t^4}{2!} + \frac{t^8}{4!} - \frac{t^{12}}{6!} + \cdots \right] dt$$

$$= 2 \left[1 - \frac{1}{10} + \frac{1}{216} - \frac{1}{9360} + \cdots \right],$$

which is exactly the same series as before.

(d) Partial integration gives

$$I = \left[2\sqrt{x} \cos x \right]_0^1 + \int_0^1 2\sqrt{x} \sin x \, dx \, .$$

The last integral can be computed numerically or by series expansion; in the latter case we obtain

$$I = 2 \cos 1 + 4 \left(\frac{1}{5 \cdot 1!} - \frac{1}{9 \cdot 3!} + \frac{1}{13 \cdot 5!} - \frac{1}{17 \cdot 7!} + \cdots \right)$$

$$= 1.809048476 \, .$$

As our second example we take $I = \int_0^1 e^{-x} \log x \, dx$. The integrand has a logarithmic singularity at the origin which can easily be neutralized. Expanding e^{-x} in a power series, we find

$$I = \int_0^1 e^{-x} \log x \, dx = \int_0^1 \left(1 - x + \frac{x^2}{2!} - \frac{x^3}{3!} + \cdots \right) \log x \, dx \, .$$

From

$$\int_0^1 x^k \log x \, dx = \left[\frac{x^{k+1}}{k + 1} \left(\log x - \frac{1}{k + 1} \right) \right]_0^1 = - \frac{1}{(k + 1)^2} \, ,$$

we easily obtain

$$I = \sum_{k=0}^{\infty} \frac{(-1)^{k+1}}{(k + 1)^2 k!} = - \left(1 - \frac{1}{4} + \frac{1}{18} - \frac{1}{96} + \cdots \right).$$

The sum of this infinite series can be found by aid of a purely numerical technique (see Chapter 11). Alternatively, using partial integration, we get

$$I = -\gamma - E_1(1) = -0.7965996 ,$$

where γ is Euler's constant (see p. 233) and

$$E_1(x) = \int_x^\infty \frac{e^{-t}}{t} \, dt .$$

As our third example we choose

$$I = \int_0^1 \frac{dx}{\sqrt{x(1 - x^2)}} .$$

The integrand is singular as $1/\sqrt{x}$ at the origin and as $1/\sqrt{2(1 - x)}$ when x is close to 1. We divide the interval into two equal parts, and make the substitution $x = t^2$ in the left and $1 - x = t^2$ in the right subinterval. After some simple calculations, we get

$$I = 2 \int_0^{1/\sqrt{2}} \frac{dt}{\sqrt{1 - t^2}} \left(\frac{1}{\sqrt{1 + t^2}} + \frac{1}{\sqrt{2 - t^2}} \right)$$

$$= 2\sqrt{2} \int_0^1 \frac{dz}{\sqrt{2 - z^2}} \left(\frac{1}{\sqrt{2 + z^2}} + \frac{1}{\sqrt{4 - z^2}} \right) .$$

Now the integrand is regular in the whole interval. In this case, however, it is possible to get rid of both singularities in one step. Putting

$$x = \frac{2u^2}{1 + u^4}$$

we obtain

$$I = 2\sqrt{2} \int_0^1 \frac{du}{\sqrt{1 + u^4}} .$$

Finally, we will briefly touch upon the case where the integration interval is infinite. In Chapter 1 we encountered an important method, namely, expansion in an asymptotic series. Another obvious method consists in making the integration interval finite by a suitable transformation. Take, for example,

$$I = \int_0^\infty \frac{e^{-z} \, dz}{ze^{-2z} + 1} .$$

Through the substitution $e^{-z} = t$, the integral is transformed into

$$I = \int_0^1 \frac{dt}{1 - t^2 \log t} ,$$

where the integrand is regular and the interval is finite.

Here we have indicated a series of methods for transforming integrals with singular integrand or with an infinite integration interval to integrals with regular integrand or finite interval. For completeness we mention that perhaps the most effective method for computation of complicated definite integrals is the residue method. However, it rests upon the theory of analytic functions and must be ignored here. Likewise, we shall not discuss the technique of differentiating or integrating with respect to a parameter. Throughout the remainder of this chapter, we shall generally treat integrals with regular integrand and mostly with finite integration interval.

10.1. Cote's formulas

We shall now treat the problem of computing $\int_a^b f(x)\,dx$ numerically, where $f(x)$ is a function whose numerical value is known in certain points. In principle we meet the same problem as in interpolation, and clearly we must replace $f(x)$ by a suitable function $P(x)$, which is usually a polynomial. If the given x-values are not equidistant, the reader should compute the interpolation polynomial explicitly and then perform the integration directly. In general, the function values are available in equidistant points, and from now on we shall preferably consider this case.

Thus we assume a constant interval length h and put

$$P(x) = \sum_{k=0}^{n} L_k(x) y_k ,$$

where

$$L_k(x) = \frac{(x - x_0)(x - x_1) \cdots (x - x_{k-1})(x - x_{k+1}) \cdots (x - x_n)}{(x_k - x_0)(x_k - x_1) \cdots (x_k - x_{k-1})(x_k - x_{k+1}) \cdots (x_k - x_n)} .$$

Here $x_k = x_0 + kh$, and further we put $x = x_0 + sh$, obtaining $dx = h\,ds$. Hence

$$L_k = \frac{s(s - 1) \cdots (s - k + 1)(s - k - 1) \cdots (s - n)}{k(k - 1) \cdots (1)(-1) \cdots (k - n)} ,$$

and we get

$$\int_{x_0}^{x_n} P(x)\,dx = nh\,\frac{1}{n}\sum_{k=0}^{n} y_k \int_0^n L_k\,ds .$$

Finally, putting $(1/n)\int_0^n L_k\,ds = C_k^n$, we can write the integration formula

$$\int_{x_0}^{x_n} P(x)\,dx = nh\sum_{k=0}^{n} C_k^n y_k = (x_n - x_0)\sum_{k=0}^{n} C_k^n y_k . \qquad (10.1.1)$$

The numbers C_k^n ($0 \leq k \leq n$) are called *Cote's numbers*. As is easily proved, they satisfy $C_k^n = C_{n-k}^n$ and $\sum_{k=0}^{n} C_k^n = 1$. Later on we shall examine what can be said about the difference between the desired value, $\int_{x_0}^{x_n} f(x)\,dx$, and the numerically computed value, $\int_{x_0}^{x_n} P(x)\,dx$.

EXAMPLE $n = 2$

$$\begin{cases} C_0^2 = \dfrac{1}{2} \displaystyle\int_0^2 \dfrac{(s-1)(s-2)}{(-1)(-2)}\, ds = \dfrac{1}{6} \,, \\[2mm] C_1^2 = \dfrac{1}{2} \displaystyle\int_0^2 \dfrac{s(s-2)}{1(-1)}\, ds = \dfrac{4}{6} \,, \\[2mm] C_2^2 = \dfrac{1}{2} \displaystyle\int_0^2 \dfrac{s(s-1)}{2 \cdot 1}\, ds = \dfrac{1}{6} \,, \end{cases}$$

$$\int_{x_0}^{x_2} f(x)\, dx \simeq (x_2 - x_0)\left(\frac{1}{6} y_0 + \frac{4}{6} y_1 + \frac{1}{6} y_2 \right). \qquad (10.1.2)$$

This is the well-known *Simpson formula.*

For $n = 1$ we find $C_0^1 = C_1^1 = \frac{1}{2}$, that is, the trapezoidal rule. It should be used only in some special cases.

Error estimates

We shall now compute the error in the Simpson formula. We suppose that $f^{IV}(x)$ exists and is continuous in the whole integration interval. Then we construct

$$F(h) = \frac{h}{3}[f(-h) + 4f(0) + f(h)] - \int_{-h}^{h} f(x)\, dx \, .$$

Differentiating with respect to h several times, we get

$$F'(h) = \frac{1}{3}[f(-h) + 4f(0) + f(h)]$$

$$+ \frac{h}{3}[f'(h) - f'(-h)] - [f(h) + f(-h)]$$

$$= -\frac{2}{3}[f(h) + f(-h)] + \frac{4}{3}f(0)$$

$$+ \frac{h}{3}[f'(h) - f'(-h)] \, .$$

We note that $F'(0) = 0$. Further,

$$F''(h) = \frac{h}{3}[f''(h) + f''(-h)] - \frac{1}{3}[f'(h) - f'(-h)] ; \qquad F''(0) = 0 \, .$$

$$F'''(h) = \frac{h}{3}[f'''(h) - f'''(-h)] ; \qquad F'''(0) = 0 \, .$$

Since f^{IV} is continuous, we can use the mean-value theorem of differential calculus on $F'''(h)$:

$$F'''(h) = \frac{2h^2}{3} f^{IV}(\theta h) ; \qquad -1 < \theta < 1 \, ,$$

and then we can define a continuous function $g(h)$ such that

$$F'''(h) = \frac{2h^2}{3} g(h) .$$ (10.1.3)

Obviously, $g(h) = f^{IV}(\theta h)$, $-1 < \theta < 1$, where, of course, $\theta = \theta(h)$ is a function of h. Integrating (10.1.3) directly, we get

$$F(h) = \int_0^h \frac{(h - t)^2}{2} \frac{2}{3} t^2 g(t) \, dt .$$ (10.1.4)

Formula (10.1.4) implies $F(0) = F'(0) = F''(0) = 0$ and further the validity of (10.1.3). Now we use the first mean-value theorem of integral calculus,

$$\int_a^b f(t)g(t) \, dt = g(\xi) \int_a^b f(t) \, dt \qquad (a < \xi < b) ,$$

provided that f and g are continuous and that f is of the same sign in the interval. These conditions are fulfilled in (10.1.4) with

$$f(t) = \frac{t^2(h - t)^2}{3} ,$$

and hence

$$F(h) = g(\theta'h) \int_0^h \frac{(h - t)^2}{2} \frac{2}{3} t^2 \, dt = \frac{h^5}{90} g(\theta'h)$$

with $0 < \theta' < 1$. Thus

$$F(h) = \frac{h^5}{90} f^{IV}(\xi) ; \qquad -h < \xi < h .$$ (10.1.5)

Putting $h = \frac{1}{2}(b - a)$, we get

$$F = \frac{(b - a)^5}{2880} f^{IV}(\xi) \simeq 3.5 \cdot 10^{-4}(b - a)^5 f^{IV}(\xi) .$$

Using (10.1.5) over n intervals, we obtain

$$F_1 + F_2 + \cdots + F_n = \frac{nh^5}{90} \frac{f^{IV}(\xi_1) + f^{IV}(\xi_2) + \cdots + f^{IV}(\xi_n)}{n} .$$

Putting for a moment $(1/n) \sum_{i=1}^n f^{IV}(\xi_i) = S$, we have with $m < f^{IV}(\xi_i) < M$ trivially $m < S < M$, and since f^{IV} is continuous, we also have $S = f^{IV}(\xi)$ with $a < \xi < b$. But $2nh = b - a$, and we obtain finally

$$F = (b - a) \frac{h^4}{180} f^{IV}(\xi) , \qquad a < \xi < b .$$ (10.1.6)

Similar error estimates can be made for the other Cote formulas. The coefficients and the error terms appear in the table below.

Table of Cote's numbers

n	N	$NC_0^{(n)}$	$NC_1^{(n)}$	$NC_2^{(n)}$	$NC_3^{(n)}$	$NC_4^{(n)}$	$NC_5^{(n)}$	$NC_6^{(n)}$	Remainder term
1	2	1	1						$8.3 \cdot 10^{-2}(b-a)^3 f''(\xi)$
2	6	1	4	1					$3.5 \cdot 10^{-4}(b-a)^5 f^{IV}(\xi)$
3	8	1	3	3	1				$1.6 \cdot 10^{-4}(b-a)^5 f^{IV}(\xi)$
4	90	7	32	12	32	7			$5.2 \cdot 10^{-7}(b-a)^7 f^{VI}(\xi)$
5	288	19	75	50	50	75	19		$3.0 \cdot 10^{-7}(b-a)^7 f^{VI}(\xi)$
6	840	41	216	27	272	27	216	41	$6.4 \cdot 10^{-10}(b-a)^9 f^{VIII}(\xi)$

For example, we have

$$\int_0^{6h} y\, dx \simeq \frac{6h}{840}[41y_0 + 216y_1 + 27y_2 + 272y_3 + 27y_4 + 216y_5 + 41y_6].$$

Choosing $y = e^x$ and $h = 0.2$, we get from this formula 2.32011 6928, compared with the exact value

$$e^{1.2} - 1 = 2.32011\ 69227\ldots.$$

The error must be less than $6.4 \cdot 10^{-10} \cdot 1.2^9 \cdot e^{1.2} = 11 \cdot 10^{-9}$ compared with the actual error $5 \cdot 10^{-9}$.

We observe that a Cote formula of order $2n + 1$ has an error term which is only slightly less than the error term of the formula of order $2n$. For this reason the even formulas predominate. In practice, the error is estimated directly only on rare occasions. Instead, we use the fact that if the interval length is doubled in a formula of order $2n$, then the error will become about 2^{2n+2} times larger. Hence, by doubling the interval in Simpson's formula, one will obtain an error which is about 16 times larger. Using two intervals to compute $\int_1^2 dx/x$, we get the result 0.69325; with one interval we get 0.69444. Denoting the exact value by I, we have approximately

$$I + 16\varepsilon = 0.69444\,,$$
$$I + \varepsilon = 0.69325\,,$$

and hence $I \simeq 0.69317$ (compared with $\log 2 = 0.69315$). The method described here is usually called *Richardson extrapolation* (or *deferred approach to the limit*); obviously, it must be used with caution and discrimination.

If we take Cote's sixth-order formula and add $((b-a)/840)\delta^6 y_3$, we obtain:

$$\int_{x_0}^{x_0+6h} y\, dx = \frac{6h}{20}[(y_0 + y_6) + 5(y_1 + y_5) + (y_2 + y_4) + 6y_3] + R'.$$

The remainder term R' is fairly complicated, but on the other hand, the coefficients are very simple. The relation is known under the name *Weddle's rule* and has been widely used, especially for hand computation.

Simpson's formula with end correction

We shall now try to improve the usual Simpson formula by also allowing derivatives in the end points. Putting

$$\int_{-h}^{+h} y \, dx \simeq h(ay_{-1} + by_0 + ay_1) + h^2(cy'_{-1} - cy'_1) \,,$$

and expanding in series we find

$$\begin{cases} 2a + b = 2 \,, \\ a - 2c = \tfrac{1}{3} \,, \\ a - 4c = \tfrac{1}{5} \,, \end{cases}$$

whence

$$a = \frac{7}{15}; \qquad b = \frac{16}{15}; \qquad c = \frac{1}{15} \,.$$

If this formula is used for several adjacent intervals, the y'-terms cancel in all interior points, and we are left with the final formula

$$\int_a^b y \, dx = \frac{h}{15} (7y_0 + 16y_1 + 14y_2 + 16y_3 + \cdots + 7y_{2n})$$

$$+ \frac{h^2}{15} (y'_0 - y'_{2n}) + O(h^6) \,,$$

where $b - a = 2nh$ and $y_0, y_1, \ldots, y_{2n}$ denote the ordinates in the points $a, a + h, \ldots, a + 2nh = b$.

EXAMPLE

$$\int_0^1 e^{-x} \, dx \simeq \frac{1}{6} (1 + 4e^{-1/2} + e^{-1}) = 0.63233368$$

(Simpson); error: $2.13 \cdot 10^{-4}$.

$$\int_0^1 e^{-x} \, dx \simeq \frac{1}{30} (7 + 16e^{-1/2} + 7e^{-1}) + \frac{1}{60} (-1 + e^{-1}) = 0.63211955$$

(Simpson with end correction); error: $-1.01 \cdot 10^{-6}$. In the latter case the error is 211 times less than in the former. A similar correction can be performed for the trapezoidal formula. The result is the same as though Euler-Maclaurin's summation formula were truncated (see Section 11.2).

   If we want to compute an integral, using three ordinates in nonequidistant points, the following modification of Simpson's formula, suggested by V. Brun, may be used:

$$\int_{x_0}^{x_0+a+b} y \, dx \simeq \frac{a + b}{6} (y_0 + 4y_1 + y_2) + \frac{b - a}{3} (y_2 - y_0) \,.$$

The Cote formulas just described are of closed type, since the end-point ordinates also enter the formulas. If this is not the case, we get open-type formulas; the coefficients can be computed as above.

Table of coefficients of open-type quadrature formulas

n	N	$NC_1^{(n)}$	$NC_2^{(n)}$	$NC_3^{(n)}$	$NC_4^{(n)}$	$NC_5^{(n)}$	$NC_6^{(n)}$	Remainder term
1	2	1	1					$2.8 \cdot 10^{-2} (b - a)^3 f''(\xi)$
2	3	2	-1	2				$3.1 \cdot 10^{-4} (b - a)^5 f^{IV}(\xi)$
3	24	11	1	1	11			$2.2 \cdot 10^{-4} (b - a)^5 f^{IV}(\xi)$
4	20	11	-14	26	-14	11		$1.1 \cdot 10^{-6} (b - a)^7 f^{VI}(\xi)$
5	1440	611	-453	562	562	-453	611	$7.4 \cdot 10^{-7} (b - a)^7 f^{VI}(\xi)$

For example, we have approximately:

$$\int_{-3h}^{+3h} y \, dx = \frac{6h}{20} (11y_{-2} - 14y_{-1} + 26y_0 - 14y_1 + 11y_2) .$$

If $\int_0^{1.2} e^{-x} \, dx$ is computed in this way, we obtain 0.6988037, compared with the exact value $1 - e^{-1.2} = 0.6988058$. The remainder term is $3.9 \cdot 10^{-6}$, compared with the actual error $2.1 \cdot 10^{-6}$.

Romberg's method

In this method we shall make use of Richardson extrapolation in a systematic way. Suppose that we start with the trapezoidal rule for evaluating $\int_a^b f(x) \, dx$. Using the interval $h_m = (b - a) \cdot 2^{-m}$, we denote the result with A_m. Since the error is proportional to h_m^2, we obtain an improved value if we form

$$B_m = A_m + \frac{A_m - A_{m-1}}{3} , \qquad m \geq 1 .$$

As a matter of fact, B_m is the same result as obtained from Simpson's formula. Now we make use of the fact that the error in B_m is proportional to h_m^4 and form

$$C_m = B_m + \frac{B_m - B_{m-1}}{15} , \qquad m \geq 2 .$$

This formula, incidentally, is identical with the Cote formula for $n = 4$. Since the error is proportional to h_m^6, we form

$$D_m = C_m + \frac{C_m - C_{m-1}}{63} , \qquad m \geq 3 .$$

The error is now proportional to h_m^8, but the formula is not of the Cote type any longer. The same process can be repeated again; we stop when two successive values are sufficiently close to each other. Further details of the method can be found in [5] and [6].

EXAMPLE. $f(x) = 1/x$; $a = 1$, $b = 2$. We find the following values:

m	A_m	B_m	C_m	D_m	E_m
0	0.75				
1	0.70833 33333	0.69444 44444			
2	0.69702 38095	0.69325 39682	0.69317 46031		
3	0.69412 18504	0.69315 45307	0.69314 79015	0.69314 74777	
4	0.69339 12022	0.69314 76528	0.69314 71943	0.69314 71831	0.69314 71819

The value of the integral is $\log 2 = 0.69314\ 71805$.

Filon's formula

As mentioned above, numerical quadrature is "easy" compared with numerical differentiation. However, there are exceptional cases, and one such case is obtained if the integrand is an oscillating function. A special quadrature formula of the Cote type for this case has been obtained by Filon:

$$\int_{x_0}^{x_{2n}} f(x) \cos tx \, dx = h\left[\alpha(th)\{f_{2n} \sin tx_{2n} - f_0 \sin tx_0\}\right.$$
$$\left. + \beta(th)C_{2n} + \gamma(th)C_{2n-1} + \frac{2}{45} th^4 S'_{2n-1}\right] - R_n \, ,$$

$$\int_{x_0}^{x_{2n}} f(x) \sin tx \, dx = h\left[\alpha(th)\{f_0 \cos tx_0 - f_{2n} \cos tx_{2n}\}\right.$$
$$\left. + \beta(th)S_{2n} + \gamma(th)S_{2n-1} + \frac{2}{45} th^4 C'_{2n-1}\right] - R_n \, .$$

Here

$$C_{2n} = \sum_{i=0}^{n} f_{2i} \cos tx_{2i} - \frac{1}{2}\left[f_{2n} \cos tx_{2n} + f_0 \cos tx_0\right],$$

$$C_{2n-1} = \sum_{i=1}^{n} f_{2i-1} \cos tx_{2i-1}, \qquad C'_{2n-1} = \sum_{i=1}^{n} f'''_{2i-1} \cos tx_{2i-1};$$

$$S_{2n} = \sum_{i=0}^{n} f_{2i} \sin tx_{2i} - \frac{1}{2}\left[f_{2n} \sin tx_{2n} + f_0 \sin tx_0\right],$$

$$S_{2n-1} = \sum_{i=1}^{n} f_{2i-1} \sin tx_{2i-1}, \qquad S'_{2n-1} = \sum_{i=1}^{n} f'''_{2i-1} \sin tx_{2i-1};$$

$$2nh = x_{2n} - x_0 .$$

The functions α, β, and γ are defined through

$$\alpha(\theta) = \frac{1}{\theta} + \frac{\sin 2\theta}{2\theta^2} - \frac{2\sin^2 \theta}{\theta^3} \, ,$$

$$\beta(\theta) = 2\left(\frac{1 + \cos^2 \theta}{\theta^2} - \frac{\sin 2\theta}{\theta^3}\right),$$

$$\gamma(\theta) = 4\left(\frac{\sin \theta}{\theta^3} - \frac{\cos \theta}{\theta^2}\right).$$

The remainder term, finally, has the form

$$R_n = \frac{1}{90} n h^5 f^{\mathrm{IV}}(\xi) + O(th^7) ; \qquad x_0 < \xi < x_{2n} .$$

10.2. Gauss' quadrature formulas

In deducing the Cote formulas, we made use of the ordinates in equidistant points, and the weights were determined in such a way that the formula obtained as high an accuracy as possible. We shall now choose abscissas without any special conditions and make the following attempt:

$$\int_a^b w(x) f(x) \, dx = A_1 f(x_1) + A_2 f(x_2) + \cdots + A_n f(x_n) + R_n . \qquad (10.2.1)$$

Here the weights A_i, as well as the abscissas x_i, are at our disposal. Further, $w(x)$ is a weight function, which will be specialized later on; so far, we only suppose that $w(x) \geq 0$. We shall use the same notations as in Lagrange's and Hermite's interpolation formulas, namely,

$$F(x) = \prod_{k=1}^n (x - x_k) ; \qquad F_k(x) = \frac{F(x)}{x - x_k} ; \qquad L_k(x) = \frac{F_k(x)}{F_k(x_k)} .$$

Now we form the Lagrangian interpolation polynomial $P(x)$:

$$P(x) = \sum_{k=1}^n L_k(x) f(x_k) , \qquad (10.2.2)$$

and find $f(x) - P(x) = 0$ for $x = x_1, x_2, \ldots, x_n$.
   Hence we can write

$$f(x) = P(x) + F(x)(\alpha_0 + \alpha_1 x + \alpha_2 x^2 + \cdots) ,$$

which inserted into (10.2.1), gives

$$\int_a^b w(x) f(x) \, dx$$

$$= \sum_{k=1}^n \left(\int_a^b w(x) L_k(x) \, dx \right) f(x_k) + \int_a^b w(x) F(x) \{ \alpha_0 + \alpha_1 x + \alpha_2 x^2 + \cdots \} \, dx$$

$$= \sum_{k=1}^n A_k f(x_k) + R_n .$$

Identifying, we obtain

$$\begin{cases} A_k = \int_a^b w(x) L_k(x) \, dx , \\ R_n = \int_a^b w(x) F(x) \sum_{r=0}^\infty \alpha_r x^r \, dx . \end{cases} \qquad (10.2.3)$$

Here the abscissas $x_1, x_2, \ldots, x_n$ are still at our disposal, but as soon as they are given, the weight constants A_k are also determined.

For determination of the abscissas we can formulate n more conditions to be fulfilled, and we choose the following:

$$\int_a^b w(x)F(x)x^r \, dx = 0 \; ; \qquad r = 0, 1, 2, \ldots, n - 1 \; . \qquad (10.2.4)$$

Since $F(x) = \prod (x - x_k)$, we get n equations which can be solved without any special difficulties, at least for small values of n. We take as example $w(x) = 1$, $a = 0$, $b = 1$, and $n = 3$, putting

$$(x - x_1)(x - x_2)(x - x_3) \equiv x^3 - s_1 x^2 + s_2 x - s_3 \; .$$

The equations are:

$$\left\{ \begin{aligned} \frac{1}{4} - \frac{s_1}{3} + \frac{s_2}{2} - s_3 &= 0 \, , \\ \frac{1}{5} - \frac{s_1}{4} + \frac{s_2}{3} - \frac{s_3}{2} &= 0 \, , \\ \frac{1}{6} - \frac{s_1}{5} + \frac{s_2}{4} - \frac{s_2}{3} &= 0 \, , \end{aligned} \right.$$

with the solution $s_1 = \frac{3}{2}$, $s_2 = \frac{3}{5}$, and $s_3 = \frac{1}{20}$. The abscissas are obtained from the equation

$$x^3 - \frac{3}{2} x^2 + \frac{3}{5} x - \frac{1}{20} = 0 \, ,$$

and we find

$$x_1 = 0.5 - \sqrt{0.15} = 0.1127 \; ;$$
$$x_2 = 0.5 \; ;$$
$$x_3 = 0.5 + \sqrt{0.15} = 0.8873 \; .$$

In order to get a theoretically more complete and more general discussion we are now going to treat the problem from another point of view. Integrating Hermite's interpolation formula (8.2.5) and (8.2.9) we get

$$\int_a^b w(x)f(x) \, dx = \sum_{k=1}^n B_k f(x_k) + \sum_{k=1}^n C_k f'(x_k) + E \, , \qquad (10.2.5)$$

where

$$\left\{ \begin{aligned} B_k &= \int_a^b w(x)[1 - 2L_k'(x_k)(x - x_k)][L_k(x)]^2 \, dx \, , \\ C_k &= \int_a^b w(x)(x - x_k)[L_k(x)]^2 \, dx \, , \\ E &= \int_a^b w(x) \frac{f^{(2n)}(\xi)}{(2n)!} [F(x)]^2 \, dx \, . \end{aligned} \right. \qquad (10.2.6)$$

First we will consider the error term, and since $w(x) \geq 0$, we can use the first mean-value theorem of integral calculus:

$$E = \frac{f^{(2n)}(\eta)}{(2n)!} \int_a^b w(x)[F(x)]^2 \, dx \; . \tag{10.2.7}$$

Here $a < \xi$, $\eta < b$ and further ξ is a function of x.

So far, the abscissas $x_1, x_2, \ldots, x_n$ are not restricted, but now we shall examine whether they can be chosen in such a way that the n constants C_k vanish. First we transform C_k [see (8.1.3)]:

$$C_k = \int_a^b w(x)[(x - x_k)L_k(x)]L_k(x) \, dx = \frac{1}{F'(x_k)} \int_a^b w(x)F(x)L_k(x) \, dx = 0 \; .$$

We can say that the n polynomials $L_k(x)$ of degree $n - 1$ must be orthogonal to $F(x)$ with the weight function $w(x)$. It is also easily understood that the n powers $1, x, x^2, \ldots, x^{n-1}$ must also be orthogonal to $F(x)$ in the same way. Then (10.2.5) transforms to

$$\int_a^b w(x)f(x) \, dx = \sum_{k=1}^n B_k f(x_k) + \frac{f^{(2n)}(\eta)}{(2n)!} \int_a^b w(x)[F(x)]^2 \, dx \; .$$

The abscissas $x_1, x_2, \ldots, x_n$ are now determined in principle through the conditions $C_k = 0$ which are equivalent to (10.2.4). Further we see that A_k in (10.2.3) can be identified with B_k in (10.2.6). Rewriting B_k we get directly

$$B_k = \int_a^b w(x)[L_k(x)]^2 \, dx - 2L_k'(x_k)C_k = \int_a^b w(x)[L_k(x)]^2 \, dx \; .$$

Thus we have the following double formula for A_k:

$$A_k = \int_a^b w(x)L_k(x) \, dx = \int_a^b w(x)[L_k(x)]^2 \, dx \; . \tag{10.2.8}$$

An important conclusion can be drawn at once from this relation: all weights A_k are > 0. Again, the condition for the validity of (10.2.7) is that the relation (10.2.4) is satisfied. Thus R_n vanishes if $f(x)$ is a polynomial of degree $\leq 2n-1$.

So far we have not specialized the weight function $w(x)$, but this is necessary if we want explicit formulas. We start with the simple case $w(x) = 1$, treated already by Gauss. The natural interval turns out to be $-1 \leq x \leq 1$. For determination of the abscissas, we have the equation

$$\int_{-1}^{+1} F(x)x^r \, dx = 0 \; ; \qquad r = 0, 1, 2, \ldots, n - 1 \; . \tag{10.2.9}$$

If n is not too large, this equation allows the direct determination of $F(x) = (x - x_1)(x - x_2) \cdots (x - x_n)$, but the calculation is done more conveniently by use of the Legendre polynomials. These are defined by

$$P_0(x) = 1 \; ; \qquad P_n(x) = \frac{1}{2^n \cdot n!} \frac{d^n}{dx^n}(x^2 - 1)^n \; .$$

Then $\int_{-1}^{+1} x^r P_n(x) \, dx = 0$ if r is such an integer that $0 \leq r < n$, because

$$\int_{-1}^{+1} x^r \frac{d^n}{dx^n} (x^2 - 1)^n \, dx$$

$$= \left[x^r \frac{d^{n-1}}{dx^{n-1}} (x^2 - 1)^n \right]_{-1}^{+1} - \int_{-1}^{+1} r x^{r-1} \frac{d^{n-1}}{dx^{n-1}} (x^2 - 1)^n \, dx$$

$$\underbrace{\qquad\qquad\qquad\qquad\qquad}_{=0}$$

$$= \cdots = (-1)^r \cdot r! \int_{-1}^{+1} \frac{d^{n-r}}{dx^{n-r}} (x^2 - 1)^n \, dx = 0 \; .$$

If $r = n$, we obtain:

$$(-1)^n \cdot n! \int_{-1}^{+1} (x^2 - 1)^n \, dx = 2 \cdot n! \int_0^1 (1 - x^2)^n \, dx$$

$$= 2n! \int_0^{\pi/2} \cos^{2n+1} \varphi \, d\varphi$$

$$= 2n! \, \frac{2n(2n - 2) \cdots 2}{(2n + 1)(2n - 1) \cdots 3}$$

$$= \frac{2^{2n+1}(n!)^3}{(2n + 1)!} \; .$$

Now we can compute

$$\int_{-1}^{+1} [P_n(x)]^2 \, dx = \int_{-1}^{+1} \left(\frac{1}{2^n n!} \frac{(2n)!}{n!} x^n + \cdots \right) \frac{1}{2^n n!} \frac{d^n}{dx^n} (x^2 - 1)^n \, dx$$

$$= \frac{(2n)!}{2^{2n}(n!)^3} \frac{2^{2n+1}(n!)^3}{(2n + 1)!} = \frac{2}{2n + 1} \, ,$$

since powers below x^n give no contributions to the integral. Hence we have the following relations:

$$\int_{-1}^{+1} x^r P_n(x) \, dx = 0 \; ; \qquad r = 0, 1, 2, \ldots, n - 1 \; ; \qquad (10.2.10)$$

$$\int_{-1}^{+1} P_m(x) P_n(x) \, dx = 0 \; ; \qquad m \neq n \; . \qquad (10.2.11)$$

$$\int_{-1}^{+1} [P_n(x)]^2 \, dx = \frac{2}{2n + 1} \; . \qquad (10.2.12)$$

The relation (10.2.11) follows from (10.2.10), since if, for example, $m < n$, $P_m(x)$ is a polynomial of degree m, where every power separately is annihilated by integration with $P_n(x)$.

Further, we see that $(x^2 - 1)^n$ is a polynomial of degree $2n$, which vanishes n times in $x = -1$ and n times in $x = +1$. Differentiating once, we obtain a polynomial of degree $2n - 1$, which vanishes $(n - 1)$ times in $x = -1$ and $(n - 1)$ times in $x = +1$, and according to Rolle's theorem, once in a point in the interval between -1 and $+1$. A similar argument can be repeated, and for each differentiation, the number of "interior" zeros increases by 1. Hence it is inferred that the polynomial $P_n(x)$ of degree n has exactly n zeros in the interval $(-1, +1)$. Thus it is clear that we can satisfy equation (10.2.9) by putting $F(x) = CP_n(x)$, where C is a constant. Hence we obtain $x_1, x_2, \ldots, x_n$ from the equation

$$P_n(x) = 0 . \tag{10.2.13}$$

The constant C is determined in such a way that the coefficient of the x^n-term is 1, that is $C = 2^n(n!)^2/(2n)!$. The first Legendre polynomials are

$$P_0(x) = 1 ; \qquad\qquad\qquad P_1(x) = x ;$$

$$P_2(x) = \frac{1}{2}(3x^2 - 1) ; \qquad\qquad P_3(x) = \frac{1}{2}(5x^3 - 3x) ;$$

$$P_4(x) = \frac{1}{8}(35x^4 - 30x^2 + 3) ; \qquad P_5(x) = \frac{1}{8}(63x^5 - 70x^3 + 15x) ; \ldots .$$

Higher polynomials can be obtained form the recursion formula

$$P_{k+1}(x) = \frac{2k+1}{k+1} xP_k(x) - \frac{k}{k+1} P_{k-1}(x) .$$

From (10.2.8) we obtain for the weights,

$$A_k = \int_{-1}^{+1} L_k(x)\, dx = \int_{-1}^{+1} [L_k(x)]^2\, dx .$$

Hence

$$A_k = \frac{1}{P_n'(x_k)} \int_{-1}^{+1} \frac{P_n(x)}{x - x_k}\, dx . \tag{10.2.14}$$

Using the properties of the Legendre polynomials, we can, after somewhat complicated calculations, transform A_k to the form

$$A_k = \frac{1}{[P_n'(x_k)]^2} \frac{2}{1 - x_k^2} . \tag{10.2.15}$$

Finer details are given, for example, in [4], p. 321.

Using (10.2.7) and (10.2.12), and changing η to ξ, we obtain the error term

$$R_n = \frac{f^{(2n)}(\xi)}{(2n)!} C^2 \int_{-1}^{+1} [P_n(x)]^2\, dx = \frac{2^{2n+1}(n!)^4}{(2n+1)[(2n)!]^3} f^{(2n)}(\xi) . \tag{10.2.16}$$

In this form, however, the error term is of limited use. Since $R_n = \alpha_n f^{(2n)}(\xi)$, halving the interval with n unchanged will give rise to a factor 2^{2n}, if we can assume that $f^{(2n)}(\xi)$ varies slowly. As before, we can apply Richardson's extrapolation to increase the accuracy.

Abscissas and weights in the Gauss-Legendre quadrature formula

n	x_k	A_k
2	$\pm$0.57735 02691 89626	1.00000 00000 00000
3	$\pm$0.77459 66692 41483	0.55555 55555 55556
	0.00000 00000 00000	0.88888 88888 88889
4	$\pm$0.86113 63115 94053	0.34785 48451 37454
	$\pm$0.33998 10435 84856	0.65214 51548 62546
5	$\pm$0.90617 98459 38664	0.23692 68850 56189
	$\pm$0.53846 93101 05683	0.47862 86704 99366
	0.00000 00000 00000	0.56888 88888 88889
6	$\pm$0.93246 95142 03152	0.17132 44923 79170
	$\pm$0.66120 93864 66265	0.36076 15730 48139
	$\pm$0.23861 91860 83197	0.46791 39345 72691

The table reproduced above has been taken from a paper by Lowan, Davids, and Levenson [1] in which values are given up to $n = 16$.

If an integral has to be computed between other limits, a suitable linear transformation must be performed:

$$\int_a^b f(x)\, dx = \frac{b-a}{2} \sum_{k=1}^{n} A_k f\left(\frac{b-a}{2} x_k + \frac{b+a}{2}\right) + R_n \, .$$

Hence $f^{(2n)}(\xi)$ will obtain a factor $[(b-a)/2]^{2n}$, and in the special case when the interval is halved, the error term will decrease by a factor 2^{2n}. For practical use this rule seems to be the most realistic one, while the error term (10.2.16) is mainly of theoretical interest. We also note that Richardson extrapolation can be performed as usual.

Next we consider the case where the weight function $w(x)$ is $=e^{-x}$. Using the same technique as before, we obtain the equations

$$\begin{cases} \int_0^\infty F(x)e^{-x}x^r\, dx = 0 & \text{for} \quad r = 0, 1, 2, \ldots, (n-1) \\ A_k = \frac{1}{F'(x_k)} \int_0^\infty \frac{F(x)e^{-x}}{x - x_k}\, dx \, . \end{cases} \tag{10.2.17}$$

Hence the formula

$$\int_0^\infty e^{-x} f(x)\, dx = \sum_{k=1}^{n} A_k f(x_k)$$

becomes exact for polynomials of degree up to $2n - 1$. The first equation in (10.2.17) fits the Laguerre polynomials

$$L_n(x) = e^x \frac{d^n}{dx^n}\left(e^{-x}x^n\right), \tag{10.2.18}$$

which satisfy the equation

$$\int_0^\infty e^{-x} x^r L_n(x)\, dx = 0 \qquad \text{for} \quad r = 0, 1, 2, \ldots, (n-1)\,. \qquad (10.2.19)$$

The abscissas $x_1, x_2, \ldots, x_n$ are obtained from the equation

$$L_n(x) = 0\,,$$

and further it can be shown that the weights and the remainder term can be written

$$A_k = \frac{(n!)^2}{x_k [L_n'(x_k)]^2}\,; \qquad R_n = \frac{(n!)^2}{(2n)!} f^{(2n)}(\xi)\,. \qquad (10.2.20)$$

A table of x_k and A_k is given below.

Abscissas and weights in the Gauss-Laguerre quadrature formula

n	x_k	A_k
2	0.58578 64376 27	0.85355 33905 93
	3.41421 35623 73	0.14644 66094 07
3	0.41577 45567 83	0.71109 30099 29
	2.29428 03602 79	0.27851 77335 69
	6.28994 50829 37	0.01038 92565 016
4	0.32254 76896 19	0.60315 41043 42
	1.74576 11011 58	0.35741 86924 38
	4.53662 02969 21	0.03888 79085 150
	9.39507 09123 01	0.00053 92947 05561
5	0.26356 03197 18	0.52175 56105 83
	1.41340 30591 07	0.39866 68110 83
	3.59642 57710 41	0.07594 24496 817
	7.08581 00058 59	0.00361 17586 7992
	12.64080 08442 76	0.00002 33699 723858

The table above has been taken from Salzer-Zucker [2], where values are given up to $n = 15$.

Last we consider the case $w = e^{-x^2}$. Then we obtain the abscissas as roots of the Hermitian polynomials, defined by

$$H_n(x) = (-1)^n e^{x^2} \frac{d^n}{dx^n} (e^{-x^2})\,. \qquad (10.2.21)$$

The weights and the remainder term are

$$\begin{cases} A_k = \dfrac{2^{n+1} \cdot n!\, \sqrt{\pi}}{[H_n'(x_k)]^2}\,, \\[2mm] R_n = \dfrac{n!\, \sqrt{\pi}}{2^n (2n)!} f^{(2n)}(\xi)\,. \end{cases} \qquad (10.2.22)$$

The values listed at the top of p. 213 have been taken from Salzer, Zucker, and Capuano [3].

Finally, we shall discuss the case $w(x) = 1/\sqrt{1 - x^2}$ in the interval $(-1, 1)$. The condition

$$\int_{-1}^{+1} \frac{1}{\sqrt{1 - x^2}} F(x) x^r \, dx = 0, \qquad r = 0, 1, 2, \ldots, (n - 1)$$

fits the Chebyshev polynomials $T_n(x) = \cos(n \arccos x)$, which will be treated in some detail in Chapter 16. Here we can compute the abscissas explicitly, since from $T_n(x) = 0$ we get $x_k = \cos\left((2k - 1)\pi/2n\right)$, $k = 1, 2, \ldots, n$. The weights take the amazingly simple form

$$A_k = \frac{\pi}{n}, \tag{10.2.23}$$

that is, all weights are equal. Hence the final formula reads

$$\int_{-1}^{+1} \frac{f(x)}{\sqrt{1 - x^2}} \, dx = \frac{\pi}{n} \sum_{k=1}^{n} f\left(\cos \frac{2k - 1}{2n} \pi\right) + \frac{2\pi}{2^{2n}(2n)!} f^{(2n)}(\xi), \tag{10.2.24}$$

where $-1 < \xi < 1$.

Summing up, we have the following Gauss formulas:

$$\int_{-1}^{+1} f(x) \, dx = 2 \sum_{k=1}^{n} \frac{f(x_k)}{(1 - x_k^2)[P_n'(x_k)]^2} + \frac{2^{2n+1}(n!)^4}{(2n + 1)[(2n)!]^3} f^{(2n)}(\xi)$$

$$x_1, x_2, \ldots, x_n \qquad \text{roots of} \qquad P_n(x) = 0.$$

$$\int_{0}^{\infty} e^{-x} f(x) \, dx = (n!)^2 \sum_{k=1}^{n} \frac{f(x_k)}{x_k [L_n'(x_k)]^2} + \frac{(n!)^2}{(2n)!} f^{(2n)}(\xi)$$

$$x_1, x_2, \ldots, x_n \qquad \text{roots of} \qquad L_n(x) = 0.$$

$$\int_{-\infty}^{+\infty} e^{-x^2} f(x) \, dx = 2^{n+1} \cdot n! \sqrt{\pi} \sum_{k=1}^{n} \frac{f(x_k)}{[H_n'(x_k)]^2} + \frac{n! \sqrt{\pi}}{2^n(2n)!} f^{(2n)}(\xi)$$

$$x_1, x_2, \ldots, x_n \qquad \text{roots of} \qquad H_n(x) = 0.$$

$$\int_{-1}^{+1} \frac{f(x)}{\sqrt{1 - x^2}} \, dx = \frac{\pi}{n} \sum_{k=1}^{n} f\left(\cos \frac{2k - 1}{2n} \pi\right) + \frac{2\pi}{2^{2n}(2n)!} f^{(2n)}(\xi).$$

We will devote a few words to the usefulness of the Gaussian formulas. It is true that for hand computation formulas of the Cote type are preferred in spite of the larger error terms. The reasons are essentially twofold: first, the abscissas in the Gaussian formulas are not equidistant, which, as a rule, necessitates interpolation. Second, the weights are not integers, and hence a time-consuming multiplication has to be performed. If an automatic computer is available, these reasons are no longer decisive, at least if the integrand has to be computed at each step. As a matter of fact, the interest in Gaussian methods has grown considerably in later years.

Abscissas and weights in the Gauss-Hermite quadrature formula

n	x_k	A_k
2	$\pm 0.70710\ 67811\ 87$	$0.88622\ 69254\ 53$
3	0	$1.18163\ 59006\ 04$
	$\pm 1.22474\ 48713\ 92$	$0.29540\ 89751\ 51$
4	$\pm 0.52464\ 76232\ 75$	$0.80491\ 40900\ 06$
	$\pm 1.65068\ 01238\ 86$	$0.08131\ 28354\ 473$
5	0	$0.94530\ 87204\ 83$
	$\pm 0.95857\ 24646\ 14$	$0.39361\ 93231\ 52$
	$\pm 2.02018\ 28704\ 56$	$0.01995\ 32420\ 591$
6	$\pm 0.43607\ 74119\ 28$	$0.72462\ 95952\ 24$
	$\pm 1.33584\ 90740\ 14$	$0.15706\ 73203\ 23$
	$\pm 2.35060\ 49736\ 74$	$0.00453\ 00099\ 0551$

10.3. Chebyshev's formulas

As is easily understood, the Cote formulas constitute a special case of the Gaussian formulas: namely, equidistant abscissas are given in advance. Now we can also explain a peculiarity in the Cote formula's error terms. For $n = 2$ (Simpson's formula) and $n = 3$, we have essentially the same error term. The reason is that in the first case, when we have three abscissas, one of them (in the middle of the interval) happens to coincide with the Gaussian value, and this is reflected as an improvement in the error term. The corresponding phenomenon is repeated for all even values of n which involve an odd number of abscissas.

We will now specialize in such a way that all weights are equal. The corresponding formulas were obtained by Chebyshev. Let $w(x)$ be a weight function and put

$$\int_{-1}^{+1} w(x)f(x)\, dx = k[\, f(x_1) + f(x_2) + \cdots + f(x_n)] + R_n \,.$$

We suppose that $f(x)$ can be expanded in a Maclaurin series:

$$f(x) = f(0) + xf'(0) + \frac{x^2}{2!} f''(0) + \cdots + \frac{x^n}{n!} f^{(n)}(0) + \frac{x^{n+1}}{(n+1)!} f^{(n+1)}(\xi) \,.$$

Inserting, we obtain

$$\int_{-1}^{+1} w(x)f(x)\, dx$$

$$= \sum_{r=0}^{n} \frac{f^{(r)}(0)}{r!} \int_{-1}^{+1} w(x)x^r\, dx + \frac{1}{(n+1)!} \int_{-1}^{+1} w(x)f^{(n+1)}(\xi)x^{n+1}\, dx \,,$$

(note that ξ is a function of x). On the other hand, we expand $f(x_1)$,

$f(x_2), \ldots, f(x_n)$ in the Maclaurin series:

$$k[f(x_1) + \cdots + f(x_n)] = k\left[nf(0) + f'(0)\sum x^r + \frac{1}{2}f''(0)\sum x_r^2 + \cdots \right.$$

$$\left. + \frac{1}{n!}f^{(n)}(0)\sum x_r^n + \frac{1}{(n+1)!}\sum f^{(n+1)}(\xi_r)x_r^{n+1}\right].$$

Identifying, we obtain the following system:

$$\left\{\begin{array}{l} kn = \displaystyle\int_{-1}^{+1} w(x)\,dx , \\[2mm] k\sum x_r = \displaystyle\int_{-1}^{+1} xw(x)\,dx , \\ \vdots \\ k\sum x_r^n = \displaystyle\int_{-1}^{+1} x^n w(x)\,dx . \end{array}\right. \qquad (10.3.1)$$

Putting $S_m = \sum x_r^m$, we get

$$S_m = \frac{1}{k}\int_{-1}^{+1} x^m w(x)\,dx = n\int_{-1}^{+1} x^m w(x)\,dx \Big/ \int_{-1}^{+1} w(x)\,dx .$$

As in (8.1.2) we introduce the function $F(z)$ and obtain:

$$F(z) = \prod_{r=1}^{n}(z - x_r) = z^n\prod_{r=1}^{n}\left(1 - \frac{x_r}{z}\right) = z^n\exp\left(\sum_{r=1}^{n}\log\left(1 - \frac{x_r}{z}\right)\right)$$

$$= z^n\exp\left(\sum_{r=1}^{n}\left(-\sum_{t=1}^{\infty}\frac{x_r^t}{tz^t}\right)\right) = z^n\exp\left(-\sum_{t=1}^{\infty}\frac{S_t}{tz^t}\right)$$

$$= z^n\exp\left(-\sum_{t=1}^{\infty}\frac{1}{tz^t}n\int_{-1}^{+1}x^t w(x)\,dx \Big/ \int_{-1}^{+1}w(x)\,dx\right)$$

$$= z^n\exp\left(n\int_{-1}^{+1}w(x)\left[-\sum_{t=1}^{\infty}\frac{x^t}{tz^t}\right]dx \Big/ \int_{-1}^{+1}w(x)\,dx\right)$$

$$= z^n\exp\left(n\int_{-1}^{+1}w(x)\log\left(1 - \frac{x}{z}\right)dx \Big/ \int_{-1}^{+1}w(x)\,dx\right).$$

If in particular $w(x) = 1$, we get

$$\int_{-1}^{+1}\log\left(1 - \frac{x}{z}\right)dx = -\int_{-1}^{+1}\sum_r \frac{x^r}{rz^r}\,dx$$

$$= -\sum_r\left[\frac{x^{r+1}}{r(r+1)z^r}\right]_{-1}^{+1} = -2\left[\frac{1}{2\cdot 3z^2} + \frac{1}{4\cdot 5z^4} + \cdots\right].$$

Our final result then takes the form

$$F(z) = z^n\exp\left(-n\left[\frac{1}{2\cdot 3z^2} + \frac{1}{4\cdot 5z^4} + \frac{1}{6\cdot 7z^6} + \cdots\right]\right). \qquad (10.3.2)$$

The left-hand side is a polynomial of degree n in z, and hence, on the right-hand side, we may include only nonnegative powers. The reason for this apparent contradiction is that we have substituted $(1/k) \int_{-1}^{+1} x^t w(x)\, dx$ for S_t even for $t > n$; that is, we have assumed our integration formula to be exact, which is the case only exceptionally. Expanding in power series we obtain

$$F(z) = z^n - \frac{n}{3!} z^{n-2} + \frac{n}{5!} \left(\frac{5n}{3} - 6 \right) z^{n-4}$$

$$- \frac{n}{7!} \left(\frac{35n^2}{9} - 42n + 120 \right) z^{n-6} + \cdots .$$

Explicitly, we find for different values of n:

$n = 1$ $F(z) = z$

$n = 2$ $F(z) = z^2 - \dfrac{1}{3}$

$n = 3$ $F(z) = z \left(z^2 - \dfrac{1}{2} \right)$

$n = 4$ $F(z) = z^4 - \dfrac{2}{3} z^2 + \dfrac{1}{45}$

$n = 5$ $F(z) = z \left(z^4 - \dfrac{5}{6} z^2 + \dfrac{7}{72} \right)$

$n = 6$ $F(z) = z^6 - z^4 + \dfrac{1}{5} z^2 - \dfrac{1}{105}$

$n = 7$ $F(z) = z \left(z^6 - \dfrac{7}{6} z^4 + \dfrac{119}{360} z^2 - \dfrac{149}{6480} \right)$

$n = 8$ $F(z) = z^8 - \dfrac{4}{3} z^6 + \dfrac{22}{45} z^4 - \dfrac{148}{2835} z^2 - \dfrac{43}{42525}$

$n = 9$ $F(z) = z \left(z^8 - \dfrac{3}{2} z^6 + \dfrac{27}{40} z^4 - \dfrac{57}{560} z^2 + \dfrac{53}{22400} \right)$

$n = 10$ $F(z) = z^{10} - \dfrac{5}{3} z^8 + \dfrac{8}{9} z^6 - \dfrac{100}{567} z^4 + \dfrac{17}{1701} z^2 - \dfrac{43}{56133}$

For $n = 1, 2, \ldots, 7$, all roots are real, but for $n = 8$, we meet the peculiar phenomenon that only two roots are real. For $n = 9$, all roots are real, but for $n = 10$, complex roots appear again. Do any values $n > 9$ exist for which the corresponding Chebyshev quadrature polynomial has only real roots? This question was left open by Chebyshev, but another Russian mathematician, Bernstein, has proved that this is not the case so long as we use the weight function $w(x) = 1$.

Another important weight function is $w(x) = 1/\sqrt{1 - x^2}$. In this case we have, as has been pointed out before, the weight $k = \pi/n$, while $F(z)$ has the form $F(z) = 2^{-n+1}T_n(z) = 2^{-n+1}\cos(n \arccos z)$, with the roots

$$x_r = \cos \frac{2r - 1}{2n} \pi .$$

Hence all roots are real, and, as a matter of fact, the formula

$$\int_{-1}^{+1} \frac{f(x)\,dx}{\sqrt{1 - x^2}} = \int_0^\pi f(\cos\theta)\,d\theta \simeq \frac{\pi}{n} \sum_{r=1}^n f\left(\cos\frac{2r - 1}{2n}\pi\right)$$

is also of Gaussian type. This means that the special nth-order Chebyshev formula above has the Gaussian precision $2n - 1$. The polynomials $T_n(z) = \cos(n \arccos z)$ are called Chebyshev polynomials and will be treated at some length in Chapter 16. Summing up, we have for $w(x) = 1$:

$$\int_{-1}^{+1} f(x)\,dx = \frac{2}{n} \sum_{r=1}^n f(x_r) + R_n ,$$

$$R_n = \frac{1}{(n + 1)!}\left[\int_{-1}^{+1} x^{n+1} f^{(n+1)}(\xi)\,dx - \frac{2}{n} \sum_{r=1}^n x_r^{n+1} f^{(n+1)}(\xi_r)\right];$$

$$-1 < \xi < 1 ; \quad -1 < \xi_r < x_r .$$

The real roots are collected in the table below:

n	x_r	n	x_r
2	$\pm 0.57735\ 02691$	6	$\pm 0.26663\ 54015$
			$\pm 0.42251\ 86538$
			$\pm 0.86624\ 68181$
3	0		
	$\pm 0.70710\ 67812$	7	0
			$\pm 0.32391\ 18105$
			$\pm 0.52965\ 67753$
4	$\pm 0.18759\ 24741$		$\pm 0.88386\ 17008$
	$\pm 0.79465\ 44723$		
		9	0
			$\pm 0.16790\ 61842$
5	0		$\pm 0.52876\ 17831$
	$\pm 0.37454\ 14096$		$\pm 0.60101\ 86554$
	$\pm 0.83249\ 74870$		$\pm 0.91158\ 93077$

EXAMPLE

The mean temperature shall be determined for a 24-hour period running from midnight to midnight, by aid of four readings. What times should be chosen?

Since we do not want to give one reading more weight than another, we choose a Chebyshev formula with $n = 4$ and find the times $12 \pm 12 \cdot 0.1876$ and $12 \pm 12 \cdot 0.7947$ or, approximately, 2.30 a.m., 9.45 a.m., 2.15 p.m., and 9.30 p.m.

10.4. Numerical quadrature by aid of differences

The problem of determining $\int_{x_0}^{x_1} y \, dx$ by use of usual forward differences can be solved by operator technique. We have

$$\int_{x_0}^{x_1} y \, dx = J_1 y_0 = \left[\frac{h\Delta}{\log(1 + \Delta)}\right] y_0 \,,$$

where $h = x_1 - x_0$ [cf. (7.1.4)]. Here we need the power-series expansion of the function $z/\log(1 + z)$. Putting

$$\frac{z}{\log(1 + z)} = \frac{1}{1 - z/2 + z^2/3 - z^3/4 + \cdots}$$

$$= 1 + b_1 z + b_2 z^2 + \cdots \,, \qquad (10.4.1)$$

we obtain the system of equations

$$\begin{cases} b_1 - \frac{1}{2} = 0 \,, \\ b_2 - \frac{1}{2}b_1 + \frac{1}{3} = 0 \,, \\ b_3 - \frac{1}{2}b_2 + \frac{1}{3}b_1 - \frac{1}{4} = 0 \,, \\ \vdots \end{cases} \qquad (10.4.2)$$

The coefficients are easily obtained recursively.

Alternatively, we can also get the desired result by integrating Newton's interpolation formula. Hence we find with $x = x_0 + ph$:

$$\int_{x_0}^{x_1} y \, dx = \int_{x_0}^{x_0+h} (1 + \Delta)^p y_0 \, dx = h \int_0^1 \sum_{k=0}^{\infty} \binom{p}{k} \Delta^k y_0 \, dp$$

$$= h \sum_{k=0}^{\infty} \frac{\Delta^k y_0}{k!} \int_0^1 p^{(k)} \, dp$$

$$= h \sum_{k=0}^{\infty} \frac{\Delta^k y_0}{k!} \int_0^1 \sum_{r=1}^{k} \alpha_r^{(k)} p^r \, dp$$

$$= h \sum_{k=0}^{\infty} \frac{1}{k!} \sum_{r=1}^{k} \frac{\alpha_r^{(k)}}{r + 1} \Delta^k y_0 \,.$$

Thus we get

$$b_k = \frac{1}{k!} \sum_{r=1}^{k} \frac{\alpha_r^{(k)}}{r + 1} \,, \qquad (10.4.3)$$

where we have used (8.5.1) and (7.3.5). Using the table of Stirling's numbers of the first kind in Section 7.3, we obtain

$$b_0 = 1 \,, \qquad b_4 = -\frac{19}{720} \,, \qquad b_8 = -\frac{33953}{3628800} \,,$$

$$b_1 = \frac{1}{2} \,, \qquad b_5 = \frac{3}{160} \,, \qquad b_9 = \frac{8183}{1036800} \,,$$

$$b_2 = -\frac{1}{12} \,, \qquad b_6 = -\frac{863}{60480} \,, \qquad b_{10} = -\frac{3250433}{479001600} \,,$$

$$b_3 = \frac{1}{24} \,, \qquad b_7 = \frac{275}{24192} \,, \qquad b_{11} = \frac{4671}{788480} \,.$$

Written explicitly, the integration formula has the form

$$\int_{x_0}^{x_1} y\,dx = h\left[y_0 + \frac{1}{2}\,\varDelta y_0 - \frac{1}{12}\,\varDelta^2 y_0 \right.$$
$$\left. + \frac{1}{24}\,\varDelta^3 y_0 - \frac{19}{720}\,\varDelta^4 y_0 + \frac{3}{160}\,\varDelta^5 y_0 - \cdots \right]. \qquad (10.4.4)$$

An analogous formula can be deduced for backward differences. A better result, however, is obtained in the following way.

$$\int_{x_0}^{x_0+nh} y\,dx = h\int_0^n E^p y_0\,dp$$
$$= h\,\frac{E^n - 1}{\log E}\,y_0 = h\left[-\frac{y_n}{\log(1 - \nabla)} - \frac{y_0}{\log(1 + \varDelta)} \right].$$

But

$$\begin{cases} -\dfrac{1}{\log(1 + \varDelta)} = -\dfrac{1}{\varDelta}\left(1 + \dfrac{1}{2}\,\varDelta - \dfrac{1}{12}\,\varDelta^2 + \dfrac{1}{24}\,\varDelta^3 - \dfrac{19}{720}\,\varDelta^4 + \cdots \right). \\[2mm] -\dfrac{1}{\log(1 - \nabla)} = \dfrac{1}{\nabla}\left(1 - \dfrac{1}{2}\,\nabla - \dfrac{1}{12}\,\nabla^2 - \dfrac{1}{24}\,\nabla^3 - \dfrac{19}{720}\,\nabla^4 - \cdots \right). \end{cases}$$

The first term in each of these expansions must be treated separately; conveniently we bring them together and obtain

$$\frac{y_n}{\nabla} - \frac{y_0}{\varDelta} = \frac{E^n y_0}{1 - E^{-1}} - \frac{y_0}{E - 1}$$
$$= \frac{E^{n+1} - 1}{E - 1}\,y_0 = y_0 + y_1 + y_2 + \cdots + y_n.$$

Hence we find

$$\int_{x_0}^{x_n} y\,dx = h\left[\frac{1}{2}\,y_0 + y_1 + y_2 + \cdots + y_{n-1} + \frac{1}{2}\,y_n - \frac{1}{12}\,(\nabla y_n - \varDelta y_0) \right.$$
$$- \frac{1}{24}\,(\nabla^2 y_n + \varDelta^2 y_0)$$
$$\left. - \frac{19}{720}\,(\nabla^3 y_n - \varDelta^3 y_0) - \frac{3}{160}\,(\nabla^4 y_n + \varDelta^4 y_0) - \cdots \right]$$
$$= h\left[\frac{1}{2}\,y_0 + y_1 + y_2 + \cdots + y_{n-1} + \frac{1}{2}\,y_n - \frac{1}{12}\,(\varDelta y_{n-1} - \varDelta y_0) \right.$$
$$- \frac{1}{24}\,(\varDelta^2 y_{n-2} + \varDelta^2 y_0)$$
$$\left. - \frac{19}{720}\,(\varDelta^3 y_{n-3} - \varDelta^3 y_0) - \frac{3}{160}\,(\varDelta^4 y_{n-4} + \varDelta^4 y_0) - \cdots \right].$$

This is Gregory's formula.

If this formula is used, for example, for $y = \pi^{-1/2}e^{-x^2}$ between $-\infty$ and $+\infty$, all differences will vanish in the limits, and we would obtain

$$\int_{-\infty}^{+\infty} y\,dx = h\left[\frac{1}{2}y_0 + y_1 + y_2 + \cdots + y_{n-1} + \frac{1}{2}y_n\right]$$

exactly, independently of h. Of course, this cannot be true, and the reason is that the remainder term does not tend to zero. Denoting the sum by S, we find

h	S	h	S
0.5	1.00000 00000 00000	0.9	1.00001 02160
0.6	1.00000 00000 02481	1	1.00010 34464
0.7	1.00000 00035	1.5	1.02489
0.8	1.00000 04014	2	1.16971

We also consider $\int_{-\infty}^{+\infty}\exp(-x^4)\,dx = \frac{1}{2}\Gamma(\frac{1}{4})$. For $h = 0.2$, we get 1.81280 49540, and for $h = 0.5$, 1.81362, while $\frac{1}{2}\Gamma(\frac{1}{4}) = 1.81280\ 49541\ 10954$. Gregory,s formula, as a rule, is not used for numerical quadrature but sometimes for summation.

We shall now consider quadrature formulas containing central differences:

$$\int_{x_0}^{x_1} y\,dx = J_0 y_{1/2} = \frac{h}{2}\varphi(\delta)(y_0 + y_1) = h\varphi(\delta)\mu y_{1/2}.$$

From (7.1.4) we have $J_0 = h\delta/2 \sinh^{-1}(\delta/2) = h\delta/U$, and hence, using (7.2.11), we have

$$\varphi(\delta) = \frac{\delta}{\mu U} = 1 - \frac{\delta^2}{12} + \frac{11\delta^4}{720} - \frac{191\delta^6}{60480} + \cdots.$$

Thus we get the following formula:

$$\int_{x_0}^{x_1} y\,dx = \frac{h}{2}(c_0 + c_2\delta^2 + c_4\delta^4 + \cdots)(y_0 + y_1). \qquad (10.4.5)$$

The constants c_{2k} are displayed in the table below.

$$c_0 = 1 \qquad\qquad\qquad = 1$$
$$c_2 = -\tfrac{1}{12} \qquad\qquad\ \ = -0.08333\ 33333$$
$$c_4 = \tfrac{11}{720} \qquad\qquad\quad = 0.01527\ 77778$$
$$c_6 = -\tfrac{191}{60480} \qquad\qquad = -0.00315\ 80688$$
$$c_8 = \tfrac{2497}{3628800} \qquad\quad = 0.00068\ 81063$$
$$c_{10} = -\tfrac{14797}{95800320} \qquad = -0.00015\ 44567$$
$$c_{12} = \tfrac{92427157}{261534873 6000} = 0.00003\ 53403$$

An alternative method is the following. We start from Everett's formula and integrate in p from 0 to 1, term by term (since $q = 1 - p$, the integration in q gives the same result):

$$y_p = \sum_{k=0}^{\infty} \binom{p+k}{2k+1} \delta^{2k} y_1 + \sum_{k=0}^{\infty} \binom{q+k}{2k+1} \delta^{2k} y_0 .$$

Treating the first term only, we get

$$\sum_{k=0}^{\infty} \binom{p+k}{2k+1} \delta^{2k} y_1 = \sum_{k=0}^{\infty} \frac{(p+k)^{(2k+1)}}{(2k+1)!} \delta^{2k} y_1$$

$$= \sum_{k=0}^{\infty} \frac{\delta^{2k} y_1}{(2k+1)!} \sum_{r=1}^{2k+1} \alpha_r^{(2k+1)} (p+k)^r .$$

Now the integration can be performed:

$$\int_0^1 (p+k)^r \, dp = \frac{(k+1)^{r+1} - k^{r+1}}{r+1} ,$$

and we obtain finally

$$c_{2k} = \frac{2}{(2k+1)!} \sum_{r=1}^{2k+1} \frac{(k+1)^{r+1} - k^{r+1}}{r+1} \alpha_r^{(2k+1)} , \qquad (10.4.6)$$

where a factor 2 has been added to compensate for the factor $h/2$ in (10.4.5); here we have further assumed that $h = 1$.

However, it is possible to obtain a formula in which the coefficients decrease much faster. Starting from $Jf(x) = \int_a^x f(t) \, dt$ [cf. (7.1.1)] and operating on both sides with

$$\frac{\mu \delta}{h} = \frac{E - E^{-1}}{2h} ,$$

we get on the left-hand side

$$\frac{1}{h} \mu \delta J = \frac{1}{h} \mu J_0 = \frac{1}{h} \mu \frac{h \delta}{2 \sinh^{-1} (\delta/2)} = \frac{\mu \delta}{U} ,$$

operating on $y = f(x)$, and on the right-hand side $(\frac{1}{2}h) \int_{x-h}^{x+h} y \, dt$. Using (7.2.10), we can write the result:

$$\int_{x_0-h}^{x_0+h} y \, dx = 2h \left(y_0 + \frac{1}{6} \delta^2 y_0 - \frac{1}{180} \delta^4 y_0 + \frac{1}{1512} \delta^6 y_0 - \cdots \right). \qquad (10.4.7)$$

This formula can be extended to larger intervals in a simple way. Especially if we truncate the expansion after the second term, we get

$$\int_{x_0-h}^{x_0+h} y \, dx \simeq 2h \left[y_0 + \frac{1}{6} \delta^2 y_0 \right] = 2h \left[y_0 + \frac{1}{6} y_1 - \frac{2}{6} y_0 + \frac{1}{6} y_{-1} \right]$$

$$= \frac{2h}{6} [y_{-1} + 4y_0 + y_1] ,$$

that is, Simpson's formula. If we truncate after the third, fourth, ... term, we do not obtain formulas of the Cote type any longer.

REFERENCES

[1] Lowan, Davids, Levenson: *Bull. Am. Math. Soc.*, **48**, 739 (1942).

[2] Salzer, Zucker: *Bull. Am. Math. Soc.*, **55**, 1004 (1949).

[3] Salzer, Zucker: Capuano: *Jour. Res.*, **48**, 11 (1952).

[4] Hildebrand: *Introduction to Numerical Analysis* (McGraw-Hill, New York, 1956).

[5] W. Romberg, "Vereinfachte numerische Integration," *Det Kgl. Norske Vid. Selsk. Forh.*, **28**, 30–36 (1955).

[6] Stiefel-Ruthishauser: "Remarques concernant l'integration numerique," *CR* (Paris), **252**, 1899–1900 (1961).

[7] V. I. Krylov: *Approximate Calculation of Integrals* (Macmillan, New York and London, 1962).

[8] L. N. G. Filon, *Proc. Roy. Soc. Edinburgh* **49**, 38–47 (1928)

[9] *NBS Handbook of Math. Functions* (1964) p. 890.

EXERCISES

1. The prime number theorem states that the number of primes in the interval $a < x < b$ is approximately $\int_a^b dx/\log x$. Use this for $a = 100$, $b = 200$, and compare with the exact value.

2. Calculate $\int_0^1 x^x \, dx$ correct to four places.

3. Find $\int_0^{\pi/2} (\cos x/(1 + x)) \, dx$ correct to four places.

4. Calculate $\int_0^{\pi/2} \sqrt{1 - 0.25 \sin^2 x} \, dx$ correct to five places.

5. A rocket is launched from the ground. Its acceleration is registered during the first 80 seconds and is given in the table below:

t (sec)	0	10	20	30	40	50	60	70	80
a (m/sec^2)	30.00	31.63	33.44	35.47	37.75	40.33	43.29	46.69	50.67

Find the velocity and the height of the rocket at the time $t = 80$.

6. Find $\int_0^1 \exp{(-\exp{(-x)})} \, dx$ correct to five places.

7. Calculate $\int_0^{\pi/2} \exp{(\sin x)} \, dx$ correct to four places.

8. Calculate $\int_0^1 ((\sin x)^{3/2}/x^2) \, dx$ correct to four places.

9. Two prime numbers p and q such that $|p - q| = 2$ are said to form a twin prime. For large values of n it is known that the number of twin primes less than n is asymptotically equal to $2C_2 \int_2^n dx/(\log x)^2$, where C_2 is a certain constant. Compute C_2 to 3 decimals using the fact that the number of twin primes less than 2^{20} is 8535.

10. Find

$$\int_0^1 \frac{(1 - e^{-x})^{1/2}}{x} \, dx$$

correct to four places.

11. Find $\int_0^1 \log x \cos x \, dx$ correct to five places.

12. Find

$$\int_0^1 \frac{dx}{\sqrt{e^x + x - 1}}$$

correct to four places.

13. Calculate the total arc length of the ellipse $x^2 + y^2/4 = 1$. The result should be accurate to six places.

14. A function $y = y(x)$ is known in equidistant points $0, \pm h, \pm 2h, \ldots$ Find an approximate value of $\int_{-2h}^{+2h} y \, dx$ which contains only y_{-1}, y_0, and y_1.

15. Find an integration formula of the type

$$\int_{-h}^{+h} y \, dx \simeq h[Ay_0 + B(y_1 + y_{-1}) + C(y_2 + y_{-2})],$$

where $y_1 = y(h)$, $y_2 = y(2h)$, and so on.

16. Find x_1 and x_2 such that the following approximate formula becomes exact to the highest possible order:

$$\int_{-1}^{+1} f(x) \, dx \simeq \frac{1}{3} [f(-1) + 2f(x_1) + 3f(x_2)].$$

17. One wants to construct a quadrature formula of the type

$$\int_0^h y \, dx = \frac{h}{2} (y_0 + y_1) + ah^2(y_0' - y_1') + R.$$

Calculate the constant a and find the order of the remainder term R.

18. In the interval $0 \le x \le \pi$, y is defined as a function of x by the relation

$$y = -\int_0^x \log \left(2 \sin \frac{t}{2}\right) dt.$$

Find the maximum value to five places.

19. Calculate $\int_0^\infty (e^{-x^2}/(1 + x^2)) \, dx$ correct to four places.

20. Find $\int_0^\infty (dx/\sqrt{e^x + x})$ correct to three places.

21. Find $\int_0^\infty e^{-x} \log x \, dx$ correct to five places.

22. Calculate numerically $\int_0^\infty (x \, dx/(e^x - 1))$ and $\int_0^\infty (x \, dx/(e^x + 1))$.

23. Find $\int_0^\infty e^{-(x+1/x)} \, dx$ correct to four places.

24. Determine abscissas and weights in the Gauss formula

$$\int_0^1 \sqrt{x} \, f(x) \, dx \simeq a_1 f(x_1) + a_2 f(x_2).$$

25. Determine abscissas and weights in the Gauss formula

$$\int_0^1 f(x) \log \frac{1}{x} \, dx \simeq A_1 f(x_1) + A_2 f(x_2).$$

26. Calculate the coefficients k_1, k_2, and k_3 in the approximate integration formula

$$\int_{-1}^{+1} f(x) \, dx \simeq k_1 f(-1) + k_2 f(1) + k_3 f(\alpha),$$

where α is a given number such that $-1 < \alpha < 1$. Then test the formula on the function $f(x) = [(5x + 13)/2]^{1/2}$ with $\alpha = -0.1$.

27. One wants to construct a quadrature formula of so-called Lobatto type:

$$\int_{-1}^{+1} f(x)\, dx = a[f(-1) + f(1)] + b[f(-\alpha) + f(\alpha)] + cf(0) + R \, .$$

Find a, b, c, and α such that R becomes of the highest possible order.

28. Determine the constants A, B, C, and D, and the remainder term R, in the following integration formula:

$$\int_{x_0}^{x_1} (x - x_0)f(x)\, dx = h^2(Af_0 + Bf_1) + h^3(Cf_0' + Df_1') + R \, ,$$

where $h = x_1 - x_0$.

29. Find a Chebyshev quadrature formula with three abscissas for the interval $0 \le x \le 1$, and the weight function $w(x) = 1/\sqrt{x}$.

30. The function $Si(x)$ is defined through $Si(x) = \int_0^x \sin t/t\, dt$. Calculate

$$\int_0^1 \frac{Si(x) - \sin x}{x^3}\, dx$$

correct to four places.

31. The integral equation $y(x) = 1 + \int_0^x f(t)y(t)\, dt$, where f is a given function, can be solved by forming a sequence of functions $y_0, y_1, y_2, \ldots$ according to

$$y_{n+1}(x) = 1 + \int_0^x f(t)y_n(t)\, dt \, .$$

Find the first five functions for $x = 0(0.25)1$, when $f(x)$ is as given in the table below. Start with $y_0 = 1$ and use Simpson's formula and Bessel's interpolation formula.

x	0	0.25	0.50	0.75	1
f	0.5000	0.4794	0.4594	0.4398	0.4207

Chapter 11

Summation

11.0. Bernoulli numbers and Bernoulli polynomials

In Chapter 7 we mentioned the Bernoulli numbers in connection with the factorials. The reason was that the Stirling numbers could be considered as special cases of the Bernoulli numbers. For the general definition it is convenient to make use of a so-called generating function. In (7.3.10) the Bernoulli numbers of order n were defined by the expansion

$$\frac{t^n}{(e^t - 1)^n} = \sum_{k=0}^{\infty} \frac{t^k}{k!} B_k^{(n)} . \tag{11.0.1}$$

The Bernoulli polynomials of order n and degree k are defined by

$$e^{xt} \frac{t^n}{(e^t - 1)^n} = \sum_{k=0}^{\infty} \frac{t^k}{k!} B_k^{(n)}(x) . \tag{11.0.2}$$

If in (11.0.2) we put $x = 0$, we obtain directly $B_k^{(n)}(0) = B_k^{(n)}$. Subsequently we shall exclusively treat the case $n = 1$, that is, Bernoulli numbers and polynomials of the first order. For brevity, we shall call them *Bernoulli numbers* and *Bernoulli polynomials*:

$$\frac{t}{e^t - 1} = \sum_{k=0}^{\infty} \frac{t^k}{k!} A_k , \tag{11.0.3}$$

$$e^{xt} \frac{t}{e^t - 1} = \sum_{k=0}^{\infty} \frac{t^k}{k!} B_k(x) . \tag{11.0.4}$$

Here we have $A_k = B_k(0)$; we use the notation A_k, since we want to reserve the notation B_k for other numbers closely associated with the numbers A_k. First we shall compute these latter numbers. From (11.0.3) we get

$$t = \sum_{k=0}^{\infty} A_k \frac{t^k}{k!} \cdot \sum_{r=1}^{\infty} \frac{t^r}{r!} = \sum_{s=1}^{\infty} \alpha_s t^s ,$$

224

where

$$\alpha_s = \sum_{i=0}^{s-1} \frac{A_i}{i!\,(s-i)!}\,.$$

Hence we have

$$s!\,\alpha_s = \sum_{i=0}^{s-1} \frac{s!}{i!\,(s-i)!}\,A_i = \sum_{i=0}^{s-1} \binom{s}{i} A_i\,.$$

But since the left-hand side was t, we have $\alpha_1 = 1$ and $\alpha_s = 0$, $s > 1$. Hence we obtain the following equation for determining A_k:

$$A_0 = 1\,;\qquad \sum_{i=0}^{s-1} \binom{s}{i} A_i = 0\,,\qquad s = 2, 3, 4,\ldots \qquad (11.0.5)$$

Written out explicitly, this is a linear system of equations which can be solved recursively without difficulty:

$$\begin{cases} A_0 = 1\,, \\ A_0 + 2A_1 = 0\,, \\ A_0 + 3A_1 + 3A_2 = 0\,, \\ A_0 + 4A_1 + 6A_2 + 4A_3 = 0\,, \\ \vdots \end{cases} \qquad (11.0.6)$$

whence $A_1 = -\frac{1}{2}$; $A_2 = \frac{1}{6}$; $A_3 = 0$; $A_4 = -\frac{1}{30}$; $A_5 = 0$; $A_6 = \frac{1}{42}$; $\ldots$ Using a symbolic notation, we can write the system (11.0.5) in a very elegant way, namely,

$$(A + 1)^s - A^s = 0\,,\qquad s = 2, 3, 4,\ldots$$

with the convention that all powers A^i should be replaced by A_i.

Now we treat (11.0.4) in a similar way and obtain

$$\sum_{r=0}^{\infty} \frac{x^r t^r}{r!} \cdot \sum_{k=0}^{\infty} A_k \frac{t^k}{k!} = \sum_{n=0}^{\infty} B_n(x) \frac{t^n}{n!}\,.$$

Identifying we get

$$B_n(x) = \sum_{r=0}^{n} \binom{n}{r} A_{n-r} x^r\,.$$

Symbolically, we can write $B_n(x) = (A + x)^n$, with the same convention $A^i \rightarrow A_i$ as before. Summing up, we have the following simple rules for the formation of Bernoulli numbers and polynomials:

$$\begin{cases} A_0 = 1\,, \\ (A + 1)^n - A^n = 0 \qquad \text{with} \qquad A^k \rightarrow A_k\,, \\ B_n(x) = (A + x)^n \qquad \text{with} \qquad A^k \rightarrow A_k\,. \end{cases} \qquad (11.0.7)$$

The first Bernoulli polynomials are

$$B_0(x) = 1 ,$$
$$B_1(x) = x - 1/2 ,$$
$$B_2(x) = x^2 - x + 1/6 ,$$
$$B_3(x) = x^3 - 3x^2/2 + x/2 ,$$
$$B_4(x) = x^4 - 2x^3 + x^2 - 1/30 ,$$
$$B_5(x) = x^5 - 5x^4/2 + 5x^3/3 - x/6 ,$$
$$B_6(x) = x^6 - 3x^5 + 5x^4/2 - x^2/2 + 1/42 .$$

Equation (11.0.4) can be written

$$e^{xt} \frac{t}{e^t - 1} = 1 + \sum_{n=1}^{\infty} B_n(x) \frac{t^n}{n!} ,$$

and differentiating with respect to x, we obtain

$$e^{xt} \frac{t}{e^t - 1} = \sum_{n=1}^{\infty} \frac{B_n'(x)}{n} \frac{t^{n-1}}{(n-1)!} = \sum_{n=1}^{\infty} B_{n-1}(x) \frac{t^{n-1}}{(n-1)!} .$$

Comparing the coefficients, we get

$$B_n'(x) = nB_{n-1}(x) . \tag{11.0.8}$$

The same relation is obtained by differentiating the symbolic equation

$$B_n(x) = (A + x)^n .$$

Now let x be an arbitrary complex number. From $(A + 1)^i = A^i$ we also have

$$\sum_{i=2}^{n} \binom{n}{i} (A + 1)^i x^{n-i} = \sum_{i=2}^{n} \binom{n}{i} A^i x^{n-i} ,$$

or

$$(A + 1 + x)^n - n(A + 1)x^{n-1} = (A + x)^n - nAx^{n-1} .$$

Hence we find the following formula, which we shall use later:

$$(A + 1 + x)^n - (A + x)^n = nx^{n-1} . \tag{11.0.9}$$

It can also be written $\Delta(A + x)^n = Dx^n$, or, since $(A + x)^n = B_n(x)$,

$$\Delta B_n(x) = Dx^n . \tag{11.0.10}$$

Conversely, if we want to find a polynomial $B_n(x)$ fulfilling (11.0.10), we have for $h = 1$:

$$B_n(x) = \frac{D}{e^D - 1} x^n = \left(\sum_{k=0}^{\infty} A_k \frac{D^k}{k!} \right) x^n = \sum_{k=0}^{n} \binom{n}{k} A_k x^{n-k} = (A + x)^n$$

in agreement with (11.0.7).

Writing (11.0.8) in the form $B_n(t) = B'_{n+1}(t)/(n+1)$ and integrating, we obtain

$$\int_x^{x+1} B_n(t)\,dt = \frac{B_{n+1}(x+1) - B_{n+1}(x)}{n+1} = \frac{\varDelta B_{n+1}(x)}{n+1} = \frac{Dx^{n+1}}{n+1} = x^n,$$

where we have also used (11.0.10). Putting $x = 0$, we find for $n = 1, 2, 3, \ldots$,

$$\int_0^1 B_n(t)\,dt = 0.$$

The relations

$$\begin{cases} B_0(t) = 1, \\ B'_n(t) = nB_{n-1}(t), \\ \int_0^1 B_n(t)\,dt = 0, \qquad (n = 1, 2, 3, \ldots), \end{cases}$$

give a simple and direct method for recursive determination of $B_n(t)$, and hence also of $A_n = B_n(0)$.

From (11.0.6) we obtained $A_3 = A_5 = 0$, and we shall now prove that $A_{2k+1} = 0$ for $k = 1, 2, 3, \ldots$ Forming

$$f(t) = \frac{t}{e^t - 1} + \frac{t}{2} = \frac{t}{2}\,\frac{e^t + 1}{e^t - 1} = \frac{t}{2}\coth\frac{t}{2},$$

we see at once that $f(-t) = f(t)$, which concludes the proof. Hence, there are only even powers of t left, and we introduce the common notation

$$B_k = (-1)^{k+1} A_{2k}; \qquad k = 1, 2, 3, \ldots. \tag{11.0.11}$$

The first Bernoulli numbers B_k are

$B_1 = 1/6$	$B_7 = 7/6$	$B_{13} = 8553103/6$
$B_2 = 1/30$	$B_8 = 3617/510$	$B_{14} = 23749461029/870$
$B_3 = 1/42$	$B_9 = 43867/798$	$B_{15} = 8615841276005/14322$
$B_4 = 1/30$	$B_{10} = 174611/330$	$B_{16} = 7709321041217/510$
$B_5 = 5/66$	$B_{11} = 854513/138$	$B_{17} = 2577687858367/6$
$B_6 = 691/2730$	$B_{12} = 236364091/2730$.	

The expansion (11.0.3) now takes the form

$$\frac{t}{e^t - 1} = 1 - \frac{t}{2} + B_1\frac{t^2}{2!} - B_2\frac{t^4}{4!} + B_3\frac{t^6}{6!} - \cdots. \tag{11.0.12}$$

Above we discussed the even function

$$\frac{t}{e^t - 1} + \frac{t}{2} = \frac{t}{2}\coth\frac{t}{2}.$$

Putting $t/2 = x$, we obtain the expansion

$$x \coth x = 1 + B_1 \frac{(2x)^2}{2!} - B_2 \frac{(2x)^4}{4!} - B_3 \frac{(2x)^6}{6!} - \cdots . \qquad (11.0.13)$$

Replacing x by ix, we get

$$x \cot x = 1 - B_1 \frac{(2x)^2}{2!} - B_2 \frac{(2x)^4}{4!} - B_3 \frac{(2x)^6}{6!} - \cdots ,$$

or

$$1 - x \cot x = \sum_{p=1}^{\infty} \frac{2^{2p}}{(2p)!} B_p x^{2p} . \qquad (11.0.14)$$

Using the identity $\tan x = \cot x - 2 \cot 2x$, we find

$$\tan x = \sum_{p=1}^{\infty} \frac{2^{2p}(2^{2p} - 1)}{(2p)!} B_p x^{2p-1} . \qquad (11.0.15)$$

Written out explicitly, formula (11.0.12) and the last two formulas take the following form:

$$\begin{cases} \dfrac{x}{e^x - 1} = 1 - \dfrac{x}{2} + \dfrac{x^2}{12} - \dfrac{x^4}{720} + \dfrac{x^6}{30240} - \dfrac{x^8}{1209600} + \dfrac{x^{10}}{47900160} - \cdots , \\[2mm] \tan x = x + \dfrac{x^3}{3} + \dfrac{2}{15} x^5 + \dfrac{17}{315} x^7 + \dfrac{62}{2835} x^9 + \dfrac{1382}{155925} x^{11} + \cdots , \\[2mm] \cot x = \dfrac{1}{x} - \dfrac{x}{3} - \dfrac{x^3}{45} - \dfrac{2x^5}{945} - \dfrac{x^7}{4725} - \dfrac{2x^9}{93555} - \cdots . \end{cases}$$

$$(11.0.16)$$

The first series converges for $|x| < 2\pi$, the second for $|x| < \pi/2$, and the last for $|x| < \pi$. A brief convergence discussion will be given later.

11.1. Sums of factorials and powers

In Chapter 7 we mentioned the fundamental properties of the factorials in the formation of differences, and according to (7.3.2) we have $\Delta p^{(n)} = np^{(n-1)}$. From this we obtained an analogous summation formula (7.3.12),

$$\sum_{p=P}^{Q} p^{(n)} = \frac{(Q + 1)^{(n+1)} - P^{(n+1)}}{n + 1} , \qquad (11.1.1)$$

valid for all integers of n except $n = -1$. As an example we first calculate

$$\sum_{p=3}^{6} p^{(3)} = \sum_{p=3}^{6} p(p - 1)(p - 2)$$

$$= \frac{7^{(4)} - 3^{(4)}}{4} = \frac{7 \cdot 6 \cdot 5 \cdot 4 - 3 \cdot 2 \cdot 1 \cdot 0}{4} = 210 ,$$

which is easily verified by direct computation. We can also compute, for example,

$$\sum_{p=1}^{4} p^{(-3)} = \sum_{p=1}^{4} \frac{1}{(p+1)(p+2)(p+3)}$$

$$= \frac{5^{(-2)} - 1^{(-2)}}{-2} = \frac{1}{2} \left(\frac{1}{2 \cdot 3} - \frac{1}{6 \cdot 7} \right) = \frac{1}{14},$$

which is also easily verified. The case $n = -1$ can be treated as a limit case; then we have to write $p^{(n)} = p!/(p-n)! = \Gamma(p+1)/\Gamma(p-n+1)$, giving a definition of the factorial also for values of n other than integers.

In Chapter 7 we showed how a power could be expressed as a linear combination of factorials by use of Stirling's numbers of the second kind. Thus we have directly a method for computing sums of powers, and we demonstrate the technique on an example.

$$\sum_{p=0}^{N} p^4 = \sum_{p=0}^{N} [p^{(1)} + 7p^{(2)} + 6p^{(3)} + p^{(4)}]$$

$$= \frac{(N+1)^{(2)}}{2} + \frac{7(N+1)^{(3)}}{3} + \frac{6(N+1)^{(4)}}{4} + \frac{(N+1)^{(5)}}{5}$$

$$= \frac{N(N+1)}{30} [15 + 70(N-1)$$

$$+ 45(N-1)(N-2) + 6(N-1)(N-2)(N-3)]$$

$$= \frac{N(N+1)(2N+1)(3N^2 + 3N - 1)}{30}.$$

For brevity we put $S_n(p) = 1^n + 2^n + \cdots + p^n$. The expressions $S_n(p)$ have been tabulated as functions of p for $n = 0(1)15$ in [4], pp. 188–189. By repeated use of the relation $\Delta B_{n+1}(x)/(n+1) = x^n$, we obtain

$$S_n(p) = \frac{B_{n+1}(p+1) - B_{n+1}(0)}{n+1}, \tag{11.1.2}$$

which is an alternative formula for computation of power sums.

Here we shall also indicate how power sums with even negative exponents can be expressed by use of Bernoulli numbers. We need the following theorem from the theory of analytic functions:

$$\pi \cot \pi x = \frac{1}{x} + \sum_{n=1}^{\infty} \left(\frac{1}{x-n} + \frac{1}{x+n} \right);$$

it is given here without proof (see, for example, [1], p. 32, or [2], p. 122).

We consider the series expansion for small values of x on both sides of the equation:

$$\pi x \cot \pi x = 1 + x \cdot \sum_{n=1}^{\infty} \left(\frac{1}{x-n} + \frac{1}{x+n} \right).$$

For the left-hand side, we use equation (11.0.14) and find that the coefficient of x^{2p} is $-(2\pi)^{2p}B_p/(2p)!$ For the right-hand side we have, apart from the constant term,

$$x \cdot \sum_{n=1}^{\infty} \left(-\frac{1}{n} \frac{1}{1-x/n} + \frac{1}{n} \frac{1}{1+x/n} \right)$$

$$= -\sum_{n=1}^{\infty} \frac{x}{n} \left\{ 1 + \frac{x}{n} + \frac{x^2}{n^2} + \frac{x^3}{n^3} + \cdots - \left(1 - \frac{x}{n} + \frac{x^2}{n^2} - \frac{x^3}{n^3} + \cdots \right) \right\}$$

$$= -2 \cdot \sum_{n=1}^{\infty} \left(\frac{x^2}{n^2} + \frac{x^4}{n^4} + \frac{x^6}{n^6} + \cdots \right).$$

Hence the coefficient of x^{2p} is $-2 \cdot \sum_{n=1}^{\infty} n^{-2p}$, and we get

$$\sum_{n=1}^{\infty} \frac{1}{n^{2p}} = \frac{(2\pi)^{2p}}{2 \cdot (2p)!} B_p. \tag{11.1.3}$$

For $p = 1, 2, 3, 4$, and 5, we obtain

$$\sum_{n=1}^{\infty} \frac{1}{n^2} = \frac{\pi^2}{6}; \qquad \sum_{n=1}^{\infty} \frac{1}{n^4} = \frac{\pi^4}{90}; \qquad \sum_{n=1}^{\infty} \frac{1}{n^6} = \frac{\pi^6}{945};$$

$$\sum_{n=1}^{\infty} \frac{1}{n^8} = \frac{\pi^8}{9450}; \qquad \sum_{n=1}^{\infty} \frac{1}{n^{10}} = \frac{\pi^{10}}{93555}.$$

On the other hand, we have no simple expressions for the sums $\sum_{n=1}^{\infty} n^{-2p-1}$; the formulas turn out to be fairly complicated. The sums $\sum_{n=1}^{\infty} n^{-p}$ have been tabulated in [3], p. 224, for $p = 2(1)100$.

From equation (11.1.3) several interesting conclusions can be drawn. First of all, we see that $B_p > 0$ for $p = 1, 2, 3, \ldots$ Further, we can easily obtain estimates of B_p for large values of p, since the left-hand side approaches 1 very rapidly, and we have asymptotically

$$B_p \sim \frac{2(2p)!}{(2\pi)^{2p}}. \tag{11.1.4}$$

This can be transformed by use of Stirling's formula:

$$B_p \sim 4\sqrt{\pi p} \left(\frac{p}{\pi e} \right)^{2p}. \tag{11.1.5}$$

Using (11.1.4) we can investigate the convergence of the series (11.0.12), where the general term for large values of p can be written $2(-1)^{p+1}(t/2\pi)^p$. Hence we have convergence for $|t| < 2\pi$.

In an analogous way, we can examine the convergence for the expansions (11.0.14) and (11.0.15). The general term behaves like $2(x/\pi)^{2p}$ and $2(2x/\pi)^{2p}$, and we have convergence for $|x| < \pi$ and $|x| < \pi/2$, respectively. On the other hand, this is quite natural, since $x \cot x$ and $\tan x$ are regular in exactly these domains.

11.2. Series with positive terms

Suppose that $f(x)$ is an analytic function and consider

$$S = f(x_0) + f(x_0 + h) + f(x_0 + 2h) + \cdots + f(x_0 + (n-1)h)$$

$$= \left(\sum_{r=0}^{n-1} E^r\right) f(x_0) = \frac{E^n - 1}{E - 1} f(x_0) \; .$$

The main idea is to consider the sum as a trapezoidal approximation of a definite integral; hence we are actually looking for the correction terms. As usual, they can be expressed in forward or central differences, or in derivatives. In the first two cases, we get the formulas by Laplace and Gauss, and in the last case we obtain the summation formula by Euler-Maclaurin.

In the Laplace case, we have

$$\frac{1}{h} \int_{x_0}^{x_n} f(x)\,dx = \frac{1}{h}(1 + E + E^2 + \cdots + E^{n-1}) J_1 f(x_0)$$

$$= \frac{E^n - 1}{E - 1} \frac{J_1}{h} f(x_0) \; ,$$

and

$$S - \frac{1}{h} \int_{x_0}^{x_n} f(x)\,dx = \frac{E^n - 1}{\varDelta}\left(1 - \frac{J_1}{h}\right) f(x_0)$$

$$= \frac{E^n - 1}{\varDelta}\left(1 - \frac{\varDelta}{\log(1 + \varDelta)}\right) f(x_0)$$

$$= -(E^n - 1)(b_1 + b_2\varDelta + b_3\varDelta^2 + \cdots) f(x_0) \; .$$

Here we have used (7.1.4) and (10.4.1). Hence we obtain

$$f(x_0) + f(x_1) + \cdots + f(x_{n-1})$$

$$= \frac{1}{h} \int_{x_0}^{x_n} f(x)\,dx - \sum_{k=1}^{m} b_k\left(\varDelta^{k-1} f(x_n) - \varDelta^{k-1} f(x_0)\right) + R_m \; . \qquad (11.2.1)$$

We give the remainder term without proof:

$$R_m = nh^{m+1} b_{m+1} f^{(m+1)}(\xi) \; ; \qquad x_0 < \xi < x_{n+m} \; .$$

In the Gaussian case we form

$$S - \frac{1}{h} \int_{x_0 - h/2}^{x_n - h/2} f(x)\,dx = \frac{E^n - 1}{E - 1}\left(1 - \frac{J_0}{h}\right) f(x_0)$$

$$= \frac{E^n - 1}{E^{1/2}\delta}\left(1 - \frac{\delta}{U}\right) f(x_0) \; .$$

The expression δ/U is easily obtained by multiplication of (7.2.9) by (7.2.10):

$$\frac{\delta}{U} = 1 + \frac{\delta^2}{24} - \frac{17\delta^4}{5760} + \frac{367\delta^6}{967680} - \frac{27859\delta^8}{464486400} + \cdots \; ,$$

and hence we have the formula

$$f(x_0) + f(x_1) + \cdots + f(x_{n-1}) = \frac{1}{h} \int_{x_0-h/2}^{x_n-h/2} f(x)\,dx$$

$$- \left(\frac{\delta}{24} - \frac{17\delta^3}{5760} + \frac{367\delta^5}{967680} - \cdots \right) \left(f\left(x_n - \frac{h}{2}\right) - f\left(x_0 - \frac{h}{2}\right) \right).$$

$$(11.2.2)$$

Alternatively, we consider

$$\frac{1}{2}\left[f(x_0 + f(x_1)) \right] - \frac{1}{h} \int_{x_0}^{x_1} f(x)\,dx$$

$$= \left(\mu - \frac{J_0}{h} \right) f_{1/2} = E^{1/2} \left(\mu - \frac{J_0}{h} \right) f_0$$

$$= E^{1/2} \mu \left(1 - \frac{\delta}{\mu U} \right) f_0$$

$$= E^{1/2} \mu \left(\frac{\delta^2}{12} - \frac{11\delta^4}{720} + \frac{191\delta^6}{60480} - \cdots \right) f_0$$

$$= \mu\delta \left(\frac{1}{12} - \frac{11\delta^2}{720} + \frac{191\delta^4}{60480} - \cdots \right)(E - 1)f_0\,.$$

By summing such relations to $\frac{1}{2}[f(x_{n-1}) + f(x_n)]$ and further adding the term $\frac{1}{2}[f(x_0) + f(x_n)]$, we obtain

$$f(x_0) + f(x_1) + \cdots + f(x_n) = \frac{1}{h} \int_{x_0}^{x_n} f(x)\,dx + \frac{1}{2}\left[f(x_0) + f(x_n) \right]$$

$$+ \frac{1}{12}\left[\mu\delta f(x_n) - \mu\delta f(x_0) \right] - \frac{11}{720}\left[\mu\delta^3 f(x_n) - \mu\delta^3 f(x_0) \right]$$

$$+ \frac{191}{60480}\left[\mu\delta^5 f(x_n) - \mu\delta^5 f(x_0) \right] - \cdots. \qquad (11.2.3)$$

Both (11.2.2) and (11.2.3) are due to Gauss.

Finally, we again form

$$f(x_0) + f(x_1) + \cdots + f(x_{n-1}) - \frac{1}{h} \int_{x_0}^{x_n} f(x)\,dx$$

$$= \frac{E^n - 1}{E - 1} \left(1 - \frac{\Delta}{U} \right) f(x_0)$$

$$= (E^n - 1) \left(\frac{1}{e^U - 1} - \frac{1}{U} \right) f(x_0)$$

$$= \left(-\frac{1}{2} + B_1 \frac{U}{2!} - B_2 \frac{U^3}{4!} + B_3 \frac{U^5}{6!} - \cdots \right) \left(f(x_n) - f(x_0) \right).$$

Adding $f(x_n)$ to both sides, we obtain the following equality:

$$f(x_0) + f(x_1) + \cdots + f(x_n)$$

$$= \frac{1}{h} \int_{x_0}^{x_n} f(x)\,dx + \frac{1}{2}[f(x_0) + f(x_n)] + \frac{h}{12}[f'(x_n) - f'(x_0)]$$

$$- \frac{h^3}{720}[f'''(x_n) - f'''(x_0)] + \frac{h^5}{30240}[f^{\mathrm{v}}(x_n) - f^{\mathrm{v}}(x_0)] - \cdots . \quad (11.2.4)$$

This is Euler-Maclaurin's well-known summation formula. If the series is truncated after the term $B_{m-1}U^{2m-3}/(2m - 2)!$, the remainder term is

$$R = \frac{nB_m h^{2m}}{(2m)!} f^{(2m)}(\xi) ; \qquad x_0 < \xi < x_n .$$

(See, for example, [4], p. 190.) If n tends to infinity, however, this remainder term is of little use. Instead it can be proved that the absolute value of the error is strictly less than twice the absolute value of the first neglected term, provided that $f^{(2m)}(x)$ has the same sign throughout the interval (x_0, x_n). Further, the error has the same sign as the first neglected term.

EXAMPLE

Euler's constant is defined by

$$\gamma = \lim_{n \to \infty} \left(1 + \frac{1}{2} + \frac{1}{3} + \cdots + \frac{1}{n} - \log n \right),$$

which can also be written

$$\gamma = 1 + \sum_{n=2}^{\infty} \left(\frac{1}{n} + \log \frac{n-1}{n} \right).$$

We have

$$\int_{10}^{\infty} \left(\frac{1}{x} - \log x + \log (x - 1) \right) dx = -1 + 9 \log 10 - 9 \log 9$$

$$= -0.05175\ 53591 .$$

The sum up to $n = 9$ is computed directly, and we get

$$
\begin{aligned}
S_9 &= 0.63174\ 36767 \\
\smallint &= -0.05175\ 53591 \\
\tfrac{1}{2} f(10) &= -0.00268\ 02578 \\
-\tfrac{1}{12} f'(10) &= -0.00009\ 25926 \\
\tfrac{1}{720} f'''(10) &= 0.00000\ 01993 \\
-\tfrac{1}{30240} f^{\mathrm{v}}(10) &= -0.00000\ 00015 \\
\hline
S &= 0.57721\ 56650
\end{aligned}
$$

The exact value of Euler's constant is $\gamma = 0.57721\ 56649\ 015\ldots$

It may well be the case that even when the integration can be performed explicitly, it is easier to use series expansion. In the example just discussed we have

$$\int_{10}^{\infty}(x^{-1} + \log(1 - x^{-1}))\,dx = -\int_{10}^{\infty}\left(\frac{1}{2x^2} + \frac{1}{3x^3} + \frac{1}{4x^4} + \cdots\right)dx$$

$$= -\left(\frac{1}{1\cdot 2\cdot 10} + \frac{1}{2\cdot 3\cdot 10^2} + \frac{1}{3\cdot 4\cdot 10^3} + \cdots\right).$$

Already from 8 terms we get the value correctly to 10 decimals.

A natural generalization arises in the case where one wishes to perform the summation at closer intervals than those used at the tabulation. Let us suppose that $y_0, y_1, \ldots, y_n$ are known function values belonging to the x-values $x_i = x_0 + ih$. Now we want to sum the y-values with intervals h/N in the independent variable, where N is an integer >1. Suppose that the operators E, Δ, and ∇ refer to the interval h; denote the desired sum by S, and put $1/N = p$. Then we get

$$S = (1 + E^p + \cdots + E^{n(N-1)p})y_0 + y_n = \frac{E^n - 1}{E^p - 1}y_0 + y_n$$

$$= \frac{y_n}{(1 - \nabla)^{-p} - 1} - \frac{y_0}{(1 + \Delta)^p - 1} + y_n\,.$$

We easily find the following series expansion:

$$\frac{x}{(1 + x)^p - 1} = \frac{1}{p}\left(1 + (1 - p)\frac{x}{2} - (1 - p^2)\frac{x^2}{12} + (1 - p^2)\frac{x^3}{24}\right.$$

$$\left. - (1 - p^2)(19 - p^2)\frac{x^4}{720} + (1 - p^2)(9 - p^2)\frac{x^5}{480} - \cdots\right).$$

Replacing x by ∇ and Δ, respectively, we find

$$S = \frac{1}{p}\left(\frac{y_n}{\nabla} - \frac{y_0}{\Delta}\right) + \frac{1}{2}(y_0 - y_n) - \frac{1}{2p}(y_0 + y_n) + y_n$$

$$- \frac{1 - p^2}{12p}(\nabla y_n - \Delta y_0) - \frac{1 - p^2}{24p}(\nabla^2 y_n + \Delta^2 y_0) - \cdots\,.$$

As was shown in Section 10.4, we have $y_n/\nabla - y_0/\Delta = y_0 + y_1 + \cdots + y_n$. Further, we make use of the relation $\nabla = E^{-1}\Delta$, from which we get $\nabla^r y_n = \Delta^r y_{n-r}$. Hence we finally obtain the formula

$$S = \frac{1}{p}(y_0 + y_1 + \cdots + y_n) - \frac{1 - p}{2p}(y_0 + y_n) - \frac{1 - p^2}{12p}(\Delta y_{n-1} - \Delta y_0)$$

$$- \frac{1 - p^2}{24p}(\Delta^2 y_{n-2} + \Delta^2 y_0) - \frac{(1 - p^2)(19 - p^2)}{720p}(\Delta^3 y_{n-3} - \Delta^3 y_0)$$

$$- \frac{(1 - p^2)(9 - p^2)}{480p}(\Delta^4 y_{n-4} + \Delta^4 y_0) - \cdots\,. \tag{11.2.5}$$

This is Lubbock's formula. For analytic functions we could equally well use Euler-Maclaurin's formula directly, but Lubbock's is quite useful for empirical functions.

We observe that

$$\lim_{p \to 0} pS = \frac{1}{h} \int_{x_0}^{x_n} y \, dx \, ,$$

and hence for $p \to 0$, the formula passes into Gregory's formula.

We can, of course, work with central differences or with derivatives; in the latter case the derivation is performed in the following way. We have

$$S - \frac{1}{p}(y_0 + y_1 + \cdots + y_n)$$

$$= \frac{E^n - 1}{E^p - 1} y_0 + y_n - \frac{1}{p} \frac{E^n - 1}{E - 1} y_0 - \frac{1}{p} y_n$$

$$= \left(\frac{1}{e^{pU} - 1} - \frac{1}{p(e^U - 1)} \right)(y_n - y_0) - \frac{1 - p}{p} y_n$$

$$= \frac{1}{pU} \left\{ \frac{U}{2} - B_1 \frac{U^2}{2!} + B_2 \frac{U^4}{4!} - B_3 \frac{U^6}{6!} + \cdots - \frac{pU}{2} + B_1 \frac{p^2 U^2}{2!} \right.$$

$$\left. - B_2 \frac{p^4 U^4}{4!} + B_3 \frac{p^6 U^6}{6!} - \cdots \right\} (y_n - y_0) - \frac{1 - p}{p} y_n \, ,$$

and hence

$$S = \frac{1}{p}(y_0 + y_1 + \cdots + y_n) - \frac{1 - p}{2p}(y_0 + y_n) - \frac{1 - p^2}{12p} h(y_n' - y_0')$$

$$+ \frac{1 - p^4}{720p} h^3(y_n''' - y_0''') - \frac{1 - p^6}{30240p} h^5(y_n^{\mathrm{V}} - y_0^{\mathrm{V}}) + \cdots . \qquad (11.2.6)$$

This is Woolhouse's formula. If we now form pS and let $p \to 0$, the formula passes into Euler-Maclaurin's formula.

11.3. Slowly converging series. Euler's transformation

Summation of slowly converging series is a nasty problem, and it is difficult to devise any general methods. In some cases the following method, suggested by Salzer, can be used. Forming a suitable number of partial sums S_n and regarding them as functions of $1/n$, one performs Lagrangian extrapolation to $1/n = 0$.

EXAMPLE

$$S = \sum_{k=1}^{\infty} k^{-2} \, ; \qquad S_n = \sum_{k=1}^{n} k^{-2} \, .$$

Putting $S_n = S(1/n)$, we obtain

$$S(0.2) \ = 1.46361111 \,,$$
$$S(0.1) \ = 1.54976773 \,,$$
$$S(0.05) = 1.59616325 \,,$$
$$S(0.04) = 1.60572341 \,.$$

Then we obtain from Lagrange's formula

$$S(0) = \tfrac{1}{12}[125 \cdot S(0.04) - 128 \cdot S(0.05) + 16 \cdot S(0.1) - S(0.2)]$$
$$= 1.64493357 \,.$$

The exact value is

$$S(0) = \frac{\pi^2}{6} = 1.644934066\cdots .$$

With $S = S_n + R_n$, we find in the special case when $R_n = A/n^2 + B/n^3 + \cdots$
that $S'_n = \tfrac{1}{3}(4S_{2n} - S_n)$ is a good approximation of S.

Another method, analogous to the technique for computing certain integrals
described in Section 10.0, consists of subtracting a known series whose convergence properties are similar to those of the given series. Lubbock's and
Woolhouse's methods may also be useful in such cases. However, we shall
now consider a more systematic transformation technique, originally due to
Euler.

We start from the power series $S = u_0 + u_1x + u_2x^2 + \cdots$, and we consider
only such values of x for which the series converges. Symbolically we have

$$S = (1 + Ex + E^2x^2 + \cdots)u_0 = \frac{1}{1 - Ex} u_0 = \frac{u_0}{1 - x - x\varDelta}$$

$$= \frac{1}{1 - x} \frac{u_0}{1 - \big(x/(1 - x)\big)\varDelta} \,,$$

and hence

$$S = \frac{1}{1 - x} \cdot \sum_{s=0}^{\infty} \left(\frac{x}{1 - x}\right)^s \varDelta^s u_0 \,. \tag{11.3.1}$$

A corresponding derivation can be made by use of central differences:

$$S = \frac{u_0}{1 - Ex} = \frac{u_0}{1 - x(1 + \delta^2/2 + \mu\delta)}$$

$$= \frac{1}{1 - x} \frac{u_0}{1 - \big(x/(1 - x)\big)(\delta^2/2) - \big(x/(1 - x)\big)\mu\delta}$$

$$= \frac{1}{1 - x} \frac{1 - \big(x/(1 - x)\big)(\delta^2/2) + \big(x/(1 - x)\big)\mu\delta}{\big[1 - \big(x/(1 - x)\big)(\delta^2/2)\big]^2 - \big(x/(1 - x)\big)^2\mu^2\delta^2} u_0$$

$$= \frac{1}{1 - x} \frac{1 - z(\delta^2/2) + z\mu\delta}{1 - z(1 + z)\delta^2} u_0 \,,$$

where we have denoted $x/(1 - x)$ by z. Hence,

$$S = \frac{1}{1 - x}\left\{\left(1 - \frac{z}{2}\,\delta^2\right)\left(1 + z(1 + z)\delta^2 + z^2(1 + z)^2\delta^4 + \cdots\right)\right.$$

$$\left. + z\mu\delta\left(1 + z(1 + z)\delta^2 + \cdots\right)\right\}u_0 \,.$$

Simplifying, we finally obtain

$$S = \frac{u_0}{1 - x} + \frac{1}{2}\frac{1 + x}{1 - x}\left(\frac{x}{(1 - x)^2}\,\delta^2 u_0 + \frac{x^2}{(1 - x)^4}\,\delta^4 u_0 + \frac{x^3}{(1 - x)^6}\,\delta^6 u_0 + \cdots\right)$$

$$+ \frac{1}{2}\frac{x}{(1 - x)^2}\left((u_1 - u_{-1}) + \frac{x}{(1 - x)^2}\,(\delta^2 u_1 - \delta^2 u_{-1})\right.$$

$$\left. + \frac{x^2}{(1 - x)^4}\,(\delta^4 u_1 - \delta^4 u_{-1}) + \cdots\right). \tag{11.3.2}$$

Naturally, we can use derivatives instead of differences. Then we obtain with the same z as before:

$$S = \frac{1}{1 - x}\,[u_0 + Q_1(z)hu_0' + Q_2(z)h^2 u_0'' + \cdots], \tag{11.3.3}$$

where we have

$$\begin{cases} Q_1(z) = z \,, \\[2mm] Q_2(z) = z^2 + \dfrac{z}{2} \,, \\[2mm] Q_3(z) = z^3 + z^2 + \dfrac{z}{6} \,, \\[2mm] Q_4(z) = z^4 + \dfrac{3}{2}\,z^3 + \dfrac{7}{12}\,z^2 + \dfrac{z}{24} \,, \\[2mm] Q_5(z) = z^5 + 2z^4 + \dfrac{5}{4}\,z^3 + \dfrac{z^2}{4} + \dfrac{z}{120} \,, \\[2mm] Q_6(z) = z^6 + \dfrac{5}{2}\,z^5 + \dfrac{13}{6}\,z^4 + \dfrac{3}{4}\,z^3 + \dfrac{31}{360}\,z^2 + \dfrac{z}{720} \,. \\[2mm] \vdots \end{cases} \tag{11.3.4}$$

The general form of these polynomials is

$$Q_n(z) = \frac{1}{n!}\sum_{i=1}^{n} z^i \sum_{k=1}^{i}(-1)^{i+k}\binom{i}{k}k^n \,.$$

EXAMPLE

Consider

$$S = \sum_{n=0}^{\infty}\frac{(-1)^n}{2n + 1} = \frac{1}{2}\sum_{n=0}^{\infty}(-1)^n\,\frac{1}{n + \frac{1}{2}} \,.$$

We use (11.3.1) for $x = -1$ and $u_n = 1/(n + \frac{1}{2}) = (n - \frac{1}{2})^{(-1)}$ [cf. (7.3.2 through 7.3.4)]. Hence

$$\Delta u_n = -(n - \tfrac{1}{2})^{(-2)} ; \quad \Delta^2 u_n = 2(n - \tfrac{1}{2})^{(-3)}; \quad \Delta^3 u_n = -2 \cdot 3(n - \tfrac{1}{2})^{(-4)} , \text{ etc.,}$$

and from this we get

$$\Delta^s u_0 = (-1)^s \frac{(s!)^2 2^{2s+1}}{(2s + 1)!} ,$$

and the final result

$$S = \sum_{s=0}^{\infty} \frac{2^{s-1}(s!)^2}{(2s + 1)!} .$$

The quotient between the $(s + 1)$- and the sth term is $(s + 1)/(2s + 3)$, that is, the convergence is now slightly better than in a geometric series with quotient $\frac{1}{2}$. The remainder term is practically equal to the last term included. Already 10 terms, corrected with a corresponding remainder term, give the result 0.7855, deviating from the correct value $\pi/4$ with only one unit in the last place. In order to obtain this accuracy by use of the original series, we would have to compute about 5000 terms.

Here we also mention a technique which is often quite useful. Consider

$$f(n) = \frac{\alpha_2}{n^2} + \frac{\alpha_3}{n^3} + \frac{\alpha^4}{n^4} + \cdots ,$$

where the constants α_r are assumed to have such properties as to make the series convergent. For $s > 1$, Riemann's ζ-function can be defined through

$$\zeta(s) = \sum_{k=1}^{\infty} k^{-s} ,$$

and this function has been tabulated accurately. If one now wants to compute $S = \sum_{n=1}^{\infty} f(n)$, we immediately find the result

$$S = \alpha_2 \zeta(2) + \alpha_3 \zeta(3) + \alpha_4 \zeta(4) + \cdots .$$

The convergence is often very slow but can be improved considerably if we write instead $S = f(1) + \alpha_2[\zeta(2) - 1] + \alpha_3[\zeta(3) - 1] + \cdots$ For facilitating such computations we give a brief table of the ζ-function when $s = 2(1)22$.

s	$\zeta(s)$	s	$\zeta(s)$	s	$\zeta(s)$
2	1.6449 3407	9	1.0020 0839	16	1.0000 1528
3	1.2020 5690	10	1.0009 9458	17	1.0000 0764
4	1.0823 2323	11	1.0004 9419	18	1.0000 0382
5	1.0369 2776	12	1.0002 4609	19	1.0000 0191
6	1.0173 4306	13	1.0001 2271	20	1.0000 0095
7	1.0083 4928	14	1.0000 6125	21	1.0000 0048
8	1.0040 7736	15	1.0000 3059	22	1.0000 0024

EXAMPLES

1. $$\sum_{k=1}^{\infty} \frac{1}{n^3+1} = \frac{1}{2} + \sum_{n=2}^{\infty} \left(\frac{1}{n^3} - \frac{1}{n^6} + \frac{1}{n^9} - \cdots \right)$$

$$= \frac{1}{2} + [\zeta(3) - 1] - [\zeta(6) - 1] + [\zeta(9) - 1] - \cdots$$

$$\simeq 0.686503 \qquad \text{(7 terms)}.$$

2. $$\sum_{n=1}^{\infty} \sin^2 \frac{1}{n} = \sin^2 1 + \frac{1}{2} \sum_{n=2}^{\infty} \left(1 - \cos \frac{2}{n} \right)$$

$$= \sin^2 1 + \sum_{n=2}^{\infty} \left(\frac{1}{n^2} - \frac{1}{3n^4} + \frac{2}{45n^6} - \frac{1}{315n^8} + \cdots \right)$$

$$= \sin^2 1 + [\zeta(2) - 1] - \frac{1}{3}[\zeta(4) - 1] + \frac{2}{45}[\zeta(6) - 1] - \cdots$$

$$\simeq 1.326324 \qquad \text{(5 terms)}.$$

11.4. Alternating series

This special case, as a matter of fact, can be treated by putting $x = -1$ in the formulas (11.3.1), (11.3.2), and (11.3.3). Nevertheless, we prefer to treat the case separately. Using ordinary differences, we find the formula

$$S = u_0 - u_1 + u_2 - \cdots = \sum_{s=0}^{\infty} \frac{(-1)^s}{2^{s+1}} \Delta^s u_0$$

$$= \frac{1}{2} \left[u_0 - \frac{1}{2} \Delta u_0 + \frac{1}{4} \Delta^2 u_0 \cdots \right]. \qquad (11.4.1)$$

If we use central differences, we get instead

$$S = \frac{1}{2} u_0 - \frac{1}{2^3}(u_1 - u_{-1}) + \frac{1}{2^5}(\delta^2 u_1 - \delta^2 u_{-1}) - \frac{1}{2^7}(\delta^4 u_1 - \delta^4 u_{-1}) + \cdots .$$

$$(11.4.2)$$

If we prefer to use derivatives, it is more convenient to make a special calculation as follows:

$$S = \frac{u_0}{1+E} = \frac{u_0}{e^U+1} = \frac{1}{U} \left[\frac{U}{e^U-1} - \frac{2U}{e^{2U}-1} \right] u_0$$

$$= \frac{1}{U} \left[1 - \frac{U}{2} + B_1 \frac{U^2}{2!} - B_2 \frac{U^4}{4!} + \cdots \right.$$

$$\left. - 1 + \frac{2U}{2} - B_1 \frac{4U^2}{2!} + B_2 \frac{16U^4}{4!} - \cdots \right] u_0$$

$$= \left[\frac{1}{2} - B_1 \frac{2^2-1}{2!} U + B_2 \frac{2^4-1}{4!} U^3 - \cdots \right] u_0 .$$

Hence we find the formula

$$S = \frac{1}{2}\left[u_0 - \frac{h}{2}\,u_0' + \frac{h^3}{24}\,u_0''' - \frac{h^5}{240}\,u_0^{\mathrm{v}} \right.$$

$$\left. + \frac{17h^7}{40320}\,u_0^{\mathrm{VII}} - \frac{31h^9}{725760}\,u_0^{\mathrm{IX}} + \frac{691h^{11}}{159667200}\,u_0^{\mathrm{XI}} - \cdots \right]. \qquad (11.4.3)$$

EXAMPLE

$$S = \left(1 - \frac{1}{3} + \frac{1}{5} - \cdots - \frac{1}{19}\right) + \left(\frac{1}{21} - \frac{1}{23} + \cdots\right) = S_1 + S_2.$$

By direct evaluation we get $S_1 = 0.76045\ 990472\ldots$ Here

$$u = \frac{1}{2x - 1}; \qquad u' = -\frac{2}{(2x-1)^2}; \qquad u''' = -\frac{2\cdot 4\cdot 6}{(2x-1)^4}\cdots,$$

with $2x - 1 = 21$. Thus

$$u = \frac{1}{21}; \qquad u' = -\frac{2}{21^2}; \qquad u''' = -\frac{2\cdot 4\cdot 6}{21^4};$$

$$u^{\mathrm{v}} = -\frac{2\cdot 4\cdot 6\cdot 8\cdot 10}{21^6}, \qquad \text{etc.,}$$

and

$$S_2 = \frac{1}{2}\left[\frac{1}{21} + \frac{1}{21^2} - \frac{2}{21^4} + \frac{16}{21^6} - \frac{272}{21^8} + \frac{7936}{21^{10}} - \cdots\right];$$

that is, $S_2 = 0.02493\ 82586\ 8\ldots$ Hence $S_1 + S_2 = 0.78539\ 81634$, compared with the exact value

$$\pi/4 = 0.78539\ 81633\ 974\ldots$$

An interesting technique has been suggested by van Wijngaarden, who transforms a series with positive terms to an alternating series. Let

$$S = u_1 + u_2 + u_3 + \cdots,$$

and put

$$\begin{cases} v_1 = u_1 + 2u_2 + 4u_4 + 8u_8 + \cdots, \\ v_2 = u_2 + 2u_4 + 4u_8 + 8u_{16} + \cdots, \\ v_3 = u_3 + 2u_6 + 4u_{12} + 8u_{24} + \cdots, \\ \quad\vdots \\ v_k = u_k + 2u_{2k} + 4u_{4k} + 8u_{8k} + \cdots, \\ \quad\vdots \end{cases}$$

Then it is easily shown that $S = v_1 - v_2 + v_3 - v_4 + \cdots$ The conditions for the validity of this transformation are quite mild; for example, it suffices that the terms u_k decrease as $k^{-1-\epsilon}$, where $\epsilon > 0$.

EXAMPLE
$$S = \sum_{k=1}^{\infty} \frac{1}{k^3} ,$$

$$v_k = \frac{1}{k^3} \left(1 + \frac{2}{2^3} + \frac{4}{4^3} + \frac{8}{8^3} + \cdots \right) = \frac{4}{3} \frac{1}{k^3} .$$

Hence we have

$$S = \frac{4}{3} \cdot \sum_{k=1}^{\infty} (-1)^{k-1} \frac{1}{k^3} .$$

REFERENCES

[1] Knopp: *Funktionentheorie*, II (Berlin, 1931).

[2] Hurwitz-Courant: *Funktionentheorie* (Berlin, 1929).

[3] Davis, Harold T.: *Tables of the Higher Mathematical Functions*, II (Indiana University, Waterman Institute for Scientific Research. Contributions. The Principia Press, Bloomington, 1935).

[4] Milne-Thompson: *The Calculus of Finite Differences* (Macmillan, London, 1933).

EXERCISES

1. Calculate

$$S = \frac{1}{1 \cdot 3} - \frac{3}{5 \cdot 7} + \frac{5}{9 \cdot 11} - \frac{7}{13 \cdot 15} + \cdots$$

correct to five places.

2. Calculate

$$S = 1 - \frac{1}{\sqrt{3}} + \frac{1}{\sqrt{5}} - \frac{1}{\sqrt{7}} + \frac{1}{\sqrt{9}} - \cdots$$

correct to four places.

3. Find to five places:

$$S = \frac{1}{\log 2} - \frac{1}{\log 3} + \frac{1}{\log 4} - \cdots$$

4. Calculate Catalan's constant $\sum_{k=0}^{\infty} (-1)^k/(2k+1)^2$ to ten places.

5. Find the sum $S = \sin 1 - \sin \frac{1}{3} + \sin \frac{1}{5} - \sin \frac{1}{7} + \cdots$ to five places.

6. Compute to five places

$$S = (2 - 1) - (\sqrt{3} - 1) + (\sqrt[3]{4} - 1) - (\sqrt[4]{5} - 1) + (\sqrt[5]{6} - 1) - \cdots$$

7. Find $e^{-1} - e^{-\sqrt{2}} + e^{-\sqrt{3}} - e^{-\sqrt{4}} + \cdots$ to four places.

8. One wants to compute $S = \sum_{n=1}^{\infty} (1/n^3)$. Putting $S_N = \sum_{n=1}^{N} (1/n^3)$ and $S = S_N + R_N$, and assuming that $R_N = 1/2N^2 + a/N^3$, one computes S_N explicitly for $N = 10$ and $N = 20$ (six places). Aided by these values, find S.

9. Compute $\sum_{n=1}^{\infty} (n^3/n!)$ by use of factorials.

10. Calculate explicitly A_n/B_n for $n = 0(1)12$, when

$$\begin{cases} A_n = A_{n-1} + (n-1)A_{n-2}, \\ B_n = B_{n-1} + (n-1)B_{n-2}, \end{cases}$$

$A_0 = 0$, $A_1 = B_0 = B_1 = 1$, and use the values for computing $\lim_{n \to \infty} A_n/B_n$ (four decimal places).

11. Find $S = \sum_{n=1}^{\infty} (1/n) \arctan (1/n)$ to five decimal places.

12. Compute $\sum_{k=1}^{\infty} k^{-5/2}$ correct to six places.

13. Compute

$$1 - \frac{1}{1+i} + \frac{1}{2} - \frac{1}{2+i} + \frac{1}{3} - \frac{1}{3+i} + \cdots$$

to four places.

14. Compute

$$\frac{1}{1-p^2} + \frac{1}{4-p^2} + \frac{1}{9-p^2} + \cdots + \frac{1}{n^2-p^2} + \cdots$$

for $p = \frac{1}{4}$ (four places).

15. Find $\prod_{k=1}^{\infty} (1 + k^{-3/2})$ to five places.

16. Show that the series

$$S = \frac{1}{1 \cdot 1} - \frac{2}{3 \cdot 3} + \frac{1}{5 \cdot 5} + \frac{1}{7 \cdot 7} - \frac{2}{9 \cdot 9}$$

$$+ \frac{1}{11 \cdot 11} + \frac{1}{13 \cdot 13} - \frac{2}{15 \cdot 15} + \frac{1}{17 \cdot 17} + \cdots$$

can be transformed to

$$S = \frac{2}{3} \sum_{n=0}^{\infty} \frac{1}{(2n+1)^2}.$$

Then compute S correct to four decimals.

17. $f(x)$ is a function which can be differentiated an infinite number of times. Further, $f(x) = u_0$; $f(x+1) = u_1$; $f(x+2) = u_2$; ... The series $S = u_0 + u_1 - u_2 - u_3 + u_4 + u_5 - \cdots$ is supposed to be convergent. Find S expressed in $f(x), f'(x), f''(x), \ldots$ up to terms of the fifth order.

18. Integrals of oscillating functions can be computed in such a way that each domain above and below the x-axis is taken separately, and the total integral appears as an alternating series. Using this method, calculate

$$\int_0^{\infty} \frac{\sin \pi x}{\log (1+x)} \, dx$$

to three decimal places.

19. Putting $z_n = \sum_{k=2}^{\infty} k^{-n}$, $n = 2, 3, 4, \ldots$, prove that $z_2/2 + z_3/3 + z_4/4 + \cdots = 1 - \gamma$, where γ is Euler's constant.

20. Find $\sum_{n=1}^{\infty} 1/\alpha_n^2$ where $\alpha_1, \alpha_2, \alpha_3, \ldots$ are the positive roots of the equation $\tan x = x$ taken in increasing order.

21. Using series expansion compute $\int_0^{\infty}(x/\sinh x)\, dx$ and $\int_0^{\infty}(x/\sinh x)^2\, dx$.

22. Using series expansion compute $\int_0^{\infty} x\, dx/(e^x - 1)$ and $\int_0^{\infty} x\, dx/(e^x + 1)$.

23. Compute $\prod_{n=1}^{\infty}(10n - 1)(10n - 9)/(10n - 5)^2$.

24. Compute $\prod_{n=1}^{\infty}(1 - q^n)^{-1}$ for $q = 0.9$.

Chapter 12

Multiple integrals

The Good Lord said to the animals:
Go out and multiply!
But the snake answered:
How could I? I am an adder!

Working with functions of several variables, one encounters essentially greater difficulties than in the case of just one variable. Already interpolation in two variables is a troublesome and laborious procedure. As in the case of one variable, one can use the operator technique and deduce interpolation coefficients for the points in a square grid. Usually one prefers to perform first a series of simple interpolations in one variable and then to interpolate the other variable by means of the new values. If, for example, $f(x_0 + nh, y_0 + vk)$ has to be computed, one can keep $y_0 + vk$ constant and compute $f(x_0 + nh, y_0 + vk)$ for $n = 0, \pm 1, \pm 2, \ldots$ (interpolation in the y-direction with different "integer" x-values). Then the desired value is obtained by one more interpolation in x.

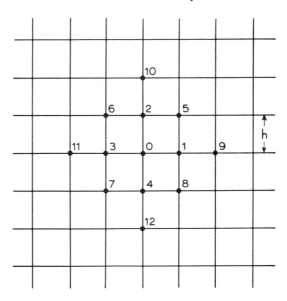

Figure 12.1

For performing differentiation and integration (which, in two dimensions, is sometimes called cubature), we conveniently use an operator technique. From now on we restrict ourselves to two independent variables and put

$$\xi = h\frac{\partial}{\partial x}; \qquad \eta = h\frac{\partial}{\partial y}. \tag{12.1}$$

Thus we are working with a square grid, where we number a few points near the origin (see Fig. 12.1) With these notations we have, for example, $f_1 = e^{\xi}f_0$; $f_2 = e^{\eta}f_0$; $f_3 = e^{-\xi}f_0$; $f_4 = e^{-\eta}f_0$; $f_5 = e^{\xi+\eta}f_0$, and so on. We shall here use the symbols ∇ and D in the following sense:

$$\begin{cases} \xi^2 + \eta^2 = h^2\left(\dfrac{\partial^2}{\partial x^2} + \dfrac{\partial^2}{\partial y^2}\right) = h^2\nabla^2 \,, \\[2ex] \xi\eta = h^2\dfrac{\partial^2}{\partial x\,\partial y} = h^2 D^2 \,. \end{cases} \tag{12.2}$$

Further, we shall consider the following three sums:

$$\begin{aligned} S_1 &= f_1 + f_2 + f_3 + f_4 \,, \\ S_2 &= f_5 + f_6 + f_7 + f_8 \,, \\ S_3 &= f_9 + f_{10} + f_{11} + f_{12} \,. \\ S_1 &= 2(\cosh\xi + \cosh\eta)f_0 \\ &= 4f_0 + (\xi^2 + \eta^2)f_0 + \tfrac{1}{12}(\xi^4 + \eta^4)f_0 + \cdots, \\ S_2 &= 4\cosh\xi\cosh\eta \cdot f_0 \\ &= 4f_0 + 2(\xi^2 + \eta^2)f_0 + \tfrac{1}{6}(\xi^4 + \eta^4)f_0 + \xi^2\eta^2 f_0 + \cdots, \\ S_3 &= 2(\cosh 2\xi + \cosh 2\eta)f_0 \\ &= 4f_0 + 4(\xi^2 + \eta^2)f_0 + \tfrac{4}{3}(\xi^4 + \eta^4)f_0 + \cdots. \end{aligned}$$

The most important differential expression is $\nabla^2 f$, and we find

$$\nabla^2 f_0 \simeq \frac{1}{h^2}(S_1 - 4f_0)\,. \tag{12.3}$$

Further, we obtain

$$4S_1 + S_2 - 20f_0 = (6h^2\nabla^2 + \tfrac{1}{2}h^4\nabla^4 + \cdots)f_0\,, \tag{12.4}$$

with the main disturbing term invariant under rotation, and

$$\nabla^2 f_0 \simeq \frac{1}{12h^2}(16S_1 - S_3 - 60f_0)\,, \tag{12.5}$$

with the error proportional to h^4.

Analogously, we find without difficulty:

$$\nabla^4 f_0 \simeq \frac{1}{h^4}(20f_0 - 8S_1 + 2S_2 + S_3)$$

with the error proportional to h^2.

We now pass to the integration (cubature), and try to compute

$$V = \int_{-h}^{+h} \int_{-h}^{+h} f(x, y) \, dx \, dy \ .$$

Then

$$f(x, y) = f(hr, hs) = e^{r\xi + s\eta} f_0 \ ; \qquad \begin{cases} dx = h \, dr \ , \\ dy = h \, ds \ , \end{cases}$$

and hence

$$V = h^2 \int_{-1}^{+1} \int_{-1}^{+1} e^{r\xi + s\eta} f_0 \, dr \, ds$$

$$= h^2 \left[\frac{e^{r\xi}}{\xi} \right]_{r=-1}^{+1} \left[\frac{e^{s\eta}}{\eta} \right]_{s=-1}^{+1} f_0 = \frac{4h^2 \cdot \sinh \xi \cdot \sinh \eta}{\xi \eta} f_0$$

$$= 4h^2 \left[1 + \frac{1}{3!} \, (\xi^2 + \eta^2) + \frac{1}{5!} \left(\xi^4 + \frac{10}{3} \, \xi^2 \eta^2 + \eta^4 \right) + \cdots \right] f_0 \ .$$

Setting for a moment $\xi^2 + \eta^2 = P$, $\xi^4 + \eta^4 = Q$, and $\xi^2 \eta^2 = R$, and adding the condition that the error term shall be proportional to ∇^4, that is, to $(\xi^2 + \eta^2)^2 = Q + 2R$ (thus invariant under rotation), we obtain

$$V = 4h^2[\alpha f_0 + \beta S_1 + \gamma S_2 + c(Q + 2R)] + \cdots .$$

Identifying, we get

$$\begin{cases} \alpha + 4\beta + 4\gamma = 1 \ , \\[2mm] \beta + 2\gamma = \dfrac{1}{6} \ , \\[2mm] \dfrac{\beta}{12} + \dfrac{\gamma}{6} + c = \dfrac{1}{120} \ , \\[2mm] \gamma + 2c = \dfrac{1}{36} \ , \end{cases} \qquad \text{whence} \qquad \begin{cases} \alpha = \dfrac{22}{45} \ , \\[2mm] \beta = \dfrac{4}{45} \ , \\[2mm] \gamma = \dfrac{7}{180} \ , \\[2mm] c = -\dfrac{1}{180} \ . \end{cases}$$

Thus we find the formula

$$V = \frac{h^2}{45} \, (88f_0 + 16S_1 + 7S_2) - \frac{h^6}{45} \, \nabla^4 f_0 + \cdots .$$

On the other hand, Simpson's formula in two directions gives

$$V = \frac{h^2}{9} \, (16f_0 + 4S_1 + S_2) + \cdots ,$$

but the error term is more complicated.

All these formulas can conveniently be illustrated by so-called computation molecules. The examples in Fig. 12.2 should need no extra explanation.

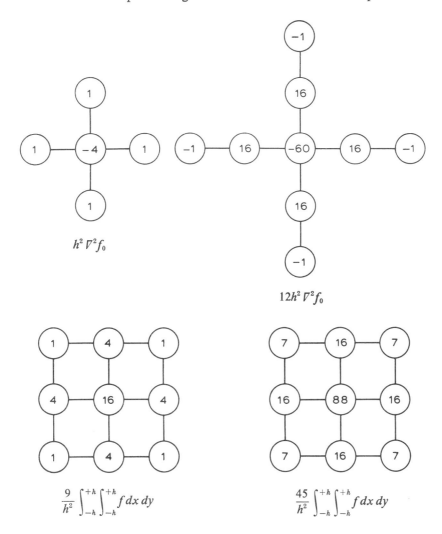

Figure 12.2

The two integration formulas are valid for integration over a square with the side $= 2h$. Integrating over a larger domain, we see that the coefficients at the corners receive contributions from four squares, while the coefficients in the midpoints of the sides get contributions from two squares. Taking out the factor 4, we obtain the open-type formulas shown in Fig. 12.3.

This standard configuration is then repeated in both directions.

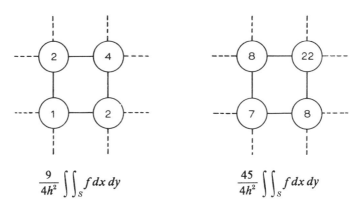

$$\frac{9}{4h^2}\int\int_S f\,dx\,dy \qquad\qquad \frac{45}{4h^2}\int\int_S f\,dx\,dy$$

Figure 12.3

The integration in higher dimensions will give rise to great difficulties. One possible way will be sketched in a later chapter, but this Monte Carlo method has serious disadvantages due to its small accuracy. Of course, it is possible, in principle at least, to construct formulas corresponding to the Cote and Gauss formulas, but they are extremely clumsy and awkward. Although some progress has been made in recent years, no completely satisfactory method has appeared so far.

On practical work with multiple integrals the most common way is perhaps first to perform as many exact integrations as possible and then to integrate each variable by itself. This may create a need for huge amounts of computing time, and the demand for accuracy has to be weighed against available computational resources.

REFERENCES

[1] G. W. Tyler: *Can. Jour. of Math.*, **3**, 393 (1953).

[2] P. C. Hammer, *Proc. of a Symposium on Numerical Approximation*, ed. E. Langer, Madison, Wisc. 1958.

[3] Thacher, *MTAC*, **11**, 189 (1957).

[4] Thacher, *Comm. ACM*, **7**, 23 (Jan. 1964).

[5] Stroud, *Ann. N. Y. Acad. Sci.*, **86**, 776 (1960).

EXERCISES

1. Compute numerically $\int\int_D (dx\,dy/(x^2 + y^2))$, where D is a square with corners $(1, 1)$, $(2, 1)$, $(2, 2)$, and $(1, 2)$.

2. Compute numerically $\int_0^\infty \int_0^\infty e^{-(x^2+y^2)^2}\,dx\,dy$ $(h = 0.5)$, and compare with the exact value.

3. For computation of $\int\int_D f(x, y)\,dx\,dy$, where D is a square with the sides parallel to the axes, one wants to use the approximation $h^2(af_0 + bS)$. Here h is the side of the

square, f_0 is the function value at the midpoint, and S is the sum of the function values at the corners. Find a and b, and compute approximately

$$\int_{-1}^{+1}\int_{-1}^{+1} e^{-(x^2+y^2)}\, dx\, dy$$

by using 13 points in all.

4. One wants to compute $\iint_C f(x, y)\, dx\, dy$, where C is a circle with radius a, by using the approximating expression $(\pi a^2/4)(f_1 + f_2 + f_3 + f_4)$. Here f_1, f_2, f_3, and f_4 are the function values at the corners of a square with the same center as the circle has. Find the side of the square when the formula is correct to the highest possible order.

5. A function $u = u(x, y)$ sufficiently differentiable is given, together with a regular hexagon with corners $P_1, P_2, \ldots, P_6$, center P_0 and side h. The function value in the point P_ν is denoted by u_ν. Putting $S = u_1 + u_2 + \cdots + u_6 - 6u_0$ and $\varDelta = \partial^2/\partial x^2 + \partial^2/\partial y^2$, show that

$$S = \tfrac{3}{2}h^2\,\varDelta u + \tfrac{3}{32}h^4\,\varDelta^2 u + O(h^6)\,,$$

where $\varDelta u$ and $\varDelta^2 u$ are taken in P_0.

Chapter 13

Difference equations

*My goodness, I'm feeling fine
said Granny, waving with her hand.
Next birthday I'll be ninety-nine
with just one year left to thousand.*

ELIAS SEHLSTEDT*

13.0. Notations and definitions

We will now work with a function $y = y(x)$, which can be defined for all real values of x, or only in certain points. In the latter case we will assume that the points are equidistant (grid points), and further we suppose that the variable x has been transformed in such a way that the interval length is 1, and the grid points correspond to integer values of x. Then we shall also use the notation y_n for $y(n)$.

We define an ordinary *difference equation* as an equation which contains an independent variable x, a dependent variable y, and one or several differences $\Delta y, \Delta^2 y, \ldots, \Delta^n y$. Of course, it is not an essential specialization to assume *forward* differences.

If we use the formula $\Delta = E - 1$, we infer that the difference equation can be trivially transformed to a relation of the type

$$F\big(x, y(x), y(x + 1), \ldots, y(x + n)\big) = 0 , \tag{13.0.1}$$

where n is called the order of the equation. As a matter of fact, an equation of this kind should be called a *recurrence equation*, but we shall continue with the name difference equation.

The treatment of general difference equations is extremely complicated, and here we can discuss only some simple special cases. First of all we assume that the equation is *linear*, that is, that F has the form

$$p_0(x)y(x + n) + p_1(x)y(x + n - 1) + \cdots + p_n(x)y(x) = g(x) . \tag{13.0.2}$$

If $g(x) = 0$, the equation is said to be homogeneous, and further we can specialize to the case when all $p_i(x)$ are constant. We will restrict ourselves mainly to this latter case. After a suitable translation, which moves the origin to the point x, such an equation can always be written

$$y_n + c_1 y_{n-1} + \cdots + c_n y_0 = (E^n + c_1 E^{n-1} + \cdots + c_n)y_0$$
$$\equiv \varphi(E)y_0 = 0 . \tag{13.0.3}$$

---

* This is a free, though not very poetic, translation of a humorous Swedish verse.

13.1. The solution concept of a difference equation

At first glance one might expect that several details from the theory of differential equations could be transferred directly to the theory of difference equations. However, in the latter case we meet difficulties of quite unexpected character and without correspondence in the former case. A fundamental question is, what is meant by a "solution of a difference equation"? In order to indicate that the question is legitimate, we consider the following example.

The equation $y_{n+1} = 3y_n + 2n - 1$ is a first-order inhomogeneous difference equation. Assuming that $y_0 = 1$, we see that y_n is uniquely determined for all *integer* values n. We find

n	0	1	2	3	4	5	6 $\cdots$	-1	-2	-3 $\cdots$
y_n	1	2	7	24	77	238	723 $\cdots$	$\frac{4}{3}$	$\frac{19}{9}$	$\frac{82}{27}$ $\cdots$

Since the solution appears only in certain points, it is called *discrete*; further, it is a *particular solution*, since it depends on a special value of y_0. Now we easily find that $y(x) = c \cdot 3^x - x$, where c is a constant, satisfies the equation. The discrete solution, just discussed, is obtained if we take $c = 1$. The solution $y = 3^x - x$ is defined for all values of x and is called a *continuous particular solution*.

One might believe $y = c \cdot 3^x - x$ to be the general solution, but it is easy to show that this is not the case. Consider, for example,

$$y(x) = 3^x(c + \tfrac{1}{3}\cos 2\pi x + \tfrac{2}{5}\cos 6\pi x - \tfrac{1}{4}\sin^2 3\pi x) - x ,$$

and we see at once that the difference equation is satisfied again. In order to clarify this point, we put $y(x) = w(x)3^x - x$ and find $w(x + 1) = w(x)$. This is a new difference equation which, however, can be interpreted directly by saying that w must be a *periodic function* with period 1. We can choose $w = w(x)$ completely arbitrarily for $0 \leq x < 1$, but the corresponding solution y does not, in general, obtain any of the properties of continuity, differentiability, and so on. If we claim such properties, suitable conditions must be imposed on $w(x)$. For continuity, $w(x)$ must be continuous, and further, $\lim_{x \to 1-0} w(x) = w(0)$.

In the remainder of this chapter, $w(x)$, $w_1(x)$, $w_2(x)$, $\ldots$ will stand for periodic functions of x with period 1. Hence the general solution of the equation which we have discussed can be written

$$y(x) = w(x) \cdot 3^x - x .$$

On the other hand, consider a relation containing the independent variable x, the dependent variable y, and a periodic function $w(x)$:

$$F(x, y(x), w(x)) = 0 . \tag{13.1.1}$$

Operating with E, we get

$$F(x + 1, y(x + 1), w(x + 1)) = F_1(x, y(x + 1), w(x)) = 0 .$$

Eliminating $w(x)$ from these two equations, we obtain

$$G(x, y(x), y(x + 1)) = 0 , \qquad (13.1.2)$$

which is a first-order difference equation.

13.2. Linear, homogeneous difference equations

We shall now discuss equation (13.0.3) in somewhat greater detail. We assume $c_n \neq 0$; otherwise, the equation would be of lower order. First we consider only the discrete particular solutions appearing when the function values $y_0, y_1, \ldots, y_{n-1}$ are given. Thus y_n can be determined from the difference equation, and then the whole procedure can be repeated with n replaced by $n + 1$. Clearly, all discrete solutions are uniquely determined from the n given parameter values.

On the other hand, putting $y = \lambda^x$, we get $E^r y = \lambda^r \cdot \lambda^x$ and

$$(\lambda^n + c_1 \lambda^{n-1} + \cdots + c_n) \lambda^x \equiv \varphi(\lambda) \cdot \lambda^x = 0 .$$

Hence $y = \lambda^x$ is a solution of the difference equation if λ satisfies the algebraic equation

$$\lambda^n + c_1 \lambda^{n-1} + \cdots + c_n = \varphi(\lambda) = 0 . \qquad (13.2.1)$$

This equation is called the *characteristic equation* of the difference equation (13.0.3).

Since our equation is linear and homogeneous, we infer that, provided u and v are solutions and a and b constants, then $au + bv$ also is a solution. Hence

$$y = A_1 \lambda_1^x + A_2 \lambda_2^x + \cdots + A_n \lambda_n^x$$

is a solution of (13.0.3) if $\lambda_1, \lambda_2, \ldots, \lambda_n$ (all assumed to be different from one another), are the roots of (13.2.1). The n parameters $A_1, A_2, \ldots, A_n$ can be determined in such a way that for $x = 0, 1, 2, \ldots, n - 1$, the variable y takes the assigned values $y_0, y_1, y_2, \ldots, y_{n-1}$. Thus we have constructed a *continuous* particular solution which coincides with the discrete solution in the grid points. As in a previous special example, we conclude that the general solution can be written

$$y_0 = w_1(x) \cdot \lambda_1^x + w_2(x) \cdot \lambda_2^x + \cdots + w_n(x) \cdot \lambda_n^x . \qquad (13.2.2)$$

If the characteristic equation has complex roots, this does not give rise to any special difficulties. For example, the equation $y(x + 2) + y(x) = 0$ has the solution

$$y = a \cdot i^x + b(-i)^x = a \cdot e^{i\pi x/2} + b \cdot e^{-i\pi x/2} = A \cos \frac{\pi x}{2} + B \sin \frac{\pi x}{2} .$$

So far, we have considered only the case when the characteristic equation has simple roots. For multiple roots we can use a limit procedure, analogous to the well-known process for differential equations. Thus we suppose that r and

$r + \varepsilon$ are roots of the characteristic equation, that is, that r^x and $(r + \varepsilon)^x$ are solutions of the difference equation. Due to the linearity,

$$\lim_{\varepsilon \to 0} \frac{(r + \varepsilon)^x - r^x}{\varepsilon} = xr^{x-1}$$

is also a solution. Since multiplication with r is trivially allowed, we conclude that we have the two solutions r^x and xr^x. If the characteristic equation has a triple root r, then also x^2r^x is a solution. Hence, the general solution of the equation $(E - r)^n y(x) = 0$, that is,

$$y(x + n) - \binom{n}{1} ry(x + n - 1) + \binom{n}{2} r^2 y(x + n - 2) - \cdots$$

$$+ (-1)^n r^n y(x) = 0 ,$$

is $y = [w_0(x) + xw_1(x) + \cdots + x^{n-1}w_{n-1}(x)]r^x$.

EXAMPLES

1. The Fibonacci numbers are defined by $y(0) = 0$; $y(1) = 1$; $y(x + 2) = y(x + 1) + y(x)$ (hence 0, 1, 1, 2, 3, 5, 8, 13, 21, 34, ...). The characteristic equation is $\lambda^2 - \lambda - 1 = 0$ with the roots $\frac{1}{2}(1 \pm \sqrt{5})$. The "general" solution is

$$y(x) = a \left(\frac{1 + \sqrt{5}}{2}\right)^x + b \left(\frac{1 - \sqrt{5}}{2}\right)^x.$$

The initial conditions give

$$\begin{cases} a + b = 0 \\ (a - b)\sqrt{5} = 2 \end{cases} \quad \text{and} \quad a = -b = \frac{1}{\sqrt{5}} .$$

Hence

$$y(x) = \frac{1}{2^{x-1}} \left[\binom{x}{1} + \binom{x}{3} 5 + \binom{x}{5} 5^2 + \binom{x}{7} 5^3 + \cdots + \binom{x}{2m + 1} 5^m + \cdots\right].$$

2. $y(x) = py(x + 1) + qy(x - 1)$ with $p + q = 1$, and $y(0) = 1$, $y_N = 0$. The characteristic equation is $\lambda = p\lambda^2 + q$, with the roots 1 and q/p. If $p \neq q$, we find the solution

$$y(x) = \frac{p^{N-x}q^x - q^N}{p^N - q^N} ;$$

if $p = q = \frac{1}{2}$, we get $y = 1 - x/N$.

We shall also indicate briefly how inhomogeneous linear difference equations can be handled. The technique is best demonstrated in a few examples.

EXAMPLES

1. $\Delta y = (4x - 2) \cdot 3^x$.
Putting $y = (ax + b) \cdot 3^x$, we get $\Delta y = (2ax + 3a + 2b) \cdot 3^x$, and identifying, we find the solution $y = (2x - 4) \cdot 3^x + w(x)$.

2. $y(x + 2) - 3y(x + 1) + 2y(x) = 6 \cdot 2^x$.

The solution of the homogeneous equation is $y = w_1(x) \cdot 2^x + w_2(x)$. Putting $y = a \cdot x \cdot 2^x$, we easily find $a = 3$. Hence the general solution is

$$y = \left(w_1(x) + 3x\right) \cdot 2^x + w_2(x) .$$

13.3. Differences of elementary functions

The equation $\varDelta f(x) = g(x)$, where $g(x)$ is a given function, was discussed in Chapter 11 in the special case when $g(x)$ is a factorial. A formal solution can be obtained in the following way. Setting $D^{-1}g(x) = \int_a^x g(t)\, dt = G(x)$, we find $(h = 1)$:

$$f(x) = \frac{g(x)}{e^D - 1} = \frac{D}{e^D - 1}\, G(x) ,$$

and hence, adding a periodic function $w(x)$, we get

$$f(x) = w(x) + G(x) - \frac{1}{2}\, g(x) + \frac{1}{6}\, \frac{g'(x)}{2!} - \frac{1}{30}\, \frac{g'''(x)}{4!}$$

$$+ \frac{1}{42}\, \frac{g^{\mathrm{V}}(x)}{6!} - \frac{1}{30}\, \frac{g^{\mathrm{VII}}(x)}{8!} + \cdots .$$

The coefficient of $g^{(2p-1)}(x)$ for large p behaves like $(-1)^{p-1} \cdot 2/(2\pi)^{2p}$ [cf. (11.1.4)], and this fact gives some information concerning the convergence of the series. However, we shall now look at the problem from a point of view which corresponds to the well-known technique for performing integration by recognizing differentiation formulas. First, we indicate a few general rules for calculating differences, and then we shall give explicit formulas for the differences of some elementary functions.

$$\begin{cases} \varDelta(u_i \pm v_i) = \varDelta u_i \pm \varDelta v_i , \\[1mm] \varDelta(u_i v_i) = u_{i+1}\, \varDelta v_i + v_i\, \varDelta u_i = u_i\, \varDelta v_i + v_{i+1}\, \varDelta u_i , \\[1mm] \varDelta\left(\dfrac{u_i}{v_i}\right) = \dfrac{v_i\, \varDelta u_i - u_i\, \varDelta v_i}{v_i v_{i+1}} . \end{cases} \tag{13.3.1}$$

$$\begin{cases} \varDelta x^{(n)} = n x^{(n-1)} , \\[1mm] \varDelta a^x = (a - 1)a^x , \\[1mm] \varDelta \log x = \log\,(1 + 1/x) , \\[1mm] \varDelta \sin px = 2 \sin\,(p/2) \cos\,(px + p/2) , \\[1mm] \varDelta \cos px = -2 \sin\,(p/2) \sin\,(px + p/2) , \\[1mm] \varDelta(u_m + u_{m+1} + \cdots + u_n) = u_{n+1} - u_m , \\[1mm] \varDelta(u_m u_{m+1} \cdots u_n) = u_{m+1} u_{m+2} \cdots u_n(u_{n+1} - u_m) . \end{cases} \tag{13.3.2}$$

Returning to the equation $\Delta f(x) = g(x)$, we will now show how a solution can be obtained when $g(x)$ is a rational function. We restrict ourselves to the case when the linear factors of the denominator are different; then $g(x)$ can be written as a sum of terms of the form $a/(x - \alpha)$. Hence the problem has been reduced to solving the equation

$$\Delta f(x) = \frac{a}{x - \alpha}.$$

Now there exists a function $\Gamma(x)$ which is continuous for $x > 0$, and takes the value $(x - 1)!$ for integer, positive x (see further Chapter 18). The following fundamental functional relation is fulfilled:

$$\Gamma(x + 1) = x\Gamma(x).$$

Differentiating, we get $\Gamma''(x + 1) = x\Gamma''(x) + \Gamma'(x)$ and hence

$$\frac{\Gamma''(x + 1)}{\Gamma'(x + 1)} = \frac{\Gamma''(x)}{\Gamma'(x)} + \frac{1}{x}.$$

The function $\Gamma''(x)/\Gamma'(x)$ is called the digamma-function and is denoted by $\psi(x)$. Thus we have the following difference formulas:

$$\begin{cases} \Delta\Gamma(x) = (x - 1)\Gamma(x), \\ \Delta\psi(x) = \dfrac{1}{x}. \end{cases} \tag{13.3.3}$$

Differentiating the second formula, we get

$$\begin{cases} \Delta\psi'(x) = -\dfrac{1}{x^2}, \\[2mm] \Delta\psi''(x) = \dfrac{2}{x^3}, \\[2mm] \Delta\psi'''(x) = -\dfrac{6}{x^4}, \\[1mm] \vdots \end{cases} \tag{13.3.4}$$

EXAMPLE

Compute $S = \sum_{k=0}^{\infty} 1/((k + \alpha)(k + \beta))$, where $0 \leq \alpha \leq \beta$. Without difficulty, we find:

$$S = \frac{1}{\beta - \alpha} \lim_{N \to \infty} \sum_{k=0}^{N-1} \left(\frac{1}{k + \alpha} - \frac{1}{k + \beta} \right)$$

$$= \frac{1}{\beta - \alpha} \lim_{N \to \infty} \{\psi(N + \alpha) - \psi(N + \beta) - \psi(\alpha) + \psi(\beta)\} = \frac{\psi(\beta) - \psi(\alpha)}{\beta - \alpha}.$$

As a matter of fact, for large x we have $\psi(x) \sim \log x - 1/2x - 1/12x^2 + \cdots$, and from this we get $\lim_{N \to \infty} \{\phi(N + \alpha) - \phi(N + \beta)\} = 0$. If $\alpha = \beta$, we find simply that $\sum_{k=0}^{\infty} 1/(k + \alpha)^2 = \phi'(\alpha)$ and, in particular, $\sum_{k=1}^{\infty} 1/k^2 = \phi'(1) = \pi^2/6$. The Γ-function, as well as the functions ϕ, ϕ', ϕ'', ..., has been tabulated by Davis (see [3] in Chapter 11).

In some cases a solution $f(x)$ of $\Delta f(x) = g(x)$ can be obtained by trying a series expansion in factorials, possibly multiplied by some other suitable functions. Although for differential equations, power-series expansions are most convenient, we find expansions in factorials more suitable for difference equations. Further details can be found in [1].

13.4. Bernoulli's method for algebraic equations

As was pointed out in Section 13.2, the solution of a linear, homogeneous difference equation with constant coefficients is closely related to a purely algebraic equation. On the other hand, we can start with the algebraic equation written in the form (13.2.1). This equation can be interpreted as the characteristic equation of the matrix C, where

$$C = \begin{bmatrix} -c_1 & -c_2 \cdots & -c_n \\ 1 & 0 \cdots & 0 \\ \vdots & & \vdots \\ 0 & 0 \cdot \cdot 1 & 0 \end{bmatrix},$$

and at the same time as the characteristic equation of the difference equation (13.0.3). Putting

$$v_1 = \begin{bmatrix} y_{n-1} \\ y_{n-2} \\ \vdots \\ y_0 \end{bmatrix} \quad \text{and} \quad v_2 = \begin{bmatrix} y_n \\ y_{n-1} \\ \vdots \\ y_1 \end{bmatrix},$$

we have $v_2 = Cv_1$, since $y_n = -c_1 y_{n-1} - c_2 y_{n-2} - \cdots - c_n y_0$ [see (13.0.3)].

Hence we have reduced the point by point formation of the solution of (13.0.3) to the usual power method for determining eigenvalues of matrices, and we have shown before that $\lim_{k \to \infty} (y_k/y_{k-1}) = \lambda_1$, where λ_1 is the numerically largest eigenvalue of C (λ_1 assumed to be real). But at the same time λ_1 is the numerically largest root of the equation (13.2.1). The discussion of the case when the equation has several roots with the same absolute value was given in Section 6.1.

EXAMPLE

The equation $\lambda^3 - 8\lambda^2 - 15\lambda + 10 = 0$ can be interpreted as the characteristic equation of the difference equation $y_{k+3} = 8y_{k+2} + 15y_{k+1} - 10y_k$. We arbitrarily set $y_0 = y_1 = 0$; $y_2 = 1$, and obtain successively $y_3 = 8$; $y_4 = 79$; $y_5 = 742$; $y_6 = 7041$; $y_7 = 66668$; $y_8 = 631539$; $y_9 = 5981922$; $y_{10} = 56661781$; $y_{11} = 536707688$; $y_{12} = 5083768999$; ... Further we have $y_6/y_5 = 9.49$; $y_8/y_7 = 9.473$;

$y_{10}/y_9 = 9.47217$; $y_{12}/y_{11} = 9.472137$, compared with the exact roots -2, $5 - \sqrt{20} = 0.527864$ and $5 + \sqrt{20} = 9.472136$. Obviously, it is essential that one of the roots be numerically much larger than any of the others. We can also find the next largest root with this method. Let λ_1 and λ_2 be the roots in question, with $|\lambda_1| > |\lambda_2|$. Then

$$\lim_{k \to \infty} \frac{y_{k+1}}{y_k} = \lambda_1 \; ; \qquad \lim_{k \to \infty} \frac{y_k y_{k+2} - y_{k+1}^2}{y_{k-1} y_{k+1} - y_k^2} = \lambda_1 \lambda_2 \; .$$

In our example we obtain

$$\frac{y_{10} y_{12} - y_{11}^2}{y_9 y_{11} - y_{10}^2} = -18.9441554 \qquad \text{and} \qquad \lambda_2 \simeq -1.999987 \; .$$

As a rule, Bernoulli's method converges slowly, so instead, one ought to use, for example, the Newton-Raphson method.

13.5. Partial difference equations

A partial difference equation contains several independent variables, and is usually written as a recurrence equation. We shall here restrict ourselves to two independent variables, which will be denoted by x and y; the dependent variable is denoted by u.

We shall discuss only linear equations with constant coefficients, and first we shall demonstrate a direct technique. Consider the equation

$$u(x + 1, y) - u(x, y + 1) = 0 \; .$$

Introducing shift operators in both directions, we get

$$\begin{cases} E_1 u(x, y) = u(x + 1, y) \; , \\ E_2 u(x, y) = u(x, y + 1) \; , \end{cases}$$

and hence we can write

$$E_1 u = E_2 u \; .$$

Previously we have found that a formal solution of $E_1 u = au$ is $u = ca^x$; here c should be a function of y only. Hence we get $u = E_2^x f(y) = f(x + y)$ [a was replaced by E_2 and c by $f(y)$].

The general solution is $u = w(x, y) f(x + y)$, where $w(x, y)$ is periodic in x and y with period 1.

Another method is due to Laplace. Here we try to obtain a generating function whose coefficients should be the values of the desired function at the grid points on a straight line parallel to one of the axes. We demonstrate the technique by an example.

We take the equation $u(x, y) = pu(x - 1, y) + qu(x, y - 1)$ with the boundary conditions $u(x, 0) = 0$ for $x > 0$; $u(0, y) = q^y$ [hence $u(0, 0) = 1$]. Introducing the generating function $\varphi_y(\xi) = u(0, y) + u(1, y)\xi + u(2, y)\xi^2 + \cdots$, we find

$$q\varphi_{y-1}(\xi) = qu(0, y - 1) + \sum_{x=1}^{\infty} qu(x, y - 1)\xi^x .$$

Further,

$$p\xi\varphi_y(\xi) = \sum_{x=1}^{\infty} pu(x - 1, y)\xi^x .$$

Adding together, we get

$$q\varphi_{y-1}(\xi) + p\xi\varphi_y(\xi) = qu(0, y - 1) + \sum_{x=1}^{\infty} u(x, y)\xi^x$$

$$= qu(0, y - 1) - u(0, y) + \varphi_y(\xi) .$$

Here we have used the initial equation; if we make use of the boundary values as well, we get

$$\varphi_y(\xi) = \left(\frac{q}{1 - p\xi}\right)^y .$$

This completes the computation of the generating function. The desired function $u(x, y)$ is now obtained as the coefficient of the ξ^x-term in the power-series expansion of $\varphi_y(\xi)$. We obtain directly:

$$\varphi_y(\xi) = q^y(1 - p\xi)^{-y} = q^y \left(1 + y \cdot p\xi + \frac{y(y + 1)}{1 \cdot 2} p^2\xi^2 + \cdots\right),$$

and hence

$$u(x, y) = \frac{y(y + 1) \cdots (y + x - 1)}{1 \cdot 2 \cdots x} p^x q^y = \binom{x + y - 1}{x} p^x q^y .$$

Still another elegant method is due to Lagrange. Here we shall only sketch the method very briefly, and for this purpose, we use the same example. We try to find a solution in the form $u = \alpha^x\beta^y$ and get the condition $\alpha\beta = p\beta + q\alpha$. Hence $u(x, y) = \alpha^x q^y(1 - p\alpha^{-1})^{-y}$ with arbitrary α is a solution. Obviously, we can multiply with an arbitrary function $\varphi(\alpha)$ and integrate between arbitrary limits, and again we obtain new solutions of the equation. Now we have $\varphi(\alpha)$, as well as the integration limits, at our disposal for satisfying also the boundary condition. The computation is performed with complex integration in a way similar to that used for solving ordinary differential equations by means of Laplace transformations. For details, see [1] and [2].

Ordinary difference equations are of great importance for the numerical solution of ordinary differential equations, and analogously, partial difference equations play an important role in connection with the numerical treatment of partial differential equations. These matters will be discussed more thoroughly in Chapters 14 and 15.

REFERENCES

[1] Milne-Thomson: *The Calculus of Finite Differences* (Macmillan, London, 1933).

[2] Uspensky: *Introduction to Mathematical Probability* (McGraw-Hill, New York, 1937).

EXERCISES

1. Solve the difference equation $\delta^2 f(x) = \alpha f(x)$. Consider particularly the cases $\alpha = -4 \sin^2 \frac{1}{2}$ and $\alpha = 4 \sinh^2 \frac{1}{2}$.

2. Solve the equation $x^4 - 17x^3 - 59x^2 + 17x + 61 = 0$, using Bernoulli's method.

3. The integral $\int_0^1 x^n e^{-x} \, dx$, where n is an integer ≥ 0, can be written in the form $a_n - b_n \cdot e^{-1}$, where a_n and b_n are integers. Prove that for $n > 1$, a_n and b_n satisfy the difference equations $a_n = na_{n-1}$; $b_n = nb_{n-1} + 1$, with $a_0 = b_0 = 1$, and find $\lim_{n \to \infty} (b_n/a_n)$.

4. $a_0, a_1,$ and a_2 are given real numbers. We form $a_n = \frac{1}{3}(a_{n-1} + a_{n-2} + a_{n-3})$ for $n = 3, 4, 5, \ldots$ and write $a_n = A_n a_0 + B_n a_1 + C_n a_2$. Further, we put $A = \lim_{n \to \infty} A_n$; $B = \lim_{n \to \infty} B_n$; $C = \lim_{n \to \infty} C_n$. Find $A, B,$ and C.

5. Solve the difference equation $u_{n+1} - 2 \cos x \cdot u_n + u_{n-1} = 0$, when

(a) $\begin{cases} u_0 = 1, \\ u_1 = \cos x, \end{cases}$ (b) $\begin{cases} u_0 = 0, \\ u_1 = 1. \end{cases}$

6. $\alpha, \beta, u_0,$ and u_1 are given numbers. For $k = 0, 1, 2, \ldots$, we have $u_{k+2} = \alpha u_{k+1} + \beta u_k$. Putting $U_k = u_k u_{k+2} - u_{k+1}^2$ show that $U_k U_{k+2} - U_{k+1}^2 = 0$.

7. The integral

$$y_n = \int_0^1 \frac{x^n}{x^2 + x + 1} \, dx$$

is given. Show that $y_{n+2} + y_{n+1} + y_n = 1/(n+1)$, and use this relation for computing y_n ($n \leq 12$); y_0 and y_1 are computed directly. Also find a value N such that $y_N = 0.04$.

8. Show that $S = \sum_{k=1}^{\infty} (1/a_k)$ is convergent, where the numbers a_k form a Fibonacci series ($a_1 = a_2 = 1$). Also show that with $R_n = \sum_{k=n+1}^{\infty} (1/a_k)$, we have $\lim_{n \to \infty} a_{n-1} R_n = 1$. Find $\frac{1}{1} + \frac{1}{1} + \frac{1}{2} + \frac{1}{3} + \frac{1}{5} + \frac{1}{8} + \frac{1}{13} + \frac{1}{21} + \cdots$ correct to six places.

9. Given a quadratic band-matrix A of order n with the elements

$$a_{ik} = \begin{cases} -1 & \text{for } |i - k| = 1, \\ 0 & \text{otherwise.} \end{cases}$$

Show that the characteristic equation $f_n(\lambda) = 0$ can be obtained from $f_k = \lambda f_{k-1} - f_{k-2}$, $f_0 = 1, f_1 = \lambda$. Find an explicit expression for $f_n(\lambda)$ and the eigenvalues of A.

10. Find all values of k for which the difference equation $y_n - 2y_{n+1} + y_{n+2} + (k^2/N^2)y_n = 0$ has nontrivial solutions such that $y_0 = y_N = 0$. Also give the form of the solutions.

11. A_n and B_n are defined through

$$\begin{cases} A_n = A_{n-1} + xB_{n-1}, \\ B_n = B_{n-1} + xA_{n-1}, \end{cases} \quad n = 1, 2, 3, \ldots, \quad A_0 = 0, \quad B_0 = 1.$$

Determine A_n/B_n as a function of n and x.

12. Solve the system of difference equations

$$\begin{cases} x_{n+1} = 7x_n + 10y_n, \\ y_{n+1} = x_n + 4y_n, \end{cases}$$

with $x_0 = 3, y_0 = 2$.

Chapter 14

Ordinary differential equations

Eppur si muove. Galilei.

14.0. Existence of solutions

An ordinary differential equation is an equation containing one independent and one dependent variable and at least one of its derivatives with respect to the independent variable; no one of the two variables need enter the equation explicitly. If the equation is of such a form that the highest (*n*th) derivative can be expressed as a function of lower derivatives and the two variables, then it is possible to replace the equation by a system of *n* first-order equations by use of a simple substitution technique. The definition of linear and homogeneous equations, as well as linear equations with constant coefficients, is trivial and should need no comment.

The discussion of a system of first-order differential equations can, in essence, be reduced to an examination of the equation

$$y' = f(x, y) . \tag{14.0.1}$$

For this reason we shall pay a great deal of attention to this equation. It is obvious that the properties of the function $f(x, y)$ are of decisive importance for the character of the solution.

We will first point out a simple geometrical fact. By equation (14.0.1) every point in the domain of definition for f is assigned one or several directions according as f is one-valued or not; only the first of these cases will be treated here. In this way it is possible to give a picture of the directions associated with the equation, and we can obtain a qualitative idea of the nature of the solutions.

We are now going to consider two important cases for $f(x, y)$: first that the function is analytic, second that it fulfills the Lipschitz condition.

If $f(x, y)$ is analytic (note that x and y may be two different complex variables), then it is an easy matter to obtain a solution of (14.0.1) by aid of Taylor's formula. The differential equation can be differentiated an arbitrary number

260

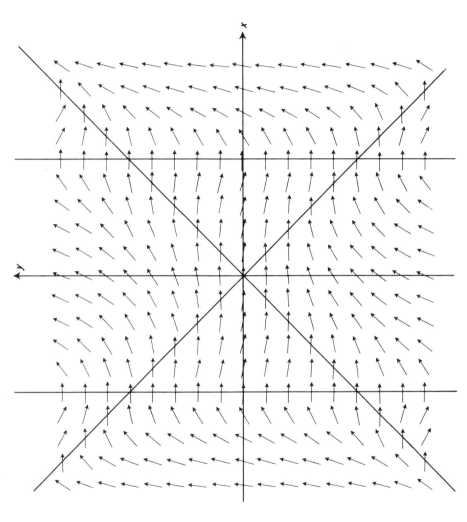

Figure 14.0.   Direction field for the differential equation $y' = (x^2 - 1)(x^2 - y^2)$.   Except the coordinate axes the lines $x = \pm 1$ and $y = \pm x$ are drawn.

of times, and hence we can obtain as many derivatives as we want. With the initial value $y = y_0$ for $x = x_0$, we get

$$y = y_0 + (x - x_0)y_0' + \frac{1}{2!}(x - x_0)^2 y_0'' + \cdots ,$$

and provided that $|x - x_0|$ is sufficiently small, the series converges and gives a unique solution of (14.0.1), with the initial condition $y = y_0$ for $x = x_0$.

Often one does not want to demand so much as analyticity from the function $f(x, y)$. If we require that $f(x, y)$ be bounded and continuous, this turns out not to be enough to guarantee a unique solution. A widely used extra condition is the so-called Lipschitz condition

$$|f(x, y) - f(x, z)| < L|y - z| , \qquad (14.0.2)$$

where L is a constant.

Now we suppose that this condition is fulfilled, and further that $f(x, y)$ is bounded and continuous within the domain under consideration: $|f(x, y)| < M$. Further we assume the initial condition $y = y_0$ for $x = x_0$. We integrate (14.0.1) between x_0 and x and obtain

$$y = y_0 + \int_{x_0}^{x} f(\xi, y)\, d\xi . \qquad (14.0.3)$$

Thus the differential equation has been transformed to an integral equation. Now we choose an initial approximation $y = y_1(x)$ which satisfies the conditions $|y_1'(x)| < M$ and $y_1(x_0) = y_0$. For example, we could choose $y_1(x) = y_0$, but in general it should be possible to find a better starting solution. Then we form a sequence of functions $y_i(x)$, $i = 2, 3, 4, \ldots$

$$y_{i+1}(x) = y_0 + \int_{x_0}^{x} f(\xi, y_i(\xi))\, d\xi . \qquad (14.0.4)$$

This equation defines Picard's method which on rare occasions is used also in practical work. We now obtain

$$|y_{i+1}(x) - y_i(x)| \leq \int_{x_0}^{x} |f(\xi, y_i(\xi)) - f(\xi, y_{i-1}(\xi))|\, d\xi$$

$$\leq L \int_{x_0}^{x} |y_i(\xi) - y_{i-1}(\xi)|\, d\xi .$$

We suppose $x_0 \leq x \leq X$, and putting $X - x_0 = h$, we obtain

$$|y_2(x) - y_1(x)| = \left| y_0 - y_1(x) + \int_{x_0}^{x} f(\xi, y_1(\xi))\, d\xi \right|$$

$$\leq |y_1(x) - y_0| + \int_{x_0}^{x} M\, d\xi \leq 2Mh = N .$$

Further, we find successively:

$$|y_3(x) - y_2(x)| \leq L \int_{x_0}^{x} N \, d\xi = NL(x - x_0) \leq NLh \, ,$$

$$|y_4(x) - y_3(x)| \leq L \int_{x_0}^{x} NL(\xi - x_0) \, d\xi = N \frac{L^2(x - x_0)^2}{2!} \leq N \frac{L^2 h^2}{2!} \, ,$$

$$\vdots$$

$$|y_{n+1}(x) - y_n(x)| \leq L \int_{x_0}^{x} N \frac{L^{n-2}(\xi - x_0)^{n-2}}{(n-2)!} \, d\xi$$

$$= N \frac{L^{n-1}(x - x_0)^{n-1}}{(n-1)!} \leq N \frac{L^{n-1} h^{n-1}}{(n-1)!} \, .$$

But

$$y_{n+1}(x) = y_1(x) + \big(y_2(x) - y_1(x)\big) + \cdots + \big(y_{n+1}(x) - y_n(x)\big) \, , \qquad (14.0.5)$$

and apart from $y_1(x)$, every term in this series is less in absolute value than the corresponding term in the series

$$N \left(1 + Lh + \frac{L^2 h^2}{2!} + \cdots + \frac{L^{n-1} h^{n-1}}{(n-1)!} \right) ,$$

which converges toward Ne^{Lh}. Thus the series (14.0.5) is absolutely and uniformly convergent toward a continuous limit function $y(x)$. The continuity follows from the fact that every partial sum has been obtained by integration of a bounded function.

Now we form

$$\left| y(x) - y_0 - \int_{x_0}^{x} f(\xi, y(\xi)) \, d\xi \right|$$

$$= \left| y(x) - y_{n+1}(x) - \int_{x_0}^{x} \{ f(\xi, y(\xi)) - f(\xi, y_n(\xi)) \} \, d\xi \right|$$

$$\leq |y(x) - y_{n+1}(x)| + L \int_{x_0}^{x} |y(\xi) - y_n(\xi)| \, d\xi \longrightarrow 0 \, ,$$

since $y_n(x)$ and $y_{n+1}(x)$ converge toward $y(x)$ when $n \to \infty$. Hence we find

$$y(x) = y_0 + \int_{x_0}^{x} f(\xi, y(\xi)) \, d\xi \, . \qquad (14.0.6)$$

Again we point out that $y(x) = \lim_{n \to \infty} y_n(x)$, where the formation law is given by (14.0.4). Differentiating (14.0.6), we get back (14.0.1).

It is easy to show that the solution is unique. Assuming that $z = z(x)$ is another solution, we obtain

$$z = y_0 + \int_{x_0}^{x} f(\xi, z(\xi)) \, d\xi \, , \qquad y_{n+1} = y_0 + \int_{x_0}^{x} f(\xi, y_n(\xi)) \, d\xi \, ,$$

and hence

$$|z - y_{n+1}| \leq \int_{x_0}^{x} |f(\xi, z) - f(\xi, y_n)| \, d\xi \leq L \int_{x_0}^{x} |z - y_n| \, d\xi \, .$$

But $|z - y_0| \leq Mh = N/2$, and hence $|z - y_1| \leq (N/2)L(x - x_0) \leq (N/2)Lh$;

$$|z - y_2| \leq \frac{N}{2} \frac{L^2(x - x_0)^2}{2!} \leq \frac{N}{2} \frac{L^2h^2}{2!} \, ,$$

and finally

$$|z - y_{n+1}| \leq \frac{N}{2} \frac{L^{n+1}h^{n+1}}{(n+1)!} \to 0 \, ,$$

when $n \to \infty$. Hence we have $z(x) \equiv y(x)$.

We can easily give examples of equations where the Lipschitz condition is not fulfilled and where more than one solution exists. The equation $y' = \sqrt{y}$ with $y(0) = 0$ has the two solutions $y = 0$ and $y = x^2/4$ for $x \geq 0$.

14.1. Classification of solution methods

In the literature one can find many examples of equations which can be solved explicitly. In spite of this fact good arguments can be given that these equations constitute a negligible minority. We shall restrict ourselves to such methods as are of interest for equations which cannot be solved explicitly in closed form. However, the methods will often be illustrated on simple equations whose exact solution is known.

If an explicit solution cannot be constructed one is usually satisfied by computing a solution in certain discrete points, as a rule equidistant in order to facilitate interpolation. On first hand, we shall consider the differential equation $y' = f(x, y)$, and we are going to use the notations $x_k = x_0 + kh$, and $y_k = $ the value of y obtained from the method [to be distinguished from the exact value $y(x_k)$]. The value y_{n+1} may then appear either as a function of just one y-value y_n, or as a function of several values $y_n, y_{n-1}, \ldots, y_{n-p}$; further, as a rule, the values x_n and h will also be present. In the first case we have a *single-step method*, in the second case a *multi-step method*.

The computation of y_{n+1} is sometimes performed through an explicit, sometimes an implicit, formula. For example,

$$y_{n+1} = y_{n-1} + 2hf(x_n, y_n)$$

is an explicit multi-step formula, while

$$y_{n+1} = y_n + \frac{h}{2} [f(x_n, y_n) + f(x_{n+1}, y_{n+1})]$$

is an implicit single-step formula. In the latter case an iterative technique is often used to determine the value of y_{n+1}.

14.2. Different types of errors

For obvious reasons error estimations play an essential role in connection with numerical methods for solution of differential equations, The errors have two different sources. First, even the simplest numerical method implies introduction of a *truncation* or *discretization error* since the exact solutions as a rule are transcendental and need an infinite number of operations, while the numerical method only resorts to a finite number of steps. Second, all operations are performed in finite precision, and hence *rounding errors* can never be avoided.

Let $y(x_n)$ be the exact solution in the point x_n, and y_n the exact value which would result from the numerical algorithm. Then the *total truncation error* ε_n is defined through

$$\varepsilon_n = y_n - y(x_n) \, . \tag{14.2.1}$$

Due to the finite precision computation we actually calculate another value $\bar{y}_n$. The rounding error ε'_n is defined through the relation

$$\varepsilon'_n = \bar{y}_n - y_n \, . \tag{14.2.2}$$

For the total error r_n we then obtain

$$|r_n| = |\bar{y}_n - y(x_n)|$$
$$= |(\bar{y}_n - y_n) + (y_n - y(x_n))| \leq |\varepsilon_n| + |\varepsilon'_n| \, .$$

There are two main problems associated with these error estimations, viz. the form of the truncation error, and the nature of the error propagation, that is, how an error introduced by, for example, round-off is propagated during later stages of the process. In the first case the situation can be improved by making h smaller; in the second case this will have no effect except in some very special cases. When one wants to investigate the truncation errors one should first study the *local truncation error*, that is, the error resulting on use of the method in just one step.

14.3. Single-step methods; Euler's method

In this section we shall to some extent discuss a method, suggested already by Euler, but almost never used in practical work. However, the method is of great interest because of its simplicity. Applied on the differential equation $y' = f(x, y)$ it is defined through

$$y_{n+1} = y_n + hf(x_n, y_n) \, ; \qquad y_0 = \alpha \, . \tag{14.3.1}$$

Geometrically the method has a very simple meaning: the wanted function curve is approximated by a polygon train where the direction of each part is determined as the function value $f(x, y)$ in its starting point. First we shall derive the local truncation error ε_1 assuming that the solution is twice continuously differentiable. Since ε_1 is the error after one step, we have

$$\varepsilon_1 = y_1 - y(x_1) = y_0 + hy_0' - y(x_0 + h) = -\tfrac{1}{2}h^2 y''(\xi) ,$$

where $x_0 < \xi < x_1$ and $y_0 = y(x_0)$. Hence the local truncation error is $O(h^2)$.
Further, we shall also estimate the total truncation error $\varepsilon_n = \varepsilon(x)$ with

$$x = x_n = x_0 + nh ,$$

assuming the Lipschitz condition $| f(x, y_2) - f(x, y_1)| \leq L|y_2 - y_1|$ and $|y''(\xi)| \leq N$:

$$\begin{cases} y_{m+1} = y_m + hf(x_m, y_m) , & m = 0, 1, 2, \ldots, n - 1 , \\ y(x_{m+1}) = y(x_m) + hf(x_m, y(x_m)) + \tfrac{1}{2}h^2 y''(\xi_m) , & x_m < \xi_m < x_{m+1} . \end{cases}$$

Subtracting and putting $\varepsilon_m = y_m - y(x_m)$, we obtain

$$\varepsilon_{m+1} = \varepsilon_m + h[f(x_m, y_m) - f(x_m, y(x_m))] - \tfrac{1}{2}h^2 y''(\xi_m) .$$

Hence

$$|\varepsilon_{m+1}| \leq (1 + hL)|\varepsilon_m| + \tfrac{1}{2}h^2 N .$$

We prefer to write this equation in the form $|\varepsilon_{m+1}| \leq A|\varepsilon_m| + B$ ($m = 0, 1,$ $2, \ldots, n - 1$) and by induction we easily prove for $A \neq 1$:

$$|\varepsilon_n| \leq A^n|\varepsilon_0| + \frac{A^n - 1}{A - 1} B .$$

But $\varepsilon_0 = 0$ and $A^n = (1 + hL)^n < e^{nhL} = e^{L(x - x_0)}$, which gives

$$|\varepsilon_n| \leq \frac{1}{2} hN \frac{e^{L(x - x_0)} - 1}{L} .$$

It is possible to prove the asymptotic formula: $\varepsilon_n \sim c_1 h + c_2 h^2 + c_3 h^3 + \cdots$, where the coefficients $c_1, c_2, c_3, \ldots$, depend upon x and the function $f(x, y)$. Starting from Euler's method, existence and uniqueness proofs can be constructed by showing that the sequence of functions which is obtained when $h \to 0$ converges toward a function $y(x)$ satisfying the differential equation.

14.4. Taylor series expansion

Assuming sufficient differentiability properties of the function $f(x, y)$ we can compute higher-order derivatives directly from the differential equation:

$$y' = f(x, y) ,$$

$$y'' = \frac{\partial f}{\partial x} + \frac{\partial f}{\partial y} y' = \frac{\partial f}{\partial x} + f \frac{\partial f}{\partial y} ,$$

$$y''' = \frac{\partial^2 f}{\partial x^2} + 2f \frac{\partial^2 f}{\partial x \partial y} + f^2 \frac{\partial^2 f}{\partial y^2} + \frac{\partial f}{\partial x} \frac{\partial f}{\partial y} + f \left(\frac{\partial f}{\partial y} \right)^2 ,$$

$$\vdots$$

After this we can compute $y(x + h) = y(x) + hy'(x) + \frac{1}{2}h^2 y''(x) + \cdots$, the series being truncated in a convenient way depending on the value of h. We demonstrate the method in an example, $y' = 1 - 2xy$. Differentiating repeatedly, we get

$$\begin{cases} y'' = -2xy' - 2y\,, \\ y''' = -2xy'' - 4y'\,, \\ \vdots \\ y^{(n+2)} = -2xy^{(n+1)} - 2(n+1)y^{(n)}\,. \end{cases}$$

Putting $\alpha_0 = y$; $\alpha_1 = hy'$; $\alpha_2 = h^2 y''/2$; $\ldots$, we obtain

$$\alpha_{n+2} = -\frac{2h}{n+2}(x\alpha_{n+1} + h\alpha_n) \qquad \text{and} \qquad y(x+h) = \alpha_0 + \alpha_1 + \alpha_2 + \cdots\,.$$

It is possible to compute

$$y'' = -2x(1 - 2xy) - 2y = (4x^2 - 2)y - 2x\,,$$
$$y''' = (-8x^3 + 4x)y + 4x^2 - 4 + 8xy = (-8x^3 + 12x)y + 4(x^2 - 1)\,, \quad \text{etc.},$$

but it should be observed that it is better *not* to construct these explicit expressions.

NUMERICAL EXAMPLE

$$x = 1\,,$$
$$y = \alpha_0 = 0.53807\ 95069 \quad \text{(taken from a table)},$$
$$h = 0.1\,.$$

$$\begin{aligned}
\alpha_0 &= \quad\ 0.53807\ 95069 \\
\alpha_1 &= -0.00761\ 59014 \\
\alpha_2 &= -0.00461\ 92049 \\
\alpha_3 &= \quad\ 0.00035\ 87197 \\
\alpha_4 &= \qquad\qquad\quad 51600 \\
\alpha_5 &= \qquad\qquad -16413 \\
\alpha_6 &= \qquad\qquad\qquad 375 \\
\alpha_7 &= \qquad\qquad\qquad\ 36 \\
\alpha_8 &= \qquad\qquad\qquad\ \ 0 \\
y(1.1) &= \quad\ 0.52620\ 66801
\end{aligned}$$

Tabulated value: $0.52620\ 66800$

A great advantage of this method is that it can be checked in a simple and effective manner. We have

$$y(x - h) = \alpha_0 - \alpha_1 + \alpha_2 - \alpha_3 + \cdots\,,$$

and this value can be compared with a previous value. In this example we find

$y(0.9) = 0.54072\ 43189$ compared with the tabulated value $0.54072\ 43187$. The differences, of course, depend on round-off errors.

The error estimates in this method must, as a rule, be made empirically. Very often the values α_n vary in a highly regular way, and moreover, larger errors would be disclosed in the back-integration checking.

The Taylor series method has a reputation of being unsuitable for numerical work, one reason being storage problems for the higher derivatives, another difficulties in computing them. The first reason hardly has any significance at all with modern computers while the difficulties in computing higher derivatives seem to have been overestimated. It should be mentioned that there are nowadays formula-manipulating programming languages in existence and thus also this part of the work can be taken over by the computer.

14.5. Runge-Kutta's method

This is one of the most widely used methods, and it is particularly suitable in cases when the computation of higher derivatives is complicated. It can be used for equations of arbitrary order by means of a transformation to a system of first-order equations. The greatest disadvantage seems to be that it is rather difficult to estimate the error, and further, the method does not offer any easy checking possibilities.

First of all we shall discuss the solution of a first-order equation; generalization to equations of arbitrary order is almost trivial. Let the equation be $y' = f(x, y)$, with starting point (x_0, y_0) and interval length h. Then we put

$$\begin{cases} k_1 = hf(x_0, y_0)\,, \\ k_2 = hf(x_0 + mh, y_0 + mk_1)\,, \\ k_3 = hf(x_0 + nh, y_0 + rk_2 + (n-r)k_1)\,, \\ k_4 = hf(x_0 + ph, y_0 + sk_2 + tk_3 + (p-s-t)k_1)\,, \\ k = ak_1 + bk_2 + ck_3 + dk_4\,. \end{cases} \qquad (14.5.1)$$

The constants should be determined in such a way that $y_0 + k$ becomes as good an approximation of $y(x_0 + h)$ as possible. By use of series expansion, one obtains, after rather complicated calculations, the following system:

$$\begin{cases} a + b + c + d = 1\,, & cmr + d(nt + ms) = \frac{1}{6}\,, \\ bm + cn + dp = \frac{1}{2}\,, & cmnr + dp(nt + ms) = \frac{1}{8}\,, \\ bm^2 + cn^2 + dp^2 = \frac{1}{3}\,, & cm^2r + d(n^2t + m^2s) = \frac{1}{12}\,, \\ bm^3 + cn^3 + dp^3 = \frac{1}{4}\,, & dmrt = \frac{1}{24}\,. \end{cases} \qquad (14.5.2)$$

We have eight equations in ten unknowns, and hence we can choose two quantities arbitrarily. If we assume that $m = n$, which seems rather natural, we find

$$m = n = \tfrac{1}{2}\,; \qquad p = 1\,; \qquad a = d = \tfrac{1}{6}\,; \qquad b + c = \tfrac{2}{3}\,;$$
$$s + t = 1\,; \qquad cr = \tfrac{1}{6}\,; \qquad rt = \tfrac{1}{2}\,.$$

If we further choose $b = c$, we get $b = c = \frac{1}{3}$ and hence

$$m = n = \tfrac{1}{2}, \qquad s = 0, \qquad a = d = \tfrac{1}{6},$$
$$p = 1, \qquad t = 1, \qquad b = c = \tfrac{1}{3},$$
$$r = \tfrac{1}{2}.$$

Thus we have the final formula system:

$$\begin{cases} k_1 = hf(x_0, y_0), \\ k_2 = hf(x_0 + \tfrac{1}{2}h, y_0 + \tfrac{1}{2}k_1), \\ k_3 = hf(x_0 + \tfrac{1}{2}h, y_0 + \tfrac{1}{2}k_2), \\ k_4 = hf(x_0 + h, y_0 + k_3). \end{cases} \qquad k = \tfrac{1}{6}(k_1 + 2k_2 + 2k_3 + k_4). \quad (14.5.3)$$

If f is independent of y, the formula passes into Simpson's rule, and it can also be shown that the local truncation error is $O(h^5)$. The total error has the asymptotic form $\varepsilon(x) \sim c_4 h^4 + c_5 h^5 + c_6 h^6 + \cdots$. The explicit formula for the error term, however, is rather complicated, and in practice one keeps track of the errors by repeating the computation with $2h$ instead of h and comparing the results; we can also obtain an improvement by aid of the usual Richardson extrapolation.

Runge-Kutta's method can be applied directly to differential equations of higher order. Taking, for example, the equation $y'' = f(x, y, y')$, we put $y' = z$ and obtain the following system of first-order equations:

$$\begin{cases} y' = z, \\ z' = f(x, y, z). \end{cases}$$

This is a special case of

$$\begin{cases} y' = F(x, y, z), \\ z' = G(x, y, z), \end{cases}$$

which is integrated through:

$$\begin{cases} k_1 = hF(x, y, z), \\ k_2 = hF(x + \tfrac{1}{2}h, y + \tfrac{1}{2}k_1, z + \tfrac{1}{2}l_1), \\ k_3 = hF(x + \tfrac{1}{2}h, y + \tfrac{1}{2}k_2, z + \tfrac{1}{2}l_2), \\ k_4 = hF(x + h, y + k_3, z + l_3), \\ k = \tfrac{1}{6}(k_1 + 2k_2 + 2k_3 + k_4), \end{cases} \qquad \begin{aligned} & l_1 = hG(x, y, z), \\ & l_2 = hG(x + \tfrac{1}{2}h, y + \tfrac{1}{2}k_1, z + \tfrac{1}{2}l_1), \\ & l_3 = hG(x + \tfrac{1}{2}h, y + \tfrac{1}{2}k_2, z + \tfrac{1}{2}l_2), \\ & l_4 = hG(x + h, y + k_3, z + l_3), \\ & l = \tfrac{1}{6}(l_1 + 2l_2 + 2l_3 + l_4). \end{aligned}$$

The new values are $(x + h, y + k, z + l)$.

NUMERICAL EXAMPLE AND COMPUTATION SCHEME

$$y'' = xy'^2 - y^2, \qquad \begin{cases} y' = z, \\ z' = xz^2 - y^2, \end{cases} \qquad \begin{aligned} & (= F(x, y, z)) \\ & (= G(x, y, z)) \end{aligned} \qquad h = 0.2.$$

Initial conditions:

$$\begin{cases} x = 0, \\ y = 1, \\ y' = 0. \end{cases}$$

x	y	$z(=F)$	$xz^2 - y^2(=G)$	k	l
0	1	0	-1	0	-0.2
0.1	1	-0.1	-0.999	-0.02	-0.1998
0.1	0.99	-0.0999	-0.979102	-0.01998	-0.1958204
0.2	0.98002	-0.1958204	-0.9527709	-0.039164	-0.1905542
0.2	0.980146	-0.196966			

The values below the line are used to initiate the next step in the integration. Note that the interval length can be changed without restrictions. On the other hand, it is somewhat more difficult to get an idea of the accuracy, as pointed out above.

It should be mentioned here that there is a whole family of Runge-Kutta methods of varying orders. For practical use, however, the classical formula has reached a dominating position by its great simplicity.

A special phenomenon, which can appear when Runge-Kutta's method is applied, deserves to be mentioned. First we take an example. The equations

$$\begin{cases} y' = -12y + 9z, \\ z' = 11y - 10z, \end{cases}$$

have the particular solution

$$\begin{cases} y = 9e^{-x} + 5e^{-21x}, \\ z = 11e^{-x} - 5e^{-21x}. \end{cases}$$

For $x \geq 1$, we have $e^{-21x} < 10^{-9}$, and one would hardly expect any difficulties. However, starting with $x = 1$, $y = 3.3111(\simeq 9e^{-1})$, $z = 4.0469(\simeq 11e^{-1})$, and $h = 0.2$, we obtain

x	y	z	$11y/9z$
1.0	3.3111	4.0469	1.0000
1.2	2.7109	3.3133	1.0000
1.4	2.2195	2.7127	1.0000
1.6	1.8174	2.2207	1.0003
1.8	1.4892	1.8169	1.0018
2.0	1.2270	1.4798	1.0134
2.2	1.0530	1.1632	1.1064
2.4	1.1640	0.6505	2.1870
2.6	2.8360	-1.3504	-2.5668

Round-off has been performed strictly according to the principles given in Chapter 1. A slight change in these principles is sufficient to produce a completely different table.

The explanation is quite simple. When we use Runge-Kutta's method, we approximate e^{-ah} by $1 - ah + \frac{1}{2}a^2h^2 - \frac{1}{6}a^3h^3 + \frac{1}{24}a^4h^4$, which is acceptable when $a = 1$ and $h = 0.2$ (the error is $< 3 \cdot 10^{-6}$); however, it is a bad approximation if $a = 21$ (6.2374 compared with $e^{-4.2} \simeq 0.014996$). In spite of the fact that proper instabilities appear only in connection with multi-step methods, the phenomenon observed here is usually called *partial instability*. A reduction of the interval length (in this case to about 0.1) would give satisfactory results. If the system of differential equations is written in the form

$$y_i' = f_i(x, y_1, y_2, \ldots, y_n) ; \qquad i = 1, 2, \ldots, n ,$$

the stability properties are associated with a certain matrix A having the elements $a_{ik} = \partial f_i / \partial y_k$. As a matter of fact, there are several matrices A since the derivatives are considered in different points, but if we assume sufficiently small intervals we can neglect this. Denoting the characteristic values of A by λ_i, we have stability for $-2.785 < h\lambda_i < 0$ if the eigenvalues are real and for $0 < |h\lambda_i| < 2\sqrt{2}$ if the eigenvalues are purely imaginary. In the case of complex eigenvalues $h\lambda_i$ must belong to a closed domain in the complex plane, mainly situated in the left half-plane (also cf. [10]). The value -2.785 is the real root not equal to 0 of the equation $1 + x + x^2/2! + x^3/3! + x^4/4! = 1$. In the example above, we had

$$A = \begin{pmatrix} -12 & 9 \\ 11 & -10 \end{pmatrix},$$

that is, $\lambda_1 = -1$ and $\lambda_2 = -21$. Consequently h must be chosen so that $21h < 2.785$, or $h < 0.1326$. Explicit calculations show, for example, that $h = 0.1$ gives perfect results.

14.6. Multi-step methods

As has been mentioned previously, a multi-step method defines the wanted value y_{n+k} as a function of several preceding values $y_{n+k-1}, y_{n+k-2}, \ldots, y_n$. In this case we have a k-step method, and if in particular $k = 1$, we are back with the single-step methods which have just been discussed. The method is *explicit* if the value can be found directly, and *implicit* if the formula contains the wanted value also on the right-hand side. We shall assume equidistant abscissas throughout, and we observe immediately that a change of the interval length h will be difficult to achieve. Further it is also obvious that a special technique is necessary for calculation of a sufficient number of initial values. Primarily we shall treat the equation $y' = f(x, y)$, later on also $y'' = f(x, y)$. For brevity the notation $f_k = f(x_k, y_k)$ will be used.

The general linear k-step method is then defined through the formula

$$\alpha_k y_{n+k} + \alpha_{k-1} y_{n+k-1} + \cdots + \alpha_0 y_n$$
$$= h[\beta_k f_{n+k} + \beta_{k-1} f_{n+k-1} + \cdots + \beta_0 f_n], \qquad n = 0, 1, 2, \ldots. \quad (14.6.1)$$

Two polynomials will be introduced, namely,

$$\begin{cases} \rho(z) = \alpha_k z^k + \alpha_{k-1} z^{k-1} + \cdots + \alpha_0, \\ \sigma(z) = \beta_k z^k + \beta_{k-1} z^{k-1} + \cdots + \beta_0. \end{cases}$$

The polynomial $\rho(z)$ at the same time is the characteristic polynomial associated with the difference equation (14.6.1) when $h = 0$. The method is now said to be *convergent* if $\lim_{h \to 0} y_n = y(x_n)$, and this must be valid even for initial values which are close to the right ones and converge to these when $h \to 0$. Note that $n \to \infty$ in such a way that nh becomes finite. Then a necessary condition for convergence is that the zeros z_i of $\rho(z)$ are such that $|z_i| \le 1$, and further that all zeros on the unit circle are simple. This condition is known as the *stability condition*. Methods for which this condition is not fulfilled are said to be *strongly unstable*.

It is easy to show that we do not get convergence if $|z_i| > 1$. To prove this we consider the equation $y' = 0$ with the initial value $y(0) = 0$ and exact solution $y(x) = 0$. Assume one root z_i such that $|z_i| = \lambda > 1$; then the solution of (14.6.1) contains a term Az_i^n (the right-hand side of the difference equation is equal to 0 since $f(x, y) = 0$). Further assume an initial value such that $A = h$ and consider y_n in the point $x = nh$. Then y_n will contain a term with the absolute value $(x/n)\lambda^n$ which tends to infinity when $h \to 0$ $(n \to \infty)$. For complete proof, see Henrici [7].

By discussing other simple equations we can obtain further conditions which must be satisfied in order to secure convergence of the method. Let us first consider the differential equation $y' = 0$ with the initial value $y(0) = 1$ and exact solution $y(x) = 1$. Since $f(x, y) = 0$ and all $y_r = 1$, we have

$$\alpha_k + \alpha_{k-1} + \cdots + \alpha_0 = 0. \quad (14.6.3)$$

Next we also consider the differential equation $y' = 1$ with the initial value $y(0) = 0$ and exact solution $y(x) = x$. Inserting this into (14.6.1) we find

$$\alpha_k y_{n+k} + \alpha_{k-1} y_{n+k-1} + \cdots + \alpha_0 y_n = h(\beta_k + \beta_{k-1} + \cdots + \beta_0).$$

Since $y_r = rh$, we obtain

$$(n + k)h\alpha_k + (n + k - 1)h\alpha_{k-1} + nh\alpha_0 = h(\beta_k + \beta_{k-1} + \cdots + \beta_0).$$

Taking (14.6.3) into account, we get

$$k\alpha_k + (k - 1)\alpha_{k-1} + \cdots + \alpha_1 = \beta_k + \beta_{k-1} + \cdots + \beta_0. \quad (14.6.4)$$

The conditions (14.6.3) and (14.6.4) can be written in a more compact form:

$$\begin{cases} \rho(1) = 0, \\ \rho'(1) = \sigma(1). \end{cases} \quad (14.6.5)$$

These relations are usually called the *consistency conditions*, and it is obvious from the derivation that they are necessary for convergence. Somewhat vaguely we can express this by saying that a method which does not satisfy the consistency conditions is mathematically wrong. On the other hand, it might well be the case that a "correct" method is strongly unstable (cf. Todd's example in a subsequent section of this chapter).

If the local truncation error is $O(h^{p+1})$, the *order* of the method is said to be p. One is clearly interested in methods which, for a given step-number k, are of highest possible order, but at the same time stability must be maintained. Dahlquist [4] has proved that the highest possible order which can be attained is $2k$, but if we also claim stability we cannot get more than $k + 1$ if k is odd, and $k + 2$ if k is even.

We shall now discuss a simple example in order to illustrate the effect of a multi-step method on the error propagation. We start from the equation

$$y' = ay , \qquad y(0) = 1 ,$$

and shall apply a method suggested by Milne:

$$y_{n+2} = y_n + \frac{h}{3} [y'_n + 4y'_{n+1} + y'_{n+2}] ,$$

that is, Simpson's formula. Also using the differential equation we obtain the following difference equation:

$$\left(1 - \frac{ah}{3}\right) y_{n+2} - \frac{4ah}{3} y_{n+1} - \left(1 + \frac{ah}{3}\right) y_n = 0 .$$

The characteristic equation is

$$\left(1 - \frac{ah}{3}\right) \lambda^2 - \frac{4ah}{3} \lambda - \left(1 + \frac{ah}{3}\right) = 0$$

with the roots $(2ah/3 \pm \sqrt{1 + a^2h^2/3})/(1 - ah/3)$ or after series expansion,

$$\lambda_1 = 1 + ah + \frac{a^2 h^2}{2} + \frac{a^3 h^3}{6} + \frac{a^4 h^4}{24} + \frac{a^5 h^5}{72} + \cdots \simeq e^{ah} + c_1 h^5 ,$$

$$\lambda_2 = -\left(1 - \frac{ah}{3} + \frac{a^2 h^2}{18} + \frac{a^3 h^3}{54} + \cdots\right) \simeq -(e^{-ah/3} + c_2 h^3) ,$$

where $c_1 = a^5/180$ and $c_2 = 2a^3/81$. The general solution of the difference equation is $y_n = A\lambda_1^n + B\lambda_2^n$, and since we are looking for the solution $y = e^{ax}$ of the differential equation we ought to choose $A = 1$, $B = 0$. We now compute approximate values of λ_1^n and λ_2^n choosing n and h in such a way that $nh = x$ where x is a given value. Then we find

$$\lambda_1^n \simeq (e^{ah} + c_1 h^5)^n = e^{anh}(1 + c_1 h^5 e^{-ah})^n = e^{ax}(1 + nc_1 h^5 e^{-ah} + \cdots)$$

$$\simeq e^{ax}(1 + c_1 x h^4) = e^{ax}(1 + \alpha x) ,$$

where $\alpha = c_1 h^4 = a^5 h^4 / 180$. In a similar way we also find

$$\lambda_2^n \simeq (-1)^n e^{-ax/3}(1 + \beta x) ,$$

where $\beta = c_2 h^2 = 2a^3 h^2 / 81$. We are now interested in the solution λ_1^n, but in a numerical calculation we can never avoid errors which will then be understood as an admixture of the other solution. However, the effect of this will be quite different in the two cases $a > 0$ and $a < 0$. For simplicity, we first assume $a = 1$. Further we assume that the parasitic solution from the beginning is represented with the fraction ε and the "correct" solution with the fraction $1 - \varepsilon$. Thus, the generated solution is

$$y_n \simeq (1 - \varepsilon)e^x(1 + \alpha x) + \varepsilon(-1)^n e^{-x/3}(1 + \beta x)$$

and the error is essentially

$$E_1 = \alpha x e^x - \varepsilon e^x + \varepsilon(-1)^n e^{-x/3} .$$

Since x is kept at a fixed value the first term depends only on the interval length $(\alpha = h^4/180)$ and can be made arbitrarily small. The second term depends on ε, that is, the admixture of the parasitic solution. Even if this fraction is small from the beginning, it can grow up successively because of accumulation of rounding errors. The last term finally represents an alternating and at the same time decreasing error which cannot cause any trouble.

We shall now give a numerical illustration and choose $h = 0.3$. Further we shall treat two alternatives for the initial value y_1 which we choose first according to the *difference equation* $(y_1 = \lambda_1)$, second according to the *differential equation* $(y_1 = \exp(0.3))$. We find

$$\lambda^2 - 4\lambda/9 = 11/9$$

and $\lambda_1 = 1.349877$; $\lambda_2 = -0.905432$. The difference equation is $9y_{n+2} = 4y_{n+1} + 11y_n$ and the results are given in the table below.

x	y_I	Error $\cdot 10^6$	$E_1^{(I)} \cdot 10^6$	y_{II}	Error $\cdot 10^6$	$E_1^{(II)} \cdot 10^6$
0	1			1	—	
0.3	1.349877	18	18	1.349859	—	0
0.6	1.822167	48	49	1.822159	40	41
0.9	2.459702	99	100	2.459676	73	74
1.2	3.320294	177	179	3.320273	156	158
1.5	4.481988	299	303	4.481948	259	262
1.8	6.050132	485	490	6.050088	441	446
2.1	8.166934	764	772	8.166864	694	703
2.4	11.024354	1178	1191	11.024270	1094	1106
2.7	14.881521	1789	1808	14.881398	1666	1686
3.0	20.088220	2683	2712	20.088062	2525	2554

For y_I we have with good approximation $\varepsilon = 0$ and hence $E_1^{(\mathrm{I})} \simeq \alpha x e^z$ with $\alpha = h^4/180 = 0.000045$. For y_II we easily find $\varepsilon = 0.000008$, and from this, E_1^II is computed. The difference between actual and theoretical values is due to the fact that the term αx is only an approximation of a whole series and also to some extent to rounding errors.

We now pass to the second case $a = -1$, that is, the differential equation $y' = -y$, $y(0) = 1$. In the same way as before we get

$$y_n \simeq (1 - \varepsilon)e^{-z}(1 + \alpha x) + \varepsilon(-1)^n e^{z/3}(1 + \beta x)\,,$$

where the signs of α and β have now been changed. Hence the error is

$$E_2 \simeq (\alpha x - \varepsilon)e^{-z} + \varepsilon(-1)^n e^{z/3}(1 + \beta x)\,.$$

Again we illustrate by a numerical example. With the same choice of interval ($h = 0.3$), the characteristic equation is $\lambda^2 + 4\lambda/11 - 9/11 = 0$ with the roots $\lambda_1 = 0.74080832$ and $\lambda_2 = -1.10444469$. Also in this case the solution is directed first according to the difference equation, that is, $y_1 = \lambda_1$, and second according to the differential equation, that is, $y_1 = \exp(-0.3)$. The difference equation has the form $11y_{n+2} = -4y_{n+1} + 9y_n$, and the results are presented in the table on p. 276.

As can be seen in the table there is good agreement between the actual and the theoretical error for the solution y_I up to about $x = 8$. Here the discrepancy between y_I and λ_1^n begins to be perceptible which indicates that ε is now not equal to 0. We also notice the characteristic alternating error which is fully developed and completely dominating for $x = 12$. Still the computed values y_I are fairly meaningful. If an error of this type appears, the method is said to be *weakly stable*. In particular it should be observed that for a given value x we can obtain any accuracy by making ε sufficiently small, and in this case, ε being equal to 0 initially, we have only to perform the calculations with sufficient accuracy. The other solution shows strong oscillations already from the beginning due to the fact that the initial value $\exp(-0.3)$ is interpreted as an admixture of the unwanted solution λ_2^n to the desired solution λ_1^n. We find

$$\varepsilon = \frac{(\lambda_1 - e^{ah})(1 - ah/3)}{2\sqrt{1 + a^2h^2/3}}\,;$$

for $a = 1$ and $h = 0.3$, we get $\varepsilon = -536 \cdot 10^{-8}$. Since $\varepsilon = O(h^5)$ the error can be brought down to a safe level by decreasing h; already a factor 4 gives $\varepsilon \sim 5 \cdot 10^{-9}$ which brings us back to the first case y_I. Again we stress the importance of steering after the *difference equation* in cases when weak stability can occur.

This case has been discussed in considerable detail because it illustrates practically all phenomena which are of interest in this connection, and the discussion of more general cases will be correspondingly facilitated. Again we choose the equation $y' = ay$, $y(0) = 1$ but now a more general method will be applied:

$$\alpha_k y_{n+k} + \alpha_{k-1} y_{n+k-1} + \cdots + \alpha_0 y_n = h(\beta_k f_{n+k} + \beta_{k-1} f_{n+k-1} + \cdots + \beta_0 f_n)\,.$$

x	$y_\mathrm{I} \cdot 10^8$	Error $\cdot 10^8$	$E_2^{(\mathrm{I})} \cdot 10^8$	$y_\mathrm{II} \cdot 10^8$	Error $\cdot 10^8$	$E_2^{(\mathrm{II})} \cdot 10^8$
0	1 0000 0000	—		1 0000 0000	—	
0.3	7408 0832	− 990	− 1000	7408 1822	—	
0.6	5487 9697	− 1467	− 1482	5487 9337	− 1826	− 1842
0.9	4065 5336	− 1630	− 1647	4065 6277	− 689	− 707
1.2	3011 7812	− 1609	− 1626	3011 7175	− 2246	− 2263
1.5	2231 1525	− 1491	− 1506	2231 2527	− 489	− 505
1.8	1652 8564	− 1325	− 1339	1652 7679	− 2210	− 2223
2.1	1224 4497	− 1146	− 1157	1224 5638	− 4	− 16
2.4	907 0826	− 969	− 980	906 9687	− 2108	− 2118
2.7	671 9743	− 808	− 816	672 1091	+ 539	+ 531
3.0	497 8042	− 665	− 672	497 6620	− 2087	− 2093
5.4	45 1551	− 107	− 110	44 8346	− 3313	− 3314
5.7	33 4510	− 87	− 86	33 8055	+ 3458	+ 3456
6.0	24 7811	− 64	− 67	24 3899	− 3976	− 3976
6.3	18 3577	− 53	− 52	18 7900	+ 4269	+ 4267
6.6	13 5999	− 38	− 40	13 1227	− 4810	− 4808
8.1	3 0340	− 14	− 11	3 8183	+ 7829	+ 7826
8.4	2 2482	− 5	− 9	1 3820	− 8667	− 8664
8.7	1 6648	− 11	− 7	2 6215	+ 9557	+ 9552
9.0	1 2341	0	− 5	1774	− 10567	− 10562
9.3	9133	− 9	− 4	2 0804	+ 11661	+ 11655
9.6	6776	+ 3	− 3	− 6114	− 12886	− 12879
12.0	628	+ 14				
12.3	439	− 16				
12.6	354	+ 17				
12.9	230	− 20				
13.2	206	+ 21				
13.5	113	− 24				
13.8	127	+ 26				
14.1	46	− 29				
14.4	83	+ 27				
14.7	7	− 34				
15.0	65	+ 35				

Putting $f = ay$ we get a difference equation with the characteristic equation

$$\rho(z) - ah\sigma(z) = 0 \; .$$

If h is sufficiently small the roots of this equation will be close to the roots z_i of the equation $\rho(z) = 0$. We assume the stability and consistency conditions to be fulfilled, and hence it is known that $|z_i| \leq 1$, $\rho(1) = 0$, and $\rho'(1) = \sigma(1)$.

Roots whose absolute values are <1 can never cause any difficulties, and we restrict ourselves to roots on the unit circle, all of which are known to be simple; among them is also $z_1 = 1$. Applying Newton-Raphson's method to the equation $\rho(z) - ah\sigma(z) = 0$ we find the root

$$z'_r = z_r + \frac{ah\sigma(z_r)}{\rho'(z_r)} + O(h^2) .$$

We now introduce the notation $\lambda_r = \sigma(z_r)/(z_r\rho'(z_r))$; for reasons which will soon become clear these quantities are called *growth parameters*. We find

$$(z'_r)^n = z_r^n\big(1 + a\lambda_r h + O(h^2)\big)^n = z_r^n\big(\exp(a\lambda_r h + c_r h^2 + O(h^3))\big)^n$$
$$= z_r^n \exp(a\lambda_r nh)\big(1 + c_r h^2 \exp(-a\lambda_r h) + O(h^3)\big)^n$$
$$\simeq z_r^n \exp(\lambda_r ax)(1 + c_r nh^2) = z_r^n \exp(\lambda_r ax)(1 + \alpha_r x) ,$$

where $\alpha_r = c_r h$ and x is supposed to be a fixed value. In particular we have

$$\lambda_1 = \frac{\sigma(1)}{\rho'(1)} = 1 ,$$

and the desired solution is obtained in the form $e^{ax}(1 + \alpha_1 x)$. Concerning the parasitic solutions we see at once that such roots z'_r, for which $|z_r| < 1$, will normally not cause any trouble (however, see the discussion below). If on the other hand $|z_r| = 1$ and $\mathrm{Re}(a\lambda_r) > 0$, we are confronted with exactly the same difficulties as have just been described (weak stability). A method which cannot give rise to weak stability is said to be *strongly stable*.

Here it should also be mentioned that the same complication as was discussed in connection with Runge-Kutta's method may appear with the present methods. For suppose that $a < 0$ and that h is chosen large enough to make $|z'_r| > 1$ in spite of the fact that $|z_r| < 1$; then the term z'^n_r will initiate a fast-growing error which can be interpreted as an unwanted parasitic solution. As has been mentioned earlier this phenomenon is called *partial instability* since it can be eliminated by choosing a smaller value of h.

At last, we shall briefly mention one more difficulty. Suppose that we are looking for a decreasing solution of a differential equation which has also an increasing solution that should consequently be suppressed. Every error introduced will be interpreted as an admixture of the unwanted solution, and it will grow quickly and independently of the difference method we are using. This phenomenon is called *mathematical* or *inherent instability*.

Finally we quote an example of strong instability which has been given by Todd. He considers the equation $y'' = -y$, and attempts a solution by $h^2 y'' = \delta^2 y - \frac{1}{12}\delta^4 y + \frac{1}{90}\delta^6 y - \cdots$. Truncating the series after two terms, he obtains with $h = 0.1$:

$$0.01 y''_n = y_{n+1} - 2y_n + y_{n-1} - \tfrac{1}{12}(y_{n+2} - 4y_{n+1} + 6y_n - 4y_{n-1} + y_{n-2})$$
$$= -0.01 y_n ,$$

or

$$y_{n+2} = 16y_{n+1} - 29.88y_n + 16y_{n-1} - y_{n-2} .$$

As initial values, he takes 0, sin 0.1, sin 0.2, and sin 0.3, rounded to 10 and 5 decimals, respectively.

x	$\sin x$	y (10 decimals)	Error	y (5 decimals)	Error
0	0	0		0	
0.1	0.09983 34166	0.09983 34166		0.09983	
0.2	0.19866 93308	0.19866 93308		0.19867	
0.3	0.29552 02067	0.29552 02067		0.29552	
0.4	0.38941 83423	0.38941 83685	262	0.38934	−8
0.5	0.47942 55386	0.47942 59960	4574	0.47819	−124
0.6	0.56464 24734	0.56464 90616	65882	0.54721	−1743
0.7	0.64421 76872	0.64430 99144	9 22272	0.40096	−24326
0.8	0.71735 60909	0.71864 22373	128 61464	−2.67357	−3.39093
0.9	0.78332 69096	0.80125 45441	1792 76345		
1.0	0.84147 09848	1.09135 22239	24988 12391		
1.1	0.89120 73601	4.37411 56871	3.48290 83270		

The explanation is quite simple. The characteristic equation of the difference equation is

$$r^4 - 16r^3 + 29.88r^2 - 16r + 1 = 0 ,$$

with two real roots $r_1 = 13.938247$ and $r_2 = 1/r_1 = 0.07174504$. The complex roots can be written $r_3 = e^{i\theta}$ and $r_4 = e^{-i\theta}$, where $\sin \theta = 0.09983347$. Thus the difference equation has the solution

$$y(n) = Ar_1^n + Br_1^{-n} + C \cos n\theta + D \sin n\theta ,$$

and for large values of n the term Ar_1^n predominates. The desired solution in our case is obtained if we put $A = B = C = 0$, $D = 1$; it becomes $y(n) = \sin (n \cdot 0.1000000556)$ instead of $\sin (n \cdot 0.1)$. In practical computation round-off errors can never be avoided, and such errors will be interpreted as admixtures of the three suppressed solutions; of these Ar_1^n represents a rapidly increasing error. The quotient between the errors in $y(1.1)$ and $y(1.0)$ is 13.93825, which is very close to r_1.

Hence the method used is very unstable, and the reason is that the differential equation has been approximated by a difference equation in an unsuitable way.

Here it should be observed that a strong instability cannot be defeated by making the interval length shorter, and as a rule such a step only makes the situation still worse.

14.7. Milne-Simpson's method

The general idea behind this method for solving the equation $y' = f(x, y)$ is first to perform an extrapolation by a coarse method, "*predictor*," to obtain an approximate value of y_{n+1}, and then to improve this value by a better formula, "*corrector*." The predictor can be obtained from an open formula of Cotes type (cf. 10.1) and we find

$$y_{n+1} - y_{n-3} = \frac{4h}{3} (2y'_{n-2} - y'_{n-1} + 2y'_n) + \frac{14}{45} h^5 y^V(\xi_1) . \qquad (14.7.1)$$

Neglecting the error term we get an approximate value of y_{n+1} with which we can compute $y'_{n+1} = f(x_{n+1}, y_{n+1})$. Then an improved value of y_{n+1} is determined from the corrector formula

$$y_{n+1} = y_{n-1} + \frac{h}{3} [y'_{n-1} + 4y'_n + y'_{n+1}] , \qquad (14.7.2)$$

with the local truncation error $-\frac{1}{90} h^5 y^V(\xi_2)$. If necessary, one can iterate several times with the corrector formula.

As before we put $y_n = y(x_n) + \varepsilon_n$ from which we get

$$y'_n = f(x_n, y_n) = f(x_n, y(x_n + \varepsilon_n)) \simeq f(x_n, y(x_n)) + \varepsilon_n \frac{\partial f}{\partial y} = y'(x_n) + \varepsilon_n \frac{\partial f}{\partial y} .$$

Assuming that the derivative $\partial f/\partial y$ varies slowly we can approximately replace it by a constant K. The corrector formula then gives the difference equation

$$\varepsilon_{n+1} = \varepsilon_{n-1} + \frac{Kh}{3} (\varepsilon_{n-1} + 4\varepsilon_n + \varepsilon_{n+1}) , \qquad (14.7.3)$$

which describes the error propagation of the method. As has been shown previously, we have weak stability if $\partial f/\partial y < 0$. If instead we use the predictor formula

$$y_{n+1} = -4y_n + 5y_{n-1} + 2h(y'_{n-1} + 2y'_n)$$

and then the corrector formula just once, it is easy to prove that we have strong stability.

14.8. Methods based on numerical integration

By formally integrating the differential equation $y' = f(x, y)$, we can transform it to an integral equation (cf. Chapter 16)

$$y(x) - y(a) = \int_a^x f(t, y(t)) \, dt .$$

The integrand contains the unknown function $y(t)$, and, as a rule, the integration cannot be performed. However, $f(t, y(t))$ can be replaced by an interpolation polynomial $P(t)$ taking the values $f_n = f(x_n, y_n)$ for $t = x_n$ (we suppose that these values have already been computed). By choosing the limits in suitable lattice points and prescribing that the graph of $P(t)$ must pass through certain points one can derive a whole series of interpolation formulas. It is then convenient to represent the interpolation polynomial by use of backward differences (cf. Newton's backward formula), and we consider the polynomial of degree q in s:

$$\varphi(s) = f_p + s\nabla f_p + \frac{s(s+1)}{1 \cdot 2} \nabla^2 f_p + \cdots + \frac{s(s+1) \cdots (s+q-1)}{q!} \nabla^q f_p.$$

Evidently $\varphi(0) = f_p$; $\varphi(-1) = f_{p-1}$; $\ldots$; $\varphi(-q) = f_{p-q}$. But $s = 0$ corresponds to $t = x_p$, $s = -1$ to $t = x_{p-1}$, and so on, and hence we must have $s = (t - x_p)/h$.

As an example we derive *Adams-Bashforth's* method, suggested as early as 1883. Then we must put

$$y_{p+1} - y_p = \int_{x_p}^{x_{p+1}} P(t)\, dt = h \int_0^1 \varphi(s)\, ds = h \sum_{r=0}^q c_r \nabla^r f_p,$$

where $c_r = (-1)^r \int_0^1 \binom{-s}{r} ds$. The first coefficients become

$$c_0 = 1; \qquad c_1 = \frac{1}{2}; \qquad c_2 = \int_0^1 \frac{s(s+1)}{2}\, ds = \frac{5}{12};$$

$$c_3 = \int_0^1 \frac{s(s+1)(s+2)}{6}\, ds = \frac{3}{8}; \qquad c_4 = \frac{251}{720};$$

$$c_5 = \frac{95}{288}; \qquad c_6 = \frac{19087}{60480}; \qquad c_7 = \frac{36799}{120960}; \qquad \cdots$$

It is obvious that a special starting procedure is needed in this case.

If the integration is performed between x_{p-1} and x_p, we get *Adams-Moulton's* method; the limits x_{p-1} and x_{p+1} gives *Nyström's* method, while x_{p-2} and x_p give *Milne-Simpson's* method which has already been treated in considerable detail. It is easy to see that Nyström's method also may give rise to weak stability.

Cowell-Numerov's method. Equations of the form $y'' = f(x, y)$, that is, not containing the first derivative, are accessible for a special technique. It has been devised independently by several authors: Cowell, Crommelin, Numerov, Störmer, Milne, Manning, and Millman. We start from the operator formula (7.2.12):

$$\frac{\delta^2}{U^2} = 1 + \frac{\delta^2}{12} - \frac{\delta^4}{240} + \cdots.$$

Hence we have

$$\delta^2 y_n = \left(1 + \frac{\delta^2}{12} - \frac{\delta^4}{240} + \cdots\right) h^2 y_n'' = h^2 \left(1 + \frac{\delta^2}{12} - \frac{\delta^4}{240} + \cdots\right) f_n,$$

or

$$y_{n+1} - 2y_n + y_{n-1} = \frac{h^2}{12}(f_{n+1} + 10f_n - f_{n-1}) - \frac{h^6}{240} y_n^{VI} + O(h^8),$$

since $\delta^4 \simeq U^4 = h^4 D^4$ and $y_n'' = f_n$. Neglecting terms of order h^6 and higher, we get the formula

$$y_{n+1} - 2y_n + y_{n-1} = \frac{h^2}{12}(f_{n+1} + 10f_n + f_{n-1}), \qquad (14.8.1)$$

with a local truncation error $O(h^6)$. In general, the formula is implicit, but in the special case when $f(x, y) = y \cdot g(x)$ we get the following explicit formula:

$$y_{n+1} = \frac{\beta_n y_n - \alpha_{n-1} y_{n-1}}{\alpha_{n+1}}, \qquad (14.8.2)$$

where $\alpha = 1 - (h^2/12)g$ and $\beta = 2 + (5h^2/6)g$.

In a special example we shall investigate the total error of the method. Let us discuss the equation $y'' = a^2 y$ leading to the following difference equation for y:

$$\left(1 - \frac{a^2 h^2}{12}\right)y_{n+1} - \left(2 + \frac{10a^2 h^2}{12}\right)y_n + \left(1 - \frac{a^2 h^2}{12}\right)y_{n-1} = 0,$$

with the characteristic equation

$$\lambda^2 - 2\left(\frac{1 + 5a^2 h^2/12}{1 - a^2 h^2/12}\right)\lambda + 1 = 0.$$

The roots are $\{1 + 5a^2 h^2/12 \pm ah(1 + a^2 h^2/6)^{1/2}\}/(1 - a^2 h^2/12)$, and after series expansion we find

$$\begin{cases} \lambda_1 = \exp(ah) + \dfrac{a^5 h^5}{480} + \dfrac{a^6 h^6}{480} + \cdots, \\[2mm] \lambda_2 = \exp(-ah) - \dfrac{a^5 h^5}{480} + \dfrac{a^6 h^6}{480} - \cdots. \end{cases} \qquad (14.8.3)$$

Hence we have

$$\lambda_1^n = \exp(anh)\left(1 + \frac{a^5 h^5}{480}\exp(-ah) + \frac{a^6 h^6}{480}\exp(-ah) + \cdots\right)^n$$

$$= \exp(ax)\left(1 + \frac{a^5 h^5}{480} + O(h^7)\right)^n = \exp(ax)[1 + \alpha_4 h^4 + O(h^6)],$$

where $\alpha_4 = a^5 x/480$ and $nh = x$. In a similar way we find

$$\lambda_2^n = \exp(-ax)\left(1 - \frac{a^5 h^5}{480} + O(h^7)\right)^n = \exp(-ax)[1 - \alpha_4 h^4 + O(h^6)].$$

This means that the total truncation error can be written

$$\varepsilon(x) \sim c_4 h^4 + c_6 h^6 + \cdots.$$

An error analysis in the general case gives the same result.

For initial value problems a summed form of the method should be preferred. First, the formula is rewritten as follows:

$$y_{n+1} - \frac{h^2}{12}f_{n+1} = y_n - \frac{h^2}{12}f_n + \left\{\left(y_n - \frac{h^2}{12}f_n\right) - \left(y_{n-1} - \frac{h^2}{12}f_{n-1}\right) + h^2f_n\right\}.$$

With the notation $z_n = y_n - (h^2/12)f_n$ and s_n for the expression within brackets we obtain:

$$
\begin{cases}
z_n = y_n - \dfrac{h^2}{12}f_n, & \text{(a)} \\[2mm]
z_{n+1} = z_n + s_n, & \text{(b)} \\[2mm]
z_n = z_{n-1} + s_{n-1}, & \text{(c)} \\[2mm]
s_n = z_n - z_{n-1} + h^2f_n. & \text{(d)}
\end{cases}
$$

Then (b) and (c) express the method while (d) is the definition of s_n. Adding (c) and (d) we find $s_n = s_{n-1} + h^2f_n$. We start the computation with a special technique (for example, series expansion) for obtaining y_1, and then we get

$$z_0 = y_0 - \frac{h^2f_0}{12}; \qquad z_1 = y_1 - \frac{h^2f_1}{12}; \qquad s_0 = z_1 - z_0;$$

$$
\begin{cases}
s_n = s_{n-1} + h^2f_n, & \\
z_{n+1} = z_n + s_n, & n = 1, 2, 3, \ldots \\
y_{n+1} - \dfrac{h^2f_{n+1}}{12} = z_{n+1}.
\end{cases}
\qquad (14.8.4)
$$

From the last relation y_{n+1} is solved, usually by iteration.

In the special case $f(x, y) = y \cdot g(x)$, the following scheme can be applied as before:

$$\alpha = 1 - \frac{h^2}{12}g, \qquad \beta = 2 + \frac{5h^2}{6}g.$$

x	y	g	α	β	αy
.	.	.	.	.	R
.	P	.	.	Q	.
.	V	.	T	.	S
.		.	.	.	
.		.	.	.	
.		.	.	.	

When we start, all values of x, g, α, and β are known, as well as the first two values of y and αy. One step in the computation comprises the following calculations:

$$
\begin{cases}
PQ - R = S, \\
\dfrac{S}{T} = V.
\end{cases}
$$

As is easily inferred, the method is extremely fast. The following disadvantages must, however, be taken into account: (1) The first step needs special attention, for example, series expansion. (2) Interval changes are somewhat difficult. (3) The derivative is not obtained during the calculation. Just as was the case with Milne-Simpson's method for the equation $y' = f(x, y)$, Cowell-Numerov's method can be interpreted as one of a whole family of methods designed for the equation $y'' = f(x, y)$. We start from the following Taylor expansion:

$$y(x + k) = y(x) + \sum_{m=1}^{n-1} \frac{k^m}{m!} y^{(m)}(x) + \int_0^1 \frac{k^n(1 - t)^{n-1}}{(n - 1)!} y^{(n)}(x + kt)\, dt \; .$$

The first mean value theorem of integral calculus transforms the integral to

$$k^n y^{(n)}(x + \theta k) \int_0^1 \frac{(1 - t)^{n-1}}{(n - 1)!}\, dt = \frac{k^n}{n!} y^{(n)}(x + \theta k)\,,$$

that is, Lagrange's remainder term. Using the formula for $n = 2$ and performing the transformation $x + kt = z$, we get

$$y(x + k) - y(x) = ky'(x) + \int_x^{x+k} (x + k - z)y''(z)\, dz \; .$$

The same relation with k replaced by $-k$ becomes

$$y(x - k) - y(x) = -ky'(x) + \int_x^{x-k} (x - k - z)y''(z)\, dz \; .$$

Replacing z by $2x - z$, putting $f(x, y(x)) = f(x)$, and adding the relations for k and $-k$ we obtain

$$y(x + k) - 2y(x) + y(x - k) = \int_x^{x+k} (x + k - z)[f(z) + f(2x - z)]\, dz \; .$$

$$(14.8.5)$$

Instead of $f(z)$ we then introduce a suitable interpolation polynomial of degree q through the points $x_p, x_{p-1}, \ldots, x_{p-q}$, and by different choices of x, k, and q we are now able to derive a whole family of methods. If we choose $x = x_p$ and $x + k = x_{p+1}$, we get *Störmer's* method:

$$y_{p+1} - 2y_p + y_{p-1} = h^2 \sum_{m=0}^{q} a_m \nabla^m f_p \,,$$

where

$$a_m = (-1)^m \int_0^1 (1 - t)\left\{\binom{-t}{m} + \binom{t}{m}\right\} dt \; .$$

One finds $a_0 = 1$, $a_1 = 0$, $a_2 = \frac{1}{12}$, $a_3 = \frac{1}{12}$, $a_4 = \frac{19}{240}$, $a_5 = \frac{3}{40}$, ...

If instead we choose $x = x_{p-1}$ and $x + k = x_p$ with $q \geq 2$ we again find Cowell-Numerov's method:

$$y_p - 2y_{p-1} + y_{p-2} = h^2 \sum_{m=0}^{q} b_m \nabla^m f_p \,,$$

where

$$b_m = (-1)^m \int_0^1 t\left\{\binom{t}{m} + \binom{2 - t}{m}\right\} dt \; .$$

The following values are obtained: $b_0 = 1$, $b_1 = -1$, $b_2 = \frac{1}{12}$, $b_3 = 0$, $b_4 = -\frac{1}{240}$, $b_5 = -\frac{1}{240}$, ...

Finally, we shall also briefly discuss the stability problems for equations of the form $y'' = f(x, y)$. First, we define the *order* of the method: if the local truncation error is $O(h^{p+2})$, then the order of the method is said to be p. This means, for example, that Cowell-Numerov's method as just discussed is of order 4. Again, we use the same notations $\rho(z)$ and $\sigma(z)$ as for the equation $y' = f(x, y)$. Our method can now be written

$$\rho(E)y_n = h^2 \sigma(E)f_n .\tag{14.8.6}$$

The *stability condition* can then be formulated as follows. A necessary condition for the convergence of the multistep method defined by (14.8.6) is that all zeros z_i of $\rho(z)$ are such that $|z_i| \leq 1$, and further for the roots on the unit circle the multiplicity is not greater than 2.

The proof is conducted by discussing the problem $y'' = 0$, $y(0) = y'(0) = 0$, with the exact solution $y(x) = 0$ (for closer details, see, for example, Henrici [7]).

In a similar way we derive the *consistency conditions*

$$\rho(1) = 0 ; \qquad \rho'(1) = 0 ; \qquad \rho''(1) = 2\sigma(1) ,\tag{14.8.7}$$

with the simple meaning that if the integration formula is developed in powers of h we must have identity in terms of the orders h^0, h^1, and h^2. For Cowell-Numerov's method we have

$$\begin{cases} \rho(z) = z^2 - 2z + 1 , \\ \sigma(z) = \frac{1}{12}(z^2 + 10z + 1) , \end{cases}$$

and hence both the stability and the consistency conditions are satisfied. In Todd's example we have $\rho(z) = z^4 - 16z^3 + 30z^2 - 16z + 1$ and $\sigma(z) = -12$. The roots of $\rho(z) = 0$ are $1, 1, 7 + 4\sqrt{3}$, and $7 - 4\sqrt{3}$, and since $7 + 4\sqrt{3} > 1$ the stability condition is not fulfilled. On the other hand, the consistency conditions are satisfied since

$$\rho(1) = \rho'(1) = 0 \qquad \text{and} \qquad \rho''(1) = -24 = 2\sigma(1) .$$

14.9. Systems of first-order linear differential equations

We denote the independent variable with t, and the n dependent variables with $x_1, x_2, \ldots, x_n$. They will be considered as components of a vector x, and hence the whole system can be written in the compact form

$$\frac{dx}{dt} = A(t)x .\tag{14.9.1}$$

Here $A(t)$ is a square matrix whose elements are functions of t. Now we assume that $A(t)$ is continuous for $0 \leq t \leq T$. Then we shall prove that the system (14.9.1) with the initial condition $x(0) = c$ has a unique solution in the interval $0 \leq t \leq T$; c is, of course, a constant vector.

The proof is essentially the same as that given in Section 14.0. Forming

$$\begin{cases} x_0 = c, \\ x_{k+1} = c + \int_0^t A(\tau) x_k \, d\tau \, ; \qquad k = 0, 1, 2, \ldots \end{cases} \tag{14.9.2}$$

we get directly

$$x_{k+1} - x_k = \int_0^t A(\tau)(x_k - x_{k-1}) \, d\tau \, ; \qquad k \geq 1 \, .$$

Since x_k are vectors, we have to work with norms instead of absolute values:

$$|x_{k+1} - x_k| \leq \int_0^t ||A(\tau)|| \cdot |x_k - x_{k-1}| \, d\tau \, .$$

Now let $M = \sup_{0 \leq \tau \leq T} ||A(\tau)||$, and we get

$$|x_{k+1} - x_k| \leq M \int_0^t |x_k - x_{k-1}| \, d\tau \, .$$

But $|x_1 - x_0| \leq \int_0^t ||A(\tau)|| \cdot |x_0| \, d\tau \leq M|c|t$, and hence

$$|x_{k+1} - x_k| \leq |c| \frac{(Mt)^{k+1}}{(k+1)!} \, ; \qquad k = 0, 1, 2, \ldots .$$

Thus the series $\sum_{k=0}^{\infty} |x_{k+1} - x_k|$ converges uniformly, since its terms are less than the corresponding terms in the exponential series. Then it is also clear that the series $\sum_{k=0}^{\infty} (x_{k+1} - x_k)$ converges uniformly, that is, $x_k \to x$ when $k \to \infty$. For the limit vector x, we have

$$x = c + \int_0^t A(\tau) x \, d\tau \, ,$$

which on differentiation gives $dx/dt = A(t)x$.

The uniqueness is proved in complete analogy to the one-dimensional case.

In the remainder of this section we shall concentrate on a very important special case, namely, the case when A is constant. Then the system has the simple form

$$\frac{dx}{dt} = Ax \, ; \qquad x(0) = x_0 \, . \tag{14.9.3}$$

We can write down the solution directly:

$$x = e^{tA} x_0 \, . \tag{14.9.4}$$

Surprisingly enough, this form is well suited for numerical work. First we choose a value of t, which is so small that e^{tA} can easily be computed. Then we use the properties of the exponential function to compute $e^{mtA} = e^{TA}$, where

$T = mt$ is the desired interval length. After this has been done, one step in the computation consists of multiplying the solution vector by the matrix e^{TA}.

It is of special interest to examine under what conditions the solution is bounded also for $t \rightarrow \infty$. We then restrict ourselves to the case when all eigenvalues are different. In this case there exists a regular matrix S such that $S^{-1}AS = D$, where D is a diagonal matrix. Hence $e^{tA} = Se^{tD}S^{-1}$, and we have directly one sufficient condition: for all eigenvalues we must have

$$\text{Re}\,(\lambda_i) \leq 0\;.$$

Problems of this kind are common in biology, where $x_1, x_2, \ldots, x_n$ (the components of the vector x) represent, for example, the amounts of a certain substance at different places in the body. This implies that $\sum_{i=1}^{n} x_1 = \text{constant}$ (for example 1), and hence $\sum_{i=1}^{n} (dx_i/dt) = 0$, which means that the sum of the elements in each column of A is zero. Thus the matrix A is *singular*, and at least one eigenvalue is zero. Often the matrix has the following form:

$$A = \begin{bmatrix} -a_{11} & a_{12} & \cdots & a_{1n} \\ a_{21} & -a_{22} & \cdots & a_{2n} \\ \vdots & & & \\ a_{n1} & a_{n2} & \cdots & -a_{nn} \end{bmatrix}$$

with $a_{ij} \geq 0$ and

$$a_{jj} = \sum_{\substack{i=1 \\ i \neq j}}^{n} a_{ij}\;.$$

From the estimate (3.3.6), we have

$$|\lambda + a_{jj}| \leq \sum_{\substack{i=1 \\ i \neq j}}^{n} a_{ij}\;,$$

or

$$|\lambda + a_{jj}| \leq a_{jj}\;. \tag{14.9.5}$$

Hence for every eigenvalue, either $\lambda = 0$ or $\text{Re}\,(\lambda) < 0$.

When $t \rightarrow \infty$, the contribution from those eigenvalues which have a negative real part will vanish, and the limiting value will be the eigenvector belonging to the eigenvalue $\lambda = 0$. This can also be inferred from the fact that the final state must be stationary, that is, $dx/dt = 0$ or $Ax = 0$.

If the problem is changed in such a way that A is singular, as before, but at least some eigenvalue has a positive real part, we have still a stationary solution which is not 0, but it is unstable. Every disturbance can be understood as an admixture of the components which correspond to eigenvalues with positive real parts, and very soon they will predominate completely.

14.10. Boundary value problems

A first-order differential equation $y' = f(x, y)$ has, in general, a function of the form $F(x, y, C) = 0$, where C is an arbitrary constant, as solution. The integral curves form a one-parameter family, where a special curve corresponds to a special choice of the constant C. Usually we specialize by assigning a function value $y = y_0$ which should correspond to the abscissa $x = x_0$. Then the integration can begin at this point, and we are solving what is called an *initial-value problem*. For equations of second and higher order, we can specialize by conditions at *several* points, and in this way we obtain a *boundary value problem*. Later we shall consider only equations of the second order; the methods can be generalized directly to equations of higher order.

For linear equations a direct technique is often successful, as will be demonstrated with an example. Find a solution of the equation $y'' = xy + x^2$ passing through the points $(0, 0)$ and $(1, 1)$. We see that $y = -x$ is a particular solution, and accordingly we put $y = Cz - x$, where $z = 0$ at the origin. Using Picard's method, we obtain

$$z = x + \frac{x^4}{12} + \frac{x^7}{12 \cdot 42} + \frac{x^{10}}{12 \cdot 42 \cdot 90} + \frac{x^{13}}{12 \cdot 42 \cdot 90 \cdot 156} + \cdots$$

$$= x + \frac{2x^4}{4!} + \frac{2 \cdot 5x^7}{7!} + \frac{2 \cdot 5 \cdot 8x^{10}}{10!} + \frac{2 \cdot 5 \cdot 8 \cdot 11x^{13}}{13!} + \cdots .$$

The boundary condition in the origin is satisfied, and we have only to choose C in such a way that the boundary condition in point $(1, 1)$ is also fulfilled. From

$$y = -x + C\left(x + \frac{x^4}{12} + \frac{x^7}{12 \cdot 42} + \cdots\right)$$

and $x = y = 1$, we get

$$C = \frac{2}{1 + 1/12 + 1/504 + \cdots} = 1.84274 .$$

Often the boundary conditions have the form

$$\begin{cases} ay_0 + by_0' = c , \\ \alpha y_1 + \beta y_1' = \gamma . \end{cases} \tag{14.10.1}$$

In nth-order equations such conditions can be imposed in n points:

$$\sum_{k=0}^{n-1} a_{ik} y_i^{(k)} = b_i ; \qquad i = 0, 1, \ldots, n - 1 . \tag{14.10.2}$$

Even under these circumstances the described technique, which is nothing but linear interpolation, can be used so long as the equation is linear. If this is not the case, we can use a trial-and-error technique. In our case $n = 2$, we guess a value of the derivative at the left boundary point, perform the integration, and observe the error at the other boundary point. The procedure is repeated, and in this way we can improve the initial value; conveniently, we use *Regula falsi* on the error at the right boundary point.

We can also approximate the derivatives of the equation by suitable difference expressions, and hence we obtain a linear system of equations. As an example, we consider the equation

$$y'' - (1 + 2/x)y + x + 2 = 0 ,$$

with the boundary conditions $y = 0$ for $x = 0$ and $y = 2$ for $x = 1$. Approximating y'' by $\delta^2 y/h^2$ and choosing $h = 0.2$, we obtain

$$y_{n+1} - (2.04 + 0.08/x)y_n + y_{n-1} + 0.04(x + 2) = 0 .$$

This results in a linear system of equations with four unknowns which are denoted as follows: $y(0.2) = y_1$, $y(0.4) = y_2$, $y(0.6) = y_3$, and $y(0.8) = y_4$. The system becomes

$$\begin{cases} 2.44y_1 - y_2 = 0.088 , \\ -y_1 + 2.24y_2 - y_3 = 0.096 , \\ -y_2 + 2.1733y_3 - y_4 = 0.104 , \\ -y_3 + 2.14y_4 = 2.112 . \end{cases}$$

We find

$$\begin{cases} y_1 = 0.2902 & \text{(exact } 0.2899) , \\ y_2 = 0.6202 & (\text{ ,, } \quad 0.6195) , \\ y_3 = 1.0030 & (\text{ ,, } \quad 1.0022) , \\ y_4 = 1.4556 & (\text{ ,, } \quad 1.4550) , \end{cases}$$

in good agreement with the exact solution $y = x(e^{x-1} + 1)$. The method described here can be improved by use of better approximations for the derivatives. It is characteristic that the resulting matrix is a band matrix. The complete system now has the form

$$Ay = b - Cy ,$$

where C contains higher differences. In the beginning we neglect Cy and compute a first approximation, also including a few points outside the interval, since we have to compute higher differences. Finally, the obtained values can be refined by the usual iteration technique.

14.11. Eigenvalue problems

Consider the following boundary value problem:

$$y'' + a^2 y = 0 , \qquad y(0) = y(1) = 0 .$$

The differential equation has the solution $y = A \cos ax + B \sin ax$. From $y(0)=0$ we get $A - 0$, while $y(1) = 0$ gives $B \sin a = 0$. If $\sin a \neq 0$ we have $B = 0$, that is, the only possible solution is the trivial one $y(x) = 0$. If, on the other hand, $\sin a = 0$, that is, if $a = n\pi$ where n is an integer, then B can be chosen arbitrarily. These special values $a^2 = n^2\pi^2$ are called *eigenvalues* and the corresponding solutions *eigenfunctions*.

A differential equation with boundary values corresponds exactly to a linear system of equations, and the situation discussed here corresponds to the case in which the right-hand side is equal to 0. If the coefficient matrix is regular, we have only the trivial solution zero, but if the matrix is singular, that is, if the determinant is equal to 0, we have an infinity of nontrivial solutions.

The eigenvalue problems play an important role in modern physics, and as a rule, the eigenvalues represent quantities which can be measured experimentally, (for example, energies). Usually the differential equation in question is written

$$\frac{d}{dx}\left(p\frac{dy}{dx}\right) - qy + \lambda\rho y = 0 , \qquad (14.11.1)$$

where p, q, and ρ are real functions of x.

The problem of solving this equation with regard to the boundary conditions

$$\begin{cases} \alpha_0 y(a) + \alpha_1 y'(a) = 0 , \\ \beta_0 y(b) + \beta_1 y'(b) = 0 , \end{cases}$$

or $y(a) = y(b)$; $p(a)y'(a) = p(b)y'(b)$ is called Sturm-Liouville's problem.

If the interval between a and b is divided into equal parts, and the derivatives are approximated by difference expressions, we obtain

$$\frac{p_r}{h^2}(y_{r-1} - 2y_r + y_{r+1}) + \frac{p'_r}{2h}(y_{r+1} - y_{r-1}) - qy_r + \lambda\rho_r y_r = 0 .$$

This can obviously be written in the form

$$(A - \lambda I)y = 0 ,$$

where A is a band matrix and y is a column vector. Nontrivial solutions exist if $\det(A - \lambda I) = 0$, and hence our eigenvalue problem has been transformed into an algebraic eigenvalue problem. However, only the lowest eigenvalues can be obtained in this way.

By the trial-and-error technique discussed above, we approach the solution via functions fulfilling the differential equation and one of the boundary conditions. But there is an interesting method by which we can advance via functions fulfilling *both* boundary conditions and simultaneously try to improve the fit of the differential equation.

In the Sturm-Liouville case we compute y_n iteratively from

$$\frac{d}{dx}\left(p(x)\frac{dy_n}{dx}\right) = q(x)y_{n-1} - \lambda_n\rho(x)y_{n-1} . \qquad (14.11.2)$$

When integrating, we obtain two integration constants which can be determined from the boundary conditions; further, λ_n is obtained from the condition $\int_a^b y_n^2 \, dx = 1$, or from some similar condition.

EXAMPLE

$$y'' + \alpha^2 y = 0 ; \qquad 0 \le x \le 2 ; \qquad y(0) = y(2) = 0 .$$

Find the lowest value of α^2 and the corresponding solution. We choose $y_0 = 2x - x^2$ fulfilling $y_0(0) = y_0(2) = 0$, $y_0(1) = 1$. Then we form $y_1, y_2, \ldots$ according to

$$y_n'' = -\alpha_n^2 y_{n-1} .$$

The integration constants are obtained from the boundary conditions and α_n^2 from the condition $y_n(1) = 1$. Hence we obtain

$$y_1'' = -\alpha_1^2 (2x - x^2) ,$$

$$y_1' = \alpha_1^2 \left(c - x^2 + \frac{x^3}{3} \right) ,$$

$$y_1 = \alpha_1^2 \left(d + cx - \frac{x^3}{3} + \frac{x^4}{12} \right) .$$

$y_1(0) = 0$ gives $d = 0$, $y_1(2) = 0$ gives $c = \frac{2}{3}$, and $y_1(1) = 1$ gives $\alpha_1^2 = 2.4$, and we therefore get $y_1 = \frac{1}{5}(8x - 4x^3 + x^4)$. The procedure is then repeated. The analytic solution is, of course, $y = \sin \alpha x$, and the condition $y_1(2) = 0$ gives the eigenvalues $\alpha = m\pi/2$, $m = 1, 2, 3, \ldots$ In our case we have, evidently, $m = 1$. For the exact solution corresponding to $m = 1$, the following relations are valid:

$$y\left(\frac{1}{3}\right) = 0.5 ; \qquad \left[y\left(\frac{1}{2}\right)\right]^2 = 0.5 ; \qquad 2y'(0) = \pi ;$$

$$\text{and} \qquad \alpha^2 = \frac{\pi^2}{4} = 2.4674011 \ldots .$$

The result is given in the following table.

n	y_n
0	$2x - x^2$
1	$(8x - 4x^3 + x^4)/5$
2	$(96x - 40x^3 + 6x^5 - x^6)/61$
3	$(2176x - 896x^3 + 112x^5 - 8x^7 + x^8)/1385$
4	$(79360x - 32640x^3 + 4032x^5 - 240x^7 + 10x^9 - x^{10})/50521$

n	$y_n(\frac{1}{3})$	$[y_n(\frac{1}{2})]^2$	$2y_n'(0)$	α_n^2
0	0.55556	0.5625	4	—
1	0.506173	0.507656	3.20	2.40
2	0.500686	0.5008895	3.1475	2.459
3	0.5000762	0.50010058	3.1422	2.4664
4	0.50000847	0.500011248	3.14166	2.46729

REFERENCES

[1] W. E. Milne: *Numerical Solution of Differential Equations* (Wiley, New York, 1953).

[2] L. Collatz: *Numerische Behandlung von Differentialgleichungen* (Berlin, Göttingen, Heidelberg, 1951).

[3] R. Bellman: *Stability Theory of Differential Equations* (McGraw-Hill, New York, 1953).

[4] G. Dahlquist: *Stability and Error Bounds in the Numerical Integration of Ordinary Differential Equations* (Diss. Uppsala, 1958).

[5] *Modern Computing Methods* (London, 1961).

[6] L. Fox: *The Numerical Solution of Two-Point Boundary Problems in Ordinary Differential Equations* (Oxford, 1957).

[7] L. Fox, Ed.: *Numerical Solution of Ordinary and Partial Differential Equations* (Oxford, 1962).

[8] P. Henrici: *Discrete Variable Methods in Ordinary Differential Equations* (New York 1962).

[9] J. Babuska, M. Práger, and E. Vitásek: *Numerical Processes in Differential Equations* (Prag, 1965).

[10] A. I. A. Karim: *Comp. Jour.* (Nov. 1966) p. 308.

EXERCISES

1. Solve the differential equation $y' = x - y^2$ by series expansion for $x = 0.2(0.2)1$. Sketch the function graphically and read off the minimum point. Initial value: $x = 0$, $y = 1$.

2. Solve the differential equation $y' = 1/(x + y)$ for $x = 0.5(0.5)2$ by using Runge-Kutta's method. Initial value: $x = 0$, $y = 1$.

3. Solve the differential equation $y'' = xy$ for $x = 0.5$ and $x = 1$ by use of series expansion and Runge-Kutta's method. Initial values: $x = 0$, $y = 0$, $y' = 1$.

4. If Runge-Kutta's method is applied to the differential equation $y' = -y$, with $y(0) = 1$ and the interval h, we obtain after n steps the value $y_n = y(nh) = [A(h)]^n$ as an approximation of the exact solution e^{-nh}. Find an algebraic expression for $A(h)$, and compute $A(h)$ and $A(h) - e^{-h}$ with less than 1% relative error for $h = 0.1, 0.2, 0.5, 1, 2$, and 5. Further, compute $y_n - e^{-nh}$ to four significant figures for $n = 100$ and $h = 0.1$.

5. Solve the differential equation $y'' = (x - y)/(1 + y^2)$ for $x = 2.4(0.2)3$ by use of Cowell-Numerov's method. Initial values: $(2, 1)$ and $(2.2, 0.8)$. Also find the coordinates of the minimum point graphically.

6. Solve the differential equation $xy'' - yy' = 0$ by Runge-Kutta's method for $x = 0.5$ and $x = 1$ ($h = 0.5$) when $y(0) = 1$ and $y''(0) = 2$ (4 decimals). Then compare with the exact solution which is of the form

$$y = \frac{ax^2 + b}{cx^2 + d}.$$

7. The differential equation $y'' = -y$ with initial conditions $y(0) = 0$ and $y(h) = k$ is solved by Numerov's method. Find the explicit solution in the simplest possible form. Then compute y_6 when $h = \pi/6$ and $k = \frac{1}{2}$.

8. The differential equation $y' = 1 + x^2 y^2$ with $y = 0$ for $x = 0$ is given. When x increases from 0 to ξ, y increases from 0 to ∞. The quantity ξ is to be determined to

three places in the following way. First we compute $y(0.8)$, using Picard's method. Then we introduce $z = 1/y$, and the differential equation for z is integrated with $h = 0.3$ by use of, for example, Runge-Kutta's method. When z is sufficiently small, we extrapolate to $z = 0$, using a series expansion.

9. The differential equation $y' = axy + b$ is given. Then we can write $y^{(n)} = P_n y + Q_n$, where P_n and Q_n are polynomials in x. Show that $z_n = P_n Q_{n+1} - P_{n+1} Q_n$ is a constant which depends on n, and compute z_n.

10. The differential equation $y'' + x^3 y = 0$ is given. For large values of x the solution y behaves like a damped oscillation with decreasing wavelength. The distance between the Nth and the $(N + 1)$-zero is denoted by z_N. Show that $\lim_{N \to \infty} z_N^5 \cdot N^3 = 8\pi^2/125$.

11. Find the solution of the differential equation $y'' + x^2 y = 0$ with the boundary conditions $y(0) = 0$, $y(1) = 1$ at the points 0.25, 0.50, and 0.75:
   (a) by approximating the differential equation with a second-order difference equation;
   (b) by performing two steps in the iteration $y''_{n+1} = -x^2 y_n$, $y_0 = x$.

12. Given the system of differential equations
$$\begin{cases} xy' + z' + \frac{1}{2}y = 0, \\ xz' - y' + \frac{1}{2}z = 0, \end{cases}$$
with the initial conditions $x = 0$, $y = 0$, $z = 1$. Show that $y^2 + z^2 = 1/\sqrt{1 + x^2}$ and find $y(\frac{1}{2})$ and $z(\frac{1}{2})$ to five decimals.

13. The following system is given
$$\begin{cases} y'' - 2xy' + z = 0, \\ z'' + 2xz' + y = 0, \end{cases}$$
with boundary values $y(0) = 0$, $y'(0) = 1$, $z(\frac{1}{2}) = 0$, and $z'(\frac{1}{2}) = -1$. Find $z(0)$ and $y(\frac{1}{2})$ to 3 decimals. Also show that $y^2 + z^2 + 2y'z'$ is independent of x.

14. The differential equation $y' = (x^2 - y^2)/(x^2 + y^2)$ is given. It has solutions which asymptotically approach a certain line $y = px$ which is also a solution of the equation. Another line $y = qx$ is perpendicular to all integral curves. Show that $pq = 1$ and find p and q to five decimals.

15. What is the largest value of the interval length h leaving all solutions of the Cowell-Numerov difference equation corresponding to $y'' + y = 0$ finite when $x \to \infty$?

16. The differential equation $y'' + ay' + by = 0$ is given with $0 < a < 2\sqrt{b}$. Find the largest possible interval length h such that no instability will occur when the equation is solved by use of the formulas
$$y_{n+1} = y_n + hy'_n; \qquad y'_{n+1} = y'_n + hy''_n.$$

17. Investigate the stability properties of the method
$$y_{n+3} = y_n + \frac{3h}{8}(y'_n + 3y'_{n+1} + 3y'_{n+2} + y'_{n+3}),$$
when applied to the equation $y' = -y$.

18. The differential equation $y'' + x^2(y + 1) = 0$ is given together with the boundary values $y(1) = y(-1) = 0$. Find approximate values of $y(0)$ and $y(\frac{1}{2})$ by assuming $y = (1 - x^2)(a + bx^2 + cx^4)$ which satisfies the boundary conditions. Use the points $x = 0$, $x = \frac{1}{2}$, and $x = 1$ (3 significant figures).

19. Determine $y(0)$ of Exercise 18 approximating by a difference equation of second order. Use $h = \frac{1}{5}$ and solve the linear system $Az = a + Dz$ iteratively by the formula

$$z_{n+1} = A^{-1}a + A^{-1}Dz_n .$$

The inverse of an $n \times n$-matrix A with elements $a_{11} = 1$, $a_{ik} = 2\delta_{ik} - \delta_{i-1,k} - \delta_{i+1,k}$ otherwise, is a matrix B with elements $b_{ik} = n + 1 - \max(i, k)$.

20. A certain eigenvalue λ of the differential equation

$$y'' - 2xy' + 2\lambda y = 0, \qquad y(0) = y(1) = 0,$$

is associated with the eigenfunction $y(x) = \sum_{n=0}^{\infty} a_n(\lambda)x^n$. Determine $a_n(\lambda)$ and also give the lowest eigenvalue with two correct decimals.

21. The differential equation $y'' + \lambda xy = 0$ is given, together with the boundary conditions $y(0) = 0$, $y(1) = 0$. Find the smallest eigenvalue λ by approximating the second derivative with the second difference for $x = \frac{1}{4}, \frac{1}{2}$, and $\frac{3}{4}$.

22. Find the smallest eigenvalue of the differential equation $xy'' + y' + \lambda xy = 0$, with the boundary conditions $y(0) = y(1) = 0$. The equation is approximated by a system of difference equations, first with $h = \frac{1}{2}$, then with $h = \frac{1}{4}$, and finally Richardson extrapolation is performed to $h = 0$ (the error is proportional to h^2).

23. Find the smallest positive eigenvalue of the differential equation $y^{IV}(x) = \lambda y(x)$ with the boundary conditions $y(0) = y'(0) = y''(1) = y'''(1) = 0$.

24. The differential equation $y'' + (1/x)y' + \lambda^2 y = 0$ is given together with the boundary conditions $y'(0) = 0$, $y(1) = 0$. Determine the lowest eigenvalue by assuming

$$y = a(x^2 - 1) + b(x^3 - 1) + c(x^4 - 1)$$

which satisfies the boundary conditions. Use the points $x = 0$, $x = \frac{1}{2}$, and $x = 1$.

25. The smallest eigenvalue λ of the differential equation $y'' + \lambda x^2 y = 0$ with boundary conditions $y(0) = y(1) = 0$ can be determined as follows. The independent variable is transformed by $x = \alpha t$ and the equation takes the form $y'' + t^2 y = 0$. This equation can be solved, for example, by Cowell-Numerov's method and, say, $h = \frac{1}{2}$ (since only the zeros are of interest we can choose $y(\frac{1}{2})$ arbitrary, for instance equal to 1). Compute λ rounded to the nearest integer by determining the first zero.

26. The differential equation $(1 + x)y'' + y' + \lambda(1 + x)y = 0$ is given together with the boundary conditions $y'(0) = 0$, $y(1) = 0$. Show that by a suitable transformation $x = \alpha t + \beta$ of the independent variable, the equation can be brought to a Bessel equation of order 0 (cf. Chapter 18). Then show that the eigenvalues λ can be obtained from $\lambda = \xi^2$ where ξ is a root of the equation

$$J_1(\xi)Y_0(2\xi) - Y_1(\xi)J_0(2\xi) = 0 .$$

It is known that $J_0'(\xi) = -J_1(\xi)$ and $Y_0'(\xi) = -Y_1(\xi)$.

Chapter 15

Partial differential equations

Le secret d'ennuyer est celui de tout dire.
VOLTAIRE.

15.0. Classification

Partial differential equations and systems of such equations appear in the description of physical processes, for example, in hydrodynamics, the theory o elasticity, the theory of electromagnetism (Maxwell's equations), and quantum mechanics. The solutions of the equations describe possible physical reactions that have to be fixed through boundary conditions, which may be of quite a different character. Here we will restrict ourselves to second-order partial differential equations, which dominate in the applications. Such equations are often obtained when systems of the kind described above are specialized and simplified in different ways.

We shall assume that the equations are linear in the second derivatives, that is, of the form

$$a\,\frac{\partial^2 u}{\partial x^2} + 2b\,\frac{\partial^2 u}{\partial x \partial y} + c\,\frac{\partial^2 u}{\partial y^2} = e\,, \qquad (15.0.1)$$

where a, b, c, and e are functions of x, y, u, $\partial u/\partial x$, and $\partial u/\partial y$. We introduce the conventional notations:

$$p = \frac{\partial u}{\partial x}\,; \qquad q = \frac{\partial u}{\partial y}\,; \qquad r = \frac{\partial^2 u}{\partial x^2}\,; \qquad s = \frac{\partial^2 u}{\partial x \partial y}\,; \qquad t = \frac{\partial^2 u}{\partial y^2}\,.$$

Then we have the relations:

$$du = \frac{\partial u}{\partial x}\,dx + \frac{\partial u}{\partial y}\,dy = p\,dx + q\,dy\,,$$

$$dp = \frac{\partial}{\partial x}\left(\frac{\partial u}{\partial x}\right)dx + \frac{\partial}{\partial y}\left(\frac{\partial u}{\partial x}\right)dy = r\,dx + s\,dy\,,$$

$$dq = \frac{\partial}{\partial x}\left(\frac{\partial u}{\partial y}\right)dx + \frac{\partial}{\partial y}\left(\frac{\partial u}{\partial y}\right)dy = s\,dx + t\,dy\,.$$

With the notations introduced, the differential equation itself can be written

$$ar + 2bs + ct = e\,.$$

When treating ordinary differential equations numerically it is customary to start in a certain point where also a number of derivatives are known. For a second-order equation it is usually sufficient to know $y_0 = y(x_0)$ and y_0'; higher derivatives are then obtained from the differential equation by successive differentiation. For a second-order partial differential equation it would be natural to conceive a similar procedure, at least for "open" problems for which the boundary conditions do not prevent this. Then it would be reasonable to replace the starting point by an initial curve along which we assume the values of u, p, and q given. With this background we first encounter the problem of determining r, s, and t in an arbitrary point of the curve. We suppose that the equation of the curve is given in parameter form:

$$x = x(\tau), \qquad y = y(\tau).$$

Since u, p, and q are known along the curve, we can simply write $u = u(\tau)$, $p = p(\tau)$, and $q = q(\tau)$, and introducing the notations $x' = dx/d\tau$ and so on, we find

$$\begin{cases} x'r + y's & = p' . \\ x's + y't = q' , \\ ar + 2bs + ct & = e . \end{cases} \tag{15.0.2}$$

From this system we can determine r, s, and t as functions of τ, provided that the coefficient determinant $D \neq 0$. We find directly

$$D = ay'^2 - 2bx'y' + cx'^2 . \tag{15.0.3}$$

It is rather surprising that the most interesting situation appears in the exceptional case $D = 0$. This equation has a solution consisting of two directions $y'/x' = dy/dx$ or rather a field of directions assigning two directions to every point in the plane. These directions then define two families of curves which are called *characteristics*. They are real if $b^2 - ac > 0$ (*hyperbolic* equation) or if $b^2 - ac = 0$ (*parabolic* equation) but imaginary if $b^2 - ac < 0$ (*elliptic* equation).

If $D = 0$ there is no solution of (15.0.2) unless one of the following three relations is fulfilled:

$$\begin{vmatrix} p' & x' & y' \\ q' & 0 & x' \\ e & a & 2b \end{vmatrix} = 0 ; \qquad \begin{vmatrix} p' & x' & 0 \\ q' & 0 & y' \\ e & a & c \end{vmatrix} = 0 ; \qquad \begin{vmatrix} p' & y' & 0 \\ q' & x' & y' \\ e & 2b & c \end{vmatrix} = 0 .$$

(Incidentally, they are equivalent if $D = 0$.) This means that one cannot prescribe arbitrary initial values (x, y, u, p, and q) on a characteristic.

As is easily understood, an equation can be elliptic in one domain, hyperbolic in another. A well-known example is gas flow at high velocities; the flow can be subsonic at some places, supersonic at others.

The following description gives the typical features of equations belonging to these three kinds, as well as additional conditions. In the *hyperbolic* case we have an open domain bounded by the x-axis between $x = 0$ and $x = a$, and the lines $x = 0$ and $x = a$ for $y \geq 0$. On the portion of the x-axis between 0 and a, the functions $u(x, 0)$ and $\partial u/\partial y$ are given as initial conditions. On each of the vertical lines a boundary condition of the form $\alpha u + \beta(\partial u/\partial x) = \gamma$ is given. In the *parabolic* case we have the same open domain, but on the portion of the x-axis only $u(x, 0)$ is given as an initial condition. Normally, $u(0, y)$ and $u(a, y)$ are also given as boundary conditions. Finally, in the *elliptic* case we have a closed curve on which u or the normal derivative $\partial u/\partial n$ (or a linear combination of both) is given; together with the equation, these conditions determine u in all interior points.

Hyperbolic equations, as a rule, are connected with oscillating systems (example: the wave equation), and parabolic equations are connected with some kind of diffusion. An interesting special case is the Schrödinger equation, which appears in quantum mechanics. Elliptic equations usually are associated with equilibrium states and especially with potential problems of all kinds.

15.1. Hyperbolic equations

We shall first give an explicit example, and we choose the wave equation in one dimension, setting the propagation velocity equal to c:

$$\frac{\partial^2 u}{\partial x^2} = \frac{1}{c^2} \frac{\partial^2 u}{\partial t^2} . \tag{15.1.1}$$

In this case we can easily write down the general solution:

$$u = f(x + ct) + g(x - ct) . \tag{15.1.2}$$

The solution contains two arbitrary functions f and g. Its physical meaning is quite simple: u can be interpreted as the superposition of two waves traveling in opposite directions. If, in particular, we choose $f(z) = -\frac{1}{2} \cos \pi z$ and $g(z) = \frac{1}{2} \cos \pi z$, we get $u = \sin \pi x \sin \pi ct$, which describes a standing wave with $u(0, t) = u(1, t) = 0$ and $u(x, 0) = 0$. Physically, the phenomenon can be realized by a vibrating string stretched between the points $x = 0$ and $x = 1$.

Often the initial conditions have the form

$$\begin{cases} u(x, 0) = f(x) , \\ \dfrac{\partial u}{\partial t}(x, 0) = g(x) , \end{cases}$$

and then we easily find the general solution

$$u(x, t) = \frac{f(x + ct) + f(x - ct)}{2} + \frac{1}{2c} \int_{x-ct}^{x+ct} g(\xi) \, d\xi . \tag{15.1.3}$$

Also for the two- and three-dimensional wave equation, explicit solutions can be given (see, for example, [3]).

Setting $y = ct$ in (15.1.1), we get the equation $\partial^2 u/\partial x^2 - \partial^2 u/\partial y^2 = 0$ with the explicit solution $u = f(x + y) + g(x - y)$. It is interesting to observe that the partial difference equation

$$u(x + 1, y) + u(x - 1, y) = u(x, y + 1) + u(x, y - 1) \qquad (15.1.4)$$

has exactly the same solution. This equation is obtained if we approximate the differential equation by second-order differences. However, if the coefficients of the equation do not have this simple form, it is, in general, impossible to give explicit solutions, and instead we must turn to numerical methods. A very general method is to replace the differential equation with a difference equation. This procedure will be exemplified later, and then we shall also treat the corresponding stability problems. At present we shall consider a method which is special for hyperbolic equations and which makes use of the properties of the *characteristics*.

First, we take a simple example. The equation $\partial^2 u/\partial x\partial y = 0$ gives the equation for the characteristics $dx\, dy = 0$, that is, $x = $ constant, $y = $ constant. The characteristics have the property that if we know u along two intersecting characteristics, then we also know u in the whole domain, where the equation is hyperbolic. In this special case we have the general solution $u = \varphi(x) + \psi(y)$. Now suppose that u is known on the line $x = x_0$, as well as on the line $y = y_0$. Then we have

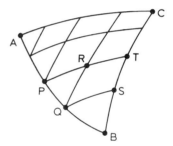

$$\begin{cases} \varphi(x) + \psi(y_0) = F(x), \\ \varphi(x_0) + \psi(y) = G(y), \end{cases}$$

where both $F(x)$ and $G(y)$ are known functions. Adding and putting $x = x_0$ and $y = y_0$, we obtain

Figure 15.1

$$\varphi(x_0) + \psi(y_0) = \tfrac{1}{2}[F(x_0) + G(y_0)],$$

where $F(x_0) = G(y_0)$, and

$$u = \varphi(x) + \psi(y) = F(x) + G(y) - \tfrac{1}{2}[F(x_0) + G(y_0)].$$

In general, the characteristics are obtained from the equation

$$ay'^2 - 2bx'y' + cx'^2 = 0$$

or

$$\frac{dy}{dx} = \frac{b \pm \sqrt{b^2 - ac}}{a};$$

these two values will be denoted by f and g (they are functions of x, y, u, p, and q). Further, consider an arc AB which is not characteristic and on which u, p, and q are given. Let PR be a characteristic of f-type, QR one of g-type.

Further, let $P = (x_1, y_1)$; $Q = (x_2, y_2)$; $R = (x_3, y_3)$. In first approximation we have

$$\begin{cases} y_3 - y_1 = f_1(x_3 - x_1) \,, \\ y_3 - y_2 = g_2(x_3 - x_2) \,. \end{cases}$$

From this system x_3 and y_3 can easily be computed. The second of the three equivalent determinantal equations can be written

$$ex'y' - ap'y' - cq'x' = 0$$

or

$$e \frac{dy}{dx} dx - a \frac{dy}{dx} dp - c \, dq = 0 \,.$$

Along PR this relation is approximated by

$$e_1(x_3 - x_1)f_1 - a_1(p_3 - p_1)f_1 - c_1(q_3 - q_1) = 0 \,,$$

and along QR by

$$e_2(x_3 - x_2)g_2 - a_2(p_3 - p_2)g_2 - c_2(q_3 - q_2) = 0 \,.$$

Since x_3 and y_3 are now approximately known, p_3 and q_3 can be solved from this system. Then it is possible to compute u_3 from

$$u_3 = u_1 + \frac{\partial u}{\partial x} dx + \frac{\partial u}{\partial y} dy \,,$$

where we approximate $\partial u/\partial x$ by $\frac{1}{2}(p_1 + p_3)$ and $\partial u/\partial y$ by $\frac{1}{2}(q_1 + q_3)$; further, dx is replaced by $x_3 - x_1$ and dy by $y_3 - y_1$. Hence we get

$$u_3 = u_1 + \frac{1}{2}(x_3 - x_1)(p_1 + p_3) + \frac{1}{2}(y_3 - y_1)(q_1 + q_3) \,.$$

From the known approximations of u_3, p_3, and q_3, we can compute f_3 and g_3. Then improved values for x_3, y_3, p_3, q_3, and u_3 can be obtained in the same way as before, but with f_1 replaced by $\frac{1}{2}(f_1 + f_3)$, g_2 by $\frac{1}{2}(g_2 + g_3)$, a_1 by $\frac{1}{2}(a_1 + a_3)$, and so on. When the value in R has been obtained, we can proceed to S, and from R and S we can reach T. It is characteristic that we can only obtain values within a domain ABC of triangular shape and with AB as "base."

We shall now in an example demonstrate how it is possible to apply an ordinary series expansion technique. Again we start from the wave equation

$$\frac{\partial^2 u}{\partial x^2} - \frac{\partial^2 u}{\partial y^2} = 0$$

with the initial conditions

$$u(x, 0) = x^2 \,; \qquad \frac{\partial u}{\partial y}(x, 0) = e^{-x} \,.$$

(Note that the straight line $y = 0$, where the initial conditions are given, is not a characteristic.) We are then going to compute $u(0.2, 0.1)$ starting from the origin. Differentiating the initial conditions, we obtain

$$\frac{\partial u}{\partial x}(x, 0) = 2x, \qquad \frac{\partial^2 u}{\partial x \partial y}(x, 0) = -e^{-x},$$

$$\frac{\partial^2 u}{\partial x^2}(x, 0) = 2, \qquad \frac{\partial^3 u}{\partial x^2 \partial y}(x, 0) = e^{-x},$$

$$\frac{\partial^3 u}{\partial x^3}(x, 0) = 0.$$

From the differential equation we get

$$\frac{\partial^2 u}{\partial y^2}(x, 0) = 2$$

and then successively:

$$\frac{\partial^3 u}{\partial x^3} - \frac{\partial^3 u}{\partial x \partial y^2} = 0; \qquad \frac{\partial^3 u}{\partial x^2 \partial y} - \frac{\partial^3 u}{\partial y^3} = 0,$$

and hence

$$\frac{\partial^3 u}{\partial x \partial y^2}(x, 0) = 0; \qquad \frac{\partial^3 u}{\partial y^3}(x, 0) = e^{-x},$$

and so on. We then finally obtain the desired value through

$$u(x, y) = \exp(xD_x + yD_y)u(0, 0)$$

$$= 0.1 + \frac{1}{2}(2 \cdot 0.04 - 0.04 + 0.02)$$

$$+ \frac{1}{6}(3 \cdot 0.004 + 0.001) + \frac{1}{24}(-4 \cdot 0.0008 - 4 \cdot 0.0002) + \cdots$$

$$= 0.13200.$$

The exact solution is

$$u = \frac{(x + y)^2 - (x - y)^2}{2} + \frac{1}{2}\int_{x-y}^{x+y} e^{-t}\, dt = x^2 + y^2 + e^{-x}\sinh y,$$

which for $x = 0.2$, $y = 0.1$ gives the value $u = 0.13201$.

As mentioned before, it is of course also possible to approximate a hyperbolic differential equation with a difference equation. From the wave equation

$$\frac{\partial^2 u}{\partial x^2} - \frac{\partial^2 u}{\partial y^2} = 0,$$

we get the difference equation

$$\frac{u_{r,s+1} - 2u_{r,s} + u_{r,s-1}}{k^2} = \frac{u_{r+1,s} - 2u_{r,s} + u_{r-1,s}}{h^2}. \qquad (15.1.5)$$

As before, the indices r and s are completely equivalent, but a difference is introduced through the initial conditions which define an integration direction, viz., toward increasing values of s. In 1928, Courant, Friedrichs, and Lewy in a famous paper [5] proved that the difference equation is stable only if $k \leq h$. In following sections we shall give much attention to stability problems.

15.2. Parabolic equations

The simplest nontrivial parabolic equation has the form

$$\frac{\partial^2 u}{\partial x^2} = \frac{\partial u}{\partial t}, \tag{15.2.1}$$

with the initial values $u(x, 0) = f(x)$, $0 \leq x \leq 1$, and the boundary conditions $u(0, t) = \varphi(t)$; $u(1, t) = \psi(t)$. This is the *heat equation* in one dimension.

We shall first solve the problem by approximating $\partial^2 u/\partial x^2$, but not $\partial u/\partial t$, by a difference expression. The portion between $x = 0$ and $x = 1$ is divided into n equal parts in such a way that $x_r = rh$ and $nh = 1$. Then we have approximately

$$\frac{\partial u_r}{\partial t} = \frac{du_r}{dt} = \frac{1}{h^2}(u_{r-1} - 2u_r + u_{r+1}).$$

Writing out the equations for $r = 1, 2, \ldots, n - 1$, we obtain

$$\begin{cases} h^2 \dfrac{du_1}{dt} = u_0 - 2u_1 + u_2, \\[2mm] h^2 \dfrac{du_2}{dt} = u_1 - 2u_2 + u_3, \\[1mm] \;\vdots \\[1mm] h^2 \dfrac{du_{n-1}}{dt} = u_{n-2} - 2u_{n-1} + u_n. \end{cases} \tag{15.2.2}$$

Hence we get a system of $n - 1$ ordinary differential equations in $u_1, u_2, \ldots, u_{n-1}$; the quantities u_0 and u_n are, of course, known functions of t:

$$\begin{cases} u_0 = \varphi(t), \\ u_n = \psi(t). \end{cases}$$

The values of $u_1, u_2, \ldots, u_{n-1}$ for $t = 0$ are $f(h), f(2h), \ldots, f((n-1)h)$. Putting

$$U = \begin{bmatrix} u_1 \\ u_2 \\ \vdots \\ u_{n-1} \end{bmatrix}; \quad V = \begin{bmatrix} u_0 \\ 0 \\ \vdots \\ 0 \\ u_n \end{bmatrix}; \quad A = \begin{bmatrix} -2 & 1 & 0 \cdots 0 & 0 \\ 1 & -2 & 1 \cdots 0 & 0 \\ \vdots & & & \vdots \\ 0 & 0 & 0 \cdots 1 & -2 \end{bmatrix},$$

we can write

$$h^2 \frac{dU}{dt} = AU + V .$$

As an example, we solve the heat equation for a thin, homogeneous rod of uniform thickness, which at time $t = 0$ has temperature $u = 0$. Further we assume $u(0, t) = 0$ and $u(1, t) = t$. We will then consider the temperature u_1 for $x = \frac{1}{4}$, u_2 for $x = \frac{1}{2}$, and u_3 for $x = \frac{3}{4}$, and we find with $h = \frac{1}{4}$:

$$\begin{cases} \dfrac{du_1}{dt} = 16(-2u_1 + u_2) , \\[2mm] \dfrac{du_2}{dt} = 16(u_1 - 2u_2 + u_3) , \\[2mm] \dfrac{du_3}{dt} = 16(u_2 - 2u_3 + t) . \end{cases}$$

Using the indefinite-coefficient method, we obtain the particular solution

$$\begin{pmatrix} u_1 \\ u_2 \\ u_3 \end{pmatrix} = \frac{1}{128} \begin{pmatrix} -5 + 32t \\ -8 + 64t \\ -7 + 96t \end{pmatrix} .$$

The solution of the homogeneous system is best obtained by putting $u_k = e^{\lambda t} = e^{16\mu t}$. This gives a homogeneous system of equations whose coefficient determinant must be zero:

$$\begin{vmatrix} -2 - \mu & 1 & 0 \\ 1 & -2 - \mu & 1 \\ 0 & 1 & -2 - \mu \end{vmatrix} = 0 .$$

The roots are

$$\mu_1 = -2 ; \qquad \mu_2 = -2 + \sqrt{2} ; \qquad \mu_3 = -2 - \sqrt{2} ,$$

and the corresponding eigenvectors are

$$\begin{pmatrix} 1 \\ 0 \\ -1 \end{pmatrix} ; \qquad \begin{pmatrix} 1 \\ \sqrt{2} \\ 1 \end{pmatrix} ; \qquad \begin{pmatrix} 1 \\ -\sqrt{2} \\ 1 \end{pmatrix} .$$

Hence the solution is

$$\begin{cases} u_1 = A \cdot e^{-32t} + B \cdot e^{-16(2-\sqrt{2})t} + C \cdot e^{-16(2+\sqrt{2})t} + \dfrac{1}{128}(-5 + 32t) , \\[2mm] u_2 = B\sqrt{2} \cdot e^{-16(2-\sqrt{2})t} - C\sqrt{2} \cdot e^{-16(2+\sqrt{2})t} + \dfrac{1}{16}(-1 + 8t) , \\[2mm] u_3 = -A \cdot e^{-32t} + B \cdot e^{-16(2-\sqrt{2})t} + C \cdot e^{-16(2+\sqrt{2})t} + \dfrac{1}{128}(-7 + 96t) ; \end{cases}$$

A, B, and C can be obtained from $u_1(0) = u_2(0) = u_3(0) = 0$.

$$A = -\frac{1}{128}; \qquad B = \frac{3 + 2\sqrt{2}}{128}; \qquad C = \frac{3 - 2\sqrt{2}}{128}.$$

For large values of t we have

$$u_1(t) \sim \frac{t}{4}; \qquad u_2(t) \sim \frac{t}{2}; \qquad u_3(t) \sim \frac{3t}{4}.$$

However, we can also approximate $\partial u/\partial t$ by a difference expression. Then it often turns out to be necessary to use smaller intervals in the t-direction. As before, we take $x_r = rh$, but now we also choose $t_s = sk$. Then we obtain the following partial difference equation:

$$\frac{u_{r-1,s} - 2u_{r,s} + u_{r+1,s}}{h^2} = \frac{u_{r,s+1} - u_{r,s}}{k}. \tag{15.2.3}$$

With $\alpha = k/h^2$, we get

$$u_{r,s+1} = \alpha u_{r-1,s} + (1 - 2\alpha)u_{r,s} + \alpha u_{r+1,s}. \tag{15.2.4}$$

Later on we shall examine for what values of α the method is stable; here we restrict ourselves to computing the truncation error. We get

$$\frac{u(x, t + k) - u(x, t)}{k} = \frac{\partial u}{\partial t} + \frac{k}{2!}\frac{\partial^2 u}{\partial t^2} + \frac{k^2}{3!}\frac{\partial^3 u}{\partial t^3} + \cdots,$$

$$\frac{u(x - h, t) - 2u(x, t) + u(x + h, t)}{h^2} = \frac{\partial^2 u}{\partial x^2} + \frac{h^2}{12}\frac{\partial^4 u}{\partial x^4} + \frac{h^4}{360}\frac{\partial^6 u}{\partial x^6} + \cdots.$$

Since u must satisfy $\partial u/\partial t = \partial^2 u/\partial x^2$, we also have, under the assumption of sufficient differentiability,

$$\frac{\partial^2 u}{\partial t^2} = \frac{\partial^4 u}{\partial x^4}; \qquad \frac{\partial^3 u}{\partial t^3} = \frac{\partial^6 u}{\partial x^6}; \cdots$$

Hence we obtain the truncation error

$$F = \frac{h^2}{2}\left(\alpha\frac{\partial^2 u}{\partial t^2} - \frac{1}{6}\frac{\partial^4 u}{\partial x^4}\right) + \frac{h^4}{6}\left(\alpha^2\frac{\partial^3 u}{\partial t^3} - \frac{1}{60}\frac{\partial^6 u}{\partial x^6}\right) + \cdots$$

$$= \frac{h^2}{2}\left(\alpha - \frac{1}{6}\right)\frac{\partial^2 u}{\partial t^2} + \frac{h^4}{6}\left(\alpha^2 - \frac{1}{60}\right)\frac{\partial^3 u}{\partial t^3} + \cdots.$$

Obviously, we get a truncation error of highest possible order when we choose $\alpha = \frac{1}{6}$, that is, $k = h^2/6$. Then we find

$$F = \frac{h^4}{540}\frac{\partial^3 u}{\partial t^3} + \cdots.$$

Now we solve the heat equation again with the initial condition $u(x, 0) = 0$

and the boundary conditions $u(0, t) = 0$, $u(1, t) = t$, choosing $h = \frac{1}{4}$ and $k = \frac{1}{96}$ corresponding to $\alpha = \frac{1}{6}$. The recursion formula takes the following simple form:

$$u(x, t) = \frac{1}{6} \left(u(x - h, t - k) + 4u(x, t - k) + u(x + h, t - k) \right) .$$

After 12 steps we have reached $t = 0.125$ and the u-values, $u_1 = 0.00540$, $u_2 = 0.01877$, $u_3 = 0.05240$, and, of course, $u_4 = 0.125$. From our previous solution, we obtain instead

$$u_1 = 0.00615 , \quad u_2 = 0.01994 , \quad \text{and} \quad u_3 = 0.05331 .$$

The exact solution of the problem is given by

$$u(x, t) = \frac{1}{6} (x^3 - x + 6xt) + \frac{2}{\pi^3} \sum_{n=1}^{\infty} \frac{(-1)^{n-1}}{n^3} e^{-n^2\pi^2 t} \sin n\pi x ,$$

which gives $u_1 = 0.00541$, $u_2 = 0.01878$, and $u_3 = 0.05240$. Hence the method with $\alpha = \frac{1}{6}$ gives an error which is essentially less than what is obtained if we use ordinary derivatives in the t-direction, corresponding to $\alpha = 0$.

The method defined by 15.2.3 or 15.2.4 gives the function values at time $t + k$ as a linear combination of three function values at time t; these are then supposed to be known. For this reason the method is said to be *explicit*. Since the time step k has the form αh^2, it is interesting to know how large values of α we can choose and still have convergence. Let us again consider the same equation

$$\frac{\partial^2 u}{\partial x^2} = \frac{\partial u}{\partial t}$$

but now with the boundary values $u(0, t) = u(1, t) = 0$ when $t \geq 0$, and with the initial condition $u(x, 0) = f(x)$, $0 \leq x \leq 1$. If $f(x)$ is continuous we can write down an analytic solution

$$u = \sum_{m=1}^{\infty} a_m \sin m\pi x \, e^{-m^2\pi^2 t} , \tag{15.2.5}$$

where the coefficients are determined from the initial condition which demands

$$f(x) = u(x, 0) = \sum_{m=1}^{\infty} a_m \sin m\pi x .$$

Thus we obtain (cf. Section 17.3):

$$a_m = 2 \int_0^1 f(x) \sin m\pi x \, dx .$$

We shall now try to produce an analytic solution also for the difference equation which we now write in the form

$$u_{r,s+1} - u_{r,s} = \alpha(u_{r-1,s} - 2u_{r,s} + u_{r+1,s}) .$$

In order to separate the variables we try with the expression $u_{r,s} = X(rh)T(sk)$ and obtain

$$\frac{T((s + 1)k) - T(sk)}{T(sk)} = \alpha \frac{X((r + 1)h) - 2X(rh) + X((r - 1)h)}{X(rh)} = -p^2 .$$

(Since the left-hand side is a function of s and the right-hand side a function of r, both must be constant; $-p^2$ is called the *separation constant*.) We notice that the function $X(rh) = \sin m\pi rh$ satisfies the equation

$$X((r + 1)h) - 2X(rh) + X((r - 1)h) = -\frac{p^2}{\alpha} X(rh) ,$$

if we choose $p^2 = 4\alpha \sin^2(m\pi h/2)$. Further we assume $T(sk) = e^{-qsk}$ and obtain $e^{-qk} = 1 - 4\alpha \sin^2(m\pi h/2)$. In this way we finally get

$$u_{r,s} = \sum_{m=1}^{M-1} b_m \sin m\pi rh \left(1 - 4\alpha \sin^2 \frac{m\pi h}{2} \right)^s . \tag{15.2.6}$$

Here $Mh = 1$ and further the coefficients b_m should be determined in such a way that $u_{r,0} = f(rh)$, which gives

$$b_m = \frac{2}{M} \sum_{n=1}^{M-1} f(nh) \sin m\pi nh .$$

(Cf. section 17.3.)

The time dependence of the solution is expressed through the factor $\{1 - 4\alpha \sin^2(m\pi h/2)\}^s$, and for small values of α we have

$$1 - 4\alpha \sin^2 \frac{m\pi h}{2} \simeq e^{-\alpha m^2 \pi^2 h^2} .$$

If this is raised to the power s we get

$$e^{-\alpha m^2 \pi^2 h^2 s} = e^{-m^2 \pi^2 k s} = e^{-m^2 \pi^2 t} ,$$

that is, the same factor as in the solution of the differential equation. If α increases, the factor will deviate more and more from the right value, and we observe that the situation becomes dangerous if $\alpha > \frac{1}{2}$ since the factor then may become absolutely > 1. Therefore, it is clear that h and k must be chosen such that $k < h^2/2$. Even if $h \to 0$ one cannot guarantee that the solution of the difference equation is close to the solution of the differential equation unless k is chosen such that $\alpha \le \frac{1}{2}$.

A numerical method is usually said to be *stable* if an error which has been introduced in one way or other keeps finite all the time. Now it is obvious that the errors satisfy the same difference equation, and hence we have proved that the stability condition is exactly $\alpha < \frac{1}{2}$. It should be pointed out, however, that the dividing line between *convergence* on one side and *stability* on

the other is rather vague. In general, convergence means that the solution
of the difference equation when $h \to 0$ and $k \to 0$ approaches the solution of
the differential equation, while stability implies that an introduced error does
not grow up during the procedure. Normally these two properties go together,
but examples have been given of convergence with an unstable method.

Before we enter upon a more systematic error investigation we shall demonstrate a technique which has been used before to illustrate the error propagation in a difference scheme. We choose the heat conduction equation which
we approximate according to (15.2.4) with $\alpha = \frac{1}{2}$:

$$u_{r,s+1} = \tfrac{1}{2}u_{r-1,s} + \tfrac{1}{2}u_{r+1,s} \,.$$

The following scheme is obtained:

	$s = 0$	$s = 1$	$s = 2$	$s = 3$	$s = 4$
$r = -4$	0	0	0	0	$\frac{1}{16}$
$r = -3$	0	0	0	$\frac{1}{8}$	0
$r = -2$	0	0	$\frac{1}{4}$	0	$\frac{4}{16}$
$r = -1$	0	$\frac{1}{2}$	0	$\frac{3}{8}$	0
$r = 0$	1	0	$\frac{2}{4}$	0	$\frac{6}{16}$
$r = 1$	0	$\frac{1}{2}$	0	$\frac{3}{8}$	0
$r = 2$	0	0	$\frac{1}{4}$	0	$\frac{4}{16}$
$r = 3$	0	0	0	$\frac{1}{8}$	0
$r = 4$	0	0	0	0	$\frac{1}{16}$

The numerators contain the binomial coefficients $\binom{s}{n}$ and the denominators
2^s; the method is obviously stable. If the same equation is approximated with
the difference equation

$$\frac{u_{r-1,s} - 2u_{r,s} + u_{r+1,s}}{h^2} = \frac{u_{r,s+1} - u_{r,s-1}}{2k} \,,$$

which has considerably smaller truncation error than (15.2.3), we get with
$\alpha = k/h^2 = \frac{1}{2}$:

$$u_{r,s+1} = u_{r,s-1} + u_{r-1,s} + u_{r+1,s} - 2u_{r,s}$$

	$s = -1$	$s = 0$	$s = 1$	$s = 2$	$s = 3$	$s = 4$
$r = -4$	0	0	0	0	0	1
$r = -3$	0	0	0	0	1	-8
$r = -2$	0	0	0	1	-6	31
$r = -1$	0	0	1	-4	17	-68
$r = 0$	0	1	-2	7	-24	89
$r = 1$	0	0	1	-4	17	-68
$r = 2$	0	0	0	1	-6	31
$r = 3$	0	0	0	0	1	-8
$r = 4$	0	0	0	0	0	1

Evidently, the method is very unstable and useless in practical computation. It is easy to give a sufficient condition for stability. Suppose that the term to be computed is given as a linear combination of known terms with coefficients $c_1, c_2, c_3, \ldots, c_n$. Then the method is stable if

$$\sum_{i=1}^{n} |c_i| \le 1 .$$

We see, for example, that (15.2.4) is stable if $\alpha \le \frac{1}{2}$.

We shall now treat the problem in a way which shows more clearly how the error propagation occurs, and we start again from the same system of equations:

$$u_{r,s+1} = \alpha u_{r-1,s} + (1 - 2\alpha)u_{r,s} + \alpha u_{r+1,s} .$$

Introducing the notation $\boldsymbol{u}_s$ for a vector with the components $u_{1,s}, u_{2,s}, \ldots, u_{M-1,s}$, we find $\boldsymbol{u}_{s+1} = A\boldsymbol{u}_s$; where

$$A = \begin{bmatrix} 1 - 2\alpha & \alpha & 0 & 0 \cdots & 0 \\ \alpha & 1 - 2\alpha & \alpha & 0 \cdots & 0 \\ 0 & \alpha & 1 - 2\alpha & \alpha \cdots & 0 \\ \vdots & & & & \vdots \\ 0 & 0 & 0 & \cdots & \alpha & 1 - 2\alpha \end{bmatrix} .$$

If we also introduce an error vector e, we get trivially

$$e_s = A^s e_0 ,$$

and we have stability if e_n stays finite when $n \rightarrow \infty$. We split the matrix A into two parts: $A = I - \alpha T$, where

$$T = \begin{bmatrix} 2 & -1 & 0 & \cdots & 0 \\ -1 & 2 & -1 & \cdots & 0 \\ \vdots & & & & \vdots \\ 0 & 0 & \cdots & -1 & 2 \end{bmatrix} ,$$

all matrices having the dimension $(M - 1) \times (M - 1)$. Supposing that T has an eigenvector x we obtain the following system of equations:

$$x_n + (\lambda - 2)x_{n+1} + x_{n+2} = 0 ,$$

with $n = 0, 1, 2, \ldots, M - 2$, and $x_0 = x_M = 0$. This can be interpreted as a difference equation with the characteristic equation

$$\mu^2 + (\lambda - 2)\mu + 1 = 0 .$$

Since T is symmetric, λ is real, and Gershgorin's theorem gives $|\lambda - 2| \le 2$, that is, $0 \le \lambda \le 4$. Hence we have the following roots of u:

$$\mu = 1 - \frac{\lambda}{2} \pm i\sqrt{\lambda - \frac{\lambda^2}{4}} .$$

But $0 \leq \lambda \leq 4$ implies $-1 \leq 1 - \lambda/2 \leq 1$ and $0 \leq \lambda - \lambda^2/4 \leq 1$, and further we have $(1 - \lambda/2)^2 + (\lambda - \lambda^2/4) = 1$. This shows that we can put

$$\begin{cases} \cos \varphi = 1 - \dfrac{\lambda}{2}, \\[2mm] \sin \varphi = \sqrt{\lambda - \dfrac{\lambda^2}{4}}, \end{cases}$$

and $\mu = e^{\pm i\varphi}$. Hence $x_n = Ae^{in\varphi} + Be^{-in\varphi}$, and $x_0 = x_M = 0$ gives:

$$\begin{cases} A + B = 0, \\ A \sin M\varphi = 0. \end{cases}$$

But $A \neq 0$ and consequently we have

$$M\varphi = m\pi, \qquad m = 1, 2, 3, \ldots, M - 1,$$

$$\lambda_n = 2(1 - \cos \varphi) = 4 \sin^2 \frac{m\pi}{2M}.$$

This shows that the eigenvalues of A are $1 - 4\alpha \sin^2 (m\pi/2M)$, and we find the stability condition:

$$-1 \leq 1 - 4\alpha \sin^2 \frac{m\pi}{2M} \leq 1,$$

that is, $\alpha \leq [2 \sin^2 (m\pi/2M)]^{-1}$. The worst case corresponds to $m = M - 1$ when the expression within brackets is very close to 2, and we conclude that the method is stable if we choose

$$\alpha \leq \tfrac{1}{2}.$$

The necessity to work with such small time-steps has caused people to try to develop other methods which allow greater steps without giving rise to instability. The best known of these methods is that of *Crank-Nicolson*. It coincides with the forward-difference method, however with the distinction that $\partial^2 u/\partial x^2$ is approximated with the *mean value* of the second difference quotients taken at times t and $t + k$. Thus the method can be described through

$$\frac{u_{r,s+1} - u_{r,s}}{k} = \frac{1}{2h^2}(u_{r+1,s} - 2u_{r,s} + u_{r-1,s} + u_{r+1,s+1} - 2u_{r,s+1} + u_{r-1,s+1}).$$

$$(15.2.7)$$

We have now three unknowns, viz., $u_{r-1,s+1}$, $u_{r,s+1}$, and $u_{r+1,s+1}$, and hence the method is *implicit*. In practical computation the difficulties are not very large since the coefficient matrix is tridiagonal.

We are now going to investigate the convergence properties of the method. As before we put

$$u_{r,s} = X(rh)T(sk)$$

and obtain

$$\frac{T[(s+1)k] - T(sk)}{T[(s+1)k] + T(sk)} = \frac{\alpha}{2} \frac{X[(r+1)h] - 2X(rh) + X[(r-1)h]}{X(rh)} = -p^2 .$$

Trying with $X(rh) = \sin m\pi rh$ we get

$$p^2 = 2\alpha \sin^2 \left(\frac{m\pi h}{2}\right),$$

and putting $T(sk) = e^{-\beta sk}$ we find

$$\frac{1 - e^{-\beta k}}{1 + e^{-\beta k}} = 2\alpha \sin^2 \left(\frac{m\pi h}{2}\right)$$

or finally

$$e^{-\beta k} = \frac{1 - 2\alpha \sin^2 (m\pi h/2)}{1 + 2\alpha \sin^2 (m\pi h/2)} .$$

We can now write down the solution of the Crank-Nicolson difference equation:

$$u_{r,s} = \sum_{m=1}^{M-1} b_m \sin m\pi rh \left(\frac{1 - 2\alpha \sin^2 (m\pi h/2)}{1 + 2\alpha \sin^2 (m\pi h/2)}\right)^s, \qquad (15.2.8)$$

where b_m has the same form as before. The shape of the solution immediately tells us that it can never grow with time irrespective of what positive values we assign to α. It is also easy to see that when $h, k \longrightarrow 0$, then (15.2.8) converges toward (15.2.5).

In order to examine the stability properties we again turn to matrix technique. We note that the system can be written

$$(2I + \alpha T)u_{s+1} = (2I - \alpha T)u_s ,$$

that is,

$$u_{s+1} = \left(I + \frac{\alpha}{2} T\right)^{-1} \left(I - \frac{\alpha}{2} T\right) u_s = Cu_s .$$

A similar relation holds for the error vector: $e_s = C^s e_0$. It is now obvious that the eigenvalues of C are of decisive importance for the stability properties of the method. We have already computed the eigenvalues of T:

$$\lambda_m = 4 \sin^2 \frac{m\pi}{2M}, \qquad m = 1, 2, 3, \ldots, M - 1 ,$$

and from this we get the eigenvalues of C (note that $Mh = 1$):

$$\frac{1 - 2\alpha \sin^2 (m\pi/2M)}{1 + 2\alpha \sin^2 (m\pi/2M)} .$$

Hence Crank-Nicolson's method is stable for all positive values of α.

We shall now also compute the truncation error of the method which is done most easily by operator technique. Let L' be the difference operator, and

$L = D_t - D_x^2$ (observe that $D_x^2 u = \partial^2 u/\partial x^2$, not $(\partial u/\partial x)^2$). We get

$$L' - L = \frac{e^{kD_t} - 1}{k} - \frac{1}{2h^2}(e^{hD_x} - 2 + e^{-hD_x})(1 + e^{kD_t}) - D_t + D_x^2$$

$$= D_t + \frac{k}{2} D_t^2 + \frac{k^2}{6} D_t^3 + \cdots$$

$$- \left(\frac{1}{2} D_x^2 + \frac{h^2}{24} D_x^4 + \cdots\right)\left(2 + kD_t + \frac{k^2}{2} D_t^2 + \cdots\right) - D_t + D_x^2$$

$$= \frac{k}{2} D_t(D_t - D_x^2) + \frac{k^2}{6} D_t^3 - \frac{k^2}{4} D_x^2 D_t^2 - \frac{h^2}{12} D_x^4 + \cdots$$

$$= O(h^2) + O(k^2),$$

since $D_t u = D_x^2 u$ if u is a solution of the differential equation. This result snows that without any risk we can choose k of the same order of magnitude as h, that is, $k = ch$, which is a substantial improvement compared with $k = \alpha h^2$.

Crank-Nicolson's method can be generalized to equations of the form

$$\frac{\partial u}{\partial t} = \frac{\partial^2 u}{\partial x^2} + \frac{\partial^2 u}{\partial y^2},$$

but usually one prefers another still more economic method, originally suggested by Peaceman and Rachford. The same technique can be used in the elliptic case, and it will be briefly described in the next section.

15.3. Elliptic equations

When one is working with ordinary differential equations, the main problem is to find a solution depending on the same number of parameters as the order of the equation. If such a solution has been found, it is usually a simple matter to adapt the parameter values to the given boundary conditions. For partial differential equations, the situation is completely different. In many cases it is relatively simple to find the general solution, but usually it is a difficult problem to adapt it to the boundary conditions. As examples we shall take the Laplace equation

$$\frac{\partial^2 u}{\partial x^2} + \frac{\partial^2 u}{\partial y^2} = 0, \tag{15.3.1}$$

and the Poisson equation

$$\frac{\partial^2 u}{\partial x^2} + \frac{\partial^2 u}{\partial y^2} = F(x, y). \tag{15.3.2}$$

In the remainder of this section, we will give much attention to these two most important equations.

The Laplace equation has the general solution

$$u = f(x + iy) + g(x - iy).$$

The problem of specializing f and g to the boundary conditions (*Dirichlet's problem*) is extremely complicated and has been the subject of extensive investigations.

We shall now consider the numerical treatment of the Poisson equation; the Laplace equation is obtained as a special case if we put $F(x, y) = 0$. According to (12.3) we have approximately.

$$\nabla^2 u(x, y) = \frac{1}{h^2} [u(x + h, y) + u(x, y + h)$$

$$+ u(x - h, y) + u(x, y - h) - 4u(x, y)] . \quad (15.3.3)$$

The equation $\nabla^2 u = F$ is then combined with boundary conditions in such a way that u is given, for example, on the sides of a rectangle or a triangle. If the boundary is of a more complex shape, we can take this into account by modifying (15.3.3). Now we get a partial difference equation which can be solved approximately by Liebmann's iteration method (15.3.4). Here index n indicates a certain iteration:

$$u_{n+1}(x, y) = \frac{1}{4} [u_{n+1}(x, y - h) + u_{n+1}(x - h, y)$$

$$+ u_n(x, y + h) + u_n(x + h, y)] - \frac{h^2}{4} F(x, y) . \quad (15.3.4)$$

We start with guessed values and iterate row by row, moving upward, and repeat the process until no further changes occur. As an example we treat the equation

$$\frac{\partial^2 u}{\partial x^2} + \frac{\partial^2 u}{\partial y^2} = \frac{1}{x^2} + \frac{1}{y^2}$$

in the interior of a triangle with vertices at the points $(1, 1)$, $(1, 2)$, and $(2, 2)$, and with the boundary values $u = x^2 - \log x - 1$ on the horizontal side, $u = 4 - \log 2y - y^2$ on the vertical side, and $u = -2 \log x$ on the oblique side. With $h = 0.2$, we obtain six interior grid points: $A(1.4, 1.2)$; $B(1.6, 1.2)$, $C(1.8, 1.2)$, $D(1.6, 1.4)$; $E(1.8, 1.4)$; and $F(1.8, 1.6)$. Choosing the starting value 0 at all these points, we obtain

A	B	C	D	E	F
−0.1155	0.2327	0.8824	−0.3541	0.3765	−0.3726
−0.0573	0.3793	1.0131	−0.2233	0.3487	−0.3795
−0.0207	0.4539	1.0248	−0.2116	0.3528	−0.3785
−0.0020	0.4644	1.0285	−0.2079	0.3549	−0.3779
+0.0006	0.4669	1.0296	−0.2068	0.3556	−0.3778
+0.0012	0.4676	1.0300	−0.2064	0.3558	−0.3777
+0.0014	0.4679	1.0301	−0.2063	0.3559	−0.3777
+0.0015	0.4679	1.0301	−0.2063	0.3559	−0.3777
+0.0012	0.4677	1.0299	−0.2065	0.3557	−0.3778

The exact solution is $u = x^2 - y^2 - \log xy$, from which the values in the last line have been computed.

The variant of Liebmann's method which has been discussed here is closely related to Gauss-Seidel's method. In general, when a discretization of the differential equation has been performed, we have an ordinary linear system of equations of high but finite order. For this reason we can follow up on the discussion of methods for such systems given in Chapter 4; in particular, this is the case for the overrelaxation method.

All these methods (Jacobi, Gauss-Seidel, SOR) are *point-iterative*, that is, of the form

$$u^{(n+1)} = Mu^{(n)} + c .\tag{15.3.5}$$

Obviously, this equation describes an *explicit* method. A natural generalization is obtained if a whole group of interrelated values (for example, in the same row) are computed simultaneously. This is called *block iteration* and evidently defines an *implicit* method. We give an example with the Laplace equation in a rectangular domain 4×3:

$\cdot 9$	$\cdot 10$	$\cdot 11$	$\cdot 12$
$\cdot 5$	$\cdot 6$	$\cdot 7$	$\cdot 8$
$\cdot 1$	$\cdot 2$	$\cdot 3$	$\cdot 4$

The coefficient matrix becomes

4	-1			-1							
-1	4	-1			-1						
	-1	4	-1			-1					
		-1	4				-1				
-1				4	-1			-1			
	-1			-1	4	-1			-1		
		-1			-1	4	-1			-1	
			-1			-1	4				-1
				-1				4	-1		
					-1			-1	4	-1	
						-1			-1	4	-1
							-1			-1	4

where all empty spaces should be filled with zeros. Using block-matrices we get the system:

$$\begin{pmatrix} D_1 & F_1 & O \\ E_1 & D_2 & F_2 \\ O & E_2 & D_3 \end{pmatrix} \begin{pmatrix} U_1 \\ U_2 \\ U_3 \end{pmatrix} = \begin{pmatrix} K_1 \\ K_2 \\ K_3 \end{pmatrix} .$$

Here, for example, U_1 is a vector with components $u_1, u_2, u_3,$ and u_4. This equation can now be solved following Jacobi, Gauss-Seidel, or SOR. It must then be observed, however, that we get a series of systems of equations for each iteration and not the unknown quantities directly.

Finally, we shall also treat alternating direction implicit ($=$ ADI) methods, and we then restrict ourselves to the Peaceman-Rachford method. First we discuss application of the method on an equation of the form

$$\frac{\partial}{\partial x}\left[A(x,y)\frac{\partial u}{\partial x}\right] + \frac{\partial}{\partial y}\left[C(x,y)\frac{\partial u}{\partial y}\right] - Fu + G = 0, \qquad (15.3.6)$$

where A, C, and F are ≥ 0.

The derivatives are approximated as follows:

$$\frac{\partial}{\partial x}\left[A(x,y)\frac{\partial u}{\partial x}\right] \simeq A\left(x + \frac{h}{2},y\right)\frac{u(x+h,y) - u(x,y)}{h^2}$$

$$- A\left(x - \frac{h}{2},y\right)\frac{u(x,y) - u(x-h,y)}{h^2}$$

and analogously for the y-derivative. Further, we introduce the notations

$$a = A\left(x + \frac{h}{2},y\right), \qquad c = A\left(x - \frac{h}{2},y\right), \qquad 2b = a + c,$$

$$\alpha = C\left(x,y + \frac{h}{2}\right), \qquad \gamma = C\left(x,y - \frac{h}{2}\right), \qquad 2\beta = \alpha + \gamma,$$

and define

$$\begin{cases} H_0 u(x,y) = a(x,y)u(x+h,y) \\ \qquad\qquad - 2b(x,y)u(x,y) + c(x,y)u(x-h,y), \\ V_0 u(x,y) = \alpha(x,y)u(x,y+h) \\ \qquad\qquad - 2\beta(x,y)u(x,y) + \gamma(x,y)u(x,y-h). \end{cases} \qquad (15.3.7)$$

The differential equation will then be approximated by the following difference equation

$$(H_0 + V_0 - Fh^2)u + Gh^2 = 0.$$

Supposing boundary conditions according to Dirichlet we construct a vector k from the term Gh^2 and the boundary values, and further we put $Fh^2 = S$. The equation is then identically rewritten in the following two forms by use of matrix notations:

$$\begin{cases} \left(\rho I - H_0 + \dfrac{S}{2}\right)u = \left(\rho I + V_0 - \dfrac{S}{2}\right)u + k, \\[2mm] \left(\rho I - V_0 + \dfrac{S}{2}\right)u = \left(\rho I + H_0 - \dfrac{S}{2}\right)u + k, \end{cases}$$

where ρ is an arbitrary parameter. Putting $H = -H_0 + S/2$, $V = -V_0 + S/2$, we now define Peaceman-Rachford's ADI-method through

$$\begin{cases} (\rho_n I + H)u^{(n+1/2)} = (\rho_n I - V)u^{(n)} + k , \\ (\rho_n I + V)u^{(n+1)} = (\rho_n I - H)u^{(n+1/2)} + k , \end{cases} \tag{15.3.8}$$

where ρ_n are so-called iteration parameters, so far at our disposal.

We now assume that H and V are symmetric and positive definite. First we restrict ourselves to the case when all ρ_n are equal $(=\rho)$. The intermediate result $u^{(n+1/2)}$ can then be eliminated and we find:

$$u^{(n+1)} = Tu^{(n)} + g ,$$

where

$$T = (\rho I + V)^{-1}(\rho I - H)(\rho I + H)^{-1}(\rho I - V) ,$$
$$g = (\rho I + V)^{-1}(\rho I - H)(\rho I + H)^{-1}k + (\rho I + V)^{-1}k .$$

The convergence speed depends on the spectral radius $\lambda(T)$, but when we try to estimate this it is suitable to construct another matrix W by a similarity transformation:

$$W = (\rho I + V)T(\rho I + V)^{-1} ,$$

that is,

$$W = \{(\rho I - H)(\rho I + H)^{-1}\} \cdot \{(\rho I - V)(\rho I + V)^{-1}\} .$$

If the eigenvalues of H and V are λ_i and μ_i we find

$$\begin{aligned} \lambda(T) = \lambda(W) &\leq ||(\rho I - H)(\rho I + H)^{-1}|| \cdot ||(\rho I - V)(\rho I + V)^{-1}|| \\ &= \lambda\{(\rho I - H)(\rho I + H)^{-1}\} \cdot \lambda\{(\rho I - V)(\rho I + V)^{-1}\} \\ &= \max_{1 \leq i \leq M} \left| \frac{\rho - \lambda_i}{\rho + \lambda_i} \right| \max_{1 \leq i \leq N} \left| \frac{\rho - \mu_i}{\rho + \mu_i} \right| . \end{aligned}$$

As H is positive definite, all $\lambda_i > 0$ and we can assume $0 < a \leq \lambda_i \leq b$. Let us first try to make

$$\max_{1 \leq i \leq M} \left| \frac{\rho - \lambda_i}{\rho + \lambda_i} \right|$$

as small as possible. It is easily understood that large values of $|(\rho - \lambda_i)/(\rho + \lambda_i)|$ are obtained if ρ is far away from λ_i, and hence our critical choice of ρ should be such that $(\rho - a)/(\rho + a) = (b - \rho)/(b + \rho)$, that is, $\rho = \rho_1 = \sqrt{ab}$. In the same way, for the other factor we ought to choose $\rho = \rho_2 = \sqrt{\alpha\beta}$ where $0 < \alpha \leq \mu_i \leq \beta$. If these two values are not too far apart we can expect that the "best" of them will give a fair solution to the whole problem. From the beginning we want to choose ρ such that $\lambda(T)$ is minimized. Instead we have split a majorant for $\lambda(T)$ into two factors which have been minimized separately since the general problem is considerably more difficult.

In the special case when we are treating the usual Laplace-equation inside a rectangular domain $A \times B$, where $A = Mh$ and $B = Nh$, H is operating only horizontally and V only vertically. The $M - 1$ points along a given horizontal line form a closed system with the same matrix R as was denoted by T in the section on parabolic equations. Hence the eigenvalues of R have the form

$$\lambda_m = 4 \sin^2 \frac{m\pi}{2M}, \qquad m = 1, 2, \ldots, M - 1.$$

With, for example, $M = 6$, $N = 4$, we get the following shape for H and V:

$$H = \begin{pmatrix} R & & \\ & R & \\ & & R \end{pmatrix}, \qquad V = \begin{bmatrix} S & & & \\ & S & & \\ & & S & \\ & & & S \\ & & & & S \end{bmatrix},$$

where

$$R = \begin{bmatrix} 2 & -1 & & & \\ -1 & 2 & -1 & & \\ & -1 & 2 & -1 & \\ & & -1 & 2 & -1 \\ & & & -1 & 2 \end{bmatrix}$$

and

$$S = \begin{pmatrix} 2 & -1 & \\ -1 & 2 & -1 \\ & -1 & 2 \end{pmatrix}.$$

In a similar way we obtain the eigenvalues of S:

$$\mu_n = 4 \sin^2 \frac{n\pi}{2N}.$$

Hence we get

$$a = 4 \sin^2 \frac{\pi}{2M}; \qquad b = 4 \sin^2 \frac{\pi[M - 1]}{2M} = 4 \cos^2 \frac{\pi}{2M};$$

$$\alpha = 4 \sin^2 \frac{\pi}{2N}; \qquad \beta = 4 \cos^2 \frac{\pi}{2N}$$

and consequently

$$\rho_1 = 2 \sin \frac{\pi}{M}, \qquad \rho_2 = 2 \sin \frac{\pi}{N}.$$

Now we can compute

$$\frac{\rho_1 - a}{\rho_1 + a} \frac{\rho_1 - \alpha}{\rho_1 + \alpha} \simeq 1 - \frac{1}{N} \frac{\pi(M^2 + N^2)}{MN},$$

and analogously

$$\frac{\rho_2 - a}{\rho_2 + a} \frac{\rho_2 - \alpha}{\rho_2 + \alpha} \simeq 1 - \frac{1}{M} \frac{\pi(M^2 + N^2)}{MN} .$$

Finally we choose the value of ρ which gives the smallest limit, that is,

$$\rho = 2 \sin \frac{\pi}{P} ,$$

where $P = \max(M, N)$. Note that $P = M$ produces the result containing $1/N$.

For the general case when different parameters are used in cyclical order, there is no complete theory, but nevertheless there are a few results which are useful in practical computation. We put $c = \min(a, \alpha)$ and $d = \max(b, \beta)$. In the case when one wants to use n parameters, Peaceman-Rachford suggest the following choice:

$$\rho_j = d\left(\frac{c}{d}\right)^{(2j-1)/2n} , \qquad j = 1, 2, \ldots, n . \tag{15.3.9}$$

Wachspress suggests instead

$$\rho_j = d\left(\frac{c}{d}\right)^{(j-1)/(n-1)} , \qquad n \geq 2 , \quad i = 1, 2, \ldots, n . \tag{15.3.10}$$

More careful investigations have shown that convergence with Wachspress parameters is about twice as fast as with Peaceman-Rachford parameters.

When the iterations are performed in practical work, this means that one solves a system of equations with tridiagonal coefficient matrices which can be done by usual Gaussian elimination. Since all methods for numerical solution of elliptic equations are based upon the solution of finite (but large) linear systems of equations, one need not, as a rule, be worried about stability. Almost exclusively an iterative technique is used, and then it is about enough that the method converges. Since the domain is closed the boundary values, so to speak, by force will keep the inner function values under control, while a similar situation is not present for hyperbolic or parabolic equations.

15.4. Eigenvalue problems

Problems which contain a parameter but which can be solved only for certain values of this parameter are called eigenvalue problems, and they play a most important role in many applications. We here restrict ourselves to demonstrating in an explicit example how such problems can be attacked with numerical methods.

We consider the equation

$$\frac{\partial^2 u}{\partial x^2} + \frac{\partial^2 u}{\partial y^2} + \lambda u = 0 , \tag{15.4.1}$$

inside a square with corners $(0, 0)$, $(3, 0)$, $(3, 3)$, and $(0, 3)$. On the sides of the square we shall have $u = 0$; then the equation describes a vibrating membrane.

We will work with two mesh sizes, $h = 1$ and $h = \frac{3}{4}$. For reasons of symmetry all values are equal in the first case, while in the second case only three

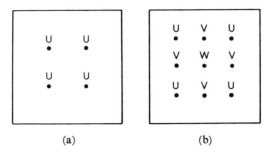

Figure 15.4 (a) (b)

different values need be considered. Starting from equation (15.3.3), we have in the first case

$$U + U + 0 + 0 - 4U + \lambda U = 0$$

and $\lambda = 2$. In the second case we get with $\mu = \frac{9}{16}\lambda$:

$$\begin{cases} 2V - 4U + \mu U = 0 \,, \\ 2U + W - 4V + \mu V = 0 \,, \\ 4V - 4W + \mu W = 0 \,. \end{cases} \tag{15.4.2}$$

The condition for a nontrivial solution is

$$\begin{vmatrix} 4 - \mu & -2 & 0 \\ -2 & 4 - \mu & -1 \\ 0 & -4 & 4 - \mu \end{vmatrix} = 0 \,. \tag{15.4.3}$$

The result is $\mu^3 - 12\mu^2 + 40\mu - 32 = 0$, with the smallest root

$$\mu = 4 - \sqrt{8} = 1.1716 \,,$$

and hence $\lambda = 2.083$. Now we know that the error is proportional to h^2, and hence we have for h^2 and λ the two pairs of values $(1, 2)$ and $(0.5625, 2.083)$. Richardson extrapolation to $h^2 = 0$ gives $\lambda = 2.190$. The exact solution $u = \sin(m\pi x/3) \sin(n\pi y/3)$ gives

$$\lambda = (m^2 + n^2) \frac{\pi^2}{9} \,. \tag{15.4.4}$$

The lowest eigenvalue is obtained for $m = n = 1$:

$$\lambda_{1,1} = \frac{2\pi^2}{9} = 2.193 \,.$$

The error is only 0.15%. Higher eigenvalues can be determined numerically, too, but then a finer mesh is necessary. We will also obtain considerably lower accuracy.

REFERENCES

[1] Sommerfeld: *Partial Differential Equations in Physics*, **6**, (Academic Press, New York, 1949).

[2] Petrovsky: *Lectures on Partial Differential Equations* (London, 1954).

[3] Beckenbach: *Modern Mathematics for the Engineer* (McGraw-Hill, New York, 1956).

[4] Richtmyer: *Difference Methods for Initial-Value Problems* [Interscience (Wiley) New York, 1957].

[5] Courant, Friedrichs, Lewy: *Math. Ann.*, **100**, 32 (1928).

[6] Forsythe and Wasow: *Finite-Difference Methods for Partial Differential Equations* (Wiley New York, 1960).

[7] J. Crank-P. Nicolson: "A Practical Method for Numerical Evaluation of Solutions of Partial Differential Equations of the Heat Conduction Type," *Proc. Camb. Phil. Soc.*, **43**, 50–67 (1947).

[8] D. W. Peaceman-H. H. Rachford, Jr.: "The Numerical Solution of Parabolic and Elliptic Differential Equations." *J. SIAM*, **3**, 28–41 (1955).

[9] E. L. Wachspress-G. J. Habetler: "An Alternating Direction Implicit Iteration Technique," *J. SIAM*, **8**, 403–424 (1960).

[10] D. Young: *Elliptic and Parabolic Partial Differential Equations* (in Survey of Numerical Analysis, ed. John Todd).

[11] G. Birkhoff-R. S. Varga: "Implicit Alternating Direction Methods," *Trans. Am. Math. Soc.*, **92**, 13–24 (1959).

[12] G. D. Smith: *Numerical Solution of Partial Differential Equations* (Oxford Univ. Press, London, 1965).

EXERCISES

1. The function $f(x, y)$ satisfies the Laplace equation

$$\frac{\partial^2 f}{\partial x^2} + \frac{\partial^2 f}{\partial y^2} = 0 \,.$$

This function $f(x, y)$ is given in the following points:

x	0	0.25	0.5	0.75	1	0	1	0
y	0	0	0	0	0	0.25	0.25	0.5
f	-1	$-\frac{3}{5}$	$-\frac{1}{3}$	$-\frac{1}{7}$	0	$-\frac{15}{17}$	$\frac{1}{65}$	$-\frac{3}{5}$

x	1	0	1	0	0.25	0.5	0.75	1
y	0.5	0.75	0.75	1	1	1	1	1
f	$\frac{1}{17}$	$-\frac{7}{25}$	$\frac{9}{73}$	0	$\frac{1}{41}$	$\frac{1}{13}$	$\frac{9}{65}$	$\frac{1}{5}$

Using Liebmann's method, find the value of the function in the nine mesh points: $x = 0.25, 0.5, 0.75$ and $y = 0.25, 0.5, 0.75$, to three places.

2. Using Liebmann's method, solve the differential equation

$$\frac{\partial^2 u}{\partial x^2} + \frac{\partial^2 u}{\partial y^2} = \frac{y}{10}$$

in a square with corners $(1, 0)$, $(3, 0)$, $(3, 2)$, and $(1, 2)$ $(h = \frac{1}{2})$. The boundary values are:

x	1.5	2	2.5	3	3	3
y	0	0	0	0.5	1	1.5
u	-0.3292	0.3000	1.1625	2.3757	1.9500	1.0833

x	2.5	2	1.5	1	1	1
y	2	2	2	1.5	1	0.5
u	-1.0564	-1.4500	-1.4775	-0.8077	-0.6500	-0.7500

3. Find an approximate value of u in the origin when u satisfies the equation

$$\frac{\partial^2 u}{\partial x^2} + 2\frac{\partial^2 u}{\partial y^2} = 1$$

and further $u = 0$ on the sides of a square with corners $(\pm 1, \pm 1)$.

4. The function $u(x, y)$ satisfies the differential equation

$$\frac{\partial^2 u}{\partial x^2} + (1 + x^2 + y^2)\frac{\partial^2 u}{\partial y^2} + 2y\frac{\partial u}{\partial y} = 0 .$$

On the sides of a square in the xy-plane with corners $(0, 0)$, $(0, 1)$, $(1, 0)$, and $(1, 1)$ we have $u(x, y) = x + y$. The square is split into smaller parts with side equal to $\frac{1}{3}$. Find the value of $u(x, y)$ in the four inner lattice points by approximating the differential equation with a difference equation so that the error becomes $O(h^2)$.

5. The equation

$$\frac{\partial^2 u}{\partial x^2} + \frac{1}{x}\frac{\partial u}{\partial x} = \frac{\partial u}{\partial t}$$

is given. Using separation determine a solution satisfying $u(0, t) = 1$, $u(1, t) = 0$, and explain how a more general solution fulfilling the same conditions can be constructed.

6. Let $u(x, t)$ denote a sufficiently differentiable function of x and t and put $u(mh, nk) = u_{m,n}$ where h and k are the mesh sizes of a rectangular grid. Show that the equation

$$(2k)^{-1}(u_{m,n+1} - u_{m,n-1}) - h^{-2}(u_{m+1,n} - u_{m,n+1} - u_{m,n-1} + u_{m-1,n}) = 0$$

gives a difference approximation of the heat equation

$$\frac{\partial u}{\partial t} - \frac{\partial^2 u}{\partial x^2} = 0$$

if $k = h^2$ and $h \to 0$, but a difference approximation of the equation

$$\frac{\partial u}{\partial t} + \frac{\partial^2 u}{\partial t^2} - \frac{\partial^2 u}{\partial x^2} = 0$$

if $k = h$ and $h \to 0$.

7. Compute the smallest eigenvalue of the equation

$$\frac{\partial^2 u}{\partial x^2} + \frac{\partial^2 u}{\partial y^2} + \lambda u = 0$$

by replacing it with a difference equation. Boundary values are $u = 0$ on two horizontal and one vertical side of a square with side 1, and with two sides along the positive x- and y-axes, and $\partial u/\partial x = 0$ on the remaining side along the y-axis. This latter condition is conveniently replaced by $u(-h, y) = u(h, y)$. Choose $h = \frac{1}{2}$ and $h = \frac{1}{3}$, extrapolating to $h^2 = 0$. Then find an analytic solution trying $u = \cos \alpha x \sin \beta y$ and compute an exact value of λ.

8. The equation

$$\frac{\partial^2 u}{\partial x^2} + \frac{\partial^2 u}{\partial y^2} + \lambda u = 0$$

is considered in a triangular domain bounded by the lines $x = 0, y = 0$, and $x + y = 1$. It is known that $u = 0$ on the boundary. The differential equation is replaced by a second-order difference equation, which is considered in a square grid with mesh size h. Two cases are treated, namely $h = \frac{1}{4}$ and $h = \frac{1}{5}$. Owing to the symmetry, in the first case only two function values, U and V, and in the second case, three function values, U, V, and W, have to be considered. Find the smallest eigenvalue in these two cases, and extrapolate to $h^2 = 0$.

9. Find the smallest eigenvalue of the equation

$$\frac{\partial^2 u}{\partial x^2} + \frac{\partial^2 u}{\partial y^2} + \lambda(x^2 + y^2)u = 0$$

for a triangular domain with corners $(-1, 0)$, $(0, 1)$, and $(1, 0)$, where we have $u = 0$ on the boundary. Use $h = \frac{1}{2}$ and $h = \frac{1}{3}$, and extrapolate to $h^2 = 0$.

10. Compute the two smallest eigenvalues of the differential equation

$$\frac{\partial^4 u}{\partial x^4} + 2 \frac{\partial^4 u}{\partial x^2 \partial y^2} + \frac{\partial^4 u}{\partial y^4} = \lambda u .$$

Boundary values are $u = 0$ and $\partial u/\partial n = 0$ (the derivative along the normal) for a square with side 4, valid on all 4 sides. Choose $h = 1$.

11. In the wave equation

$$\frac{\partial^2 u}{\partial x^2} - \frac{\partial^2 u}{\partial t^2} = 0$$

the left-hand side is approximated by the expression $h^{-2}[\delta_t^2 - (1 + \alpha \delta_t^2)\delta_x^2]u$. Compute the truncation error!

Chapter 16

Integral Equations

16.0. Definition and classification

An equation containing an integral where an unknown function to be determined enters under the integral sign is called an *integral equation*. If the desired function is present only in the integral, the equation is said to be of the *first kind*; if the function is also present outside the integral, the equation is said to be of the *second kind*. If the function to be determined only occurs linearly, the equation is said to be *linear*, and we shall restrict ourselves to such equations. If the limits in the integral are constant, the equation is said to be of *Fredholm's* type; if the upper limit is variable we have an equation of *Volterra's* type.

In the following the unknown function will be denoted by $y(x)$. We then suppose that $y(t)$ is integrated with another function $K(x, t)$ (with t as integration variable); the function $K(x, t)$ is called the *kernel* of the equation. While the function $y(x)$ as a rule has good regularity properties (continuity or differentiability), it is rather common that the kernel $K(x, t)$ has some discontinuity or at least discontinuous derivatives. The equation is still said to be of Fredholm's type if the limits are constant and the kernel quadratically integrable over the domain. If $K(x, t)$ is singular in the considered domain, or if one or both of the limits is infinite, the integral equation is said to be *singular*. If the equation has a term which does not contain the wanted function $y(x)$, the equation is said to be *inhomogeneous*, otherwise *homogeneous*. Often a parameter λ is placed as a factor in front of the integral, and it may then happen that the equation can be solved only for certain values of λ (eigenvalues), or that there is no solution at all. In the examples below, the terminology just described is illustrated.

$$f(x) = \int_a^b K(x, t)y(t) \, dt , \qquad \text{Fredholm, 1. kind, inhomogeneous,}$$

$$y(x) = f(x) + \lambda \int_a^b K(x, t)y(t) \, dt , \qquad \text{Fredholm, 2. kind, inhomogeneous,}$$

$$y(x) = \lambda \int_a^b K(x, t)y(t) \, dt , \qquad \text{Fredholm, 2. kind, homogeneous,}$$

$$y(x) = f(x) + \lambda \int_a^x K(x, t)y(t) \, dt , \qquad \text{Volterra, 2. kind, inhomogeneous,}$$

$$f(x) = \int_a^x K(x, t)y(t) \, dt , \qquad \text{Volterra, 1. kind, inhomogeneous.}$$

Again it is pointed out that the types referred to here do not in any way cover all possibilities. The reason that just these have been chosen is their strong dominance in the applications.

The fundamental description of the integral equations was performed by Fredholm and Volterra, but important contributions to the theory have been made by Hilbert and Schmidt. An extensive literature exists within this field, but a rather small amount has been published on numerical methods. For obvious reasons, it is only possible to give a very rhapsodic presentation. Primarily we shall discuss direct methods based on linear systems of equations, but also iterative methods will be treated. Further, special kernels (symmetric, degenerate) and methods related to them, will be mentioned briefly. In order not to make the description unnecessarily heavy we shall assume that all functions used have such properties as to make the performed operations legitimate.

16.1. Fredholm's inhomogeneous equation of the second kind

Here we shall primarily study Fredholm's inhomogeneous equation of the second kind:

$$y(x) = f(x) + \lambda \int_a^b K(x, t)y(t) \, dt , \qquad (16.1.1)$$

since the treatment of other types in many respects can be related and traced back to this one. The section between a and b is divided into n equal parts, each of length h. We introduce the notations

$$x_r = a + rh , \qquad y(x_r) = y_r , \qquad f(x_r) = f_r , \qquad K(x_r, t_s) = K_{rs} .$$

The integral is then replaced by a finite sum essentially according to the trapezoidal rule, and then the following system of equations appears:

$$Ay = f ,$$

where

$$
\begin{cases}
A = \begin{bmatrix} 1 - \lambda h K_{11} & - \lambda h K_{12} & \cdots & - \lambda h K_{1n} \\ - \lambda h K_{21} & 1 - \lambda h K_{22} & \cdots & - \lambda h K_{2n} \\ \vdots & & & \\ - \lambda h K_{n1} & - \lambda h K_{n2} & \cdots & 1 - \lambda h K_{nn} \end{bmatrix} , \\[2em]
y = \begin{bmatrix} y_1 \\ y_2 \\ \vdots \\ y_n \end{bmatrix} , \qquad f = \begin{bmatrix} f_1 \\ f_2 \\ \vdots \\ f_n \end{bmatrix} .
\end{cases}
\qquad (16.1.2)
$$

Fredholm himself gave this system careful study. Of special interest is the limiting value of $\Delta = \det A$ when $h \to 0$ (and $n \to \infty$). This limiting value is called *Fredholm's determinant* and is usually denoted by $D(\lambda)$. Applying

Cramer's rule we can write down the solution in explicit form according to

$$y_s = \frac{1}{\Delta} \sum_{r=1}^{n} f_r \Delta_{rs} \,,$$

where, as usual, Δ_{rs} is the algebraic complement of the (r, s)-element in the matrix A. Expanding Δ in powers of λ we find

$$\Delta = 1 - \lambda h \sum_i K_{ii} + \frac{\lambda^2 h^2}{2!} \sum_{i,j} \begin{vmatrix} K_{ii} & K_{ij} \\ K_{ji} & K_{jj} \end{vmatrix} - \cdots$$

and after passage to the limit $h \to 0$:

$$D(\lambda) = 1 - \lambda \int_a^b K(t, t)\, dt + \frac{\lambda^2}{2!} \int\int_a^b \begin{vmatrix} K(t_1, t_1) & K(t_1, t_2) \\ K(t_2, t_1) & K(t_2, t_2) \end{vmatrix} dt_1\, dt_2$$

$$- \frac{\lambda^3}{3!} \int\int\int_a^b \begin{vmatrix} K(t_1, t_1) & K(t_1, t_2) & K(t_1, t_3) \\ K(t_2, t_1) & K(t_2, t_2) & K(t_2, t_3) \\ K(t_3, t_1) & K(t_3, t_2) & K(t_3, t_3) \end{vmatrix} dt_1\, dt_2\, dt_3 + \cdots .$$

A similar expansion can be made for Δ_{rs} and after division by h and passage to the limit we get

$$D(x, t; \lambda) = \lambda K(x, t) - \lambda^2 \int_a^b \begin{vmatrix} K(x, t) & K(x, p) \\ K(p, t) & K(p, p) \end{vmatrix} dp$$

$$+ \frac{\lambda^3}{2!} \int\int_a^b \begin{vmatrix} K(x, t) & K(x, t_1) & K(x, t_2) \\ K(t_1, t) & K(t_1, t_1) & K(t_1, t_2) \\ K(t_2, t) & K(t_2, t_1) & K(t_2, t_2) \end{vmatrix} dt_1\, dt_2 - \cdots ,$$

where $x = a + sh$, $t = a + rh$; however, $D(x, x; \lambda) = D(\lambda)$. Introducing the notation $H(x, t; \lambda) = D(x, t; \lambda)/D(\lambda)$ we can write the solution in the form

$$y(x) = f(x) + \int_a^b H(x, t; \lambda) f(t)\, dt \,. \tag{16.1.3}$$

The function $H(x, t; \lambda)$ is usually called the *resolvent* of the integral equation. Obviously, as a rule λ must have such a value that $D(\lambda) \neq 0$. On the other hand, if we are considering the homogeneous Fredholm equation of second kind, obtained by making $f(x) = 0$ in (16.1.1), then λ must be a root of the equation $D(\lambda) = 0$ in order to secure a nontrivial solution. Hence, in this case we have an eigenvalue problem. The technique which led to the solution (16.1.3) in some simple cases can be used also in practice, and we give an example of this. Consider the equation

$$y(x) = f(x) + \lambda \int_0^1 (x + t) y(t)\, dt \,.$$

We compute $D(\lambda)$ and $D(x, t; \lambda)$ noticing that all determinants of higher order than the second vanish:

$$D(\lambda) = 1 - \lambda - \frac{\lambda^2}{12} \,; \qquad D(x, t; \lambda) = \lambda \left[\left(1 - \frac{\lambda}{2} \right) (x + t) + \lambda \left(xt + \frac{1}{3} \right) \right] .$$

Thus we find the solution:

$$y(x) = f(x) + \frac{\lambda}{1 - \lambda - \lambda^2/12} \int_0^1 \left[\left(1 - \frac{\lambda}{2}\right)(x + t) + \lambda\left(xt + \frac{1}{3}\right) \right] f(t)\, dt ,$$

which can obviously be written $y(x) = f(x) + Ax + B$, where A and B are constants depending on λ and the function $f(t)$. The solution exists unless $\lambda^2 + 12\lambda = 12$, that is, $\lambda = -6 \pm 4\sqrt{3}$.

We shall now study an iterative technique and again start from (16.1.1) constructing a series of functions $y_0, y_1, y_2, \ldots$, according to

$$y_{n+1}(x) = f(x) + \lambda \int_a^b K(x, t) y_n(t)\, dt ; \qquad y_0(x) = 0$$

(cf. Picard's method for ordinary differential equations). If this sequence of functions converges toward a limiting function, this is also a solution of the integral equation. As is easily understood, the method is equivalent to a series expansion in λ (Neumann series):

$$y = f(x) + \lambda\varphi_1(x) + \lambda^2\varphi_2(x) + \cdots .$$

Inserting this and identifying coefficients for different powers of λ we get:

$$\varphi_1(x) = \int_a^b K(x, t)f(t)\, dt ,$$

$$\varphi_2(x) = \int_a^b K(x, t)\varphi_1(t)\, dt = \int_a^b K_2(x, t)f(t)\, dt ,$$

$$\varphi_3(x) = \int_a^b K(x, t)\varphi_2(t)\, dt = \int_a^b K_3(x, t)f(t)\, dt ,$$

$$\vdots$$

Here

$$K_2(x, t) = \int_a^b K(x, z)K(z, t)\, dz$$

and more generally

$$K_n(x, t) = \int_a^b K_r(x, z)K_{n-r}(z, t)\, dz , \qquad K_1(x, t) = K(x, t) .$$

The functions $K_2, K_3, \ldots$ are called *iterated kernels*. It can be shown that the series converges for $\lambda < \|K\|^{-1}$, where

$$\|K\| = \iint_a^b K^2(x, y)\, dx\, dy .$$

Thus we obtain the final result:

$$y(x) = f(x) + \int_a^b [\lambda K_1(x, t) + \lambda^2 K_2(x, t) + \lambda^3 K_3(x, t) + \cdots] f(t)\, dt$$

or

$$y(x) = f(x) + \int_a^b H(x, t; \lambda)f(t)\, dt ,$$

where

$$H(x, t; \lambda) = \lambda K_1(x, t) + \lambda^2 K_2(x, t) + \lambda^3 K_3(x, t) + \cdots$$

as before stands for the resolvent.

The described method is well suited for numerical treatment on a computer. We shall work an example analytically with this technique. Consider the equation

$$y(x) + f(x) + \lambda \int_0^1 \exp(x - t)y(t)\, dt \,.$$

Thus we have $K(x, t) = \exp(x - t)$ and consequently

$$K_2(x, t) = \int_0^1 e^{x-z} e^{z-t}\, dz = e^{x-t} = K(x, t)$$

and all iterated kernels are equal to $K(x, t)$. Hence we obtain

$$y = f(x) + \int_0^1 e^{x-t} f(t)(\lambda + \lambda^2 + \lambda^3 + \cdots)\, dt$$

or, if $|\lambda| < 1$,

$$y = f(x) + \frac{\lambda}{1 - \lambda} e^x \int_0^1 e^{-t} f(t)\, dt \,.$$

In this special case we can attain a solution in a still simpler way. From the equation follows

$$y(x) = f(x) + \lambda e^x \int_0^1 e^{-t} y(t)\, dt = f(x) + C\lambda e^x \,.$$

Inserting this solution again for determining C, we find

$$f(x) + C\lambda e^x = f(x) + \lambda \int_0^1 e^{x-t} [f(t) + C\lambda e^t]\, dt \,,$$

giving

$$C = \int_0^1 e^{-t} f(t)\, dt + C\lambda$$

and we get back the same solution as before. However, it is now sufficient that $\lambda \neq 1$. If $\lambda = 1$ no solution exists unless $\int_0^1 e^{-t} f(t)\, dt = 0$; if so, there are an infinity of solutions $y = f(x) + Ce^x$ with arbitrary C.

This last method obviously depends on the fact that the kernel e^{x-t} could be written $e^x e^{-t}$ making it possible to move the factor e^x outside the integral sign. If the kernel decomposes in this way so that $K(x, t) = F(x)G(t)$ or, more generally, so that

$$K(x, t) = \sum_{i=1}^n F_i(x)G_i(t) \,,$$

the kernel is said to be *degenerate*. In this case, as a rule the equation can be treated in an elementary way. Let us start from

$$y(x) = f(x) + \lambda \int_a^b \left[\sum_{i=1}^n F_i(x)G_i(t) \right] y(t)\, dt = f(x) + \sum_{i=1}^n F_i(x) \int_a^b G_i(t)y(t)\, dt \ .$$

These last integrals are all constants:

$$\int_a^b G_i(t)y(t)\, dt = c_i$$

and so we get:

$$y(x) = f(x) + \lambda \sum_{i=1}^n c_i F_i(x) \ .$$

The constants c_i are determined by insertion into the equation:

$$c_i = \int_a^b G_i(t) \left[f(t) + \lambda \sum_{k=1}^n c_k F_k(t) \right] dt \ .$$

With the notations

$$\int_a^b G_i(t)F_k(t)\, dt = a_{ik} \qquad \text{and} \qquad \int_a^b G_i(t)f(t)\, dt = b_i \ ,$$

we get a linear system of equations for the c_i:

$$(I - \lambda A)c = b \ ,$$

where $A = (a_{ik})$ and b and c are column vectors. This system has a unique solution if $D(\lambda) = \det (I - \lambda A) \neq 0$. Further, we define $D(x, t; \lambda)$ as a determinant of the $(n + 1)$th order:

$$D(x, t; \lambda) = \begin{vmatrix} 0 & -F_1(x) & -F_2(x) \cdots & -F_n(x) \\ G_1(t) & 1 - \lambda a_{11} & -\lambda a_{12} \cdots & -\lambda a_{1n} \\ G_2(t) & -\lambda a_{21} & 1 - \lambda a_{22} \cdots & -\lambda a_{2n} \\ \vdots & & & \\ G_n(t) & -\lambda a_{n1} & -\lambda a_{n2} \cdots & 1 - \lambda a_{nn} \end{vmatrix} \ .$$

Then we can also write

$$y(x) = f(x) + \frac{\lambda}{D(\lambda)} \int_a^b D(x, t; \lambda)f(t)\, dt = f(x) + \lambda \int_a^b H(x, t; \lambda)f(t)\, dt \ ,$$

with the resolvent as before represented in the form

$$H(x, t; \lambda) = \frac{D(x, t; \lambda)}{D(\lambda)} \ .$$

We now treat a previous example by this technique:

$$y(x) = f(x) + \lambda \int_0^1 (x + t)y(t)\, dt \ .$$

Putting

$$\int_0^1 y(t)\, dt = A \,, \qquad \int_0^1 ty(t)\, dt = B \,, \qquad \int_0^1 f(t)\, dt = C \,, \qquad \int_0^1 tf(t)\, dt = D \,,$$

the constants C and D are known in principle, while A and B have to be computed. Now, $y(x) = f(x) + A\lambda x + B\lambda$ and $y(t) = f(t) + A\lambda t + B\lambda$ which inserted into the equation gives

$$A = C + \frac{1}{2}\, A\lambda + B\lambda \,,$$

$$B = D + \frac{1}{3}\, A\lambda + \frac{1}{2}\, B\lambda \,.$$

Solving for A and B we get

$$A = \frac{C(1 - \tfrac{1}{2}\lambda) + D\lambda}{\varDelta} \,, \qquad B = \frac{C\lambda/3 + D(1 - \tfrac{1}{2}\lambda)}{\varDelta} \,,$$

where we have supposed $\varDelta = 1 - \lambda - \tfrac{1}{12}\lambda^2 \neq 0$. Hence we get as before

$$y(x) = f(x) + \frac{\lambda}{1 - \lambda - \tfrac{1}{12}\lambda^2} \int_0^1 \left[x\left(1 - \frac{\lambda}{2}\right) + \lambda tx + \frac{\lambda}{3} + t\left(1 - \frac{\lambda}{2}\right) \right] f(t)\, dt \,.$$

16.2. Fredholm's homogeneous equation of the second kind

As mentioned earlier the case $f(x) \equiv 0$, that is, when we have a homogeneous equation, gives rise to an eigenvalue problem. The eigenvalues are determined, for example, from the equation $D(\lambda) = 0$, or, if we are approximating by a finite linear system of equations, in the usual way from a determinantal condition. There exists an extensive theory for these eigenvalue problems, but here we shall only touch upon a few details which have exact counterparts in the theory of Hermitian or real symmetric matrices. First we prove that if the kernel is real and symmetric, that is, $K(x, t) = K(t, x)$, then the eigenvalues are real. For we have $y(x) = \lambda \int_a^b K(x, t)y(t)\, dt$. Multiplying by $y^*(x)$ and integrating from a to b we get

$$\int_a^b y^*(x)y(x)\, dx = \lambda \iint_a^b y^*(x)K(x, t)y(t)\, dx\, dt \,.$$

The left-hand integral is obviously real and so is the right-hand integral since its conjugate value is

$$\iint_a^b y(x)K(x, t)y^*(t)\, dx\, dt = \iint_a^b y(x)K(t, x)y^*(t)\, dx\, dt \,,$$

which goes over into the original integral if the integration variables change places. This implies that λ also must be real.

Next we shall prove that two eigenfunctions $y_i(x)$ and $y_k(x)$ corresponding to two different eigenvalues λ_i and λ_k are orthogonal. For

$$y_i(x) = \lambda_i \int_a^b K(x, t)y_i(t)\, dt \ ,$$

$$y_k(x) = \lambda_k \int_a^b K(x, t)y_k(t)\, dt \ .$$

Now multiply the first equation by $\lambda_k y_k(x)$ and the second by $\lambda_i y_i(x)$, subtract and integrate:

$$(\lambda_k - \lambda_i)\int_a^b y_i(x)y_k(x)\, dx = \lambda_i\lambda_k\left[\iint_a^b y_k(x)K(x, t)y_i(t)\, dx\, dt\right.$$

$$\left. - \iint_a^b y_i(x)K(x, t)y_k(t)\, dx\, dt\right] \ .$$

Since $K(x, t) = K(t, x)$ and the variables x and t can change places, the two double integrals are equal, and as $\lambda_i \neq \lambda_k$ we get

$$\int_a^b y_i(x)y_k(x)\, dx = 0 \ . \tag{16.2.1}$$

We shall also try to construct a solution of the homogeneous equation

$$y(x) = \lambda \int_a^b K(x, t)y(t)\, dt \ .$$

To do this we start from the matrix A in (16.1.2). If we choose the elements of column j and multiply by the algebraic complements of column k, the sum of these products, as is well known, equals zero. Hence we have

$$(1 - \lambda h K_{jj})\Delta_{jk} - \lambda h K_{kj}\Delta_{kk} - \sum_{i \neq j, k} \lambda h K_{ij}\Delta_{ik} = 0 \ ,$$

where the products (j) and (k) have been written out separately. Dividing by h (note that Δ_{jk} and Δ_{ik} as distinguished from Δ_{kk} contain an extra factor h) we get after passage to the limit $h \to 0$:

$$D(x_k, x_j; \lambda) - \lambda K(x_k, x_j)D(\lambda) - \lambda \int_a^b K(t, x_j)D(x_k, t; \lambda)\, dt = 0 \ .$$

Replacing x_k by x and x_j by z, we get finally

$$D(x, z; \lambda) - \lambda K(x, z)D(\lambda) = \lambda \int_a^b K(t, z)D(x, t; \lambda)\, dt \ . \tag{16.2.2}$$

Expanding by rows instead of columns we find in a similar way

$$D(x, z; \lambda) - \lambda K(x, z)D(\lambda) = \lambda \int_a^b K(x, t)D(t, z; \lambda)\, dt \ . \tag{16.2.3}$$

These two relations are known as *Fredholm's first* and *second* relation.

Now suppose that $\lambda = \lambda_0$ is an eigenvalue, that is, $D(\lambda_0) = 0$. Fredholm's second relation then takes the form:

$$D(x, z; \lambda_0) = \lambda_0 \int_a^b K(x, t)D(t, z; \lambda) \, dt ,$$

independently of z. Hence we can choose $z = z_0$ arbitrarily, and under the assumption that $D(x, z; \lambda_0)$ is not identically zero, we have the desired solution apart from a trivial constant factor:

$$y(x) = D(x, z_0; \lambda_0) . \tag{16.2.4}$$

EXAMPLE

$$y(x) = \lambda \int_0^1 (x + t)y(t) \, dt .$$

The eigenvalues are obtained from $D(\lambda) = 0$, that is,

$$\lambda^2 + 12\lambda = 12 , \qquad \lambda = -6 \pm 4\sqrt{3} .$$

As before we have $D(x, t; \lambda) = \lambda[(1 - \tfrac{1}{2}\lambda)(x + t) + \lambda(xt + \tfrac{1}{3})]$, and we can choose for example, $t = 0$. One solution is consequently

$$y(x) = \lambda\left[\left(1 - \frac{\lambda}{2}\right)x + \frac{\lambda}{3}\right] .$$

Taking $\lambda = 4\sqrt{3} - 6$ we have apart from irrelevant constant factors

$$y = x\sqrt{3} + 1 .$$

If we choose other values for t we get, as is easily understood, the same result.

Another method can also be used. The equation shows that the solution must have the form $y(x) = Ax + B$, and this implies

$$Ax + B = \lambda \int_0^1 (x + t)(At + B) \, dt$$

which gives the same result.

16.3. Fredholm's equation of the first kind

The equation which we shall study very superficially in this section has the form

$$f(x) = \int_a^b K(x, t)y(t) \, dt . \tag{16.3.1}$$

A parameter λ can of course be brought together with the kernel. In general, this equation has no solution as can be realized from the following example:

$$\sin x = \int_0^1 e^{x-t}y(t) \, dt , \qquad (= ce^x) .$$

However, let us suppose that $f(x)$, $K(x, t)$, and $y(t)$ can be expanded in a series of orthonormalized functions u_i:

$$\begin{cases} f(x) = \sum_{i=1}^{\infty} b_i u_i(x) \, , \\[2mm] K(x, t) = \sum_{i,k=1}^{\infty} a_{ik} u_i(x) u_k(t) \, , \\[2mm] y(t) = \sum_{i=1}^{\infty} c_i u_i(t) \, . \end{cases} \qquad (16.3.2)$$

The functions $u_i(x)$ being orthonormalized over the interval (a, b) means that

$$\int_a^b u_i(x) u_k(x) \, dx = \delta_{ik} \, . \qquad (16.3.3)$$

After insertion into (16.3.1) we obtain:

$$\sum_{i=1}^{\infty} b_i u_i(x) = \int_a^b \sum_{i,k=1}^{\infty} a_{ik} u_i(x) u_k(t) \sum_{r=1}^{\infty} c_r u_r(t) \, dt = \sum_{i,k=1}^{\infty} a_{ik} c_k u_i(x) \, .$$

Identifying we get an infinite linear system of equations

$$\sum_{k=1}^{\infty} a_{ik} c_k = b_i \, ,$$

where the coefficients c_k should be determined from the known quantities a_{ik} and b_i. If the summation is truncated at the value n we have the system

$$Ac = b \qquad (16.3.4)$$

with well-known properties regarding existence of solutions.

Another possibility is to approximate (16.3.1) directly by a suitable quadrature formula giving rise to a linear system of equations:

$$\sum_{j=0}^{n} a_j K_{ij} y_j = f_i \, . \qquad (16.3.5)$$

However, there is an apparent risk that one might not observe that no solution exists because of the approximations which have been made.

16.4. Volterra's equations

First we are going to discuss Volterra's equation of the second kind

$$y(x) = f(x) + \lambda \int_a^x K(x, t) y(t) \, dt \, . \qquad (16.4.1)$$

Obviously, it can be considered as a special case of Fredholm's equation if we define the kernel so that $K(x, t) = 0$ when $x < t < b$. However, there are some advantages in discussing the case separately, in about the same way as for triangular systems of equations compared with the general case. For simplicity we assume $a = 0$, further the interval length h, and decide to use the trapezoidal rule (more sophisticated methods can also easily be applied). Then we obtain

$$
\begin{aligned}
y_0 &= f_0 \,, \\
y_1 &= f_1 + \lambda h[\tfrac{1}{2}K_{10}y_0 + \tfrac{1}{2}K_{11}y_1] \,, \\
y_2 &= f_2 + \lambda h[\tfrac{1}{2}K_{20}y_0 + \ K_{21}y_1 + \tfrac{1}{2}K_{22}y_2] \,, \\
&\ \ \vdots
\end{aligned}
\tag{16.4.2}
$$

and we can determine the desired values $y_0, y_1, y_2, \ldots$ successively. This method is very well adapted for use on automatic computers.

As is the case for Fredholm's equation we can also construct a Neumann series with iterated kernels. The technique is exactly the same, and there is no reason to discuss it again. However, there is one important difference: if $K(x, t)$ and $f(x)$ are real and continuous, then the series converges for all values of λ.

EXAMPLE

$$
y(x) = x + \int_0^x (t - x)y(t)\, dt \,.
$$

1. The trapezoidal rule with interval length $= h$ gives

$$
y_0 = 0
$$

$$
y_1 = \ h \qquad\qquad\qquad y_1 - \sin\ h = \frac{h^3}{6} + \cdots
$$

$$
y_2 = 2h - \ h^3 \qquad\qquad y_2 - \sin 2h = \frac{2h^3}{6} + \cdots
$$

$$
y_3 = 3h - 4h^3 + h^5 \qquad y_3 - \sin 3h = \frac{3h^3}{6} + \cdots
$$

$$
\vdots
$$

Already Simpson's formula supplemented by the $\tfrac{3}{8}$-rule gives a substantial improvement.

2. Using successive approximations we get

$$
y_0 = x \,,
$$

$$
y_1 = x + \int_0^x (t - x)t\, dt = x - \frac{x^3}{6} \,,
$$

$$
y_2 = x + \int_0^x (t - x)\left(t - \frac{t^3}{6}\right) dt = x - \frac{x^3}{6} + \frac{x^5}{120} \,,
$$

$$
\vdots
$$

and it is easy to show that we obtain the whole series expansion for $\sin x$.

3. Differentiating the equation, we find

$$y'(x) = 1 - \int_0^x y(t)\, dt\ ; \qquad y'(0) = 1\ .$$

$$y''(x) = -y(x) \qquad \text{with the solution} \qquad y = A \cos x + B \sin x\ .$$

But since $y(0) = 0$ and $y'(0) = 1$ the result is $y = \sin x$.

Very briefly we shall also comment upon Volterra's equation of the first kind

$$f(x) = \int_a^x K(x, t) y(t)\, dt\ . \tag{16.4.3}$$

Differentiation with respect to x gives

$$f'(x) = K(x, x) y(x) + \int_a^x \frac{\partial K(x, t)}{\partial x} y(t)\, dt\ ,$$

and assuming $K(x, x) \neq 0$ in the considered interval we can divide by $K(x, x)$ obtaining a Volterra equation of the second kind. If necessary, the procedure can be repeated. It is also possible to use the same difference method as before, the first function value being determined from the relation above:

$$y(a) = \frac{f'(a)}{K(a, a)}\ .$$

A special equation of Volterra type is the Abelian integral equation

$$f(x) = \int_0^x (x - t)^{-\alpha} y(t)\, dt \qquad (0 < \alpha < 1)\ . \tag{16.4.4}$$

Writing instead $f(z) = \int_0^z (z - t)^{-\alpha} y(t)\, dt$, dividing by $(x - z)^{1-\alpha}$, and integrating in z between the limits 0 and x, we obtain

$$\int_0^x \frac{f(z)}{(x - z)^{1-\alpha}}\, dz = \int_0^x \frac{dz}{(x - z)^{1-\alpha}} \int_0^z \frac{y(t)\, dt}{(z - t)^{\alpha}}\ .$$

If z is a vertical and t a horizontal axis, the integration is performed over a triangle with corners in the origin, in $(0, x)$, and in (x, x), and reversing the order of integration we find

$$\int_0^x \frac{f(z)}{(x - z)^{1-\alpha}}\, dz = \int_0^x y(t)\, dt \int_t^x \frac{dz}{(z - t)^{\alpha} (x - z)^{1-\alpha}}\ .$$

The last integral is transformed through $z - t = (x - t)\xi$ and goes over into

$$\int_0^1 \xi^{-\alpha}(1 - \xi)^{\alpha-1}\, d\xi = \Gamma(\alpha)\Gamma(1 - \alpha) = \frac{\pi}{\sin \pi\alpha}\ ,$$

that is,

$$\int_0^x y(t)\, dt = \frac{\sin \pi\alpha}{\pi} \int_0^x \frac{f(z)}{(x - z)^{1-\alpha}}\, dz\ .$$

Differentiating this relation we obtain the desired solution

$$y(x) = \frac{\sin \alpha \pi}{\pi} \frac{d}{dx} \int_0^x \frac{f(z)}{(x - z)^{1-\alpha}} dz .$$
(16.4.5)

Abel's initial problem was concerned with the case $\alpha = \frac{1}{2}$ and arose when he tried to determine tautochronous curves, that is, curves having the property that the time which a particle needed to slide along the curve to a given end point is a given function of the position of the starting point.

16.5. Connections between differential and integral equations

There exists a fundamental relationship between linear differential equations and integral equations of Volterra's type. Let us start from the equation

$$\frac{d^n u}{dx^n} + a_1(x) \frac{d^{n-1} u}{dx^{n-1}} + \cdots + a_n(x)u = F(x) ,$$
(16.5.1)

where the coefficients are continuous and obey the initial conditions

$$u(0) = c_0 , \quad u'(0) = c_1 , \quad \ldots , \quad u^{(n-1)}(0) = c_{n-1} .$$

Putting $d^n u/dx^n = y$ we find

$$D^{-1}y = \int_0^x y(t) \, dt \qquad \text{(defining the operator } D^{-1}) ,$$

$$D^{-2}y = D^{-1}(D^{-1}y) = \int_0^x (x - t)y(t) \, dt \qquad \text{(by partial integration)} ,$$

$$\vdots$$

$$D^{-n}y = D^{-1}(D^{-n+1}y) = \frac{1}{(n - 1)!} \int_0^x (x - t)^{n-1}y(t) \, dt .$$

By integrating the equation $d^n u/dx^n = y$ successively and taking the initial conditions into account, we get

$$\frac{d^{n-1} u}{dx^{n-1}} = c_{n-1} + D^{-1}y ,$$

$$\frac{d^{n-2} u}{dx^{n-2}} = c_{n-1}x + c_{n-2} + D^{-2}y ,$$

$$\vdots$$

$$u = c_{n-1} \frac{x^{n-1}}{(n - 1)!} + c_{n-2} \frac{x^{n-2}}{(n - 2)!} + \cdots + c_1 x + c_0 + D^{-n}y .$$

These values are now inserted into (16.5.1) and the following equation results:

$$y(x) = f(x) + \int_0^x K(x, t)y(t) \, dt ,$$
(16.5.2)

where

$$f(x) = F(x) - a_1(x)c_{n-1} - a_2(x)(c_{n-1}x + c_{n-2}) - \cdots$$

$$- a_n(x)\left(c_{n-1}\frac{x^{n-1}}{(n-1)!} + c_{n-2}\frac{x^{n-2}}{(n-2)!} + \cdots + c_1x + c_0\right) \quad (16.5.3)$$

and

$$K(x, t) = -\left[a_1(x) + a_2(x)(x - t) + a_3(x)\frac{(x-t)^2}{2!} + \cdots \right.$$

$$\left. + a_n(x)\frac{(x-t)^{n-1}}{(n-1)!}\right]. \quad (16.5.4)$$

Thus we can say that the integral equation describes the differential equation together with its initial conditions.

EXAMPLES

1. $u'' + u = 0$; $u(0) = 0$, $u'(0) = 1$. One finds

$$y(x) = -x + \int_0^x (t - x)y(t)\, dt\,,$$

with the solution $y(x) = -\sin x$ (note that $y = u''$).

2. $u'' - 2xu' + u = 0$; $u(0) = 1$, $u'(0) = 0$. Hence

$$y(x) = -1 + \int_0^x (x + t)y(t)\, dt\,.$$

REFERENCES

[1] Lovitt: *Linear Integral Equations* (McGraw-Hill, New York, 1924).

[2] Mikhlin: *Integral Equations* (Pergamon, New York, 1957).

[3] Tricomi: *Integral Equations* (Interscience Publishers, New York, 1957).

[4] Hilbert: *Grundzüge einer allgemeinen Theorie der linearen Integralgleichungen* (Teubner, Leipzig, 1912).

[5] Petrovsky: *Vorlesungen über die Theorie der Integralgleichungen*, (Physica-Verlag, Würzburg, 1953).

EXERCISES

1. Compute $D(\lambda)$ and $D(x, t; \lambda)$ when $K(x, t) = xt$, $a = 0$, $b = 1$.

2. Using the result in Exercise 1 solve the integral equation

$$y(x) = \frac{1}{\sqrt{x}} + \int_0^1 xty(t)\, dt\,.$$

3. Find the solution of the Fredholm integral equation

$$y(x) = x^2 + \lambda \int_0^1 (x^2 + t^2)y(t)\, dt\,.$$

4. Determine the eigenvalues of the homogeneous integral equation

$$y(x) = \lambda \int_0^1 (x + t + xt)y(t) \, dt \; .$$

5. Find an approximate solution of the integral equation

$$y(x) = x + \int_0^1 \sin (xt)y(t) \, dt \; ,$$

by replacing $\sin (xt)$ by a degenerate kernel consisting of the first three terms in the Maclaurin expansion.

6. Show that the integral equation $y(x) = 1 - x + \int_0^1 K(x, t)y(t) \, dt$ where

$$K(x, t) = \begin{cases} t(1 - x) & \text{for} \quad t \leq x \\ x(1 - t) & \text{for} \quad t > x \end{cases}$$

is equivalent to the boundary value problem $d^2y/dx^2 + y = 0; \; y(0) = 1; \; y(1) = 0$.

7. Find the eigenvalues of the integral equation $y(x) = \lambda \int_0^{\pi/2} \cos (x + t)y(t) \, dt$.

8. Solve the integral equation $y(x) = x + \int_0^x xt^2 y(t) \, dt$.

9. Solve the integral equation $y(x) = x + \int_0^x ty(t) \, dt$.

10. Determine eigenvalues and eigenfunctions to the problem $y(x) = \lambda \int_0^1 K(x, t)y(t) \, dt$ where $K(x, t) = \min (x, t)$.

11. Find the solution of the integral equation $3x^2 + 5x^3 = \int_0^x (x + t)y(t) \, dt$.

12. Using a numerical technique solve the integral equation

$$y(x) = 1 + \int_0^x \frac{y(t) \, dt}{x + t + 1}$$

for $x = 0(0.2)1$. Apply the trapezoidal rule.

13. Find an integral equation corresponding to $u' - u = 0; \; u(0) = u'(0) = 1$.

14. Find an integral equation corresponding to $u''' - 3u'' - 6u' + 8u = 0; \; u(0) = u'(0) = u''(0) = 1$.

Chapter 17

Approximation

17.0. Different types of approximation

In the following discussion we shall consider the case where we want to represent empirical data with a certain type of function and the case where an exact function whose values are uniquely determined in all points is to be represented by a finite or an infinite number of functions. The former case occurs frequently in several applied sciences, for example, biology, physics, chemistry, medicine, and social sciences; the representation obtained is used to get a deeper insight into the interrelations, to follow changes which might occur, and sometimes also for prognostic purposes. The latter case is of great theoretical interest, but it has also a practical aspect in connection with the computation of functions, especially on automatic computers.

In both cases we must define in some way how the deviations between actual and approximate values should be measured. Let $u_1, u_2, \ldots, u_n$ be the given values, and $v_1, v_2, \ldots, v_n$ the values obtained from the approximation. Putting $v_i - u_i = w_i$, we can regard the errors w_i as components of a vector W. Generally, we want to find an approximation that makes the components of W small. The most natural measure is some of the vector norms (see Section 3.6), and in this way we obtain the principle that the approximation should be chosen in such a way that the norm of the error vector (taken in some of the vector norms) is *minimized*. Which norm we should choose must be decided individually by aid of qualitative arguments. If we choose the Euclidean norm, we obtain the *least-square method*, which is very popular with many people. If, instead, we choose the maximum norm, we get the Chebyshev approximation technique, which has aroused a great deal of interest in recent years.

If we are dealing with the approximation of functions which are defined in all points, we can measure the deviation in the whole interval, and the Euclidean norm should be interpreted as the square root of the integral of the square of the deviations. For the maximum norm we measure the deviations in all local extrema, including the end points.

It is hardly surprising that polynomial approximation plays a very essential role when one wants to approximate a given function $f(x)$. One might expect that the usual Lagrangian interpolation polynomials with suitably chosen interpolation points (nodes) would be satisfactory, but this turns out to be a dangerous way. As a matter of fact, Runge has shown that the Lagrangian interpolation polynomials constructed for the function $1/(x^2 + 1)$ in the interval $[-5, 5]$ with uniformly distributed nodes give rise to arbitrary large deviations for increasing degree n.

With this background Weierstrass' theorem is indeed remarkable. This famous theorem states the following. If $f(x)$ is a continuous function in the interval $[a,b]$, then to each $\varepsilon > 0$ there exists a polynomial $P(x)$ such that $|f(x) - P(x)| < \varepsilon$ for all $x \in [a, b]$. For proof see, for example, [11] or [12].

17.1. Least-square polynomial approximation

We suppose that we have an empirical material in the form of $(n + 1)$ pairs of values $(x_0, y_0), (x_1, y_1), \ldots, (x_n, y_n)$, where the experimental errors are associated with the y-values only. Then we seek a polynomial

$$y = y_m(x) = a_0 + a_1 x + \cdots + a_m x^m ,$$

fitting the given points as well as possible. If $m = n$, the polynomial is identical with the Lagrangian interpolation polynomial; if $m < n$, we shall determine the coefficients by minimizing

$$S = \sum_{j=0}^{n} (a_0 + a_1 x_j + \cdots + a_m x_j^m - y_j)^2 , \qquad (17.1.1)$$

$$\frac{\partial S}{\partial a_k} = 2 \sum_{j=0}^{n} (a_0 + a_1 x_j + \cdots + a_m x_j^m - y_j) x_j^k = 0 , \qquad k = 0, 1, 2, \ldots, m ,$$

or

$$a_0 \sum x_j^k + a_1 \sum x_j^{k+1} + \cdots + a_m \sum x_j^{k+m} = \sum x_j^k y_j .$$

With the notations $s_k = \sum_j x_j^k$ and $v_k = \sum_j x_j^k y_j$, we get the system

$$\begin{cases} s_0 a_0 + s_1 a_1 + \cdots + s_m a_m = v_0 , \\ s_1 a_0 + s_2 a_1 + \cdots + s_{m+1} a_m = v_1 , \\ \vdots \\ s_m a_0 + s_{m+1} a_1 + \cdots + s_{2m} a_m = v_m . \end{cases} \qquad (17.1.2)$$

These equations are usually called *normal equations*.

The coefficient matrix P is symmetric, and we shall show that the system has a solution which makes S a minimum.

The problem is conveniently solved by use of matrices. We put

$$A = \begin{bmatrix} 1 & x_0 & x_0^2 & \cdots & x_0^m \\ 1 & x_1 & x_1^2 & \cdots & x_1^m \\ \vdots & & & & \\ 1 & x_n & x_n^2 & \cdots & x_n^m \end{bmatrix}, \quad a = \begin{bmatrix} a_0 \\ a_1 \\ \vdots \\ a_m \end{bmatrix}, \quad v = \begin{bmatrix} v_0 \\ v_1 \\ \vdots \\ v_m \end{bmatrix}, \quad y = \begin{bmatrix} y_0 \\ y_1 \\ \vdots \\ y_n \end{bmatrix}.$$

Here A is of type $(n + 1, m + 1)$, where $n > m$ and further $P = A^T A$. We shall minimize

$$S = |Aa - y|^2 = (Aa - y)^T (Aa - y)$$
$$= a^T A^T Aa - a^T A^T y - y^T Aa + y^T y = a^T A^T Aa - 2a^T A^T y + y^T y ;$$

here we observe that $y^T Aa$ is a pure number, and hence

$$y^T Aa = (y^T Aa)^T = a^T A^T y .$$

Putting $P = A^T A$ we shall first prove that P is nonsingular. This is equivalent to stating that the homogeneous system of equations $Pa = 0$ has the only solution $a = 0$. On the other hand, this is obvious since $Pa = 0$ implies

$$a^T Pa = a^T A^T Aa = (Aa)^T Aa = 0 ,$$

and hence we would have $Aa = 0$. This last system written out in full has the following form:

$$\begin{cases} a_0 + a_1 x_0 + \cdots + a_m x_0^m = 0 , \\ \vdots \\ a_0 + a_1 x_n + \cdots + a_m x_n^m = 0 , \end{cases}$$

which means that the polynomial $y_m(x) = a_0 + a_1 x + \cdots + a_m x^m$ vanishes for $n + 1$ $(n > m)$ different x-values $x_0, x_1, \ldots, x_n$. This is possible only if all coefficients $a_0, a_1, \ldots, a_m$ vanish, that is, $a = 0$. We have then proved that $Pa = 0$ implies $a = 0$, that is, P cannot be singular, which means that P^{-1} exists. Putting $v = A^T y$, we can write

$$S = (a - P^{-1} v)^T P(a - P^{-1} v) + y^T y - v^T P^{-1} v$$
$$= [A(a - P^{-1} v)]^T [A(a - P^{-1} v)] + y^T y - v^T P^{-1} v .$$

Now we have to choose a so that S is minimized which is achieved if

$$A(a - P^{-1} v) = 0 .$$

This equation is multiplied from the left, first by A^T, then by P^{-1} which gives

$$a = P^{-1} v .$$

As is easily seen (17.1.2) can be written $Pa = v$, and if we know that P is nonsingular we find the same solution. For large values of m, the system is highly ill-conditioned and gives rise to considerable difficulties. Later we will show how these difficulties can be overcome to some extent.

EXAMPLE

A quadratic function $y = a_0 + a_1 x + a_2 x^2$ should be fitted with the following data:

x	8	10	12	16	20	30	40	60	100
y	0.88	1.22	1.64	2.72	3.96	7.66	11.96	21.56	43.16

j	x_j	x_j^2	x_j^3	x_j^4	y_j	$x_j y_j$	$x_j^2 y_j$
0	8	64	512	4096	0.88	7.04	56.32
1	10	100	1000	10000	1.22	12.20	122.00
2	12	144	1728	20736	1.64	19.68	236.16
3	16	256	4096	65536	2.72	43.52	696.32
4	20	400	8000	160000	3.96	79.20	1584.00
5	30	900	27000	810000	7.66	229.80	6894.00
6	40	1600	64000	2560000	11.96	478.40	19136.00
7	60	3600	216000	12960000	21.56	1293.60	77616.00
8	100	10000	1000000	100000000	43.16	4316.00	431600.00
	296	17064	1322336	116590368	94.76	6479.44	537940.80

Hence we get

$$s_0 = 9 \, ,$$
$$s_1 = 296 \, , \qquad v_0 = 94.76 \, ,$$
$$s_2 = 17064 \, , \qquad v_1 = 6479.44 \, ,$$
$$s_3 = 1322336 \, , \qquad v_2 = 537940.80 \, ,$$
$$s_4 = 116590368 \, ,$$

and the system of equations becomes

$$\begin{cases} 9a_0 + 296a_1 + 17064a_2 = 94.76 \, , \\ 296a_0 + 17064a_1 + 1322336a_2 = 6479.44 \, , \\ 17064a_0 + 1322336a_1 + 116590368a_2 = 537940.80 \, . \end{cases}$$

After some computations we find that

$$\begin{cases} a_0 = -1.919 \, , \\ a_1 = 0.2782 \, , \\ a_2 = 0.001739 \, . \end{cases}$$

With these coefficients we get the following function values:

x	8	10	12	16	20	30	40	60	100
y	0.42	1.04	1.67	2.98	4.24	7.99	11.99	21.04	43.30

A quite common special case arises when data have to be represented by straight lines, and we shall consider two such cases separately.

1. In the first case we suppose that the errors in the x-values can be neglected compared with the errors in the y-values, and then it is natural to minimize the sum of the squares of the vertical deviations. We number the points from 1 to n:

$$S = \sum_{i=1}^{n} (kx_i + l - y_i)^2 \,,$$

$$\frac{\partial S}{\partial l} = 2 \cdot \sum_i (kx_i + l - y_i) = 0 \,,$$

$$\frac{\partial S}{\partial k} = 2 \cdot \sum_i (kx_i + l - y_i)x_i = 0 \,.$$

With the notations

$$\sum_i x_i = nx_0 = s_1 \,, \qquad\qquad \sum_i y_i = ny_0 = t_1 \,,$$

$$\sum_i x_i^2 = s_2 \,, \quad \sum_i x_i y_i = v_1 \,, \quad \sum_i y_i^2 = t_2 \,;$$

$$\begin{cases} A = s_2 - nx_0^2 \,, \\ B = v_1 - nx_0 y_0 \,, \\ C = t_2 - ny_0^2 \,, \end{cases}$$

we obtain $l = y_0 - kx_0$, which means that the center of gravity lies on the desired line. Further,

$$s_2 k + nlx_0 - v_1 = 0 \,;$$

that is,

$$k = \frac{v_1 - nx_0 y_0}{s_2 - nx_0^2} = \frac{B}{A} \qquad \text{and} \qquad l = y_0 - \frac{B}{A} x_0 \,.$$

2. Here we suppose that the x-values as well as the y-values are subject to errors of about the same order of magnitude, and then it is more natural to minimize the sum of the squares of the perpendicular distances to the line. Thus

$$S = \frac{1}{1 + k^2} \sum_{i=1}^{n} (kx_i + l - y_i)^2 \,;$$

$$\frac{\partial S}{\partial l} = 0 \qquad \text{gives, as before} \quad l = y_0 - kx_0 \,;$$

$$\frac{\partial S}{\partial k} = 0 \qquad \text{gives} \quad (1 + k^2) \cdot \sum_i (kx_i + l - y_i)x_i = k \sum_i (kx_i + l - y_i)^2 \,.$$

After simplification, we get

$$k^2(v_1 - nx_0 y_0) + k(s_2 - nx_0^2 - t_2 + ny_0^2) - (v_1 - nx_0 y_0) = 0$$

or

$$k^2 + \frac{A - C}{B} k - 1 = 0 .$$

From this equation we obtain two directions at right angles, and in practice there is no difficulty deciding which one gives a minimum.

Here we also mention that overdetermined systems, that is, systems where the number of equations is larger than the number of unknowns, can be "solved" by aid of the least-squares method. Let the system be $Ax = b$, where A is of type (m, n) with $m > n$; then we minimize

$$S = (Ax - b)^T(Ax - b) .$$

In this way we are back at the problem just treated.

17.2. Polynomial approximation by use of orthogonal polynomials

As has been mentioned, solving the normal equations gives rise to considerable numerical difficulties, even for moderate values of the degree m of the approximation polynomial. Forsythe [1] expresses this fact in the following striking way: "When $m \geq 7$ or 8, however, one begins to hear strange grumblings of discontent in the computing laboratory." Approximation with orthogonal polynomials might improve the situation although there is evidence of numerical difficulties even in this case. The method is not limited to this case but can be used for many other purposes, for example, for numerical quadrature and for solving ordinary, as well as partial, differential equations [3].

Suppose that the $n + 1$ points $(x_0, y_0), (x_1, y_1), \ldots, (x_n, y_n)$ are given. We seek a polynomial $y = Q(x)$ of degree m $(m < n)$ such that

$$S = \sum_{i=0}^{n} [y_i - Q(x_i)]^2 = \min . \tag{17.2.1}$$

The polynomial $Q(x)$ shall be expressed in the form

$$Q(x) = \sum_{j=0}^{m} a_j P_j(x) , \tag{17.2.2}$$

where $P_j(x)$ is a polynomial of degree j fulfilling the relations

$$\sum_{i=0}^{n} P_j(x_i) P_k(x_i) = \delta_{jk} . \tag{17.2.3}$$

The problem is now to determine the polynomials $P_j(x)$ as well as the unknown coefficients a_j. We differentiate S with respect to a_j and get

$$\frac{\partial S}{\partial a_j} = -2 \sum_{i=0}^{n} (y_i - Q(x_i))P_j(x_i)$$

$$= -2 \sum_{i=0}^{n} y_i P_j(x_i) + 2 \sum_{i=0}^{n} P_j(x_i) \cdot \sum_{k=0}^{m} a_k P_k(x_i)$$

$$= -2 \sum_{i=0}^{n} y_i P_j(x_i) + 2 \sum_{k=0}^{m} a_k \sum_{i=0}^{n} P_j(x_i)P_k(x_i)$$

$$= 2a_j - 2 \sum_{i=0}^{n} y_i P_j(x_i) .$$

The condition $\partial S / \partial a_j = 0$ gives the desired coefficients:

$$a_j = \sum_{i=0}^{n} y_i P_j(x_i) . \tag{17.2.4}$$

Then we have to determine the polynomials $P_j(x)$, which we express in the following way:

$$P_j(x) = \sum_{k=0}^{j} a_{jk} x^k . \tag{17.2.5}$$

Conversely, the powers x^j can be written as linear combinations of the polynomials:

$$x^j = \sum_{k=0}^{j} \alpha_{jk} P_k(x) = \alpha_{jj} P_j(x) + \sum_{k=0}^{j-1} \alpha_{jk} P_k(x) . \tag{17.2.6}$$

Squaring, putting $x = x_i$, and summing over i, we obtain

$$\sum_{i=0}^{n} x_i^{2j} = \alpha_{jj}^2 + \sum_{k=0}^{j-1} \alpha_{jk}^2$$

or

$$\alpha_{jj}^2 = \sum_{i=0}^{n} x_i^{2j} - \sum_{k=0}^{j-1} \alpha_{jk}^2 . \tag{17.2.7}$$

Multiplying (17.2.6) by $P_r(x)$, putting $x = x_i$, and summing over i, we get

$$\sum_{i=0}^{n} x_i^j P_r(x_i) = \sum_{i=0}^{n} \sum_{k=0}^{j} \alpha_{jk} P_k(x_i) P_r(x_i) = \alpha_{jr}$$

or

$$\alpha_{jk} = \sum_{i=0}^{n} x_i^j P_k(x_i) . \tag{17.2.8}$$

From (17.2.6) we further obtain

$$P_j(x) = \left\{ x^j - \sum_{k=0}^{j-1} \alpha_{jk} P_k(x) \right\} \Big/ \alpha_{jj} . \tag{17.2.9}$$

For $j = 0$ we get $P_0(x) = 1/\alpha_{00}$ and $\alpha_{00}^2 = n + 1$. The next steps come in the following order:

$$\alpha_{10}, \alpha_{11}, P_1; \quad \alpha_{20}, \alpha_{21}, \alpha_{22}, P_2; \ldots; \quad \alpha_{m0}, \alpha_{m1}, \ldots, \alpha_{mm}, P_m .$$

We obtain α_{jk}, $j > k$, from (17.2.8), α_{jj} from (17.2.7), and P_j from (17.2.9). Then we determine $a_0, a_1, \ldots, a_m$ from (17.2.4), and by now we also know $Q(x)$. Obviously, we do not have to fix the degree m of the polynomial $Q(x)$ from the beginning; instead m can be increased successively, which is a great advantage. Then we can follow how S decreases; as is easily found, we have

$$S = \sum_{i=0}^{n} y_i^2 - 2 \sum_{i=0}^{n} y_i \sum_{j=0}^{m} a_j P_j(x_i) + \sum_{i=0}^{n} \sum_{j,k=0}^{m} a_j a_k P_j(x_i) P_k(x_i) ,$$

and by use of (17.2.3) and (17.2.4),

$$S = \sum_{i=0}^{n} y_i^2 - \sum_{j=0}^{m} a_j^2 . \tag{17.2.10}$$

EXAMPLE

Find a straight line passing as close as possible to the points (x_0, y_0), (x_1, y_1), $\ldots$, (x_n, y_n).

Putting $s_r = \sum_i x_i^r$ and $v_r = \sum_i x_i^r y_i$, we obtain

$$\alpha_{00} = \sqrt{s_0} ; \qquad P_0 = \frac{1}{\sqrt{s_0}} ;$$

$$\alpha_{10} = \sum_i x_i P_0(x_i) = \frac{s_1}{\sqrt{s_0}} ; \qquad \alpha_{11}^2 = s_2 - \frac{s_1^2}{s_0} ;$$

$$P_1 = \frac{x - \alpha_{10} P_0}{\alpha_{11}} = \frac{x - s_1/s_0}{\sqrt{s_2 - s_1^2/s_0}} ;$$

$$a_0 = \sum_i y_i P_0(x_i) = \frac{v_0}{\sqrt{s_0}} ; \qquad a_1 = \sum_i y_i P_1(x_i) = \frac{v_1 - s_1 v_0/s_0}{\sqrt{s_2 - s_1^2/s_0}} ;$$

$$Q(x) = a_0 P_0(x) + a_1 P_1(x) = \frac{v_0}{s_0} + \frac{v_1 - s_1 v_0/s_0}{s_2 - s_1^2/s_0} \left(x - \frac{s_1}{s_0} \right) ;$$

or

$$Q(x) = \frac{s_2 v_0 - s_1 v_1 + (s_0 v_1 - s_1 v_0) x}{s_0 s_2 - s_1^2} .$$

Naturally, this is exactly what is obtained if we solve the system

$$\begin{cases} s_0 a_0 + s_1 a_1 = v_0 , \\ s_1 a_0 + s_2 a_1 = v_1 . \end{cases}$$

in the usual way.

An obvious generalization is the case when the points should be counted with certain preassigned weights. The derivations are completely analogous and will not be carried through here. Closer details are given in [2].

Another natural generalization is the approximation of a function $y = y(x)$ with $\sum_r a_r \varphi_r(x)$, where the functions φ_r form an orthonormal system over the interval $a \leq x \leq b$:

$$\int_a^b \varphi_r(x)\varphi_s(x)\, dx = \delta_{rs} . \qquad (17.2.11)$$

The best-known example is the trigonometrical functions in the interval $(0, 2\pi)$; this case will be treated separately. Adding a weight function $w(x)$ is an almost trivial complication.

We now want to determine the coefficients a_r from the condition

$$S = \int_a^b \left[y - \sum_{r=0}^n a_r \varphi_r(x) \right]^2 dx = \min . \qquad (17.2.12)$$

Differentiating with respect to a_s we get

$$\frac{\partial S}{\partial a_s} = 2 \int_a^b \left[y - \sum_{r=0}^n a_r \varphi_r(x) \right] [-\varphi_s(x)]\, dx .$$

Putting $\partial S/\partial a_s = 0$, we obtain, by use of (17.2.11),

$$a_s = \int_a^b y \cdot \varphi_s(x)\, dx . \qquad (17.2.13)$$

Analogous to (17.2.10), we also find that

$$S_{\min} = \int_a^b y^2\, dx - \sum_{r=0}^n a_r^2 . \qquad (17.2.14)$$

Since $S \geq 0$ we have

$$\sum_{r=0}^n a_r^2 \leq \int_a^b y^2\, dx ,$$

known as *Bessel's inequality*. If

$$\sum_{r=0}^\infty a_r^2 = \int_a^b y^2\, dx ,$$

the function system is said to be *complete* (*Parseval's relation*).

EXAMPLE

We consider the interval $(-1, 1)$, and according to (10.2.11) and (10.2.12), we can use the Legendre polynomials $P_n(x)$ multiplied by the normalization factor $\sqrt{n + \frac{1}{2}}$:

$$\varphi_n(x) = \sqrt{n + \frac{1}{2}} \cdot \frac{1}{2^n \cdot n!} \frac{d^n}{dx^n} [(x^2 - 1)^n] .$$

Thus $\varphi_0(x) = 1/\sqrt{2}$; $\varphi_1(x) = x\sqrt{\frac{3}{2}}$; $\varphi_2(x) = (3x^2 - 1)\sqrt{\frac{5}{8}}$; ... Now suppose that we want to approximate $y = x^4$ with $a_0\varphi_0(x) + a_1\varphi_1(x) + a_2\varphi_2(x)$. Using

(17.2.13), we get

$$a_0 = \int_{-1}^{+1} \frac{x^4}{\sqrt{2}} \, dx = \frac{\sqrt{2}}{5} \, ,$$

$$a_1 = \int_{-1}^{+1} \sqrt{\frac{3}{2}} \, x^5 \, dx = 0 \, ,$$

$$a_2 = \int_{-1}^{+1} x^4 (3x^2 - 1) \sqrt{\frac{5}{8}} \, dx = \frac{8}{35} \sqrt{\frac{5}{2}} \, ,$$

$$a_0 \varphi_0(x) + a_1 \varphi_1(x) + a_2 \varphi_2(x) = \frac{6x^2}{7} - \frac{3}{35} \, .$$

If we put conventionally

$$S = \int_{-1}^{+1} (x^4 - a - bx^2)^2 \, dx \, ,$$

and seek the minimum of S, we get the system

$$\begin{cases} a + \dfrac{b}{3} = \dfrac{1}{5} \\[2mm] \dfrac{a}{3} + \dfrac{b}{5} = \dfrac{1}{7} \end{cases}$$

Hence $a = -\frac{3}{35}$ and $b = \frac{6}{7}$.

17.3. Approximation with trigonometric functions

We shall now study the case when a function is to be approximated by trigonometric functions relating to (17.2.1) or (17.2.12). Defining

$$\delta(m, n) = \begin{cases} 1 & \text{if } m \text{ is divisible by } n \, , \\ 0 & \text{otherwise} \, , \end{cases}$$

we found our discussion on the formulas

$$\sum_{r=0}^{n-1} e^{rj\alpha i} \cdot e^{-rk\alpha i} = n\delta(j - k, n) \, , \qquad (17.3.1)$$

$$\int_0^{2\pi} e^{ijx} \cdot e^{-ikx} \, dx = 2\pi \delta_{jk} \, . \qquad (17.3.2)$$

Here we have put $2\pi/n = \alpha$; the proof of the relations is trivial. Now we can deduce corresponding relations in trigonometric form:

$$\alpha \sum_{r=0}^{n-1} \cos rj\alpha \cos rk\alpha = \frac{\alpha}{2} \sum_{r=0}^{n-1} \{\cos r(j + k)\alpha + \cos r(j - k)\alpha\}$$

$$= \frac{\alpha}{2} \operatorname{Re} \left\{ \sum_{r=0}^{n-1} [e^{r(j+k)\alpha i} + e^{r(j-k)\alpha i}] \right\}$$

$$= \pi[\delta(j + k, n) + \delta(j - k, n)] \, . \qquad (17.3.3)$$

Other relations are proved in a similar way. Thus we have

$$\begin{cases} \sum_{r=0}^{n-1} \sin rj\alpha \sin rk\alpha = \pi[\delta(j-k,n) - \delta(j+k,n)] , \\ \sum_{r=0}^{n-1} \sin rj\alpha \cos rk\alpha = 0 . \end{cases} \qquad (17.3.4)$$

If in these relations we let $\alpha \rightarrow 0$, we obtain ($j, k \neq 0$):

$$\int_0^{2\pi} \cos jx \cos kx\, dx = \int_0^{2\pi} \sin jx \sin kx\, dx = \pi\delta_{jk} , \qquad (17.3.5)$$

$$\int_0^{2\pi} \sin jx \cos kx\, dx = 0 . \qquad (17.3.6)$$

The orthogonality at summation is the basis for the harmonic analysis, while the orthogonality at integration plays a decisive role in the theory of Fourier series.

First we shall try to generate a series in $\cos rj\alpha$ and $\sin rj\alpha$, which in the points $x_j = j\alpha$ takes the values y_j, and we find

$$\begin{cases} y_j = \frac{1}{2} a_0 + \sum_{r=1}^{(n-1)/2} (a_r \cos rj\alpha + b_r \sin rj\alpha) ; \qquad n \text{ odd} , \\ y_j = \frac{1}{2}\left(a_0 + (-1)^j a_{n/2}\right) \\ \qquad + \sum_{r=1}^{n/2-1} (a_r \cos rj\alpha + b_r \sin rj\alpha) ; \qquad n \text{ even} . \end{cases} \qquad (17.3.7)$$

The coefficients are obtained from

$$a_r = \frac{2}{n} \sum_{j=0}^{n-1} y_j \cos rj\alpha ; \qquad b_r = \frac{2}{n} \sum_{j=0}^{n-1} y_j \sin rj\alpha . \qquad (17.3.8)$$

Similarly we can represent all periodic, absolutely integrable functions by a Fourier series:

$$f(x) = \frac{1}{2} a_0 + \sum_{r=1}^{\infty} (a_r \cos rx + b_r \sin rx) . \qquad (17.3.9)$$

In this case we get the coefficients from

$$\begin{cases} a_r = \frac{1}{\pi} \int_0^{2\pi} f(x) \cos rx\, dx ; \\ b_r = \frac{1}{\pi} \int_0^{2\pi} f(x) \sin rx\, dx . \end{cases} \qquad (17.3.10)$$

Here we cannot treat questions relating to the convergence of the series, if the series converges to the initial function, or special phenomena which may appear (Gibbs' phenomenon); instead we refer to [4], [5], or [6].

EXAMPLES

1. Expand the function $y = f(x)$ in a Fourier series, where

$$f(x) = \begin{cases} +1, & 0 < x < \pi, \\ -1, & \pi < x < 2\pi. \end{cases}$$

We find

$$a_n = \frac{1}{\pi} \int_0^\pi \cos nx\, dx - \frac{1}{\pi} \int_\pi^{2\pi} \cos nx\, dx = 0,$$

$$b_n = \frac{1}{\pi} \int_0^\pi \sin nx\, dx - \frac{1}{\pi} \int_\pi^{2\pi} \sin nx\, dx = \frac{2}{\pi n}[1 - (-1)^n],$$

and hence

$$b_n = \begin{cases} 0 & n \text{ even}, \\ \dfrac{4}{\pi n} & n \text{ odd}. \end{cases}$$

Thus we obtain

$$\frac{\pi}{4} = \sin x + \frac{\sin 3x}{3} + \frac{\sin 5x}{5} + \cdots, \qquad 0 < x < \pi.$$

2.

$$f(x) = \begin{cases} \dfrac{\pi}{4} - \dfrac{x}{2}, & 0 \le x \le \pi, \\ -\dfrac{3\pi}{4} + \dfrac{x}{2}, & \pi \le x \le 2\pi. \end{cases}$$

We have

$$a_n = \frac{1}{\pi} \int_0^\pi \left(\frac{\pi}{4} - \frac{x}{2}\right) \cos nx\, dx + \frac{1}{\pi} \int_\pi^{2\pi} \left(-\frac{3\pi}{4} + \frac{x}{2}\right) \cos nx\, dx$$

$$= \begin{cases} 0 & n \text{ even}, \\ \dfrac{2}{\pi n^2} & n \text{ odd}. \end{cases}$$

Further, we easily find $b_n = 0$, and hence

$$\frac{\pi}{4} - \frac{x}{2} = \frac{2}{\pi}\left(\cos x + \frac{\cos 3x}{3^2} + \frac{\cos 5x}{5^2} + \cdots\right); \qquad 0 \le x \le \pi.$$

For $x = 0$ we obtain

$$\frac{\pi^2}{8} = 1 + \frac{1}{3^2} + \frac{1}{5^2} + \cdots.$$

When a function is represented as a trigonometric series, this means that only certain discrete frequencies have been used. By passing to the integral form, we actually use *all* frequencies. The principal point is Fourier's integral theorem, which states a complete reciprocity between the amplitude and frequency functions:

$$\begin{cases} F(\nu) = \displaystyle\int_{-\infty}^{+\infty} f(x)e^{-2\pi i\nu x}\, dx\,, \\[2mm] f(x) = \displaystyle\int_{-\infty}^{+\infty} F(\nu)e^{2\pi i\nu x}\, dv\,. \end{cases} \qquad (17.3.11)$$

In practical work these integrals are evaluated numerically, and hence we must require that $f(x)$ and $F(\nu)$ vanish with a suitable strength at infinity. If $f(x)$ is known at a sufficient number of equidistant points, then $F(\nu)$ can be computed by numerical quadrature:

$$F(\nu) = \int_0^\infty g(x) \cos 2\pi\nu x\, dx - i \cdot \int_0^\infty h(x) \sin 2\pi\nu x\, dx\,,$$

with $g(x) = f(x) + f(-x)$; $h(x) = f(x) - f(-x)$. In general, $F(\nu)$ will be complex: $F(\nu) = G(\nu) - iH(\nu)$, and conveniently we compute the absolute value

$$\varphi(\nu) = \big(G(\nu)^2 + H(\nu)^2\big)^{1/2}\,.$$

If the computation is performed for a sufficient number of ν-values, one can quite easily get a general idea of which frequencies ν are dominating in the material $f(x)$. Another method is autoanalysis. In this case one forms convolution integrals of the form $\int f(x)f(x - t)\, dx$; if there is a pronounced periodicity in the material, then for certain values of t the factors $f(x)$ and $f(x - t)$ will both be large, and hence the integral will also become large. It can easily be shown that if $f(x)$ contains oscillations with amplitude a_n, the convolution integral will contain the same oscillations but with amplitude a_n^2 instead.

17.4. Approximation with exponential functions

The problem of analyzing a sum of exponential functions apparently is very close to the problem just treated. Suppose that we have a function $f(x)$ which we try to approximate by

$$f(x) = a_1 \cdot e^{\lambda_1 x} + a_2 \cdot e^{\lambda_2 x} + \cdots + a_n \cdot e^{\lambda_n x}\,. \qquad (17.4.1)$$

We assume that n is known and shall try to compute $a_1, a_2, \ldots, a_n$, and $\lambda_1, \lambda_2, \ldots, \lambda_n$. We observe that $f(x)$ satisfies a certain differential equation:

$$y^{(n)} + A_1 y^{(n-1)} + \cdots + A_n y = 0$$

with presently unknown coefficients $A_1, A_2, \ldots, A_n$. We compute $y', y'', \ldots, y^{(n)}$ numerically in n different points and obtain a linear system of equations from which the coefficients can be obtained. Last, $\lambda_1, \lambda_2, \ldots, \lambda_n$ are obtained as roots of the algebraic equation $\lambda^n + A_1\lambda^{n-1} + \cdots + A_n = 0$, and we get the coefficients $a_1, a_2, \ldots, a_n$ by use of the least-squares method.

At first glance this method seems to be clear and without complications, but in practical work we meet tremendous difficulties. The main reason is the numerical differentiation which deteriorates the accuracy and thus leads to uncertain results. Since the data, which very often come from biological, physical, or chemical experiments, are usually given to, at most, four significant digits, it is easily seen that an approximation with three exponential functions must already become rather questionable. For example, Lanczos has shown that the three functions

$$f_1(x) = 0.0951 \cdot e^{-x} + 0.8607 \cdot e^{-3x} + 1.5576 \cdot e^{-5x},$$
$$f_2(x) = 0.305 \cdot e^{-1.58x} + 2.202 \cdot e^{-4.45x},$$
$$f_3(x) = 0.041 \cdot e^{-0.5x} + 0.79 \cdot e^{-2.73x} + 1.68 \cdot e^{-4.96x},$$

approximate the same data for $0 \leq x \leq 1.2$ to two places.

A better method, which has also been discussed by Lanczos [4], is the following, originating with Prony. We assume that the function $y = f(x)$ is given in equidistant points with the coordinates $(x_0, y_0), (x_1, y_1), \ldots, (x_n, y_n)$, where $x_r = x_0 + rh$. We will approximate $f(x)$ by $a_1 e^{\lambda_1 x} + a_2 e^{\lambda_2 x} + \cdots + a_m e^{\lambda_m x}$. Putting $c_r = a_r e^{\lambda_r x_0}$ and $v_r = e^{h\lambda_r}$, we obtain for $r = 0, 1, 2, \ldots, m$ the equations:

$$\begin{cases} c_1 & + c_2 & + \cdots + c_m & = y_0, \\ c_1 v_1 & + c_2 v_2 & + \cdots + c_m v_m & = y_1, \\ \vdots \\ c_1 v_1^m & + c_2 v_2^m & + \cdots + c_m v_m^m & = y_m. \end{cases} \qquad (17.4.2)$$

Forming

$$(v - v_1)(v - v_2) \cdots (v - v_m) \equiv v^m + s_1 v^{m-1} + \cdots + s_m \qquad (17.4.3)$$
$$= \varphi(v),$$

multiplying in turn by $s_m, s_{m-1}, \ldots, s_1, s_0 = 1$, and adding, we obtain

$$\varphi(v_1)c_1 + \varphi(v_2)c_2 + \cdots + \varphi(v_m)c_m$$
$$= s_m y_0 + s_{m-1} y_1 + \cdots + s_1 y_{m-1} + s_0 y_m = 0,$$

since $\varphi(v_r) = 0$, $r = 1, 2, \ldots, m$. Normally, m is, of course, substantially smaller than n, and further it is clear that we get a new equation if we shift the origin a distance h to the right. If this is repeated, we get the following system:

$$\begin{cases} y_{m-1} s_1 & + y_{m-2} s_2 & + \cdots + y_0 s_m & = -y_m, \\ y_m s_1 & + y_{m-1} s_2 & + \cdots + y_1 s_m & = -y_{m+1}, \\ \vdots \\ y_{n-1} s_1 & + y_{n-2} s_2 & + \cdots + y_{n-m} s_m & = -y_n. \end{cases} \qquad (17.4.4)$$

Thus we have $n - m + 1$ equations in m unknowns, and normally $n - m + 1 > m$; hence $s_1, s_2, \ldots, s_m$ can be determined by means of the least-squares method. Then we get $v_1, v_2, \ldots, v_m$ from (17.4.3) and $\lambda_r = \log v_r / h$. Finally, we get

$c_1, c_2, \ldots, c_m$ from (17.4.2) (since there are $m + 1$ equations but only m unknowns, for example, the last equation can be left out), and $a_r = c_r \cdot v_r^{-x_0/h}$.

17.5. Approximation with Chebyshev polynomials

The Chebyshev polynomials are defined by the relation

$$T_n(x) = \cos(n \arccos x), \tag{17.5.1}$$

where it is natural to put $T_{-n}(x) = T_n(x)$. Well-known trigonometric formulas give at once $T_{n+m}(x) + T_{n-m}(x) = 2T_n(x)T_m(x)$. For $m = 1$ we get in particular

$$T_{n+1}(x) = 2xT_n(x) - T_{n-1}(x). \tag{17.5.2}$$

Since $T_0(x) = 1$ and $T_1(x) = x$, we can successively compute all $T_n(x)$. Putting, for a moment, $x = \cos\theta$, we get $y = T_n(x) = \cos n\theta$ and, further,

$$\frac{dy}{dx} = y' = \frac{n \sin n\theta}{\sin\theta}$$

and

$$y'' = \frac{-n^2 \cos n\theta + n \sin n\theta \cot\theta}{\sin^2\theta} = -\frac{n^2 y}{1 - x^2} + \frac{xy'}{1 - x^2}.$$

Thus the polynomials $T_n(x)$ satisfy the differential equation

$$(1 - x^2)y'' - xy' + n^2 y = 0. \tag{17.5.3}$$

As is inferred directly, the other fundamental solution of (17.5.3) is the function $S_n(x) = \sin(n \arccos x)$. In particular we have $S_0(x) = 0$ and $S_1(x) = \sqrt{1 - x^2}$. With $U_n(x) = S_n(x)/\sqrt{1 - x^2}$, we have in the same way as for (17.5.2):

$$U_{n+1}(x) = 2xU_n(x) - U_{n-1}(x). \tag{17.5.4}$$

The first polynomials are

$$
\begin{array}{ll}
T_0(x) = 1 & U_0(x) = 0 \\
T_1(x) = x & U_1(x) = 1 \\
T_2(x) = 2x^2 - 1 & U_2(x) = 2x \\
T_3(x) = 4x^3 - 3x & U_3(x) = 4x^2 - 1 \\
T_4(x) = 8x^4 - 8x^2 + 1 & U_4(x) = 8x^3 - 4x \\
T_5(x) = 16x^5 - 20x^3 + 5x & U_5(x) = 16x^4 - 12x^2 + 1
\end{array}
$$

From (17.3.5) we have directly

$$\int_{-1}^{+1} T_m(x)T_n(x) \cdot \frac{dx}{\sqrt{1 - x^2}} = \begin{cases} 0, & m \neq n, \\ \dfrac{\pi}{2}, & m = n \neq 0, \\ \pi, & m = n = 0, \end{cases} \tag{17.5.5}$$

that is, the polynomials $T_n(x)$ are orthogonal with the weight function $1/\sqrt{1 - x^2}$.

Chebyshev discovered the following remarkable property of the polynomials $T_n(x)$. For $-1 \leq x \leq 1$, we consider all polynomials $p_n(x)$ of degree n and with the coefficient of x^n equal to 1. Putting $\alpha_n = \sup_{-1 \leq x \leq 1} |p_n(x)|$, we seek the polynomial $p_n(x)$ for which α_n is as small as possible. The desired polynomial is then $2^{-(n-1)} T_n(x)$, as will become clear from the following discussion. First we observe that $T_n(x) = \cos(n \arccos x) = 0$, when

$$x = \xi_r = \cos \frac{(2r + 1)\pi}{2n}, \qquad r = 0, 1, 2, \ldots, n - 1,$$

and that $T_n(x) = (-1)^r$ for $x = x_r = \cos r\pi/n$, $r = 0, 1, 2, \ldots, n$. Now suppose that $|p_n(x)| < 2^{-(n-1)}$ everywhere in the interval $-1 \leq x \leq 1$. Then we have

$$\begin{cases} 2^{-(n-1)} T_n(x_0) - p_n(x_0) > 0, \\ 2^{-(n-1)} T_n(x_1) - p_n(x_1) < 0, \\ \vdots \end{cases}$$

that is, the polynomial $f_{n-1}(x) = 2^{-(n-1)} T_n(x) - p_n(x)$ of degree $(n - 1)$ would have an alternating sign in $(n + 1)$ points $x_0, x_1, \ldots, x_n$. Hence $f_{n-1}(x)$ would have n roots in the interval, which is possible only if $f_{n-1}(x) \equiv 0$, that is, $p_n(x) \equiv 2^{-(n-1)} T_n(x)$.

Suppose that we want to approximate a function $f(x)$ with a Chebyshev series:

$$f(x) = \tfrac{1}{2} c_0 + c_1 T_1(x) + c_2 T_2(x) + \cdots + c_{n-1} T_{n-1}(x) + R_n(x). \tag{17.5.6}$$

In analogy to (17.3.3), if j, $k < n$ and at least one of them different from zero, we have

$$\sum_{r=0}^{n-1} T_j(\xi_r) T_k(\xi_r) = \frac{n}{2} \delta_{jk}. \tag{17.5.7}$$

The general formula is

$$\sum_{r=0}^{n-1} T_j(\xi_r) T_k(\xi_r) = \frac{n}{2} [(-1)^{(j+k)/2n} \delta(j + k, 2n) + (-1)^{(j-k)/2n} \delta(j - k, 2n)].$$

To determine the coefficients c_k in (17.5.6), we can use either (17.5.5) or (17.5.7). The latter might be preferable, since the integration is often somewhat difficult:

$$c_k = \frac{2}{n} \cdot \sum_{r=0}^{n-1} f(\xi_r) T_k(\xi_r) = \frac{2}{n} \sum_{r=0}^{n-1} f(\xi_r) \cos \frac{(2r + 1)k\pi}{2n}. \tag{17.5.8}$$

In particular we observe that c_k depends upon n, which means that we cannot just add another term if the expansion has to be improved; all coefficients change if we alter the degree n. If we use (17.5.5), we obtain the limiting values of c_k when $n \to \infty$.

The remainder term R_n can be expressed as

$$R_n = c_n T_n(x) + c_{n+1} T_{n+1}(x) + \cdots,$$

and if the convergence is sufficiently fast, we have $R_n \simeq c_n T_n(x)$. In this case the error oscillates between $-c_n$ and $+c_n$.

If the function f is such that we know a power-series expansion which converges in the interval, a direct method can be recommended. We have

$$1 = T_0 , \qquad\qquad x^4 = \tfrac{1}{8}(3T_0 + 4T_2 + T_4) ,$$
$$x = T_1 , \qquad\qquad x^5 = \tfrac{1}{16}(10T_1 + 5T_3 + T_5) ,$$
$$x^2 = \tfrac{1}{2}(T_0 + T_2) , \qquad x^6 = \tfrac{1}{32}(10T_0 + 15T_2 + 6T_4 + T_6) ,$$
$$x^3 = \tfrac{1}{4}(3T_1 + T_3) , \qquad x^7 = \tfrac{1}{64}(35T_1 + 21T_3 + 7T_5 + T_7) ,$$
$$\vdots$$

The general coefficient in the expansion $x^k = \sum c_n T_n$ becomes

$$c_n = \frac{2}{\pi} \cdot \int_{-1}^{+1} x^k T_n(x) \frac{dx}{\sqrt{1 - x^2}} = \frac{2}{\pi} \int_0^\pi \cos^k \varphi \cos n\varphi \, d\varphi .$$

If k is odd, n takes the values $1, 3, 5, \ldots, k$, and if k is even, n takes the values $0, 2, 4, \ldots, k$. The integrand can be written

$$\frac{(e^{i\varphi} - e^{-i\varphi})^k}{2^k} \frac{e^{ni\varphi} + e^{-ni\varphi}}{2}$$

$$= \frac{e^{ki\varphi} + \binom{k}{1} e^{(k-2)i\varphi} + \binom{k}{2} e^{(k-4)i\varphi} + \cdots + e^{-ki\varphi}}{2^k} \frac{e^{ni\varphi} + e^{-ni\varphi}}{2} .$$

The term $e^{(k-2r)i\varphi} \cdot e^{ni\varphi}$ becomes 1 if $r = (n + k)/2$, and analogously, we get 1 also for the term $e^{(2r-k)i\varphi} \cdot e^{-ni\varphi}$ with the same r. When we integrate, all other terms vanish, and we are left with

$$c_n = 2^{-k+1} \binom{k}{(n + k)/2} . \qquad\qquad (17.5.9)$$

If $k = 0$ we have $c_n = 0$ except when $n = 0$, since $c_0 = 1$.

By use of these formulas, a power-series expansion can be replaced with an expansion in Chebyshev polynomials. This latter expansion is truncated in such a way that our requirements with respect to accuracy are fulfilled, and then the terms can be rearranged to a polynomial. Clearly, the coefficients in this polynomial differ from the corresponding coefficients in the original expansion.

EXAMPLES

1. $f(x) = \arcsin x$. Since $f(x)$ is an odd function, we get $c_{2k} = 0$. For the other coefficients we find

$$c_{2k+1} = \frac{2}{\pi} \int_{-1}^{+1} \arcsin x \cdot \cos \left((2k + 1) \arccos x\right) \frac{dx}{\sqrt{1 - x^2}} .$$

Putting $x = \cos \varphi$, we obtain

$$c_{2k+1} = \frac{2}{\pi} \int_0^\pi \left(\frac{\pi}{2} - \varphi \right) \cdot \cos (2k + 1) \varphi \, d\varphi$$

$$= \frac{4}{\pi} \frac{1}{(2k + 1)^2} \cdot$$

Thus

$$\arcsin x = \frac{4}{\pi} \left[T_1(x) + \frac{T_3(x)}{9} + \frac{T_5(x)}{25} + \cdots \right].$$

For $x = 1$ we have $T_1 = T_3 = T_5 = \cdots = 1$, and again we get the formula

$$\frac{\pi^2}{8} = 1 + \frac{1}{9} + \frac{1}{25} + \cdots .$$

Expansion of other elementary functions in Chebyshev series gives rise to rather complicated integrals, and we will not treat this problem here (cf. [7]).

2. If we expand e^{-x} in power series and then express the powers in Chebyshev polynomials, we find

$$e^{-x} = 1.266065 \, 877752 T_0$$
$$- 1.130318 \, 207984 T_1 + 0.271495 \, 339533 T_2$$
$$- 0.044336 \, 849849 T_3 + 0.005474 \, 240442 T_4$$
$$- 0.000542 \, 926312 T_5 + 0.000044 \, 977322 T_6$$
$$- 0.000003 \, 198436 T_7 + 0.000000 \, 199212 T_8$$
$$- 0.000000 \, 011037 T_9 + 0.000000 \, 000550 T_{10}$$
$$- 0.000000 \, 000025 T_{11} + 0.000000 \, 000001 T_{12} .$$

This expansion is valid only for $-1 \leq x \leq 1$. If we truncate the series after the T_5-term and rearrange in powers of x, we obtain

$$e^{-x} \simeq 1.000045 - 1.000022x + 0.499199x^2$$
$$- 0.166488x^3 + 0.043794x^4 - 0.008687x^5 .$$

The error is essentially equal to $0.000045 T_6$, whose largest absolute value is 0.000045. If, instead, we truncate the usual power series after the x^5-term, we get

$$e^{-x} \simeq 1 - x + 0.5x^2 - 0.166667x^3 + 0.041667x^4 - 0.008333x^5 ,$$

where the error is essentially $x^6/720$; the maximum error is obtained for $x = -1$ and amounts to $1/720 + 1/5040 + \cdots = 0.001615$, that is, 36 times as large as in the former case. The error curves are represented in Fig. 17.5.* For small

---

* For convenience, the latter curve is reversed.

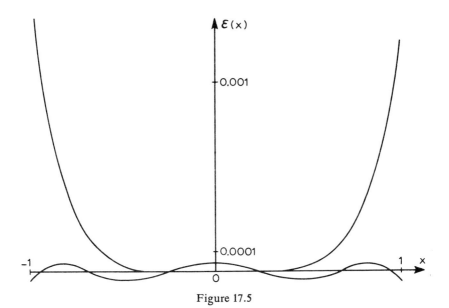

Figure 17.5

values of x the truncated power series is, of course, superior. On the other hand, the Chebyshev series has the power of distributing the error over the whole interval.

When Chebyshev series are to be used on a computer, a recursive technique is desirable. Such a technique has been devised by Clenshaw [9] for the computation of

$$f(x) = \sum_{k=0}^{n} c_k T_k(x) ,$$

where the coefficients c_k as well as the value x are assumed to be known. We form a sequence of numbers $a_n, a_{n-1}, \ldots, a_0$ by means of the relation

$$a_k - 2xa_{k+1} + a_{k+2} = c_k , \qquad (17.5.10)$$

where $a_{n+1} = a_{n+2} = 0$. Insertion in $f(x)$ gives

$$f(x) = \sum_{k=0}^{n-2} (a_k - 2xa_{k+1} + a_{k+2})T_k(x) + (a_{n-1} - 2xa_n)T_{n-1}(x) + a_n T_n(x)$$

$$= \sum_{k=0}^{n-2} (T_k - 2xT_{k+1} + T_{k+2})a_{k+2} + a_0 T_0 + a_1(T_1 - 2xT_0) .$$

Using (17.5.2) we find

$$f(x) = a_0 T_0 + a_1(T_1 - 2xT_0)$$

or

$$f(x) = a_0 - a_1 x . \qquad (17.5.11)$$

17.6. Approximations with continued fractions

A finite continued fraction is an expression of the form

$$b_0 + \cfrac{a_1}{b_1 + \cfrac{a_2}{b_2 + \cfrac{a_3}{b_3 + \cfrac{\ddots}{\quad \cfrac{a_{n-1}}{b_{n-1} + \cfrac{a_n}{b_n}}}}}} \tag{17.6.1}$$

A convenient notation for this expression is the following:

$$b_0 + \frac{a_1|}{|b_1} + \frac{a_2|}{|b_2} + \cdots + \frac{a_n|}{|b_n} = \frac{A_n}{B_n}. \tag{17.6.2}$$

A_r and B_r are polynomials in the elements a_s and b_s, $s = 0, 1, \ldots, r$. As is easily found, the following recursion formulas are valid:

$$\begin{cases} A_r = b_r A_{r-1} + a_r A_{r-2}, \\ B_r = b_r B_{r-1} + a_r B_{r-2}. \end{cases} \tag{17.6.3}$$

Here $A_{-1} = 1$, $B_{-1} = 0$, $A_0 = b_0$, and $B_0 = 1$. The formula is proved by induction, taking into account that A_{r+1}/B_{r+1} is obtained from A_r/B_r if b_r is replaced by $b_r + a_{r+1}/b_{r+1}$.

Generalization to infinite continued fractions is obvious; such a fraction is said to be convergent if $\lim_{n \to \infty} A_n/B_n = \xi$ exists.

Continued fractions containing a variable z are often quite useful for representation of functions; in particular, such expansions often have much better convergence properties than, for example, the power-series expansions. Space limitations make it impossible to give a description of the relevant theory, and so we restrict ourselves to a few examples.

$$e^z = 1 + \frac{z|}{|1} - \frac{z|}{|2} + \frac{z|}{|3} - \frac{z|}{|2} + \frac{z|}{|5} - \frac{z|}{|2} + \frac{z|}{|7} - \frac{z|}{|2} + \frac{z|}{|9} - \cdots,$$

$$\int_0^\infty \frac{e^{-u}\,du}{z+u} = \frac{1|}{|z} + \frac{1|}{|1} + \frac{1|}{|z} + \frac{2|}{|1} + \frac{2|}{|z} + \frac{3|}{|1} + \frac{3|}{|z} + \cdots,$$

$$e^x \int_x^\infty \frac{e^{-u}}{u}\,du = \frac{1|}{|x+1} - \frac{1|}{|x+3} - \frac{4|}{|x+5} - \frac{9|}{|x+7} - \frac{16|}{|x+9} - \cdots.$$

Details on the theory and application of continued fractions can be found, for example, in [10] and [11].

17.7. Approximations with rational functions

A finite continued fraction containing a variable z can be written A_n/B_n, where A_n and B_n are polynomials in z, and hence A_n/B_n is also a rational function. Here we encounter the problem of finding rational approximations of the form $P(x)/Q(x)$ to a given function $f(x)$ in an interval $[a, b]$. Then it is first necessary to prescribe how the deviation between the function and the approximation should be measured. If this is to be done in the least squares sense, then one still does not know very much. One could also demand conditions of the form

$$\left| f(x) - \frac{P_m(x)}{Q_n(x)} \right| \leq M \, |x|^{\nu} \, ,$$

where the coefficients of P_m and Q_n should be chosen so as to make ν as large as possible. In typical cases this implies that the Taylor expansions of $f(x)$ and $P_m(x)/Q_n(x)$ should coincide as far as possible. This type of approximation was introduced by Padé (cf. Exercise 4).

Most common, however, are Chebyshev approximations, and in this respect there are several theoretical results. For example, under quite general conditions there exists a "best" approximation characterized by the property that all maximal deviations are equal but with alternating sign (cf. [13] and [14]). In the examples we shall specialize to the same degree in numerator and denominator. Nearest to hand are the following two representations:

$$f(x) \simeq \frac{a_0 + a_1 x + \cdots + a_m x^m}{b_0 + b_1 x + \cdots + b_m x^m} \, , \tag{17.7.1}$$

$$f(x) \simeq \frac{\alpha_0 + \alpha_1 x^2 + \cdots + \alpha_m x^{2m}}{\beta_0 + \beta_1 x^2 + \cdots + \beta_m x^{2m}} \, . \tag{17.7.2}$$

The coefficients are determined by a semiempirical but systematic trial-and-error technique. The main idea is to adjust the zeros of the function $f(x) - P(x)/Q(x)$ until all maximal deviations are numerically equal.

Without specializing we can put $b_0 = \beta_0 = 1$. The second formula should be used if $f(x)$ is an even or an odd function; in the latter case it should first be multiplied by a suitable (positive or negative) odd power of x. It is interesting to compare the errors of the different types of approximations; naturally the type depends on the function to be approximated. However, an overall qualitative rule is the following. Let A, B, C, and D stand for power series, Chebyshev polynomial, continued fraction, and rational function, respectively, and let the signs $>$ and $\gg$ stand for "better" and "much better." Then we have, on comparing approximations with the same number of parameters,

$$D > C \gg B > A \, .$$

There are exceptions to this general rule, but this relation seems to be the typical one.

Last, we give a few numerical examples of rational approximation, all of them constructed in the Chebyshev sense; ε is the maximum error, and if nothing else is mentioned, the interval is $0 \leq x \leq 1$. Also cf. [12].

$$e^{-x} \simeq \frac{1.0000000007 - 0.4759358618x + 0.0884921370x^2 - 0.0065658101x^3}{1 + 0.5240642207x + 0.1125548636x^2 + 0.0106337905x^3},$$

$$\varepsilon = 7.34 \cdot 10^{-10}.$$

$$\log z \simeq \frac{-0.6931471773 + 0.0677412133x + 0.5297501385x^2 + 0.0956558162x^3}{1 + 1.3449644663x + 0.4547729177x^2 + 0.0286818192x^3},$$

$$x = 2z - 1; \quad \tfrac{1}{2} \leq z \leq 1; \quad \varepsilon = 3.29 \cdot 10^{-9}.$$

$$\frac{\arctan x}{x}$$

$$\simeq \frac{0.9999999992 + 1.1303754276x^2 + 0.2870044785x^4 + 0.0089472229x^6}{1 + 1.4637086496x^2 + 0.5749098994x^4 + 0.0506770959x^6},$$

$$\varepsilon = 7.80 \cdot 10^{-10}.$$

$$\frac{\sin x}{x} \simeq \frac{1 - 0.1335639326x^2 + 0.0032811761x^4}{1 + 0.0331027317x^2 + 0.0004649838x^4},$$

$$\varepsilon = 4.67 \cdot 10^{-11}.$$

$$\cos x \simeq \frac{0.9999999992 - 0.4558922221x^2 + 0.0205121130x^4}{1 + 0.0441077396x^2 + 0.0008996261x^4},$$

$$\varepsilon = 7.55 \cdot 10^{-10}.$$

The errors were computed before the coefficients were rounded to ten places.

REFERENCES

[1] George E. Forsythe: "Generation and Use of Orthogonal Polynomials for Data-fitting with a Digital Computer," *J. Soc. Indust. Appl. Math.*, **5** (June, 1957).

[2] P. G. Guest: *Numerical Methods of Curve Fitting* (Cambridge, 1961).

[3] F. L. Alt: *Advances in Computers* **2** (Academic Press, New York, 1961).

[4] C. Lanczos: *Applied Analysis* (London, 1957).

[5] N. K. Bari: *Trigonometriceskije Rjady* (Moscow, 1961).

[6] Hardy-Rogosinski: *Fourier Series* (Cambridge, 1944).

[7] Lance: *Numerical Methods for High-Speed Computers* (London, 1960).

[8] Hastings: *Approximations for Digital Computers* (Princeton University Press, Princeton, 1955).

[9] Clenshaw: MTAC (9), 118 (1955).

[10] O. Perron: *Die Lehre von den Kettenbrüchen* (Leipzig and Berlin, 1929).

[11] H. S. Wall: *Analytic Theory of Continued Fractions* (Van Nostrand, New York, 1948).

[12] Fröberg: *BIT*, **1**, 4 (1961).

[13] N. I. Achieser: *Theory of Approximation* (Ungar, New York, 1956).

[14] E. W. Cheney: *Introduction to Approximation Theory* (McGraw-Hill, New York, 1966).

EXERCISES

1. The following data x and y are subject to errors of the same order of magnitude:

x	1	2	3	4	5	6	7	8
y	3	3	4	5	5	6	6	7

Find a straight-line approximation, using the least-squares method.

2. The points $(2, 2)$, $(5, 4)$, $(6, 6)$, $(9, 9)$, and $(11, 10)$ should be approximated by a straight line. Perform this, assuming
(a) that the errors in the x-values can be neglected;
(b) that the errors in the x- and y-values are of the same order of magnitude.

3. Approximate e^{-x} for $0 \le x \le 1$ by $a - bx$ in the Chebyshev sense.

4. For small values of x, we want to approximate the function $y = e^{2x}$, with

$$f(x) = \frac{1 + ax + bx^2 + cx^3}{1 - ax + bx^2 - cx^3},$$

Find a, b, and c.

5. For $|x| < 1$, there exists an expansion of the form

$$e^{-x} = (1 + a_1 x)(1 + a_2 x^2)(1 + a_3 x^3) \cdots$$

Determine the constants $a_1, a_2, \ldots, a_6$.

6. For $0 \le x \le \pi/2$, the function $\cos x$ is to be approximated by a second-degree polynomial $y = ax^2 + bx + c$ such that $\int_0^{\pi/2} (y - \cos x)^2\, dx$ is minimized. Find a, b, and c.

7. The radiation intensity from a radioactive source is given by the formula $I = I_0 e^{-\alpha t}$. Determine the constants α and I_0 using the data below.

t	0.2	0.3	0.4	0.5	0.6	0.7	0.8
I	3.16	2.38	1.75	1.34	1.00	0.74	0.56

8. y is a function of x, given by the data below. It should be represented in the form $Ae^{-ax} + Be^{-bx}$. Determine the constants a, b, A, and B.

x	0.4	0.5	0.6	0.7	0.8	0.9	1.0	1.1
y	2.31604	2.02877	1.78030	1.56513	1.37854	1.21651	1.07561	0.95289

9. The function y of x given by the data below should be represented in the form $y = e^{-ax} \sin bx$. Find a and b.

x	0	0.2	0.4	0.6	0.8	1.0	1.2	1.4	1.6
y	0	0.15398	0.18417	0.16156	0.12301	0.08551	0.05537	0.03362	0.01909

10. The function $y(t)$ is a solution of the differential equation $y' + ay = 0$, $y(0) = b$. Using the method of least squares, find a and b if the following values are known:

t	0.1	0.2	0.3	0.4	0.5
$y(t)$	80.4	53.9	36.1	24.2	16.2

11. The function $y = 1/x$ should be approximated in the interval $1 \leq x \leq 2$ by two different straight lines. One line is used in $1 \leq x \leq a$ and another in $a \leq x \leq 2$. Each line intersects with the curve in two points, and the two lines intersect in $x = a$. All five maximum deviations are numerically equal. Find a and the maximum deviation.

12. The parabola $y = f(x) = x^2$ is to be approximated by a straight line $y = g(x)$ in the Chebyshev sense in a certain interval I so that the deviations in the two endpoints and the maximum deviations in I all are numerically $= \varepsilon$. Compute B/A when

$$A = \int_I (g(x) - f(x))\, dx \quad \text{and} \quad B = \int_I |g(x) - f(x)|\, dx \, .$$

13. Approximate $x/(e^x - 1)$ by a polynomial of lowest possible degree so that the absolute value of the error is $< 5 \times 10^{-4}$ when $x \in [0, 1]$.

14. Find a third-degree polynomial $P(x)$ which approximates the function $f(x) = \cos \pi x$ in $0 \leq x \leq 1$ so that $P(0) = f(0)$; $P(1) = f(1)$; $P'(0) = f'(0)$; $P'(1) = f'(1)$. Then calculate $\max_{0 \leq x \leq 1} |P(x) - f(x)|$ to 3 significant figures.

15. The function e^{-x} should be approximated by $ax + b$ so that $|e^{-x} - ax - b| \leq 0.005$. The constants a and b are to be chosen so that this accuracy is attained for an interval $[0, c]$ as large as possible. Find a, b, and c to three decimals.

16. The function F is defined through

$$F(x) = \int_0^x \exp\left(-\frac{t^2}{2}\right) dt \, .$$

Determine a polynomial $P(x)$ of lowest possible degree so that $|F(x) - P(x)| \leq 10^{-4}$ for $|x| \leq 1$.

17. Expand $|\sin x|$ in a Fourier series!

18. Expand the function $\cos x$ in a series of Chebyshev polynomials.

Chapter 18

Special Functions

18.0. Introduction

In this chapter we shall treat a few of the most common special functions. Primarily we shall choose such functions as are of great importance from a theoretical point of view, or in physical and technical sciences. Among these functions there are two main groups, one associated with simple linear homogeneous differential equations, usually of second order, and one with more complicated functions. In the latter case a rather thorough knowledge of the theory of analytic functions is indispensable for a comprehensive and logical treatment. However, it does not seem to be reasonable to assume such a knowledge, and for this reason it has been necessary to keep the discussion at a more elementary level, omitting some proofs and other rather vital parts. This is especially the case with the gamma function, but also for other functions with certain formulas depending on, for example, complex integration and residue calculus. Since the gamma and beta functions are of such a great importance in many applications, a rather careful description of their properties is desirable even if a complete foundation for the reasons mentioned cannot be given.

18.1. The Gamma Function

For a long time a function called the factorial function has been known and is defined for positive integer values of n through $n! = 1 \cdot 2 \cdot 3 \cdots n$. With this starting point, one has tried to find a more general function which coincides with the factorial function for integer values of the argument. Now we have trivially

$$n! = n \cdot (n - 1)! \qquad n = 2, 3, 4, \ldots,$$

and if we prescribe the validity of this relation also for $n = 1$ we find $0! = 1$. For different (mostly historical) reasons, one prefers to modify the functional equation slightly, and instead we shall consider functions $g(z)$ satisfying

$$g(z + 1) = zg(z)$$

for arbitrary complex z, and the condition $g(n + 1) = n!$ for $n = 0, 1, 2, \ldots$

359

Taking logarithms and differentiating twice we get

$$\frac{d^2}{dz^2} \log g(z + 1) = -\frac{1}{z^2} + \frac{d^2}{dz^2} \log g(z) .$$

From $g(z + n + 1) = (z + n)(z + n - 1) \cdots zg(z)$, we find in a similar way for $n = 0, 1, 2, \ldots,$

$$\frac{d^2}{dz^2} \log g(z) = \sum_{k=0}^{n} \frac{1}{(z + k)^2} + \frac{d^2}{dz^2} \log g(z + n + 1) .$$

Now we put

$$f(z) = \sum_{k=0}^{\infty} \frac{1}{(z + k)^2}$$

and find immediately

$$f(z) = \sum_{k=0}^{n} \frac{1}{(z + k)^2} + f(z + n + 1) .$$

It now seems natural indeed to identify $f(z)$ with $d^2/dz^2 \log g(z)$, and in this way define a function $\Gamma(z)$ through

$$\frac{d^2}{dz^2} \log \Gamma(z) = \sum_{n=0}^{\infty} \frac{1}{(z + n)^2} \qquad (18.1.1)$$

with

$$\Gamma(z + 1) = z\Gamma(z) ; \qquad \Gamma(1) = 1 . \qquad (18.1.2)$$

Integrating (18.1.1) from 1 to $z + 1$, we find

$$\frac{d}{dz} \log \Gamma(z + 1) = \Gamma'(1) + \sum_{n=1}^{\infty} \left(\frac{1}{n} - \frac{1}{z + n} \right) .$$

One more integration from 0 to z gives

$$\log \Gamma(z + 1) = z \cdot \Gamma'(1) + \sum_{n=1}^{\infty} \left\{ \frac{z}{n} - \log \left(1 + \frac{z}{n} \right) \right\} .$$

Putting $z = 1$ we get because of $\Gamma(2) = 1 \cdot \Gamma(1) = 1$:

$$-\Gamma'(1) = \sum_{n=1}^{\infty} \left\{ \frac{1}{n} - \log \left(1 + \frac{1}{n} \right) \right\}$$

$$= \lim_{n \to \infty} \left(1 + \frac{1}{2} + \frac{1}{3} + \cdots + \frac{1}{n} - \log n \right) = \gamma .$$

Here, as usual, γ is Euler's constant. Hence we obtain the representation

$$\Gamma(z) = \frac{1}{z} e^{-\gamma z} \prod_{n=1}^{\infty} e^{z/n} \bigg/ \left(1 + \frac{z}{n} \right) . \qquad (18.1.3)$$

If we use $\sum_{n=1}^{\infty}\{1/n - \log(1 + 1/n)\}$ instead of Euler's constant we get the equivalent formula

$$\Gamma(z) = \frac{1}{z} \prod_{n=1}^{\infty} \frac{(1 + 1/n)^z}{(1 + z/n)} . \tag{18.1.4}$$

From this formula we see that the Γ-function has poles (simple infinities) for $z = 0, -1, -2, \ldots$ Another representation which is frequently used is the following:

$$\Gamma(z) = \int_0^{\infty} e^{-t}t^{z-1}\, dt , \qquad \text{Re } z > 0 . \tag{18.1.5}$$

By the aid of partial integration it is easily verified that (18.1.2) is satisfied.

We are now going to derive some important formulas for the gamma function. First we consider $\Gamma(1 - z)$:

$$\Gamma(1 - z) = -z\Gamma(-z) = e^{\gamma z} \prod_{n=1}^{\infty} \left(1 - \frac{z}{n}\right)^{-1} e^{-z/n} .$$

Multiplying with $\Gamma(z)$ we get

$$\Gamma(z)\Gamma(1 - z) = z^{-1} \prod_{n=1}^{\infty} \left(1 - \frac{z^2}{n^2}\right)^{-1} .$$

Let us now perform a very superficial investigation of the function

$$z \prod_{n=1}^{\infty} \left(1 - \frac{z^2}{n^2}\right) .$$

Then we find

$$z(1 - z^2)\left(1 - \frac{z^2}{4}\right)\left(1 - \frac{z^2}{9}\right) \cdots = z - z^3 \left(1 + \frac{1}{4} + \frac{1}{9} + \cdots\right)$$

$$+ z^5 \left(1 \cdot \frac{1}{4} + 1 \cdot \frac{1}{9} + \cdots + \frac{1}{4} \cdot \frac{1}{9} + \frac{1}{4} \cdot \frac{1}{16} + \cdots\right) - \cdots .$$

The first coefficient is $\pi^2/6$ while the second can be written

$$\frac{1}{2}\Big\{\left(1 + \frac{1}{4} + \frac{1}{9} + \cdots\right)\left(1 + \frac{1}{4} + \frac{1}{9} + \cdots\right)$$

$$- 1 - \frac{1}{4 \cdot 4} - \frac{1}{9 \cdot 9} - \cdots\Big\} = \frac{1}{2}\left(\frac{\pi^4}{36} - \frac{\pi^4}{90}\right) = \frac{\pi^4}{120} .$$

Hence the leading three terms in the series are

$$z - \frac{\pi^2 z^3}{6} + \frac{\pi^4 z^5}{120} ,$$

coinciding with the three leading terms in the series expansion for $\sin \pi z/\pi$. In the theory of analytic functions it is strictly proved that

$$z \prod_{n=1}^{\infty} \left(1 - \frac{z^2}{n^2}\right) = \frac{\sin \pi z}{\pi} .$$

(Note that we have the same zeros on both sides!) This gives the important result

$$\Gamma(z)\Gamma(1 - z) = \frac{\pi}{\sin \pi z} . \qquad (18.1.6)$$

In particular we find for $z = \frac{1}{2}$ that

$$[\Gamma(\tfrac{1}{2})]^2 = \pi \qquad \text{and} \qquad \Gamma(\tfrac{1}{2}) = \sqrt{\pi} ,$$

since $\Gamma(x) > 0$ when $x > 0$. If the formula is used for purely imaginary argument $z = iy$, we get

$$\Gamma(iy)\Gamma(1 - iy) = -iy\Gamma(iy)\Gamma(-iy) = \frac{\pi}{\sin \pi iy} = \frac{\pi}{i \sinh \pi y} ,$$

and hence

$$|\Gamma(iy)|^2 = \frac{\pi}{y \sinh \pi y} , \qquad (18.1.7)$$

since $|\Gamma(iy)| = |\Gamma(-iy)|$. By the aid of (18.1.3) we can also compute the argument using the well-known relations $\arg(z_1 z_2) = \arg z_1 + \arg z_2$; $\arg(x + iy) = \arctan(y/x)$; and $\arg e^{i\alpha} = \alpha$. In this way the following formula is obtained:

$$\arg \Gamma(iy) = \sum_{n=1}^{\infty} \left\{ \frac{y}{n} - \arctan \left(\frac{y}{n} \right) \right\} - \gamma y - \left(\frac{\pi}{2} \right) \text{sign}(y) . \qquad (18.1.8)$$

By use of the functional relation we can then derive formulas for the absolute value and the argument of $\Gamma(N - iy)$ when N is an integer.

We shall now construct a formula which can be used for computation of the Γ-function in points close to the real axis, and in particular for real values of the argument. We start from (18.1.4) in the form

$$z(z + 1)\Gamma(z) = \Gamma(z + 2) = \exp \left\{ \sum_{n=2}^{\infty} \left[z \log \left(1 + \frac{1}{n - 1} \right) - \log \left(1 + \frac{z}{n} \right) \right] \right\} .$$

But

$$-\log \left(1 + \frac{z}{n} \right) = -\frac{z}{n} + \sum_{k=2}^{\infty} \frac{(-1)^k}{k} \left(\frac{z}{n} \right)^k .$$

If we now perform first the summation over n from 2 to N for

$$z \left[\log \left(1 + \frac{1}{n - 1} \right) - \frac{1}{n} \right] ,$$

we obtain

$$z \left[\log 2 + \log \frac{3}{2} + \log \frac{4}{3} + \cdots + \log \frac{N}{N - 1} - \frac{1}{2} - \frac{1}{3} - \cdots - \frac{1}{N} \right]$$

$$= z \left(\log N - \frac{1}{2} - \frac{1}{3} - \cdots - \frac{1}{N} \right) .$$

When $N \to \infty$ this expression tends to $z(1 - \gamma)$, and we get

$$\Gamma(z + 2) = \exp\left\{z(1 - \gamma) + \sum_{n=2}^{\infty}\sum_{k=2}^{\infty}\frac{(-1)^k}{k}\left(\frac{z}{n}\right)^k\right\}.$$

Now the summation order in n and k is reversed (which is allowed if $|z| < 2$) and using the notation

$$\zeta(k) = \sum_{n=1}^{\infty} n^{-k}, \qquad k > 1,$$

we obtain finally

$$\Gamma(z + 2) = \exp\left\{z(1 - \gamma) + \sum_{k=2}^{\infty}\frac{(-1)^k}{k}[\zeta(k) - 1]z^k\right\}. \qquad (18.1.9)$$

For $z = -1$, we find the well-known identity

$$\gamma + \frac{z_2}{2} + \frac{z_3}{3} + \frac{z_4}{4} + \cdots = 1,$$

with $z_k = \zeta(k) - 1$ (also cf. Ex. 19 of Chapter 11). Formula (18.1.9) should only be used when $|z|$ is small, preferably for real values of x such that $-1 \le x \le 1$.

If one wants to compute the value of the Γ-function for an arbitrary complex argument z, an asymptotic formula constructed by Stirling should be used. For, say, 10-digit accuracy it is suitable to require $|z| \ge 15$ and $\mathrm{Re}\, z \ge 0$. Other values of z can be mastered if Stirling's formula is used on $z + N$ (N positive integer) and the functional relation then is applied N times.

We shall now derive Stirling's formula starting from Euler-McLaurin's summation formula. Suppose $f(x) = \log x$ and $h = 1$ and let M and N be large positive integers, $N > M$. Since

$$f^{(2r+1)}(x) = (2r)!\, x^{-(2r+1)},$$

we obtain the following asymptotic relation:

$$\sum_{n=M}^{N} \log n = \int_{M}^{N} \log x\, dx + \frac{1}{2}[\log M + \log N]$$

$$- \frac{1}{12}\left[\frac{1}{M} - \frac{1}{N}\right] + \frac{1}{360}\left[\frac{1}{M^3} - \frac{1}{N^3}\right] - \frac{1}{1260}\left[\frac{1}{M^5} - \frac{1}{N^5}\right] + \cdots$$

$$+ (-1)^{r+1}\frac{B_{r+1}}{(2r + 1)(2r + 2)}\left[\frac{1}{M^{2r+1}} - \frac{1}{N^{2r+1}}\right] + \cdots .$$

Applying

$$\sum_{n=M}^{N} \log n = \log N! - \log(M - 1)!$$

and

$$\int_{M}^{N} \log x\, dx = N \log N - N - M \log M + M$$

and collecting terms in M and N we find

$$\log M! - \left(M + \frac{1}{2}\right) \log M + M - \frac{1}{12M} + \frac{1}{360M^3} - \cdots$$

$$= \log N! - \left(N + \frac{1}{2}\right) \log N + N - \frac{1}{12N} + \frac{1}{360N^3} - \cdots,$$

where both sides must be considered as asymptotic expansions. Now let M and N tend to infinity and truncate the two series just before the terms $1/12M$ and $1/12N$. Then we see at once that the remainder must be a constant K:

$$K = \lim_{n \to \infty} \{\log n! - (n + \tfrac{1}{2}) \log n + n\}.$$

Hence we obtain the asymptotic expansion

$$\log n! = K + \left(n + \frac{1}{2}\right) \log n - n + \frac{1}{12n} - \frac{1}{360n^3} + \cdots.$$

However, we have still to compute the value of K. To do this we again apply the same expansion, but now for $2n$:

$$\log (2n)! = K + \left(2n + \frac{1}{2}\right) \log (2n) - 2n + \frac{1}{24n} - \frac{1}{2880n^3} + \cdots$$

or

$$\log (2n)! - 2 \log n! = -K + \left(2n + \frac{1}{2}\right) \log 2$$

$$- \frac{1}{2} \log n - \frac{1}{8n} + \frac{1}{192n^3} - \cdots.$$

The left-hand side can be estimated by considering the integral $I_m = \int_0^{\pi/2} \sin^m x \, dx$ for non-negative integer values of m. By partial integration we find, when $m \geq 2$,

$$I_m = \frac{m-1}{m} I_{m-2},$$

with $I_0 = \pi/2$ and $I_1 = 1$. Hence we have

$$I_{2n} = \frac{2n-1}{2n} \frac{2n-3}{2n-2} \cdots \frac{1}{2} \frac{\pi}{2} = \frac{(2n)!}{2^{2n}(n!)^2} \frac{\pi}{2}$$

and

$$I_{2n+1} = \frac{2n}{2n+1} \frac{2n-2}{2n-1} \cdots \frac{2}{3} = \frac{2^{2n}(n!)^2}{(2n)! \, (2n+1)}.$$

Further we have for $n \geq 1$

$$\int_0^{\pi/2} \sin^{2n+1} x \, dx < \int_0^{\pi/2} \sin^{2n} x \, dx < \int_0^{\pi/2} \sin^{2n-1} x \, dx,$$

that is, $I_{2n+1} < I_{2n} < I_{2n-1}$. From this follows

$$\frac{I_{2n+1}}{I_{2n-1}} < \frac{I_{2n}}{I_{2n-1}} < 1 \qquad \text{or} \qquad \frac{2n}{2n+1} < \frac{[(2n)!]^2 n}{2^{4n}(n!)^4} \pi < 1 \,.$$

Hence we finally get

$$\lim_{n \to \infty} \frac{(2n)! \sqrt{n}}{2^{2n}(n!)^2} = \frac{1}{\sqrt{\pi}} \,,$$

which is the well-known formula by Wallis. For large values of n we have

$$\log (2n)! - 2 \log n! \cong 2n \log 2 - \tfrac{1}{2} \log (n\pi)$$

and comparing with our previous expression we get when $n \to \infty$

$$K = \tfrac{1}{2} \log 2\pi \,.$$

The derivation has been performed for positive integers, but the formula can be continued to arbitrary complex values z sufficiently far away from the singular points $z = 0, -1, -2, \ldots$ We now present *Stirling's formula* in the following shape ($\log \Gamma(n) = \log n! - \log n$):

$$\log \Gamma(z) = \frac{1}{2} \log 2\pi + \left(z - \frac{1}{2} \right) \log z - z$$

$$+ \frac{1}{12z} - \frac{1}{360z^3} + \frac{1}{1260z^5} - \cdots$$

$$+ \frac{(-1)^{n-1}B_n}{2n(2n-1)z^{2n-1}} + R_{n+1} \,, \qquad (18.1.10)$$

where

$$|R_{n+1}| \leq \frac{B_{n+1}}{(2n+1)(2n+2)|z|^{2n+1}(\cos (\varphi/2))^{2n+1}} \qquad \text{and} \qquad \varphi = \arg z \,.$$

For the Γ-function itself we have

$$\Gamma(z) \sim \sqrt{2\pi} z^{z-1/2} e^{-z} \left[1 + \frac{1}{12z} + \frac{1}{288z^2} - \frac{139}{51840z^3} - \frac{71}{2488320z^4} + \cdots \right] \,.$$

The formulas can be used only if $|\arg z| < \pi$, but it is advisable to keep to the right half-plane and apply the functional relation when necessary. Often the logarithmic derivative of the Γ-function is defined as a function of its own, the so-called ψ-function or the digamma function. From the functional relation for the Γ-function we have directly $\psi(z + 1) - \psi(z) = 1/z$. Further from (18.1.3) we obtain:

$$\psi(z) = -\gamma + \sum_{n=0}^{\infty} \left(\frac{1}{n+1} - \frac{1}{z+n} \right). \qquad (18.1.11)$$

As is the case for the Γ-function the ψ-function also has poles for $z = -n$, $n = 0, 1, 2, 3, \ldots$ The function values for positive integer arguments can easily be computed:

$$\psi(1) = -\gamma\,; \qquad \psi(2) = 1 - \gamma\,; \qquad \psi(3) = 1 + \tfrac{1}{2} - \gamma\,;$$
$$\psi(4) = 1 + \tfrac{1}{2} + \tfrac{1}{3} - \gamma\,, \qquad \text{and so on.}$$

It is immediately recognized that for large values of n we have $\psi(n) \sim \log n$. An asymptotic expansion can be obtained from (18.1.10):

$$\psi(z) \sim \log z - \frac{1}{2z} - \frac{1}{12z^2} + \frac{1}{120z^4} - \frac{1}{252z^6} + \cdots.$$

Due to the relation $\Delta\psi(z) = 1/z$ the ψ-function is of some importance in the theory of difference equations (cf. Section 13.3).

The derivative of the ψ-function can also be computed from (18.1.5) which allows differentiation under the integral sign:

$$\psi'(z) = \int_0^\infty e^{-t} t^{z-1} \log t \; dt\,.$$

For $z = 1$ and $z = 2$ we find in particular

$$\psi'(1) = -\gamma = \int_0^\infty e^{-t} \log t \; dt\,,$$

$$\psi'(2) = 1 - \gamma = \int_0^\infty t e^{-t} \log t \; dt\,.$$

As an example we shall now show how the properties of the Γ-function can be used for computation of infinite products of the form $\prod_{n=1}^\infty u_n$ where u_n is a rational function of n. First we can split u_n in factors in such a way that the product can be written

$$P = \prod_{n=1}^\infty \frac{A(n - a_1)(n - a_2) \cdots (n - a_r)}{(n - b_1)(n - b_2) \cdots (n - b_s)}\,.$$

A necessary condition for convergence is $\lim_{n \to \infty} u_n = 1$ which implies $A = 1$ and $r = s$. The general factor then gets the form

$$P_n = \left(1 - \frac{a_1}{n}\right) \cdots \left(1 - \frac{a_r}{n}\right)\left(1 - \frac{b_1}{n}\right)^{-1} \cdots \left(1 - \frac{b_r}{n}\right)^{-1}$$

$$= 1 - \frac{\sum a_i - \sum b_i}{n} + O(n^{-2})\,.$$

But a product of the form $\prod_{n=1}^\infty (1 - a/n)$ is convergent if and only if $a = 0$ (it $a > 0$ the product tends to 0, if $a < 0$ to ∞). Hence we must have $\sum a_i = \sum b_i$

and we may add the factor $\exp\{n^{-1}(\sum a_i - \sum b_i)\}$ without changing the value of P. In this way we get

$$P = \prod_{n=1}^{\infty} \left(1 - \frac{a_1}{n}\right) e^{a_1/n} \cdots \left(1 - \frac{a_r}{n}\right) e^{a_r/n} \left(1 - \frac{b_1}{n}\right)^{-1} e^{-b_1/n} \cdots \left(1 - \frac{b_r}{n}\right)^{-1} e^{-b_r/n} .$$

But according to (18.1.3) we have

$$\prod_{n=1}^{\infty} \left(1 - \frac{z}{n}\right) e^{z/n} = \frac{e^{\gamma z}}{[-z\Gamma(-z)]} = \frac{e^{\gamma z}}{\Gamma(1-z)}$$

and from this we get the final result:

$$P = \prod_{i=1}^{r} \frac{\Gamma(1 - b_i)}{\Gamma(1 - a_i)} . \tag{18.1.12}$$

18.2. The Beta Function

Many integrals can be brought to the form

$$B(x, y) = \int_0^1 t^{x-1}(1 - t)^{y-1} dt . \tag{18.2.1}$$

(Unfortunately, capital Beta cannot be distinguished from an ordinary capital B.) This relation defines the beta function $B(x, y)$ which, obviously, is a function of two variables x and y. For convergence of the integral we must claim Re $x > 0$, Re $y > 0$. First it will be shown that the beta function can be expressed in terms of gamma functions. Starting from (18.2.1) we perform the transformation $t = v/(1 + v)$ obtaining

$$B(x, y) = \int_0^{\infty} v^{x-1}(1 + v)^{-x-y} dv . \tag{18.2.2}$$

Multiplying by $\Gamma(x + y)$ and observing that

$$\int_0^{\infty} e^{-(1+v)\xi}\xi^{x+y-1} d\xi = (1 + v)^{-x-y}\Gamma(x + y) ,$$

which follows through the transformation $(1 + v)\xi = z$, we obtain

$$\Gamma(x + y)B(x, y) = \int_0^{\infty} dv \cdot v^{x-1} \int_0^{\infty} e^{-(1+v)\xi}\xi^{x+y-1} d\xi$$

$$= \int_0^{\infty} d\xi e^{-\xi}\xi^{x+y-1} \int_0^{\infty} v^{x-1}e^{-v\xi} dv .$$

The inner integral has the value $\xi^{-x}\Gamma(x)$ and this gives

$$\Gamma(x + y)B(x, y) = \Gamma(x) \int_0^{\infty} e^{-\xi}\xi^{y-1} d\xi = \Gamma(x)\Gamma(y) .$$

Hence we get the desired formula

$$B(x, y) = \frac{\Gamma(x)\Gamma(y)}{\Gamma(x + y)} .$$
(18.2.3)

We can now make use of this relation in deriving another important formula. If the integration interval in 18.2.2 is divided into two parts we obtain

$$B(x, y) = \int_0^1 + \int_1^\infty \frac{v^{x-1}}{(1 + v)^{x+y}} \, dv .$$

The second integral is transformed through $v = 1/t$ and hence

$$B(x, y) = \int_0^1 \frac{v^{x-1} + v^{y-1}}{(1 + v)^{x+y}} \, dv .$$

Putting $y = x$ we get

$$B(x, x) = \frac{\Gamma(x)^2}{\Gamma(2x)} = 2 \int_0^1 \frac{v^{x-1}}{(1 + v)^{2x}} \, dv .$$

If we put $v = (1 - t)/(1 + t)$, the integral is transformed to $2^{2-2x} \int_0^1 (1 - t^2)^{x-1} \, dt$, and then $t^2 = z$ gives

$$\frac{\Gamma(x)^2}{\Gamma(2x)} = \frac{1}{2^{2x-1}} \int_0^1 z^{-1/2}(1 - z)^{x-1} \, dz = \frac{B(x, \frac{1}{2})}{2^{2x-1}} .$$

From this we finally get

$$\Gamma(2x) = \pi^{-1/2} 2^{2x-1} \Gamma(x)\Gamma(x + \tfrac{1}{2}) .$$
(18.2.4)

This duplication formula is initially due to Legendre.

EXAMPLE

$$\int_0^{\pi/2} \sin^{2x} t \, \cos^{2y} t \, dt = \frac{\Gamma(x + \frac{1}{2})\Gamma(y + \frac{1}{2})}{2\Gamma(x + y + 1)} , \qquad x > -\tfrac{1}{2}, \quad y > -\tfrac{1}{2} .$$
(18.2.5)

This important formula is obtained if we put $\sin^2 t = z$.

18.3. Some functions defined by definite integrals

In this section we shall very briefly touch upon still a few functions other than the gamma and beta functions (which could also be brought here since they can be defined through definite integrals). The *error integral* is used extensively in statistics and probability theory and is defined through

$$\text{Erfc}(x) = \frac{2}{\sqrt{\pi}} \int_0^x e^{-t^2} \, dt ,$$
(18.3.1)

where Erfc is an abbreviation for Error function. For large values of x the function converges quickly toward 1 and then an asymptotic expansion can be used conveniently:

$$\sqrt{\pi}\,(1 - \text{Erfc}\,(x)) \sim \frac{e^{-x^2}}{x}\left(1 - \frac{1}{2x^2} + \frac{1 \cdot 3}{(2x^2)^2} - \frac{1 \cdot 3 \cdot 5}{(2x^2)^3} + \cdots\right). \qquad (18.3.2)$$

The exponential integrals are defined through

$$Ei(x) = \int_{-\infty}^{x} \frac{e^t}{t}\,dt\,, \qquad x > 0\,, \qquad (18.3.3)$$

$$E_1(x) = \int_{x}^{\infty} \frac{e^{-t}}{t}\,dt\,, \qquad x > 0\,. \qquad (18.3.4)$$

In the first integral the integrand is singular in the origin and the integral should be interpreted as the Cauchy principal value

$$\lim_{\varepsilon \to 0} \int_{-\infty}^{-\varepsilon} + \int_{\varepsilon}^{x} \frac{e^t}{t}\,dt\,.$$

Nearly related are the *sine* and *cosine integrals*

$$Si(x) = \int_{0}^{x} \frac{\sin t}{t}\,dt\,, \qquad (18.3.5)$$

$$Ci(x) = -\int_{x}^{\infty} \frac{\cos t}{t}\,dt = \gamma + \log x + \int_{0}^{x} \frac{\cos t - 1}{t}\,dt\,. \qquad (18.3.6)$$

The *complete elliptic integrals* are defined through

$$K(k) = \int_{0}^{\pi/2} \frac{dt}{\sqrt{1 - k^2 \sin^2 t}} = \int_{0}^{1} \frac{dx}{\sqrt{(1 - x^2)(1 - k^2 x^2)}}\,. \qquad (18.3.7)$$

$$E(k) = \int_{0}^{\pi/2} \sqrt{1 - k^2 \sin^2 t}\,dt = \int_{0}^{1} \sqrt{\frac{1 - k^2 x^2}{1 - x^2}}\,dx\,. \qquad (18.3.8)$$

A large variety of definite integrals can be reduced to this form (cf. [7]).

EXAMPLES

$$I = \int_{0}^{\infty} \frac{dt}{\sqrt{1 + t^4}} = \int_{0}^{1} + \int_{1}^{\infty} \frac{dt}{\sqrt{1 + t^4}} = 2\int_{0}^{1} \frac{dt}{\sqrt{1 + t^4}}$$

(put $t = u^{-1}$ in the second integral).

Then transform through $t = \tan(x/2)$:

$$I = \int_{0}^{\pi/2} \frac{dx}{\sqrt{\sin^4(x/2) + \cos^4(x/2)}} = \int_{0}^{\pi/2} \frac{dx}{\sqrt{1 - \frac{1}{2}\sin^2 x}} = K\left(\frac{1}{\sqrt{2}}\right)$$

$$= 1.85407\ 46773\,.$$

18.4. Functions defined through linear homogeneous differential equations

First we shall very briefly discuss how solutions of linear differential equations can be determined by series expansion. In general, one makes the attempt

$$y = \sum_{n=r}^{\infty} a_n x^n \,,$$

where the numbers a_n so far represent unknown coefficients; also the starting level r remains to be established. By insertion into the differential equation we get relations between adjacent coefficients through the condition that the total coefficient for each power x^n must vanish. In the beginning of the series we obtain a special condition for determining r (the *index equation*).

EXAMPLE

$$y'' + y = 0 \,.$$

We suppose that the series starts with the term x^r. Introducing this single term and collecting terms of lowest possible degree we are left with the expression $r(r-1)x^{r-2}$. The index equation is $r(r-1) = 0$ and we get two possible values of r: $r = 0$ and $r = 1$.

(a) $r = 0$.

$$y = \sum_{n=0}^{\infty} a_n x^n \,.$$

$$y'' = \sum_{n=2}^{\infty} n(n-1)a_n x^{n-2} = \sum_{n=0}^{\infty} (n+1)(n+2)a_{n+2} x^n \,.$$

Hence we obtain the relation $(n+1)(n+2)a_{n+2} + a_n = 0$, and choosing $a_0 = 1$ we find successively

$$a_2 = -\frac{1}{1 \cdot 2}; \qquad a_4 = \frac{1}{1 \cdot 2 \cdot 3 \cdot 4}; \qquad \cdots$$

and the solution

$$y = 1 - \frac{x^2}{2!} + \frac{x^4}{4!} - \frac{x^6}{6!} + \cdots = \cos x \,.$$

(b) $r = 1$. Here we have the same relation for the coefficients, and choosing $a_1 = 1$ we find

$$y = x - \frac{x^3}{3!} + \frac{x^5}{5!} - \frac{x^7}{7!} + \cdots = \sin x \,.$$

The general solution is $y = A \cos x + B \sin x$.

Similar series expansions cannot generally be established for all linear differential equations. If the equation has the form $y'' + P(x)y' + Q(x)y = 0$, $P(x)$ and $Q(x)$ as a rule must not contain powers of degree lower than -1 and -2, respectively. The equations we are now going to discuss are all of great importance in physics and other sciences, and the condition mentioned here will be satisfied for all these equations.

18.5. Bessel functions

We shall now treat the differential equation

$$\frac{d^2u}{dx^2} + \frac{1}{x}\frac{du}{dx} + \left(1 - \frac{\nu^2}{x^2}\right)u = 0 . \tag{18.5.1}$$

Putting $u = x^r$ we find the index equation $r(r - 1) + r - \nu^2 = 0$ with the two roots $r = \pm\nu$. Trying the expansion $u = \sum_{k=\nu}^{\infty} a_k x^k$, we find

$$a_{k+2} = -\frac{a_k}{(k + 2 + \nu)(k + 2 - \nu)} .$$

We now start with $k = \nu$ and $a_\nu = 1$ obtaining

$$\begin{cases} a_{\nu+2} = -\dfrac{1}{2(2\nu + 2)} , \\[2mm] a_{\nu+4} = \dfrac{1}{2 \cdot 4(2\nu + 2)(2\nu + 4)} , \\[2mm] \quad\vdots \\[2mm] a_{\nu+2k} = \dfrac{(-1)^k}{2^{2k} \cdot k! \,(\nu + 1)(\nu + 2) \cdots (\nu + k)} = \dfrac{(-1)^k \Gamma(\nu + 1) \cdot 2^\nu}{2^{\nu+2k} \cdot k! \,\Gamma(\nu + k + 1)} . \end{cases}$$

The factor $2^\nu \Gamma(\nu + 1)$ is constant and can be removed, and so we define a function $J_\nu(x)$ through the formula

$$J_\nu(x) = \sum_{k=0}^{\infty} \frac{(-1)^k (x/2)^{\nu+2k}}{k! \,\Gamma(\nu + k + 1)} . \tag{18.5.2}$$

The function $J_\nu(x)$ is called a *Bessel function of the first kind*. Replacing ν by $-\nu$ we get still another solution which we denote by $J_{-\nu}(x)$, and a linear combination of these two will give the general solution. However, if ν is an integer n, the terms corresponding to $k = 0, 1, \ldots, n - 1$ will disappear since the gamma function in the denominator becomes infinite. Then we are left with

$$J_{-n}(x) = \sum_{k=n}^{\infty} \frac{(-1)^k (x/2)^{-n+2k}}{k! \,\Gamma(-n + k + 1)} = \sum_{\nu=0}^{\infty} \frac{(-1)^n (-1)^\nu (x/2)^{-n+2n+2\nu}}{(n + \nu)! \,\Gamma(\nu + 1)} = (-1)^n J_n(x) ,$$

where we have put $k = n + \nu$. Note that $(n + \nu)! = \Gamma(n + \nu + 1)$ and $\Gamma(\nu + 1) = \nu!$. As soon as ν is an integer we only get one solution by direct series expansion, and we must try to find a second solution in some other way. A very general technique is demonstrated in the following example. Consider the differential equation $y'' + \alpha^2 y = 0$ with the solutions $y_1 = e^{\alpha x}$, $y_2 = e^{-\alpha x}$.

If $\alpha = 0$ we get $y_1 = y_2 = 1$ and one solution has got lost. It is now clear that also $y_0 = (y_2 - y_1)/2\alpha$ is a solution, and this holds true even if $\alpha \to 0$. Since $\lim_{\alpha \to 0} y_0 = x$, the general solution becomes $y = A + Bx$.

Along the same lines we now form

$$Y_n(x) = \frac{1}{\pi} \lim_{\nu \to n} \frac{J_\nu(x) - (-1)^n J_{-\nu}(x)}{\nu - n}$$

or

$$Y_n(x) = \frac{1}{\pi} \lim_{\nu \to n} \left\{ \frac{J_\nu(x) - J_n(x)}{\nu - n} - (-1)^n \frac{J_{-\nu}(x) - J_{-n}(x)}{\nu - n} \right\}$$

$$= \frac{1}{\pi} \left\{ \frac{\partial J_\nu(x)}{\partial \nu} - (-1)^n \frac{\partial J_{-\nu}(x)}{\partial \nu} \right\}_{\nu = n} .$$

We shall now prove that $Y_n(x)$ satisfies the Bessel equation

$$x^2 \frac{d^2 u}{dx^2} + x \frac{du}{dx} + (x^2 - \nu^2)u = 0 .$$

Differentiating with respect to ν, and with u replaced by $J_\nu(x)$ and $J_{-\nu}(x)$, we get

$$\begin{cases} x^2 \dfrac{d^2}{dx^2} \dfrac{\partial J_\nu(x)}{\partial \nu} + x \dfrac{d}{dx} \dfrac{\partial J_\nu(x)}{\partial \nu} + (x^2 - \nu^2) \dfrac{\partial J_\nu(x)}{\partial \nu} = 2\nu J_\nu(x) , \\[2mm] x^2 \dfrac{d^2}{dx^2} \dfrac{\partial J_{-\nu}(x)}{\partial \nu} + x \dfrac{d}{dx} \dfrac{\partial J_{-\nu}(x)}{\partial \nu} + (x^2 - \nu^2) \dfrac{\partial J_{-\nu}(x)}{\partial \nu} = 2\nu J_{-\nu}(x) . \end{cases}$$

We multiply the second equation by $(-1)^n$ and subtract:

$$\left(x^2 \frac{d^2}{dx^2} + x \frac{d}{dx} + x^2 - \nu^2 \right) \left(\frac{\partial J_\nu(x)}{\partial \nu} - (-1)^n \frac{\partial J_{-\nu}(x)}{\partial \nu} \right)$$

$$= 2\nu \big(J_\nu(x) - (-1)^n J_{-\nu}(x) \big) .$$

Now letting $\nu \to n$ the right-hand side will tend to zero, and we find after multiplication with $1/\pi$:

$$x^2 \frac{d^2 Y_n(x)}{dx^2} + x \frac{dY_n(x)}{dx} + (x^2 - n^2) Y_n(x) = 0 ;$$

that is, $Y_n(x)$ satisfies the Bessel differential equation.

It is also possible to define a function $Y_\nu(x)$ even if ν is not an integer:

$$Y_\nu(x) = \frac{J_\nu(x) \cos \nu\pi - J_{-\nu}(x)}{\sin \nu\pi} ,$$

and by use of l'Hospital's rule, for example, one can easily establish that

$$\lim_{\nu \to n} Y_\nu(x) = Y_n(x) .$$

The general solution can now be written in a more consistent form:

$$u = A J_\nu(x) + B Y_\nu(x) .$$

The Y-functions are usually called *Bessel functions of the second kind* or *Neumann functions.* In particular we find for $n = 0$:

$$Y_0(x) = \frac{2}{\pi} \left(\frac{\partial J_\nu(x)}{\partial \nu} \right)_{\nu=0} = \frac{2}{\pi} \left\{ \frac{\partial}{\partial \nu} \sum_{k=0}^{\infty} \frac{(-1)^k (x/2)^{\nu+2k}}{k!\, \Gamma(\nu + k + 1)} \right\}_{\nu=0}$$

$$= \frac{2}{\pi} \left\{ \sum_{k=0}^{\infty} \frac{(-1)^k (x/2)^{\nu+2k}}{k!\, \Gamma(\nu + k + 1)} \left(\log \frac{x}{2} - \frac{\partial}{\partial \nu} \log \Gamma(\nu + k + 1) \right) \right\}_{\nu=0}.$$

Hence

$$Y_0(x) = \frac{2}{\pi} \sum_{k=0}^{\infty} \frac{(-1)^k (x/2)^{2k}}{(k!)^2} \left(\log \frac{x}{2} - \phi(k + 1) \right). \qquad (18.5.5)$$

If $n > 0$ the calculations get slightly more complicated and we only give the final result:

$$Y_n(x) = \frac{2}{\pi} \log \frac{x}{2} J_n(x) - \frac{1}{\pi} \left(\frac{x}{2} \right)^{-n} \sum_{k=0}^{n-1} \frac{(n - k - 1)!}{k!} \left(\frac{x}{2} \right)^{2k}$$

$$- \frac{1}{\pi} \sum_{k=0}^{\infty} \frac{(-1)^k (x/2)^{n+2k}}{k!\, (n + k)!} \left(\phi(k + 1) + \phi(n + k + 1) \right). \qquad (18.5.6)$$

We are now going to prove a general relation for linear, homogeneous second-order equations. Consider the equation

$$y'' + P(x)y' + Q(x)y = 0 ,$$

and suppose that two linearly independent solutions y_1 and y_2 are known:

$$\begin{cases} y_1'' + P(x)y_1' + Q(x)y_1 = 0 , \\ y_2'' + P(x)y_2' + Q(x)y_2 = 0 . \end{cases}$$

The first equation is multiplied by y_2, the second by $-y_1$, and then the equations are added:

$$\frac{d}{dx} (y_1'y_2 - y_1y_2') + P(x)(y_1'y_2 - y_1y_2') = 0 ,$$

and hence we have $y_1'y_2 - y_1y_2' = \exp\{-\int P(x)\, dx\}$. For the Bessel equation we have $P(x) = 1/x$, and using this we get

$$J_\nu'(x)J_{-\nu}(x) - J_\nu(x)J_{-\nu}'(x) = \frac{C}{x} .$$

The constant C can be determined from the first terms in the series expansions:

$$J_\nu(x) = \frac{(x/2)^\nu}{\Gamma(\nu + 1)} \left(1 + O(x^2) \right) ; \qquad J_\nu'(x) = \frac{(x/2)^{\nu-1}}{2\Gamma(\nu)} \left(1 + O(x^2) \right) .$$

In this way we obtain

$$J_\nu'(x)J_{-\nu}(x) - J_\nu(x)J_{-\nu}'(x) = \frac{1}{x} \left\{ -\frac{1}{\Gamma(\nu + 1)\Gamma(-\nu)} + \frac{1}{\Gamma(\nu)\Gamma(-\nu + 1)} \right\}$$

$$+ O(x)$$

and $C = 2 \sin \nu\pi/\pi$. If $\nu = n$ is an integer we get $C = 0$, that is, $J_n(x)$ and $J_{-n}(x)$ are proportional since $J_n'/J_n = J_{-n}'/J_{-n}$.

Relations of this kind are called Wronski-relations and are of great importance. An alternative expression can be constructed from J_ν and Y_ν and we find immediately

$$J_\nu'(x)Y_\nu(x) - J_\nu(x)Y_\nu'(x) = -\frac{2}{\pi x}.$$

In connection with numerical computation of functions such relations offer extremely good checking possibilities.

Recursion formulas

Between Bessel functions of adjacent orders there are certain simple linear relationships from which new functions can be computed by aid of old ones. First we shall prove that

$$J_{\nu-1}(x) + J_{\nu+1}(x) = \frac{2\nu}{x} J_\nu(x) . \qquad (18.5.7)$$

Choosing the term (k) from the first expansion, $(k-1)$ from the second, and (k) from the third, we obtain the same power $(x/2)^{\nu+2k-1}$. The coefficients become

$$\frac{(-1)^k}{k!\,\Gamma(\nu+k)}, \qquad \frac{(-1)^{k-1}}{(k-1)!\,\Gamma(\nu+k+1)}, \qquad \text{and} \qquad \frac{(-1)^k\nu}{k!\,\Gamma(\nu+k+1)},$$

and it is easily seen that the sum of the first two is equal to the third. However, we must also investigate the case $k = 0$ separately by comparing powers $(x/2)^{\nu-1}$, since there is no contribution from the second term. The coefficients are $1/\Gamma(\nu)$ and $\nu/\Gamma(\nu+1)$ and so the whole formula is proved. In a similar way we can prove

$$J_{\nu-1}(x) - J_{\nu+1}(x) = 2J_\nu'(x) . \qquad (18.5.8)$$

Eliminating either $J_{\nu-1}(x)$ or $J_{\nu+1}(x)$ one can construct other relations:

$$\begin{cases} J_\nu'(x) + \dfrac{\nu}{x} J_\nu(x) = J_{\nu-1}(x) , \\[2mm] J_\nu'(x) - \dfrac{\nu}{x} J_\nu(x) = -J_{\nu+1}(x) . \end{cases}$$

A corresponding set of formulas can be obtained also for the functions $Y_\nu(x)$. In particular we find for $\nu = 0$:

$$J_0'(x) = -J_1(x) ; \qquad Y_0'(x) = -Y_1(x) .$$

Integral representations

Consider the Bessel equation in the special case $\nu = 0$:

$$xu'' + u' + xu = 0 .$$

If the coefficients had been constant, we could have tried $u = e^{\lambda x}$ leading to a number of discrete particular solutions from which the general solution would have been obtained as a linear combination. When the coefficients are functions of x we cannot hope for such a simple solution, but we still have a chance to try a solution which takes care of the properties of the exponential functions. Instead of discrete "frequencies" we must expect a whole band of frequencies ranging over a suitable interval, and the summation must be replaced by integration. In this way we are led to the attempt

$$u = \int_a^b \phi(\alpha)e^{\alpha x}\, d\alpha .$$

We now have large funds at our disposal consisting of the function $\phi(\alpha)$ and the limits a and b. The differential equation is satisfied if

$$\int_a^b \{(\alpha^2 + 1)\phi(\alpha)xe^{\alpha x} + \alpha\phi(\alpha)e^{\alpha x}\}\, d\alpha = 0 .$$

Observing that $xe^{\alpha x} = (\partial/\partial\alpha)e^{\alpha x}$, we can perform a partial integration:

$$[(\alpha^2 + 1)\phi(\alpha)e^{\alpha x}]_a^b - \int_a^b \{(\alpha^2 + 1)\phi'(\alpha) + \alpha\phi(\alpha)\}e^{\alpha x}\, d\alpha = 0 .$$

We can now waste part of our funds by claiming that the integrated part shall be zero; this is the case if we choose $a = -i$, $b = i$ (also $-\infty$ will work if $x > 0$). Then the integral is also equal to zero, and we spend still a part of our funds by claiming that the integrand shall be zero:

$$\frac{d\phi}{\phi} + \frac{\alpha\, d\alpha}{\alpha^2 + 1} = 0$$

or $\phi(\alpha) = 1/\sqrt{\alpha^2 + 1}$, since a constant factor can be neglected here. In this way we have obtained

$$u = \int_{-i}^{+i} \frac{e^{\alpha x}}{\sqrt{\alpha^2 + 1}}\, d\alpha .$$

Putting $\alpha = i\beta$ and neglecting a trivial factor i we get

$$u = \int_{-1}^{+1} \frac{e^{i\beta x}}{\sqrt{1 - \beta^2}}\, d\beta .$$

Writing $e^{i\beta x} = \cos\beta x + i\sin\beta x$, we can discard the second term since the integral of an odd function over the interval $(-1, 1)$ will disappear. Putting $\beta = \cos\theta$, we finally get $u = \int_0^\pi \cos(x\cos\theta)\, d\theta$.
    We now state that

$$J_0(x) = \frac{1}{\pi} \int_0^\pi \cos(x\cos\theta)\, d\theta . \tag{18.5.9}$$

This follows because

$$\frac{1}{\pi}\int_0^\pi \cos(x\cos\theta)\,d\theta = \frac{2}{\pi}\int_0^{\pi/2}\sum_{k=0}^\infty \frac{(-1)^k x^{2k}\cos^{2k}\theta}{(2k)!}\,d\theta$$

$$= \frac{2}{\pi}\sum_{k=0}^\infty \frac{(-1)^k x^{2k}}{(2k)!}\,\frac{\Gamma(\tfrac12)\Gamma(k+\tfrac12)}{2\Gamma(k+1)}$$

$$= \frac{1}{\pi}\sum_{k=0}^\infty \frac{(-1)^k x^{2k}\pi\Gamma(k+\tfrac12)}{2k\cdot 2^{2k-1}\Gamma(k)\Gamma(k+\tfrac12)\cdot\Gamma(k+1)}$$

$$= \sum_{k=0}^\infty \frac{(-1)^k(x/2)^{2k}}{(k!)^2} = J_0(x)\,,$$

where we have made use of $(2k)! = 2k\Gamma(2k)$ and the duplication formula $(18.2.4)$ for $\Gamma(2k)$.

The technique we have worked with here is a tool which can often be used on a large variety of problems in numerical analysis and applied mathematics; it is known under the name *Laplace transformation*.

We shall now by other means derive a similar integral representation for the Bessel function $J_\nu(x)$. The general term in the series expansion can be written:

$$\frac{(-1)^k(x/2)^{\nu+2k}}{k!\,\Gamma(\nu+k+1)} = \frac{(-1)^k(x/2)^\nu}{\Gamma(\nu+\tfrac12)\Gamma(\tfrac12)}\,\frac{x^{2k}}{(2k)!}\,\frac{\Gamma(\nu+\tfrac12)\Gamma(k+\tfrac12)}{\Gamma(\nu+k+1)}$$

$$= \frac{(-1)^k(x/2)^\nu}{\Gamma(\nu+\tfrac12)\Gamma(\tfrac12)}\,\frac{x^{2k}}{(2k)!}\int_0^1 t^{\nu-1/2}(1-t)^{k-1/2}\,dt\,,$$

where again we have made use of formula $(18.2.4)$. Summing the whole series, we now get

$$J_\nu(x) = \frac{(x/2)^\nu}{\Gamma(\nu+\tfrac12)\Gamma(\tfrac12)}\int_0^1 t^{\nu-1/2}\left\{\sum_{k=0}^\infty \frac{(-1)^k x^{2k}(1-t)^{k-1/2}}{(2k)!}\right\}dt\,.$$

After the transformation $t = \sin^2\theta$, we obtain

$$J_\nu(x) = \frac{(x/2)^\nu}{\Gamma(\nu+\tfrac12)\Gamma(\tfrac12)}\int_0^{\pi/2}\sin^{2\nu-1}\theta\left\{\sum_{k=0}^\infty \frac{(-1)^k x^{2k}\cos^{2k-1}\theta}{(2k)!}\right\}2\sin\theta\cos\theta\,d\theta$$

or

$$J_\nu(x) = \frac{(x/2)^\nu}{\Gamma(\nu+\tfrac12)\Gamma(\tfrac12)}\int_0^\pi \sin^{2\nu}\theta\cos(x\cos\theta)\,d\theta\,. \tag{18.5.10}$$

The factor 2 has been used for extending the integration interval from $(0,\pi/2)$ to $(0,\pi)$. Putting $\nu = 0$ we get back our preceding formula $(18.5.9)$.

The formula which has just been derived can be used, for example, in the case when ν is half-integer, that is, $\nu = n + \tfrac12$ where n is integer. It is then suitable to perform the transformation $t = \cos\theta$;

$$J_{n+1/2}(x) = \frac{(x/2)^{n+1/2}}{n!\sqrt{\pi}}\int_{-1}^{+1}(1-t^2)^n\cos xt\,dt\,.$$

This integral can be evaluated in closed form, and one finds, for example,

$$J_{1/2}(x) = \sqrt{\frac{2}{\pi x}} \sin x \; ; \qquad\qquad J_{-1/2}(x) = \sqrt{\frac{2}{\pi x}} \cos x \, ,$$

$$J_{3/2}(x) = \sqrt{\frac{2}{\pi x}} \left(\frac{\sin x}{x} - \cos x \right) ; \qquad J_{-3/2}(x) = - \sqrt{\frac{2}{\pi x}} \left(\frac{\cos x}{x} + \sin x \right) .$$

Finally we quote a few useful summation formulas:

$$\begin{cases} J_0(x) + 2J_2(x) + 2J_4(x) + \cdots = 1 \, , \\ J_0(x) - 2J_2(x) + 2J_4(x) - \cdots = \cos x \, , \\ 2J_1(x) - 2J_3(x) + 2J_5(x) - \cdots = \sin x \, . \end{cases} \qquad (18.5.11)$$

The first of these formulas has been used for numerical calculation of Bessel functions by use of (18.5.7) (see Section 1.3). For proof of the relations, cf. Ex. 12 at the end of this chapter.

Numerical values of Bessel functions can be found in several excellent tables. In particular we mention here a monumental work from Harvard Computation Laboratory [8]. In 12 huge volumes $J_n(x)$ is tabulated for $0 \leq x \leq 100$ and for $n \leq 135$; for higher values of n we have $|J_n(x)| < \frac{1}{2} \cdot 10^{-10}$ in the given interval for x. There is also a rich literature on the theory of Bessel functions, but we restrict ourselves to mentioning the standard work by Watson [5].

18.6. Modified Bessel functions

We start from the differential equation

$$x^2 \frac{d^2u}{dx^2} + x \frac{du}{dx} - (x^2 + \nu^2)u = 0 \, . \qquad (18.6.1)$$

If we put $x = -it$ the equation goes over into the usual Bessel equation, and for example, we have the solutions $J_\nu(ix)$ and $J_{-\nu}(ix)$. But

$$J_\nu(ix) = \sum_{k=0}^{\infty} \frac{(-1)^k (x/2)^{\nu+2k} i^{2k+\nu}}{k! \, \Gamma(\nu + k + 1)} = e^{\nu \pi i/2} \sum_{k=0}^{\infty} \frac{(x/2)^{\nu+2k}}{k! \, \Gamma(\nu + k + 1)} .$$

Thus $e^{-\nu \pi i/2} J_\nu(ix)$ is real for real values of x, and we now define

$$I_\nu(x) = \sum_{k=0}^{\infty} \frac{(x/2)^{\nu+2k}}{k! \, \Gamma(\nu + k + 1)} , \qquad (18.6.2)$$

where the function $I_\nu(x)$ is called a *modified Bessel function*. In a similar way we define $I_{-\nu}(x)$. When ν is an integer we encounter the same difficulties as before, and a second solution is defined through

$$K_\nu(x) = \frac{\pi}{2} \frac{I_{-\nu}(x) - I_\nu(x)}{\sin \pi \nu} .$$

For integer values of $\nu = n$ the limiting value is taken instead, and we obtain

$$K_n(x) = \frac{1}{2} \sum_{k=0}^{n-1} \frac{(-1)^k (n-k-1)! \, (x/2)^{-n+2k}}{k!} + (-1)^{n+1} \sum_{k=0}^{\infty} \frac{(x/2)^{n+2k}}{k! \, (n+k)!}$$

$$\times \left[\log \frac{x}{2} - \frac{1}{2} \psi(k+1) - \frac{1}{2} \psi(n+k+1) \right]. \qquad (18.6.3)$$

In exactly the same way as for $J_\nu(x)$ we can derive an integral representation for $I_\nu(x)$:

$$I_\nu(x) = \frac{(x/2)^\nu}{\Gamma(\tfrac{1}{2}) \Gamma(\nu + \tfrac{1}{2})} \int_0^\pi \sin^{2\nu} \theta \cosh (x \cos \theta) \, d\theta . \qquad (18.6.4)$$

Also for $K_\nu(x)$ there exists a simple integral representaion:

$$K_\nu(x) = \int_0^\infty \exp(-x \cosh t) \cosh \nu t \, dt . \qquad (18.6.5)$$

In particular, if $\nu = \tfrac{1}{2}$ the integral can be evaluated directly and we get

$$K_{1/2}(x) = \sqrt{\pi/2x} \, e^{-x} .$$

The modified Bessel functions also satisfy simple summation formulas:

$$\begin{cases} I_0(x) - 2I_2(x) + 2I_4(x) - \cdots = 1 , \\ I_0(x) + 2I_1(x) + 2I_2(x) + \cdots = e^x , \\ I_0(x) - 2I_1(x) + 2I_2(x) - \cdots = e^{-x} . \end{cases} \qquad (18.6.6)$$

There are excellent tables also for the functions $I_n(x)$ and $K_n(x)$ [9].

18.7. Spherical harmonics

Many physical situations are governed by the *Laplace* equation

$$\Delta \psi = \frac{\partial^2 \psi}{\partial x^2} + \frac{\partial^2 \psi}{\partial y^2} + \frac{\partial^2 \psi}{\partial z^2} = 0$$

or by the *Poisson* equation $\Delta \psi + \kappa \psi = 0$ where, as a rule, a unique and finite solution is wanted. Very often it is suitable to go over to polar coordinates:

$$\begin{cases} x = r \sin \theta \cos \varphi , \\ y = r \sin \theta \sin \varphi , \\ z = r \cos \theta . \end{cases}$$

If now $\kappa = \kappa(r)$ is a function of the radius only, the variables can be separated if we assume that ψ can be written $\psi = f(r) Y(\theta, \varphi)$. The following equation is then obtained for Y:

$$\frac{1}{\sin \theta} \frac{\partial}{\partial \theta} \left(\sin \theta \frac{\partial Y}{\partial \theta} \right) + \frac{1}{\sin^2 \theta} \frac{\partial^2 Y}{\partial \varphi^2} + \lambda Y = 0 ,$$

where λ is the separation constant (cf. Chapter 15). One more separation $Y(\theta, \varphi) = u(\theta)v(\varphi)$ gives

$$\frac{d^2v}{d\varphi^2} + \beta v = 0 .$$

The uniqueness requirement implies $v(\varphi + 2\pi) = v(\varphi)$ and hence $\beta = m^2$, $m = 0, \pm 1, \pm 2, \ldots$ The remaining equation then becomes

$$\frac{1}{\sin \theta} \frac{d}{d\theta} \left(\sin \theta \frac{du}{d\theta} \right) - \frac{m^2 u}{\sin^2 \theta} + \lambda u = 0 .$$

We shall here only take the case $m = 0$ into account. After the transformation $\cos \theta = x$, we obtain

$$\frac{d}{dx} \left[(1 - x^2) \frac{du}{dx} \right] + \lambda u = 0 . \tag{18.7.1}$$

The index equation is $r(r - 1) = 0$ and further we get from $u = \sum a_r x^r$:

$$a_{r+2} = \frac{r(r + 1) - \lambda}{(r + 1)(r + 2)} a_r .$$

For large values of r the constant λ can be neglected and we find approximately $(r + 2)a_{r+2} \simeq ra_r$, that is, $a_r \sim C/r$. But the series

$$1 + \frac{x^2}{2} + \frac{x^4}{4} + \cdots \qquad \text{and} \qquad x + \frac{x^3}{3} + \frac{x^5}{5} + \cdots$$

are both divergent when $x = 1$ while we are looking for a finite solution. The only way out of this difficulty is that λ has such a value that the series expansion breaks off, that is, $\lambda = n(n + 1)$ where n is an integer. In this case the solutions are simply polynomials, the so-called *Legendre polynomials*. Already in connection with the Gaussion integration formulas we have touched upon some properties of the spherical harmonics (see Section 10.2). First, we shall now prove that the polynomials defined by Rodrigues' formula

$$P_n(x) = \frac{1}{2^n n!} \frac{d^n}{dx^n} (x^2 - 1)^n \tag{18.7.2}$$

are identical with the Legendre polynomials by showing that they satisfy (18.7.1) with $\lambda = n(n + 1)$. Putting $y = (x^2 - 1)^n$ we have $y' = 2nx(x^2 - 1)^{n-1}$ and hence by repeated differentiation:

$$(x^2 - 1)y' \quad = 2nxy ,$$
$$(x^2 - 1)y'' \quad = 2(n - 1)xy' \; + 2ny ,$$
$$(x^2 - 1)y''' \quad = 2(n - 2)xy'' \; + 2[n + (n - 1)]y' ,$$
$$(x^2 - 1)y^{IV} \quad = 2(n - 3)xy''' + 2[n + (n - 1) + (n - 2)]y'' ,$$
$$\vdots$$
$$(x^2 - 1)y^{(n+1)} = 2(n - n)xy^{(n)} + 2[n + (n - 1) + (n - 2) + \cdots + 2 + 1]y^{(n-1)} .$$

Thus we obtain $(x^2 - 1)y^{(n+1)} = n(n + 1)y^{(n-1)}$ and differentiating once more

$$\frac{d}{dx}\left[(1 - x^2)\frac{dP_n}{dx}\right] + n(n + 1)P_n = 0 .$$

In Section 10.2 the first polynomials were given explicitly, and further it was proved that

$$\int_{-1}^{+1} x^r P_n(x)\, dx = 0 , \qquad r = 0, 1, 2, \ldots, n - 1 , \qquad (18.7.3)$$

$$\int_{-1}^{+1} P_m(x)P_n(x)\, dx = \frac{2}{2n + 1}\delta_{mn} . \qquad (18.7.4)$$

In the case $m \neq n$, a simpler proof can be of some interest in view of the fact that the general idea for the proof is applicable in many similar situations. We start from the equations

$$(1 - x^2)P_m'' - 2xP_m' + m(m + 1)P_m = 0 ,$$
$$(1 - x^2)P_n'' - 2xP_n' + n(n + 1)P_n = 0 .$$

Multiplying the first equation by P_n and the second by $-P_m$ and adding the equations, we obtain

$$(1 - x^2)(P_n P_m'' - P_m P_n'') - 2x(P_n P_m' - P_m P_n') + (m - n)(m + n + 1)P_m P_n = 0 .$$

Integrating from -1 to $+1$ and observing that the first two terms can be written $(d/dx)\{(1 - x^2)(P_n P_m' - P_m P_n')\}$ we get the result

$$\int_{-1}^{+1} P_m(x)P_n(x)\, dx = 0 \qquad \text{provided that } m \neq n .$$

We are now going to look for another solution to Eq. (18.7.1), and we should then keep in mind that the factor $(1 - x^2)$ enters in an essential way. Hence it becomes natural to try with the following expression

$$u = \frac{1}{2}\log\frac{1 + x}{1 - x}P_n - W_n ,$$

since the derivative of

$$\frac{1}{2}\log\frac{1 + x}{1 - x}$$

is exactly $1/(1 - x^2)$. One further finds

$$u' = \frac{1}{2}\log\frac{1 + x}{1 - x}P_n' + \frac{1}{1 - x^2}P_n - W_n' ,$$

$$u'' = \frac{1}{2}\log\frac{1 + x}{1 - x}P_n'' + \frac{2}{1 - x^2}P_n' + \frac{2x}{(1 - x^2)^2}P_n - W_n'' ,$$

and after insertion:

$$(1 - x^2)W_n'' - 2xW_n' + n(n + 1)W_n = 2P_n'$$

It is easy to see that the equation is satisfied if W_n is a polynomial of degree $n - 1$, and by the aid of indeterminate coefficients the solution W_n can be computed explicitly. For $n = 3$, for example, we have

$$P_3 = \frac{1}{2}(5x^3 - 3x) ; \qquad 2P_3' = 15x^2 - 3 ;$$

and hence

$$(1 - x^2)W_3'' - 2xW_3' + 12W_3 = 15x^2 - 3 .$$

Putting $W_3 = \alpha x^2 + \beta x + \gamma$ we get $\alpha = \frac{5}{2}$, $\beta = 0$, and $\gamma = -\frac{2}{3}$. In this way we obtain $W_3 = \frac{5}{2}x^2 - \frac{2}{3}$ and the complete solution

$$Q_3(x) = \frac{1}{4}(5x^3 - 3x) \log \frac{1 + x}{1 - x} - \frac{5}{2}x^2 + \frac{2}{3} .$$

In an analogous way we find

$$Q_0(x) = \frac{1}{2} \log \frac{1 + x}{1 - x} ,$$

$$Q_1(x) = \frac{x}{2} \log \frac{1 + x}{1 - x} - 1 ,$$

$$Q_2(x) = \frac{1}{4}(3x^2 - 1) \log \frac{1 + x}{1 - x} - \frac{3}{2}x .$$

For integer values $n > 0$ we can derive a number of recursion formulas:

$$\begin{cases} (2n + 1)xP_n = (n + 1)P_{n+1} + nP_{n-1} , \\ P_{n+1}' + P_{n-1}' = 2xP_n' + P_n , \\ (2n + 1)P_n = P_{n+1}' - P_{n-1}' , \\ P_{n+1}' = (n + 1)P_n + xP_n' , \\ P_{n-1}' = -nP_n + xP_n' . \end{cases} \tag{18.7.5}$$

These formulas are conveniently proved by complex integration (cf. [2]), but with some difficulty they can also be proved in an elementary way (cf. Ex. 9 below). Similar formulas are also valid for Q_n.

Finally we also give two integral representations:

$$P_n(x) = \frac{1}{\pi} \int_0^\pi (x + \sqrt{x^2 - 1} \cos t)^n \, dt , \tag{18.7.6}$$

$$Q_n(x) = \int_0^\infty (x + \sqrt{x^2 - 1} \cosh t)^{-n-1} \, dt . \tag{18.7.7}$$

REFERENCES

[1] Rainville, *Special Functions*.

[2] Whittaker-Watson, *A Course of Modern Analysis*.

[3] *Handbook of Mathematical Functions*, AMS 55, NBS Washington.

[4] Lösch-Schoblik, *Die Fakultät* (Teubner, Leipzig, 1951).

[5] G. N. Watson, *A Treatise on the Theory of Bessel Functions.*

[6] Magnus-Oberhettinger, *Formeln und Sätze für die speziellen Funktionen der Mathematischen Physik* (Springer, 1948).

[7] F. Oberhettinger and W. Magnus, *Anwendung der elliptischen Funktionen in Physik und Technik* (Springer, 1949).

[8] *Table of Bessel Functions*, Vol. I-XII, Harvard Comp. Laboratory, Cambridge, Mass. 1947–1951.

[9] *Bessel Functions*, British Assoc. for the Advancement of Science, Cambridge Univ. Press, Part I 1950, Part II 1952.

[10] Hj. Tallqvist, "Sechsstellige Tafeln der 16 ersten Kugelfunktionen $P_n(x)$"; *Acta Soc. Sci. Fenn.*, A II, No. 4 (1937).

EXERCISES

1. Show that

$$\int_0^1 x^x \, dx = \sum_{n=1}^\infty (-1)^{n+1} n^{-n} .$$

2. Show that

$$\int_0^\infty \left(1 + \frac{x^2}{s}\right)^{-(s+1)/2} dx = \frac{\sqrt{\pi s}}{2} \, \Gamma\left(\frac{s}{2}\right) \Big/ \Gamma\left(\frac{s+1}{2}\right) , \qquad s > 0 .$$

Also find the numerical values when $s = 4$.

3. Compute $\int_0^1 x \sqrt[3]{1 - x^3} \, dx$ exactly.

4. Show that when $n > 1$,

$$\int_0^\infty \frac{dx}{1 + x^n} = \int_0^1 \frac{dx}{\sqrt[n]{1 - x^n}} ,$$

and compute the exact values.

5. Show that

$$\int_0^\pi \exp(2 \cos x) \, dx = \pi \sum_{n=0}^\infty (n!)^{-2} .$$

6. Compute the exact value of the double integral

$$\int_0^1 \int_0^1 x^{13/3}(1 - x^2)^{1/2} y^{1/6}(1 - y)^{1/2} \, dx \, dy .$$

7. Compute

$$\prod_{n=1}^\infty \frac{(10n - 1)(10n - 9)}{(10n - 5)^2} .$$

8. Find

$$\int_{-1}^{+1} x^n P_n(x) \, dx .$$

9. Using Rodrigues' formula, show that $(2n + 1)P_n(x) = P'_{n+1}(x) - P'_{n+1}(x)$. Then use the formula for computing $\int_0^1 P_n(x) \, dx$ when n is odd.

10. Using Laplace transformation find a solution in integral form to the differential equation $xy'' - y = 0$.

11. Derive the general solution of the differential equation $xy'' - y = 0$ by performing the transformations $\xi^2 = 4x$ and $y = \xi z$. Then use the result to give the exact value of the integral obtained in Ex. 10.

12. Expand $\cos(x \sin \theta)$ in a Fourier series after cosine functions and $\sin(x \sin \theta)$ after sine functions. Which formulas are obtained for $\theta = 0$ and $\theta = \pi/2$?

13. Find the area T_p bounded by the curve $x^p + y^p = 1$, $p > 0$, and the positive x- and y-axes. Then compute $\lim_{p \to \infty} p^2(1 - T_p)$.

Chapter 19

The Monte Carlo method

Im ächten Manne ist ein Kind versteckt;
das will spielen.   NIETZSCHE.

19.0. Introduction

The Monte Carlo method is an artificial sampling method which can be used for solving complicated problems in analytic formulation and for simulating purely statistical problems. The method is being used more and more in recent years, especially in those cases where the number of factors included in the problem is so large that an analytical solution is impossible. The main idea is either to construct a stochastic model which is in agreement with the actual problem analytically, or to simulate the whole problem directly. In both cases an element of randomness has to be introduced according to well-defined rules. Then a large number of trials or plays is performed, the results are observed, and finally a statistical analysis is undertaken in the usual way. The advantages of the method are, above everything, that even very difficult problems can often be treated quite easily, and desired modifications can be applied without too much trouble. Warnings have been voiced that the method might tempt one to neglect to search for analytical solutions as soon as such solutions are not quite obvious. The disadvantages are the poor precision and the large number of trials which are necessary. The latter is, of course, not too important, since the calculations are almost exclusively performed on automatic computers.

19.1. Random numbers

Random numbers play an important role in applications of the Monte Carlo method. We do not intend to go into strict definitions since this would require a rather elaborate description.

First we are reminded of a few fundamental concepts from the theory of probability. For an arbitrary value x the *distribution function $F(x)$* associated with a stochastic variable X gives the probability that $X \leq x$, that is

$$F(x) = P(X \leq x) . \qquad (19.1.1)$$

A distribution function is always nondecreasing, and further $\lim_{x \to -\infty} F(x) = 0$, $\lim_{x \to +\infty} F(x) = 1$.

We define a series of random numbers from the distribution $F(x)$ as a sequence of independent observations of X. In a long series of this kind the relative amount of numbers $\leq x$ is approximately equal to $F(x)$. If, for example, the distribution is such that $F(0) = 0.2$, then in a long series of random numbers

384

approximately 20% of the numbers should be negative or zero while the rest should be positive.

The distribution is said to be of continuous type if we can represent $F(x)$ in the form

$$F(x) = \int_{-\infty}^{x} f(t) \, dt \; . \tag{19.1.2}$$

The function $f(x) = F'(x)$ is called the *frequency function* and is always ≥ 0. A long series of random numbers from the corresponding distribution will have the property that the relative amount of random numbers in a small interval $(x, x + h)$ is approximately equal to $hf(x)$.

In the following we are only going to deal with distributions of continuous type. In this connection the rectangular distribution is of special importance; it is defined by

$$f(x) = \begin{cases} 0, & x < 0 \, , \\ 1, & 0 < x < 1 \, , \\ 0, & x > 1 \, . \end{cases}$$

The values for $x = 0$ and $x = 1$ can be put equal to 0 or 1 depending on the circumstances. The corresponding distribution function becomes

$$F(x) = \begin{cases} 0, & x < 0 \, , \\ x, & 0 \leq x \leq 1 \, , \\ 1, & x > 1 \, . \end{cases}$$

The reason that the rectangular distribution plays such an important role for the Monte Carlo method is that random numbers from other distributions can be constructed from random numbers of this simple type. If y is a random number with rectangular distribution we get a random number from the distribution $F(x)$ by solving x from the equation

$$F(x) = y \; . \tag{19.1.3}$$

This relation means that the two numbers will cut off the same amount of the area between the frequency functions and the abscissa axes. Later on we shall provide an example of this.

We shall now discuss how rectangular random numbers can be prepared, and then we shall distinguish between physical and mathematical methods. In practically all applications of the Monte Carlo method, large amounts of random numbers are needed, and an essential requirement is that they should be quickly obtainable. The following arrangement is near at hand. A suitable number of binary counters are controlled by radioactive sources or by the noise from vacuum tubes. In the latter case we have an output voltage V fluctuating around a mean value V_0. Each time $V = V_0$, the binary counter gets an impulse which shifts the counter from 0 to 1, or vice versa. When a random number is needed, all counters are locked, and every one of them gives one binary digit of the desired number. Clearly, the numbers must not be taken

out so often that two consecutive numbers are correlated. Even if the probabilities for 0 ($=a$) and for 1 ($=b$) are different, we obtain a useful technique
by disregarding the combinations aa and bb, and by letting ab and ba represent
0 and 1, respectively.

The "mathematical" methods are remarkable because we cannot prove that
the generated numbers are random. As a matter of fact, we obtain a finite series
of numbers which come back periodically at certain intervals. However, if the
period is large enough and, further, a number of statistical tests are satisfied,
then we can use these so-called pseudorandom numbers. Mainly, the following
three methods have been discussed, viz., the midsquare method, the Fibonacci
method, and the power method. The midsquare method is best illustrated by
an example. Consider the interval $0 < z < 10^4$ and form with an arbitrary z_0:

$$z_0 = 1\ 2\ 3\ 4 \qquad z_0^2 = 0\ 1\ \boxed{5\ 2\ 2\ 7}\ 5\ 6$$

$$z_1 = 5\ 2\ 2\ 7 \qquad z_1^2 = 2\ 7\ \boxed{3\ 2\ 1\ 5}\ 2\ 9$$

$$z_2 = 3\ 2\ 1\ 5 \qquad z_2^2 = 1\ 0\ \boxed{3\ 3\ 6\ 2}\ 2\ 5$$

$$z_3 = 3\ 3\ 6\ 2 \qquad z_3^2 = 1\ 1\ \boxed{3\ 0\ 3\ 0}\ 4\ 4$$

$$z_4 = 3\ 0\ 3\ 0 \qquad z_4^2 = 0\ 9\ \boxed{1\ 8\ 0\ 9}\ 0\ 0$$

$$z_5 = 1\ 8\ 0\ 9 \qquad \vdots$$

The procedure stops if the result 0000 should appear, and the probability for
this cannot be neglected. Thus the method must
be rejected.

The Fibonacci method seems to give good
results when some precautions are taken. However, the power method seems to be most widely
used. The following process gives acceptable
results:

$$\begin{cases} x_i \equiv 23x_{i-1} \bmod 2^{39} + 1\,, \\ z_i = 2^{-39}x_i\,. \end{cases}$$

Here x_i is integer, and z_i is pseudorandom. A
detailed account of different methods for generating random numbers together with a description of the properties of the methods has been
given by Birger Jansson [4].

As has been mentioned before, we can obtain
random numbers from any distribution if we
have rectangular random numbers at our disposal. For this purpose we make use of equation

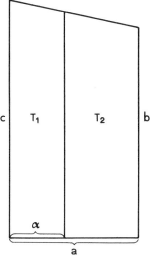

Figure 19.1

(19.1.3), and we shall show how the method works for the normal distribution. The frequency function is $f(x) = (2\pi)^{-1/2} e^{-x^2/2}$ and hence we have to solve the equation

$$(2\pi)^{-1/2} \int_{-\infty}^{x} e^{-t^2/2}\, dt = \xi,$$

where ξ is a random number from the rectangular distribution. By the aid of a table we seek out $N - 1$ values $x_1, x_2, \ldots, x_{N-1}$ such that

$$(2\pi)^{-1/2} \int_{x_r}^{x_{r+1}} e^{-t^2/2}\, dt = \frac{1}{N} \qquad \text{for} \quad r = 0, 1, \ldots, N - 1,$$

where $x_0 = -\infty$ and $x_N = +\infty$. Thus the area between the curve $y = (2\pi)^{-1/2} e^{-x^2/2}$ and $y = 0$ has been divided into N equal parts by use of the vertical lines $x = x_r$. For a given rectangular ξ we determine a value r such that $r/N \le \xi < (r + 1)/N$, that is, $r = \text{entier}\,(N\xi)$. Hence the desired normal random number falls between x_r and x_{r+1}. Putting $z = N\xi - r$, we have to find an abscissa which, together with x_r, delimits the fraction z of the segment number $r + 1$.

For sufficiently large N the segments can be approximated by a trapezoid, except for the first and the last segments, where additional prescriptions must be given. Here we neglect this complication and find with $a = x_{r+1} - x_r$; $b = f(x_{r+1})$; $c = f(x_r)$; $T_1/(T_1 + T_2) = z$, and using elementary geometrical calculations, that

$$\alpha = \frac{ac}{c - b}\left[1 \pm \sqrt{1 - z\left(1 - \frac{b^2}{c^2}\right)}\right].$$

Since α must be 0 when $z = 0$, we have to choose the minus sign. Hence we obtain the approximate formula

$$x = x_r + \alpha.$$

Against this method the following objection, first made by von Neumann, can be raised: that we choose a number completely at random and then, with the utmost care, perform a series of complex operations on it. However, it is possible to reach some kinds of distributions in a different way. For example, the sum of a sufficiently large number of rectangularly distributed numbers has normal distribution. The frequency function for the sum of three rectangular numbers is composed of three parabolic arcs:

$$
\begin{aligned}
y &= x^2/2 && \text{for} \quad 0 \le x \le 1,\\
y &= -x^2 + 3x - 3/2 && \text{for} \quad 1 \le x \le 2,\\
y &= (3 - x)^2/2 && \text{for} \quad 2 \le x \le 3.
\end{aligned}
$$

Here we are already surprisingly close to the normal Gaussian curve.

A simple and elegant method is the following. Take two rectangular random numbers x and y and construct

$$\begin{cases} u = (-2 \log x)^{1/2} \cos 2\pi y , \\ v = (-2 \log x)^{1/2} \sin 2\pi y . \end{cases}$$

Then u and v are normal random numbers with mean 0 and standard deviation 1.

Next we assume that we want random numbers θ in the interval $(-1, 1)$ with the frequency function $f(\theta) = (1/\pi)(1 - \theta^2)^{-1/2}$. Then the distribution function is $F(\theta) = (1/\pi)(\arc \sin \theta + \pi/2)$, where $\arc \sin \theta$ is determined in such a way that $-\pi/2 \leq \arc \sin \theta \leq \pi/2$. Hence the number θ can be determined from a random number ξ with uniform distribution in $(0, 1)$ through the equation $(1/\pi)(\arc \sin \theta + \pi/2) = \xi$, that is, $\theta = -\cos \pi\xi = \cos \pi(1 - \xi)$. The same thing can be attained if we choose *two* rectangular numbers x and y. Then we form $X = 2x - 1$ and $Y = 2y - 1$ and test whether $X^2 + Y^2 < 1$; otherwise the numbers are rejected. Thus the point (X, Y) will lie inside a circle with radius 1. Then $z = X/\sqrt{X^2 + Y^2}$ is a number from the desired distribution. Alternatively, we see that $(X^2 - Y^2)/(X^2 + Y^2)$ has the same distribution; the latter formulation should be preferred, since no square roots are present.

The *discrimination technique* is also very useful. Let $z = \min(x, y)$, where x and y are rectangular in $(0, 1)$. We easily find that z has the frequency function $2 - 2z$ for $0 \leq z \leq 1$ and 0 otherwise. Other examples of the same technique can be found in such references as [1], the contribution by von Neumann.

In the discussion as to whether physical or mathematical methods should be preferred, quite a few arguments may be raised. The physical method needs special equipment and yields nonreproducible results at rather low speed. The fact that the results cannot be reproduced is an advantage in some situations, but it has the drawback that the computation cannot be checked.

> *"Would you tell me, please, which way I ought to walk from here?"*
> *"That depends a good deal on where you want to get to,"* said the Cat.
> *"I don't much care where—"* said Alice.
> *"Then it doesn't matter which way you walk,"* said the Cat.   LEWIS CARROLL.

19.2. Random walks

A random walk can be defined as a game with the following rules. A domain is divided into a square pattern. One player starts with a particle in the origin, draws a random number which decides in what direction the particle should be moved, and moves the particle to the nearest point in this direction. After that a new random number is taken.

In the following example we shall assume that the probabilities for a movement to the four closest points are all equal to $\frac{1}{4}$. We denote by $P(x, y, t)$ the probability that the particle after t moves is at the point (x, y). Then we have

$$P(x, y, t + 1) = \tfrac{1}{4}[P(x - 1, y, t) + P(x, y - 1, t)$$
$$+ P(x + 1, y, t) + P(x, y + 1, t)] . \qquad (19.2.1)$$

This equation can be rewritten

$$P(x, y, t + 1) - P(x, y, t) = \tfrac{1}{4}[P(x + 1, y, t) - 2P(x, y, t) + P(x - 1, y, t)$$
$$+ P(x, y + 1, t) - 2P(x, y, t) + P(x, y - 1, t)] ,$$

and we find an obvious similarity with equation (15.2.3), or rather with its two-dimensional equivalent. Our difference equation evidently approximates the two-dimensional heat equation

$$\frac{\partial P}{\partial t} = C\left(\frac{\partial^2 P}{\partial x^2} + \frac{\partial^2 P}{\partial y^2}\right).$$

Now we suppose that a random-walk process is performed in a limited domain and in such a way that the particle is absorbed when it reaches the boundary. At the same time, a certain profit V is paid out; its amount depends on the boundary point at which the particle is absorbed. Boundary points will be denoted by (x_r, y_r) and interior points by (x, y). At each interior point we have a certain probability $P(x, y, x_r, y_r)$ that the particle which starts at the point (x, y) will be absorbed at the boundary point (x_r, y_r). The expected profit is obviously

$$u(x, y) = \sum_r P(x, y, x_r, y_r)V(x_r y_r) . \qquad (19.2.2)$$

We find immediately that $u(x, y)$ satisfies the difference equation

$$u(x, y) = \tfrac{1}{4}[u(x + 1, y)$$
$$+ u(x - 1, y) + u(x, y + 1) + u(x, y - 1)] , \qquad (19.2.3)$$

with $u(x_r, y_r) = V(x_r, y_r)$. This equation is a well-known approximation of the Laplace equation,

$$\frac{\partial^2 u}{\partial x^2} + \frac{\partial^2 u}{\partial y^2} = 0 .$$

In a simple example we will show how this equation can be solved by the random-walk technique. As our domain we choose a square with the corners $(0, 0)$, $(4, 0)$, $(4, 4)$, and $(0, 4)$, including integer grid points. As boundary values we have $u = 0$ for $x = 0$; $u = x/(x^2 + 16)$ for $y = 4$; $u = 4/(16 + y^2)$ for $x = 4$; and $u = 1/x$ for $y = 0$. We shall restrict ourselves to considering

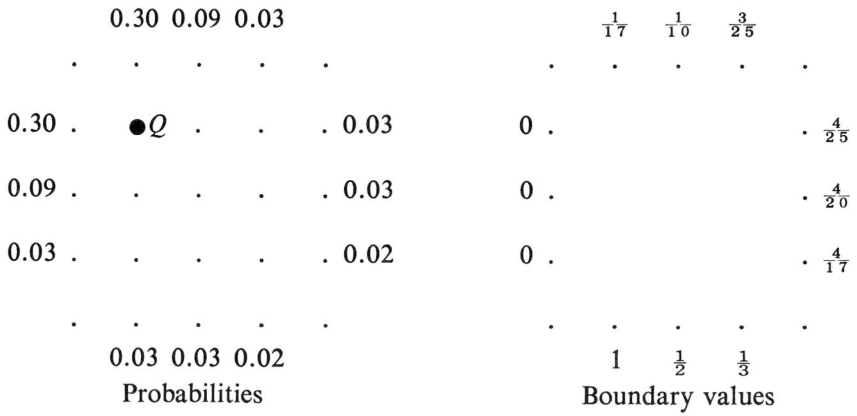

0.30 0.09 0.03 $\frac{1}{17}$ $\frac{1}{10}$ $\frac{3}{25}$

.   .   .   .

0.30 .   ●Q .   .   . 0.03   0 .   . $\frac{4}{25}$

0.09 .   .   .   . 0.03   0 .   . $\frac{4}{20}$

0.03 .   .   .   . 0.02   0 .   . $\frac{4}{17}$

.   .   .   .

0.03 0.03 0.02   1 $\frac{1}{2}$ $\frac{1}{3}$
Probabilities   Boundary values

Figure 19.2

the interior point $Q(1, 3)$. From this point we can reach the point $(1, 4)$ in one step with probability $\frac{1}{4}$. But we can also reach it in three steps in two different ways with total probability $2/4^3 = 1/32$, and in five steps in ten different ways with total probability $10/4^5 = 5/512$, and so on. Adding all these probabilities, we obtain about 0.30. In a similar way the other probabilities have been computed (see Fig. 19.2).

Insertion in (19.2.2) gives $u(Q) = 0.098$, in good agreement with the value 0.1 obtained from the exact solution $u = x/(x^2 + y^2)$.

Clearly, the method can easily be programmed for a computer, and then the probabilities are obtained by the random-walk technique. It is also evident that the method can hardly compete with direct methods if such can be applied.

19.3. Computation of definite integrals

Computation of multidimensional integrals is an extremely complicated problem, and an acceptable conventional technique still does not exist. Here the Monte Carlo method is well suited, even if the results are far from precise. For simplicity we demonstrate the technique on one-dimensional integrals and consider

$$I = \int_0^1 f(x)\, dx ; \qquad 0 \leq f(x) \leq 1 . \tag{19.3.1}$$

We choose N number pairs (x_i, y_i) with rectangular distribution and define z_i through

$$z_i = \begin{cases} 0 & \text{if } y_i > f(x_i), \\ 1 & \text{if } y_i < f(x_i), \end{cases}$$

(the case $y_i = f(x_i)$ can be neglected). Putting $n = \sum_i z_i$, we have $n/N \simeq I$.

Somewhat more precisely, we can write

$$I = \frac{n}{N} + O(N^{-1/2}) . \tag{19.3.2}$$

Obviously, the accuracy is poor; with 100 pairs we get a precision of the order of $\pm 5\%$, and the traditional formulas, for example, Simpson's formula, are much better. In many dimensions we still have errors of this order of magnitude, and since systematic formulas (so far as such formulas exist in higher dimensions) are extremely difficult to manage, it might well be the case that the Monte Carlo method compares favorably, at least if the number of dimensions is ≥ 6.

It is even more natural to consider the integral as the mean value of $f(\xi)$, where ξ is rectangular, and then estimate the mean value from

$$I \simeq \frac{1}{N} \sum_{i=1}^{N} f(\xi_i) . \tag{19.3.3}$$

This formula can easily be generalized to higher dimensions.

Equation (19.3.1) can be rewritten in the following way:

$$I = \int_0^1 \frac{f(x)}{g(x)} g(x) \, dx , \tag{19.3.4}$$

and hence I can be interpreted as the mean value of $f(\xi)/g(\xi)$, where ξ is a random number with the frequency $g(\xi)$. Thus

$$I \simeq \frac{1}{N} \sum_{i=1}^{N} \frac{f(\xi_i)}{g(\xi_i)} . \tag{19.3.5}$$

where ξ_i is a random number with the frequency $g(\xi)$. Conveniently, the function $g(\xi)$ is chosen in such a way that it does not deviate too much from the function to be integrated.

19.4. Simulation

The main idea in simulation is to construct a stochastic model of the real events, and then, by aid of random numbers or random walk, play a large number of games. The results are then analyzed statistically, as usual. This technique can be used for widely different purposes, for example, in social sciences, in physics, in operational analysis problems such as combat problems, queuing problems, industrial problems, and so on. An example which is often quoted is the case when neutrons pass through a metallic plate. The energy and direction distributions are known, and these quantities are chosen accordingly. Further, the probabilities for reflection and absorption are known, and hence we can trace the history of each neutron. When a sufficient number of events has been obtained, the whole process can be analyzed in detail.

Many phenomena which one wants to study in this way occur with very low frequency. The probability that the neutron passes through the plate might not be more than 10^{-6}, and naturally we need a very large number of trials to get enough material. In situations like this some kind of "importance sampling" is often used. This means that intentionally one directs the process toward the interesting but rare cases and then compensates for this by giving each trial correspondingly less weight. Exactly the same technique was used in a preceding discussion where random numbers were chosen with a frequency function agreeing as far as possible with the function to be integrated. Under certain circumstances when it is difficult to state exactly whether a case is interesting, one can with probability $\frac{1}{2}$ stop a trial which seems to be unsuccessful; if it does continue, its weight is doubled. In the same way trials which seem to be promising are split into two whose weights are halved at the same time.

The tricks which have been sketched here are of great practical importance, and in many cases they can imply great savings of computing time and even more make the use of the Monte Carlo method possible.

Apart from these indications, we cannot go into detail about the different applications. Instead we will content ourselves with a simple though important feature. Suppose that we have a choice among N different events with the probabilities $P_1, P_2, \ldots, P_N$, where $\sum_{r=1}^{N} P_r = 1$. We form so-called accumulated probabilities Q_r according to

$$Q_r = \sum_{i=1}^{r} P_i . \tag{19.4.1}$$

Choosing a rectangular random number ξ, we seek a value of r such that

$$Q_{r-1} < \xi < Q_r . \tag{19.4.2}$$

Then the event number r should be chosen. We can interpret the quantities P_i as distances with total length 1, and if the pieces are placed one after another, the point ξ will fall on the corresponding piece with probability P_i.

REFERENCES

[1] National Bureau of Standards, "Monte Carlo Method" *Appl. Math. Series*, No. 12 (Washington, 1951).

[2] Beckenbach: *Modern Mathematics for the Engineer* (McGraw-Hill, New York, 1956).

[3] Kendall-Smith: *Table of Random Numbers* (Cambridge, 1946).

[4] B. Janson: *Random Number Generators* (Stockholm, 1966).

[5] K. D. Tocher: *The Art of Simulation* (London, 1963).

EXERCISES

1. A distribution with the frequency function $f(x) = 0$ when $x < 0$ and $f(x) = \lambda e^{-\lambda x}$ when $x \geq 0$ ($\lambda > 0$) is called an exponential distribution. Show how a random number out of this distribution can be obtained from a random number of the rectangular distribution.

2. Construct a normal random number with mean value m and standard deviation σ from a normal random number ξ with mean value 0 and standard deviation 1 [frequency functions $\sigma^{-1}(2\pi)^{-1/2} \exp\{-(x-m)^2/2\sigma^2\}$ and $(2\pi)^{-1/2} \exp(-x^2/2)$].

3. The following method has been suggested for obtaining pseudorandom numbers. Starting from such a value x_1 that $0 < x_1 < 1$ and $x_1 \neq k/4$, $k = 0(1)4$, we form a sequence $x_1, x_2, x_3, \ldots$, where $x_{n+1} = 4x_n(1-x_n)$. Using this method, compute

$$\int_0^1 \frac{dx}{1+x} \simeq \frac{1}{10} \sum_{i=1}^{10} \frac{1}{1+x_i}$$

with $x_1 = 0.120$.

4. Using a table of random numbers with rectangular distribution (for example, Reference 3), take three such numbers and denote them with x_i, y_i, and z_i, where $x_i > y_i > z_i$. Form 20 such triples, compute $X = \frac{1}{20} \sum x_i$; $Y = \frac{1}{20} \sum y_i$; $Z = \frac{1}{20} \sum z_i$, and compare with the theoretical values, which should also be computed.

5. Two random numbers ξ_1 and ξ_2, with rectangular distribution, are chosen. Then x_1, x_2, and x_3 are formed by

$$x_1 = 1 - \sqrt{\xi_1}; \qquad x_2 = \sqrt{\xi_1} \cdot (1 - \xi_2); \qquad x_3 = \sqrt{\xi_1} \cdot \xi_2.$$

Finally, these numbers are ordered in decreasing series and renamed: $z_1 > z_2 > z_3$.
   (a) Find the mean values of x_1, x_2, and x_3.
   (b) Find the mean values of z_1, z_2, and z_3.
[*Hint:* Each triple can be represented as a point inside the triangle $x > 0$, $y > 0$, $z > 0$, in the plane $x + y + z = 1$; note that $x_1 + x_2 + x_3 = z_1 + z_2 + z_3 = 1$.]
   (c) Using a table of random numbers, construct 20 triples, form x_i and z_i, and compare the mean values with the theoretical results.

6. Two salesmen, A and B, are trying to sell their products to a group of 20 people. A is working three times as hard as B, but unfortunately his product, the dietetic drink "Moonlight Serenade," is inferior and nobody is willing to become a customer until he has received three free samples. "Lady Malvert," the brand offered by B, is much better, and any person offered this brand will accept it at once. All 20 people are very conservative, and when one of them has accepted a brand, he is not going to change. In 50 trials simulate the process described here. In each trial, two random numbers should be used, one for determining whether A or B is going to work, and one for choosing the person he will try to get as a customer.

Chapter 20

Linear programming

Le mieux est l'ennemi du bien. VOLTAIRE.

20.0. Introduction

Linear programming is a subject characterized by one main problem: to seek the maximum or the minimum of a linear expression when the variables of the problem are subject to restrictions in the form of certain linear equalities or inequalities. Problems of this kind are encountered when we have to exploit limited resources in an optimal way. Production and transport problems, which play an important role in industry, are of special significance in this respect.

It is rather surprising that a problem category like this one came into the limelight only during the last decades. However, the background is that during World War II, mathematical methods were used for planning portions of military activities in an optimal manner, especially in England and the United States. The methods developed during this period were then taken over by civilian industry, and in this way the theory of linear progamming developed.

20.1. The simplex method

We shall start by discussing a simple example. Find the maximum value of $y = 7x_1 + 5x_2$ under the conditions

$$\begin{cases} x_1 + 2x_2 \leq 6, & x_1 \geq 0, \\ 4x_1 + 3x_2 \leq 12, & x_2 \geq 0. \end{cases}$$

The point (x_1, x_2) must lie inside or on the boundary of the domain marked with lines in the figure. For different values of y the equation

$$7x_1 + 5x_2 = y$$

represents parallel straight lines, and in order to get useful solutions we must have the lines pass through the domain or the boundary. In this way it is easy to see that we only have to look at the values in the corners $(0, 0)$, $(0, 3)$, $(1.2, 2.4)$, and $(3, 0)$. We get $y = 0$, 15, 20.4, and 21 respectively, and hence $y_{max} = 21$ for $x_1 = 3$ and $x_2 = 0$.

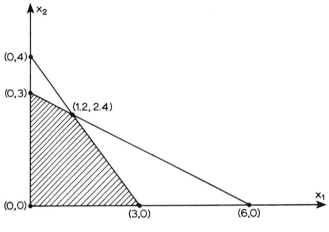

Figure 20.1

We shall now formulate the main problem. We assume that we have n variables $x_1, x_2, \ldots, x_n$ subject to the conditions

$$\begin{cases} a_{11}x_1 + a_{12}x_2 + \cdots + a_{1n}x_n = b_1, \\ a_{21}x_1 + a_{22}x_2 + \cdots + a_{2n}x_n = b_2, \qquad m < n, \\ \vdots \\ a_{m1}x_1 + a_{m2}x_2 + \cdots + a_{mn}x_n = b_m, \end{cases}$$

and further that $x_i \geq 0$, $i = 1, 2, \ldots, n$. Find the minimum value of

$$v = c_1x_1 + c_2x_2 + \cdots + c_nx_n.$$

Using matrix notations, we get the alternative formulation: Minimize $y = c^T x$ under the conditions $x \geq 0$; $Ax = b$. Often the conditions $Ax = b$ are given as inequalities, but by adding so-called slack variables, the inequalities are transformed to equalities. Hence $a_{i1}x_1 + a_{i2}x_2 + \cdots + a_{in}x_n \leq b_i$ is replaced by

$$a_{i1}x_1 + a_{i2}x_2 + \cdots + a_{in}x_n + x_{n+i} = b_i, \qquad i = 1, 2, \ldots, m,$$

and then we can also replace $m + n$ by n.

By choosing the signs of the coefficients in a suitable way, we can always formulate the problem as has just been described with all $b_i > 0$. Clearly, the domain defined by the secondary conditions is a convex hyperpolyhedron, which can be either limited or unlimited. In some cases the conditions may be such that this domain vanishes; then the problem has no solution. For different values of y, the equation $y = c_1x_1 + c_2x_2 + \cdots + c_nx_n$ defines a family of hyperplanes; for suitable values of y they have points in common with the hyper-polyhedron. When y decreases, the intersection changes, and it is easily inferred that in general there exists a position where the intersection has contracted to a point (or possibly a straight line), and this represents the solution. As a rule, the solution is a corner of the domain, and it is now possible to construct a method which implies that we start at one corner and then proceed succes-sively to other corners, simultaneously watching that y must decrease all the time. This method, which has been invented by Dantzig, is known as the simplex method (a simplex is a hyperpolyhedron with $n + 1$ corners in n dimensions, for example, a point, a limited straight line, a triangle, a tetra-hedron, etc.).

Before discussing the solution technique we shall define a few important concepts. Each vector x satisfying the equation $Ax = b$ is called a *solution*. If no component is negative, the vector x is said to be a *feasible solution*. In general, none of these is the *optimal* solution which we are looking for. If m columns of the matrix A can be chosen in such a way that the determinant is not equal to 0, these vectors are said to form a *basis* and the corresponding variables are called *basic variables*. If all other variables are put equal to 0, the system can be solved, but naturally we cannot guarantee that the solution is feasible. If some of the basic variables vanish, the solution is said to be *degenerate*.

In the sequel we suppose that all slack variables needed have already been introduced so that we obtain a linear system of m equations in n unknowns $(m < n)$. As already mentioned, the system represents m hyperplanes and normally the optimal solution is a corner in the corresponding hyperpoly-hedron. In the typical case we have m equations with m slack variables and hence only $n - m$ proper variables. Thus the hyperplanes belong to an $(n - m)$-dimensional space. Normally, a corner is obtained as the intersection of $n - m$ hyperplanes, and for each hyperplane one slack variable will vanish, that is, we have in total $n - m$ variables equal to zero. Since the coordinate planes $x_i = 0$ may contribute in forming corners, proper variables may also vanish in a corner, but the total number of variables not equal to 0 will be the same. In degenerate cases *more* than $n - m$ hyperplanes may pass through the same corner, but the conclusion is still unchanged. Hence, in an optimal solution at most m variables are positive. The reasoning can easily be generalized to cover also the case when a number of equalities needing no slack variables are present already from the beginning.

For a description of the method, we introduce the column vectors

$$P_1 = \begin{pmatrix} a_{11} \\ a_{21} \\ \vdots \\ a_{m1} \end{pmatrix}; \quad P_2 = \begin{pmatrix} a_{12} \\ a_{22} \\ \vdots \\ a_{m2} \end{pmatrix}; \quad \ldots; \quad P_n = \begin{pmatrix} a_{1n} \\ a_{2n} \\ \vdots \\ a_{mn} \end{pmatrix}; \quad P_0 = \begin{pmatrix} b_1 \\ b_2 \\ \vdots \\ b_m \end{pmatrix}, \quad P_0 > 0.$$

Then we have to minimize $c^T x$ under the secondary conditions $x \geq 0$;

$$x_1 P_1 + x_2 P_2 + \cdots + x_n P_n = P_0 .$$

We assume that we know a feasible solution x with components $x_1, x_2, \ldots, x_m$ $(x_{m+1} = x_{m+2} = \cdots = x_n = 0)$; $x_i \geq 0$, $i = 1, 2, \ldots, m$. Then we have

$$x_1 P_1 + x_2 P_2 + \cdots + x_m P_m = P_0 , \tag{20.1.1}$$

and the corresponding y-value is

$$c_1 x_1 + c_2 x_2 + \cdots + c_m x_m = y_0 . \tag{20.1.2}$$

The vectors $P_1, P_2, \ldots, P_m$ are supposed to be linearly independent (in practice, they are often chosen in such a way that they belong to the slack variables, and consequently they become unit vectors), and then $P_{m+1}, P_{m+2}, \ldots, P_n$ can be expressed as linear combinations of the base vectors:

$$P_j = x_{1j} P_1 + x_{2j} P_2 + \cdots + x_{mj} P_m , \qquad j = 1, 2, \ldots, n . \tag{20.1.3}$$

Further, we define y_j by

$$y_j = c_1 x_{1j} + c_2 x_{2j} + \cdots + c_m x_{mj} . \tag{20.1.4}$$

If for some value j the condition $y_j - c_j > 0$ is satisfied, then we can find a better solution. Multiplying (20.1.3) by a number p and subtracting from (20.1.1), we obtain

$$(x_1 - p x_{1j}) P_1 + (x_2 - p x_{2j}) P_2 + \cdots$$
$$+ (x_m - p x_{mj}) P_m + p P_j = P_0 . \tag{20.1.5}$$

Analogously, from (20.1.2) and (20.1.4),

$$(x_1 - p x_{1j}) c_1 + (x_2 - p x_{2j}) c_2 + \cdots$$
$$+ (x_m - p x_{mj}) c_m + p c_j = y_0 - p(y_j - c_j) . \tag{20.1.6}$$

If the coefficients for $P_1, P_2, \ldots, P_m$, P_j are all ≥ 0, then we have a new feasible solution with the corresponding y-value $y = y_0 - p(y_j - c_j) < y_0$, since $p > 0$ and $y_j - c_j > 0$. If for a fixed value of j at least one $x_{ij} > 0$, then the largest p-value we can choose is

$$p = \min_i \frac{x_i}{x_{ij}} > 0 ,$$

and if the problem has not degenerated, this condition determines i and p. The coefficient of P_i now becomes zero, and further we have again P_0 represented as a linear combination of m base vectors. Finally, we have also reached a lower value of y. The same process is repeated and continued until *either* all $y_j - c_j < 0$, *or* for some $y_j - c_j > 0$, we have all $x_{ij} \leq 0$. In the latter case we can choose p as large as we want, and the minimum value is $-\infty$.

Before we can proceed with the next step, all vectors must be expressed by aid of the new base vectors. Suppose that the base vector P_i has to be replaced by the base vector P_k. Our original basis was $P_1, P_2, \ldots, P_i, \ldots, P_m$, and our new basis is to be $P_1, P_2, \ldots, P_{i-1}, P_{i+1}, \ldots, P_m, P_k$. Now we have

$$\begin{cases} P_0 = x_1 P_1 + x_2 P_2 + \cdots + x_i P_i + \cdots + x_m P_m, \\ P_k = x_{1k} P_1 + x_{2k} P_2 + \cdots + x_{ik} P_i + \cdots + x_{mk} P_m, \\ P_j = x_{1j} P_1 + x_{2j} P_2 + \cdots + x_{ij} P_i + \cdots + x_{mj} P_m. \end{cases}$$

From the middle relation we solve P_i, which is inserted into the two others:

$$\begin{cases} P_0 = x_1' P_1 + \cdots + x_{i-1}' P_{i-1} + x_k' P_k + x_{i+1}' P_{i+1} + \cdots + x_m' P_m, \\ P_j = x_{1j}' P_1 + \cdots + x_{i-1,j}' P_{i-1} + x_{kj}' P_k + x_{i+1,j}' P_{i+1} + \cdots + x_{mj}' P_m, \end{cases}$$

where $x_r' = x_r - (x_i/x_{ik})x_{rk}$ for $r = 1, 2, \ldots, i-1, i+1, \ldots, m$, and $x_k' = x_i/x_{ik}$. Analogously, $x_{rj}' = x_{rj} - (x_{ij}/x_{ik})x_{rk}$ for $r \neq i$ and $x_{kj}' = x_{ij}/x_{ik}$.
Now we have

$$\begin{aligned} y_j' - c_j &= x_{1j}' c_1 + \cdots + x_{kj}' c_k + \cdots + x_{mj}' c_m - c_j \\ &= \sum_r x_{rj}' c_r - c_j = \sum_{r \neq i} \left(x_{rj} - \frac{x_{ij}}{x_{ik}} x_{rk} \right) c_r + \frac{x_{ij}}{x_{ik}} c_k - c_j \\ &= \sum_r x_{rj} c_r - x_{ij} c_i - \frac{x_{ij}}{x_{ik}} \left(\sum_r x_{rk} c_r - x_{ik} c_i - c_k \right) - c_j \\ &= y_j - c_j - \frac{x_{ij}}{x_{ik}} (y_k - c_k), \end{aligned}$$

and further

$$\begin{aligned} y_0' &= c_1 x_1' + \cdots + c_k x_k' + \cdots + c_m x_m' \\ &= \sum_{r \neq i} \left(x_r - \frac{x_i}{x_{ik}} x_{rk} \right) c_r + \frac{x_i}{x_{ik}} c_k \\ &= \sum_r c_r x_r - c_i x_i - \frac{x_i}{x_{ik}} \left(\sum_r x_{rk} c_r - x_{ik} c_i - c_k \right) \\ &= y_0 - \frac{x_i}{x_{ik}} (y_k - c_k). \end{aligned}$$

EXAMPLE

Seek the maximum of $y = -5x_1 + 8x_2 + 3x_3$ under the conditions

$$\begin{cases} 2x_1 + 5x_2 - x_3 \leq 1, \\ -3x_1 - 8x_2 + 2x_3 \leq 4, \\ -2x_1 - 12x_2 + 3x_3 \leq 9, \\ x_1 \geq 0, \quad x_2 \geq 0, \quad x_3 \geq 0. \end{cases}$$

This problem is fairly difficult to solve by use of the direct technique, and for this reason we turn to the simplex method. We rewrite the problem in standard form as follows.

Find the minimum value of

$$y = 5x_1 - 8x_2 - 3x_3$$

when

$$\begin{cases} 2x_1 + 5x_2 - x_3 + x_4 & = 1, \\ -3x_1 - 8x_2 + 2x_3 \quad + x_5 & = 4, \qquad x_i \geq 0. \\ -2x_1 - 12x_2 + 3x_3 \quad\quad + x_6 = 9, \end{cases}$$

We use the following scheme:

			5	-8	-3	0	0	0
Basis	c	P_0	P_1	P_2	P_3	P_4	P_5	P_6
P_4	0	1	2	5	-1	1	0	0
P_5	0	4	-3	-8	2	0	1	0
P_6	0	9	-2	-12	3	0	0	1
		0	-5	**8**	3	0	0	0

Over the vectors $P_1, \ldots, P_6$ one places the coefficients c_j; the last line contains y_0 and $y_j - c_j$; thus

$$0 \cdot 1 + 0 \cdot 4 + 0 \cdot 9 = 0;$$
$$0 \cdot 2 + 0 \cdot (-3) + 0 \cdot (-2) - 5 = -5,$$

and so on. Now we choose a positive number in the last line (8), and in the corresponding column a positive number (5) (both numbers are printed in bold-face type). In the basis we shall now exchange P_4 and P_2, that is, we have $i = 4, k = 2$. The first line (P_4) holds an exceptional position, and in the reduction all elements (except c) are divided by 5; the new value of c becomes -8, belonging to the vector P_2.

			5	-8	-3	0	0	0
Basis	c	P_0	P_1	P_2	P_3	P_4	P_5	P_6
P_2	-8	$\frac{1}{5}$	$\frac{2}{5}$	1	$-\frac{1}{5}$	$\frac{1}{5}$	0	0
P_5	0	$\frac{28}{5}$	$\frac{1}{5}$	0	2/5	$\frac{8}{5}$	1	0
P_6	0	$\frac{57}{5}$	$\frac{14}{5}$	0	$\frac{3}{5}$	$\frac{12}{5}$	0	1
		$-\frac{8}{5}$	$-\frac{41}{5}$	0	23/5	$-\frac{8}{5}$	0	0

The other elements are obtained in the following way:

$$9 - \tfrac{1}{5}(-12) = \tfrac{57}{5} \; ;$$

the numbers 9, 1, 5, and -12 stand in the corners of a rectangle. Analogously, $-2 - \tfrac{2}{5}(-12) = \tfrac{14}{5}$, and so on. Note also that the elements in the last line can be obtained in this way, which makes checking simple. In the next step we must choose $k = 3$, since $\tfrac{23}{5} > 0$, but we have two possibilities for the index i, namely, $i = 5$ and $i = 6$, corresponding to the elements $\tfrac{2}{5}$ and $\tfrac{3}{5}$, respectively. But $\tfrac{28}{5} \div \tfrac{2}{5} = 14$ is less than $\tfrac{57}{5} \div \tfrac{3}{5} = 19$, and hence we must choose $i = 5$. This reduction gives the result

Basis	c	P_0	5 P_1	-8 P_2	-3 P_3	0 P_4	0 P_5	0 P_6
P_2	-8	3	$\tfrac{1}{2}$	1	0	1	$\tfrac{1}{2}$	0
P_3	-3	14	$\tfrac{1}{2}$	0	1	4	$\tfrac{5}{2}$	0
P_6	0	3	$\tfrac{5}{2}$	0	0	0	$-\tfrac{3}{2}$	1
		-66	$-\tfrac{21}{2}$	0	0	-20	$-\tfrac{23}{2}$	0

Since all numbers in the last line are negative or zero, the problem is solved and we have $y_{\min} = -66$ for

$$\begin{cases} x_2 = 3 \,, \\ x_3 = 14 \,, \\ x_6 = 3 \,, \end{cases}$$

which gives

$$\begin{cases} x_1 = 0 \,, \\ x_4 = 0 \,, \\ x_5 = 0 \,. \end{cases}$$

We have now exclusively treated the case when all unit vectors entered the secondary conditions without special arrangements. If this is not the case, we can master the problem by introducing an artificial basis. Suppose that we have to minimize $c_1 x_1 + \cdots + c_n x_n$ under the conditions

$$\begin{cases} a_{11}x_1 + \cdots + a_{1n}x_n = b_1 \,, \\ \quad \vdots \\ a_{m1}x_1 + \cdots + a_{mn}x_n = b_m \,, \end{cases}$$

and $x_i \geq 0 \ (i = 1, 2, \ldots, n)$.

Instead we consider the problem of minimizing

$$c_1 x_1 + \cdots + c_n x_n + w x_{n+1} + w x_{n+2} + \cdots + w x_{n+m}$$

under the conditions

$$\begin{cases} a_{11}x_1 + \cdots + a_{1n}x_n + x_{n+1} & = b_1 , \\ a_{21}x_1 + \cdots + a_{2n}x_n \qquad\quad + x_{n+2} & = b_2 , \\ \vdots \\ a_{m1}x_1 + \cdots + a_{mn}x_n \qquad\qquad\quad + x_{n+m} = b_m , \end{cases}$$

and $x_i \geq 0$ $(i = 1, 2, \ldots, n + m)$.

Then we let w be a large positive number, which need not be specified. In this way the variables $x_{n+1}, \ldots, x_{n+m}$ are, in fact, eliminated from the secondary conditions, and we are back to our old problem. The expressions $y_j - c_j$ now become linear functions of w; the constant terms are written as before in a special line, while the coefficients of w are added in an extra line below. Of these, the largest positive coefficient determines a new base vector, and the old one, which should be replaced, is chosen along the same lines as before. An eliminated base vector can be disregarded in the following computations.

EXAMPLE

Minimize

$$y = x_1 + x_2 + 2x_3$$

under the conditions

$$\begin{cases} x_1 + x_2 + x_3 \leq 9 , \\ 2x_1 - 3x_2 + 3x_3 = 1 , \qquad x_i \geq 0 . \\ -3x_1 + 6x_2 - 4x_3 = 3 , \end{cases}$$

First we solve the problem by conventional methods. In the inequality we add a slack variable x_4 and obtain

$$x_1 + x_2 + x_3 + x_4 = 9 .$$

Regarding x_4 as a known quantity, we get

$$\begin{cases} x_1 = \tfrac{1}{2}(13 - 3x_4) , \\ x_2 = \tfrac{1}{4}(13 - x_4) , \\ x_3 = \tfrac{3}{4}(-1 + x_4) . \end{cases}$$

Hence

$$y = x_1 + x_2 + 2x_3 = \tfrac{1}{4}(33 - x_4) .$$

The conditions $x_1 \geq 0$, $x_2 \geq 0$, $x_3 \geq 0$ together give the limits $1 \leq x_4 \leq \tfrac{13}{3}$. Since y is going to be minimized, x_4 should be chosen as large as possible, that is $x_4 = \tfrac{13}{3}$. Thus we get

$$x_1 = 0 ; \qquad x_2 = \tfrac{13}{6} ; \qquad x_3 = \tfrac{5}{2} , \qquad \text{and} \qquad y_{\min} = \tfrac{43}{6} .$$

Now we pass to the simplex method.

			1	1	2	0	w	w
Basis	c	P_0	P_1	P_2	P_3	P_4	P_5	P_6
P_4	0	9	1	1	1	1	0	0
P_5	w	1	2	-3	3	0	1	0
P_6	w	3	-3	6	-4	0	0	1
		0	-1	-1	-2	0	0	0
		4	-1	3	-1	0	0	0
P_4	0	$\frac{17}{2}$	$\frac{3}{2}$	0	$\frac{5}{3}$	1	0	—
P_5	w	$\frac{5}{2}$	$\frac{1}{2}$	0	1	0	1	—
P_2	1	$\frac{1}{2}$	$-\frac{1}{2}$	1	$-\frac{2}{3}$	0	0	—
		$\frac{1}{2}$	$-\frac{3}{2}$	0	$-\frac{8}{3}$	0	0	—
		$\frac{5}{2}$	$\frac{1}{2}$	0	1	0	0	—
P_4	0	$\frac{13}{3}$	$\frac{2}{3}$	0	0	1	—	—
P_3	2	$\frac{5}{2}$	$\frac{1}{2}$	0	1	0	—	—
P_2	1	$\frac{13}{6}$	$-\frac{1}{6}$	1	0	0	—	—
		$\frac{43}{6}$	$-\frac{1}{6}$	0	0	0		

Hence we get $x_4 = \frac{13}{3}$; $x_3 = \frac{5}{2}$; $x_2 = \frac{13}{6}$, and $y_{\min} = \frac{43}{6}$, exactly as before.

Here we mention briefly the existence of the *dual* counterpart of a linear programming problem. Assume the following primary problem: Find a vector x such that $c^T x = \min$ under the conditions $x \geq 0$, $Ax = b$. Then the *dual unsymmetric* problem is the following: Find a vector y such that

$$b^T y = \max$$

under the condition $A^T y \leq c$. Here we do not require that y be ≥ 0. The following theorem has been proved by Dantzig and Orden: If one of the problems has a finite solution, then the same is true for the other problem, and further

$$\min c^T x = \max b^T y \,.$$

Alternatively for the primary problem: Find a vector x such that $c^T x = \min$ under the conditions $Ax \geq b$ and $x \geq 0$. We then have the following *dual symmetric* problem: Find a vector y such that $b^T y = \max$ under the conditions $A^T y \leq c$ and $y \geq 0$. The theorem just mentioned is valid also in this case. For proof we refer to Reference 1.

Among possible complications we have already mentioned degeneration. This is not quite unusual but on the other hand not very difficult to master. An obvious measure is perturbation as suggested by Charnes. The vector b is replaced by another vector b':

$$b' = b + \sum_{k=1}^{N} \varepsilon^k P_k,$$

where N is the total number of vectors including possible artificial base vectors. Here ε can be understood as a small positive number which need not be specified closer, and when the solution has been obtained, $\varepsilon = 0$.

Under very special circumstances, "cycling" may appear. This means that after exchange of base vectors sooner or later we return to the same combination in spite of the fact that the simplex rules have been strictly obeyed. Only a few particularly constructed cases are known, and cycling does never seem to occur in practice.

Der Horizont vieler Menschen ist ein Kreis mit Radius Null—und das nennen sie ihren Standpunkt.

20.2. The transportation problem

In many cases, we have linear programming problems of a special kind with a very simple structure, and among these the transportation problem occupies a dominant position. The problem can be formulated in the following way: An article is produced by m producers in the quantities $a_1, a_2, \ldots, a_m$, and it is consumed by n consumers in the quantities $b_1, b_2, \ldots, b_n$. To begin with, we assume that $\sum a_i = \sum b_j$. The transportation cost from producer i to consumer k is c_{ik} per unit, and we search for the quantities x_{ik} which should be delivered from i to k so that the total transportation cost will be as small as possible. The problem can be solved by the conventional simplex method, but usually one prefers a less involved iterative technique, introduced by Hitchcock.

In the usual simplex method, when we are dealing with m equations and n variables, the solution is, in general, a corner of a hyperpolyhedron. The solution contains at least $n - m$ variables which are zero. In our case the number of equations is $m + n - 1$ (namely, $\sum_j x_{ij} = a_i$ and $\sum_i x_{ij} = b_j$; however, we must take the identity $\sum a_i = \sum b_j$ into account), and the number of variables is mn. Thus a feasible solution must not contain more than $m + n - 1$ non-zero elements.

We now formulate the problem mathematically. Find such numbers $x_{ij} \geq 0$ that

$$f = \sum_{i=1}^{m} \sum_{j=1}^{n} c_{ij} x_{ij} = \min \qquad (20.2.1)$$

under the conditions

$$\sum_{j=1}^{n} x_{ij} = a_i , \tag{20.2.2}$$

$$\sum_{i=1}^{m} x_{ij} = b_j , \tag{20.2.3}$$

$$\sum_{i=1}^{m} a_i = \sum_{j=1}^{n} b_j . \tag{20.2.4}$$

In order to make the discussion easier we shall consider a special case; the conclusions which can be drawn from this are then generalized without difficulty, Suppose $m = 3$ and $n = 4$ and write down the coefficient matrix A for (20.2.2) and (20.2.3):

$$A = \begin{pmatrix} 1 & 1 & 1 & 1 & 0 & 0 & 0 & 0 & 0 & 0 & 0 & 0 \\ 0 & 0 & 0 & 0 & 1 & 1 & 1 & 1 & 0 & 0 & 0 & 0 \\ 0 & 0 & 0 & 0 & 0 & 0 & 0 & 0 & 1 & 1 & 1 & 1 \\ 1 & 0 & 0 & 0 & 1 & 0 & 0 & 0 & 1 & 0 & 0 & 0 \\ 0 & 1 & 0 & 0 & 0 & 1 & 0 & 0 & 0 & 1 & 0 & 0 \\ 0 & 0 & 1 & 0 & 0 & 0 & 1 & 0 & 0 & 0 & 1 & 0 \\ 0 & 0 & 0 & 1 & 0 & 0 & 0 & 1 & 0 & 0 & 0 & 1 \end{pmatrix},$$

We have $m + n$ rows and mn columns; the first m rows correspond to (20.2.2) and the last n to (20.2.3). Since the sum of the first m rows is equal to the sum of the last n rows the rank is at most $m + n - 1$. We shall denote the columns of A by p_{ij}, which should correspond to the variables x_{ij} taken in the order $x_{11}, x_{12}, x_{13}, x_{14}, x_{21}, x_{22}, x_{23}, x_{24}, x_{31}, x_{32}, x_{33}, x_{34}$. Let us now compare two vectors p_{ij} and p_{rs}. We see at once that if $i = r$, then the first m components coincide; and if $j = s$, then the last n components coincide. This observation gives us simple means to examine linear dependence for vectors p_{ij}. If we form a cyclic sequence where two adjacent vectors alternately coincide in row-index and column-index, then the vectors must become linearly dependent. For example, we have $p_{12} - p_{32} + p_{34} - p_{24} + p_{23} - p_{13} = 0$. This fact is of great importance when a feasible initial solution is chosen; as has already been observed it must contain $m + n - 1$ elements not equal to 0. If the vectors p_{ij} are arranged in matrix form, we get the picture:

$$\begin{pmatrix} p_{11} & p_{12} & p_{13} & p_{14} \\ p_{21} & p_{22} & p_{23} & p_{24} \\ p_{31} & p_{32} & p_{33} & p_{34} \end{pmatrix}$$

or, using the example above,

From this discussion it is clear that a feasible initial solution cannot be chosen in such a way that $x_{ij} \neq 0$ in points (i, j) which are corners in a closed polygon with only horizontal and vertical sides. For this would imply that the determinant corresponding to a number, possibly all, of the variables x_{ij} which are not equal to 0, would become zero because the vectors related to these x_{ij} are linearly dependent, and such an initial solution cannot exist.

EXAMPLES

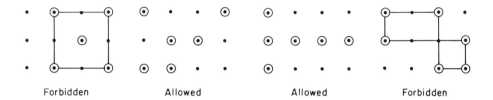

| Forbidden | Allowed | Allowed | Forbidden |

Thus we start by constructing a feasible solution satisfying the following conditions:

1. exactly $m + n - 1$ of the variables x_{ij} shall be positive, the others zero;

2. the boundary conditions (20.2.2) and (20.2.3) must be satisfied;

3. $x_{ij} \neq 0$ in such points that no closed polygon with only horizontal and vertical sides can appear.

How this construction should be best performed in practice will be discussed later.

First we determine α_i and β_j in such a way that $c_{ij} = \alpha_i + \beta_j$ for all such indices i and j that $x_{ij} > 0$; *one* value, for example, α_1, can be chosen arbitrarily. Then we *define* auxiliary quantities (also called fictitious transportation costs) $k_{ij} = \alpha_i + \beta_j$ in all remaining cases. One could, for example, imagine that the transport passes over some central storage which would account for the transport cost being split into two parts, one depending on the producer and one depending on the consumer. Thus we get

$$f = \sum_{i,j} c_{ij} x_{ij} = \sum_{i,j} (c_{ij} - k_{ij}) x_{ij} + \sum_i \alpha_i \sum_j x_{ij} + \sum_j \beta_j \sum_i x_{ij}$$
$$= \sum_{i,j} (c_{ij} - k_{ij}) x_{ij} + \sum_i a_i \alpha_i + \sum_j b_j \beta_j . \qquad (20.2.5)$$

But $c_{ij} = k_{ij}$ for all $x_{ij} \neq 0$, and hence we obtain

$$f = \sum_i a_i \alpha_i + \sum_j b_j \beta_j . \qquad (20.2.6)$$

If for some i, j we have $c_{ij} < k_{ij}$, it is possible to find a better solution. Suppose that we move a quantity ε to the place (i, j). Then the total cost will decrease with $\varepsilon(k_{ij} - c_{ij})$. This fact suggests the following procedure. We search for the minimum of $c_{ij} - k_{ij}$, and if this is < 0, we choose ε as large as possible with regard to the conditions (20.2.2) and (20.2.3). Then we calculate new values α_i, β_j, and k_{ij} and repeat the whole procedure. When all $c_{ij} \geq k_{ij}$, we have attained an optimal solution; in exceptional cases several such solutions can exist.

The technique is best demonstrated on an example. We suppose that a certain commodity is produced in three factories in quantities of 8, 9, and 13 units, and it is used by four consumers in quantities of 6, 7, 7, and 10 units. The transportation costs are given in the following table:

$$(c_{ij}) = \begin{pmatrix} 3 & 8 & 9 & 16 \\ 6 & 11 & 14 & 9 \\ 5 & 13 & 10 & 12 \end{pmatrix}.$$

We start by constructing a feasible solution. To the left of the x_{ij}-matrix, which is so far unknown, we write down the column a_i, and above the matrix we write the row b_j. Then we fill in the elements, one by one, in such a way that the conditions (20.2.2) and (20.2.3) are not violated. First we take the element with the lowest cost, in this case (1, 1), and in this place we put a number as large as possible (6).

	(6)0	7	7	10
(8)2	6			
8	0			
13	0			

Then we get zeros in the first column, while the row sum 8 is not fully exploited. From among the other elements in the first row, we choose the one with the lowest c_{ij}, in this case (1, 2), and in this place we put a number as large as possible (2):

	0	(7)5	7	10
(2)0	6	2	0	0
9	0			
13	0			

In a similar way we obtain in the next step:

	0	5	7	(10)1
0	6	2	0	0
(9)0	0	0	0	9
13	0			

In the last step we have no choice, and hence we obtain

$$(x_{ij}) = \begin{pmatrix} 6 & 2 & 0 & 0 \\ 0 & 0 & 0 & 9 \\ 0 & 5 & 7 & 1 \end{pmatrix}.$$

We find $f = \sum c_{ij}x_{ij} = 262$. Then we compute α_i and β_j, choosing arbitrarily $\alpha_1 = 0$, and we get without difficulty $\beta_1 = 3$; $\beta_2 = 8$; $\alpha_3 = 5$; $\beta_3 = 5$; $\beta_4 = 7$; $\alpha_2 = 2$. These values are obtained from the elements $(1, 1)$, $(1, 2)$, $(2, 4)$, $(3, 2)$, $(3, 3)$, and $(3, 4)$ of (c_{ij}); they correspond to the x_{ij}-elements which are not zero. From $k_{ij} = \alpha_i + \beta_j$, we easily get

$$(k_{ij}) = \begin{pmatrix} 3 & 8 & 5 & 7 \\ 5 & 10 & 7 & 9 \\ 8 & 13 & 10 & 12 \end{pmatrix}; \qquad (c_{ij} - k_{ij}) = \begin{pmatrix} 0 & 0 & 4 & 9 \\ 1 & 1 & 7 & 0 \\ -3 & 0 & 0 & 0 \end{pmatrix}.$$

Thus we have $c_{31} - k_{31} < 0$, and it should be possible to reduce the transportation cost by moving as much as possible to this place. Hence, we modify (x_{ij}) in the following way:

$$(x_{ij}) = \begin{pmatrix} 6 - \varepsilon & 2 + \varepsilon & 0 & 0 \\ 0 & 0 & 0 & 9 \\ \varepsilon & 5 - \varepsilon & 7 & 1 \end{pmatrix}.$$

The elements which can be affected are obtained if we start at the chosen place (i, j), where $c_{ij} - k_{ij} < 0$, and draw a closed polygon with only horizontal and vertical sides, and with all corners in elements that are not zero.

Since all x_{ij} must be ≥ 0, and the number of nonzero elements should be $m + n - 1 = 6$, we find $\varepsilon = 5$ and

$$(x_{ij}) = \begin{pmatrix} 1 & 7 & 0 & 0 \\ 0 & 0 & 0 & 9 \\ 5 & 0 & 7 & 1 \end{pmatrix}.$$

As before, we determine α_i, β_j, and k_{ij} and obtain

$$(c_{ij} - k_{ij}) = \begin{pmatrix} 0 & 0 & 1 & 6 \\ 4 & 4 & 7 & 0 \\ 0 & 3 & 0 & 0 \end{pmatrix},$$

which shows that the solution is optimal. The total transportation cost becomes $f = 247$.

We see that the technique guarantees an integer solution if all initial values also are integers. Further we note that the matrix $(c_{ij} - k_{ij})$ normally contains $m + n - 1$ zeros at places corresponding to actual transports. If there is, for example, one more zero we have a degenerate case which allows for several optimal solutions. A transport can be moved to this place without the solution ceasing to be optimal. Combining two different optimal solutions we can form an infinite number of optimal solutions. In this special case we can actually construct solutions with *more* than $m + n - 1$ elements not equal to 0, and, of course, there are also cases when we have less than $m + n - 1$ elements not equal to 0.

We shall now briefly indicate a few complications that may arise. So far we have assumed that the produced commodities are consumed, that is, there is no overproduction. The cases of over- and underproduction can easily be handled by introducing fictitious producers and consumers. For example, if we have a fictitious consumer, we make the transportation costs to him equal to zero. Then the calculation will show who is producing for the fictitious consumer; in other words, the overproduction is localized to special producers. If we have a fictitious producer, we can make the transportation costs from him equal to zero, as in the previous case. We give an example to demonstrate the technique. Suppose that two producers manufacture 7 and 9 units, while four consumers ask for 2, 8, 5, and 5 units. Obviously, there is a deficit of 4 units, and hence we introduce a fictitious producer. We assume the following transportation cost table:

$$(c_{ij}) = \begin{pmatrix} 0 & 0 & 0 & 0 \\ 3 & 6 & 10 & 12 \\ 7 & 5 & 4 & 8 \end{pmatrix}.$$

We easily find

$$(x_{ij}) = \begin{pmatrix} 0 & 0 & 0 & 4 \\ 2 & 5 & 0 & 0 \\ 0 & 3 & 5 & 1 \end{pmatrix} \quad \text{and} \quad (k_{ij}) = \begin{pmatrix} -6 & -3 & -4 & 0 \\ 3 & 6 & 5 & 9 \\ 2 & 5 & 4 & 8 \end{pmatrix}.$$

Hence

$$(c_{ij} - k_{ij}) = \begin{pmatrix} 6 & 3 & 4 & 0 \\ 0 & 0 & 5 & 3 \\ 5 & 0 & 0 & 0 \end{pmatrix}.$$

Hence, the solution is optimal, and the fourth customer, who asked for 5 units, does not obtain more than 1.

As a rule, $m + n - 1$ elements of our transport table should not be zero. In the initial construction of a feasible solution or in the following reduction, it sometimes happens that we get an extra zero. This difficulty can be overcome if we observe that all quantities are continuous functions of the values a_i, b_j, and c_{ij}. Hence, we simply replace one of the zeros by δ, where we put $\delta = 0$ in all arithmetic calculations. The only use we have for this δ is that the rule requiring that $m + n - 1$ elements shall not be zero still holds, and that the formation of the closed polygon is made possible.

As a rule, $m + n - 1$ elements of our tranport table should differ from zero. However, there are two cases when this requirement cannot be met. It may sometimes happen that a remaining production capacity and a remaining consumer demand vanish simultaneously while the transport table is constructed. But also on transformation of the table as described above extra zeros may appear. This difficulty can be overcome by the same perturbation technique as has been used before, and a quantity δ (some authors use $+$ instead) is placed in the table to obey the polygon rule, whereas δ is put equal to zero in all arithmetic calculations. This is motivated by the fact that the desired minimum value is a continuous function of the quantities a_i, b_j, and c_{ij}, and hence degeneration can be avoided by a slight change of some of them. The quantity (or quantities) δ are used also on computation of k_{ij}.

EXAMPLE

$$
\begin{array}{c|ccc}
 & 3 & 5 & 5 \\
\hline
5 & 2 & 6 & 11 \\
5 & 4 & 10 & 16 \\
3 & 4 & 16 & 14 \\
\end{array}
$$

To begin, we get:

$$
\begin{array}{c|ccc}
 & 3 & 5 & 5 \\
\hline
5 & 3 & 2 & \\
5 & & 3 & 2 \\
3 & & & 3 \\
\end{array}
\qquad
(k_{ij}) = \begin{pmatrix} 2 & 6 & 12 \\ (6) & 10 & 16 \\ 4 & 8 & 14 \end{pmatrix}, \qquad f = 122 .
$$

Obviously we ought to move as much as possible (4 units) to the (2, 1)-place, but then we get zeros in both (1, 1) and (2, 2). Let us therefore put, for example, the (1, 1)-element $= 0$ and the (2, 2)-element $= \delta$:

$$
\begin{array}{c|ccc}
 & 3 & 5 & 5 \\
\hline
5 & & 5 & \\
5 & 3 & 8 & 2 \\
3 & & & 3 \\
\end{array}
\qquad
(k_{ij}) = \begin{pmatrix} 0 & 6 & (12) \\ 4 & 10 & 16 \\ 2 & 8 & 14 \end{pmatrix}, \qquad f = 116 .
$$

Now we move as much as possible (2 units) to the (1, 3)-place and get:

$$
\begin{array}{c|ccc}
 & 3 & 5 & 5 \\
\hline
5 & & 3 & 2 \\
5 & 3 & 2 & \\
3 & & & 3 \\
\end{array}
\qquad
(k_{ij}) = \begin{pmatrix} 0 & 6 & 11 \\ 4 & 10 & 15 \\ 3 & 9 & 14 \end{pmatrix}, \qquad f = 114 .
$$

This is the optimal solution.

20.3. Quadratic, integer, and dynamic programming

Linear programming is, in fact, only a special case (though a very important one), and generalizations in different directions are possible. Near at hand is the possibility of minimizing a quadratic expression instead of a linear one, adhering to the secondary linear conditions (equalities or inequalities). This problem is known as quadratic programming (see [2] and [5]).

If we have a process which occurs in several stages, where each subprocess is dependent on the strategy chosen, we have a dynamic programming problem. The theory of dynamic programming is essentially due to Bellman [3]. As an illustration we give the following example, taken from [2].

We have n machines of a certain kind at our disposal, and these machines can perform two different kinds of work. If z machines are working in the first way, commodities worth $g(z)$ are produced and if z machines are work-ing in the second way, commodities worth $h(z)$ are produced. However, the machines are partly destroyed, and in the first case, $a(z)$ machines are left over and in the second, $b(z)$ machines. Here, a, b, g, and h are given functions. We assign x_1 machines for the first job, and $y_1 = n - x_1$ machines for the second job. After one stage we are left with $n_2 = a(x_1) + b(y_1)$ machines, of which we assign x_2 for the first job and y_2 for the second job. After N stages the total value of the produced goods amounts to

$$f = \sum_{i=1}^{N} [g(x_i) + h(y_i)],$$

with

$$
\begin{aligned}
& x_i + y_i = n_i, && n_1 = n, \\
& a(x_i) + b(y_i) = n_{i+1}, && i = 1, 2, \ldots, N-1, \\
& 0 \leq x_i \leq n_i, && i = 1, 2, \ldots, N.
\end{aligned}
$$

The problem is to maximize f. In particular, if the functions are linear, we have again a linear programming problem.

Let $f_N(n)$ be the maximum total value when we start with n machines and work in N stages using an optimal policy. Then we have

$$f_1(n) = \max_{0 \leq x \leq n} [g(x) + h(n - x)],$$

$$f_k(n) = \max_{0 \leq x \leq n} \{g(x) + h(n - x) + f_{k-1}[a(x) + b(n - x)]\}; \qquad k > 1.$$

In this way the solution can be obtained by use of a recursive technique.

Last, we also mention that in certain programming problems, all quantities must be integers (integer programming). However, a closer account of this problem falls outside the scope of this book.

Finally we shall illustrate the solution technique in a numerical example simultaneously containing elements of dynamic and integer programming. Suppose that a ship is to be loaded with different goods and that every article is available only in units with definite weight and definite value. The problem is now to choose goods with regard to the weight restrictions (the total weight being given) so that the total value is maximized. Let the number of articles be N, the weight capacity z, and further the value, weight, and number of units of article i be v_i, w_i and x_i. Then we want to maximize

$$L_N(x) = \sum_{i=1}^{N} x_i v_i,$$

under the conditions $\sum_{i=1}^{N} x_i w_i \leq z$ with x_i integer and ≥ 0. Defining

$$f_N(z) = \max_{\{x_i\}} L_N(x),$$

we shall determine the maximum over combinations of x_i-values satisfying the conditions above. We can now derive a functional relation as follows. Let us first choose an arbitrary value x_N leaving a remaining weight capacity $z - x_N w_N$.

By definition, the best value we can get from this weight is $f_{N-1}(z - x_N w_N)$. Our choice of x_N gives the total value $x_N v_N + f_{N-1}(z - x_N w_N)$, and hence we must choose x_N so that this value is maximized. From this we get the fundamental and typical relationship

$$f_N(z) = \max_{x_N} \{x_N v_N + f_{N-1}(z - x_N w_N)\},$$

with $0 \leq x_N \leq$ entier (z/w_N). The initial function is trivially

$$f_1(z) = v_1 \cdot \text{entier } \frac{z}{w_1}.$$

Below the solution in the case $N = 5$, $z = 20$ is derived for the following values of v_i and w_i:

i	v_i	w_i	v_i/w_i
1	9	4	2.25
2	13	5	2.60
3	16	6	2.67
4	20	9	2.22
5	31	11	2.82

The computation can be performed through successive tabulation of $f_1, f_2, \ldots, f_5$ for $z = 1, 2, 3, \ldots, 20$. The results are presented in the following table.

z	f_1	f_2	f_3	f_4	f_5
1	0	0	0	0	0
2	0	0	0	0	0
3	0	0	0	0	0
4	9	9	9	9	9
5	9	13	13	13	13
6	9	13	16	16	16
7	9	13	16	16	16
8	18	18	18	18	18
9	18	22	22	22	22
10	18	26	26	26	26
11	18	26	29	29	31
12	27	27	32	32	32
13	27	31	32	32	32
14	27	35	35	35	35
15	27	39	39	39	40
16	36	39	42	42	44
17	36	40	45	45	47
18	36	44	48	48	48
19	36	48	48	48	49
20	45	52	52	52	53

The maximum value 53 is attained for $x_1 = 1$, $x_2 = 1$, $x_3 = 0$, $x_4 = 0$, and $x_5 = 1$.

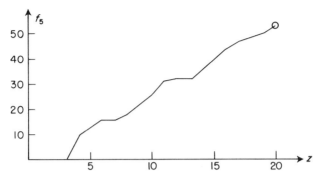

Figure 20.2. Maximum value as a function of weight capacity.

REFERENCES

[1] Gass: *Linear Programming* (McGraw-Hill, New York, 1958).

[2] Vajda: *Mathematical Programming* (London, 1961).

[3] Bellman: *Dynamic Programming* (Princeton University Press, Princeton, 1957).

[4] Charnes-Cooper: *Management Models and Industrial Applications of Linear Programming* (Wiley, New York, 1961).

[5] Künzi-Krelle: *Nichtlineare Programmierung* (Berlin, 1962).

[6] Ph. Wolfe: *Recent Developments in Nonlinear Programming; Advances in Computers*, Vol. 3 (Academic Press, New York, 1962).

[7] G. Hadley: *Linear Programming* (Addison-Wesley, Reading, Mass., 1962).

EXERCISES

1. Find the maximum of $y = x_1 - x_2 + 2x_3$ when

$$\begin{cases} x_1 + x_2 + 3x_3 + x_4 \le 5, \\ x_1 + x_3 - 4x_4 \le 2, \\ x_i \ge 0. \end{cases}$$

2. Find the minimum of $y = 5x_1 - 4x_2 + 3x_3$ when

$$\begin{cases} 2x_1 + x_2 - 6x_3 = 20, \\ 6x_1 + 5x_2 + 10x_3 \le 76, \\ 8x_1 - 3x_2 + 6x_3 \le 50, \\ x_i \ge 0. \end{cases}$$

3. Find the minimum of $f = \lambda x_1 - x_2$ as a function of $\lambda(-\infty < \lambda < \infty)$, when

$$\begin{cases} x_1 + x_2 \le 6, \\ x_1 + 2x_2 \le 10, \\ x_i \ge 0. \end{cases}$$

4. Find the maximum of $f = -1 + x_2 - x_3$ when

$$\begin{cases} x_4 = x_1 - x_2 + x_3 \,, \\ x_5 = 2 \ - x_1 - x_3 \,, \\ x_i \geq 0 \,. \end{cases}$$

5. Find the minimum of $f = x_1 + x_4$ when

$$\begin{cases} 2x_1 + 2x_2 + \ x_3 \qquad\quad \leq 7 \,, \\ 2x_1 + \ x_2 + 2x_3 \qquad\quad \leq 4 \,, \\ \qquad\quad x_2 + \qquad\quad x_4 \geq 1 \,, \\ \qquad\quad x_2 + \ x_3 + x_4 = 3 \,, \\ \qquad\qquad\qquad\qquad\quad x_i \geq 0 \,. \end{cases}$$

6. Find the minimum of $f = 4x_1 + 2x_2 + 3x_3$ when

$$\begin{cases} 2x_1 + \qquad\quad 4x_3 \geq 5 \,, \\ 2x_1 + 3x_2 + \ x_3 \geq 4 \,, \\ \qquad\qquad\qquad x_i \geq 0 \,. \end{cases}$$

7. Maximize $f = 2x_1 + x_2$ when

$$\begin{cases} x_1 - \ x_2 \leq 2 \,, \\ x_1 + \ x_2 \leq 6 \,, \\ x_1 + 2x_2 \leq \alpha \,, \\ \qquad\quad x_i \geq 0 \,. \end{cases}$$

The maximum value of f should be given as a function of α when $0 \leq \alpha \leq 12$.

8. Minimize $f = -3x + y - 3z$ under the conditions

$$\begin{cases} -x + 2y + \ z \leq 0 \,, \\ 2x - 2y - 3z = 9 \,, \\ \ x - \ y - 2z \geq 6 \,, \end{cases}$$

and $x \geq 0$, $y \geq 0$, $-\infty < z < \infty$.

9. The following linear programming problem is given: Maximize $c^T x$ under the conditions $Ax = b$, $x \geq d \geq 0$. Show how this problem can be transformed to the following type: Maximize $g^T y$ under the conditions $Fy = f$, $y \geq 0$, where the matrix F is of the same type (m, n) as the matrix A. Also solve the following problem: Maximize $z = 3x_1 + 4x_2 + x_3 + 7x_4$ when

$$\begin{cases} 8x_1 + 3x_2 + 4x_3 + \ x_4 \leq 42 \,, \\ \qquad\quad 6x_2 + \ x_3 + 2x_4 \leq 20 \,, \\ x_1 + 4x_2 + 5x_3 + 2x_4 \leq 37 \,, \end{cases}$$

and $x_1 \geq 2$; $x_2 \geq 1$; $x_3 \geq 3$; $x_4 \geq 4$.

10. A manufacturer uses three raw products, a, b, c, priced at 30, 50, 120 \$/lb, respectively. He can make three different products, A, B, and C, which can be sold at 90, 100, and 120 \$/lb, respectively. The raw products can be obtained only in limited quantities, namely, 20, 15, and 10 lb/day. Given: 2 lb of a plus 1 lb of b plus 1 lb of c will yield

4 lb of A; 3 lb of a plus 2 lb of b plus 2 lb of c will yield 7 lb of B; 2 lb of b plus 1 lb of c will yield 3 lb of C. Make a production plan, assuming that other costs are not influenced by the choice among the alternatives.

11. A mining company is taking a certain kind of ore from two mines, A and B. The ore is divided into three quality groups, a, b, and c. Every week the company has to deliver 240 tons of a, 160 tons of b, and 440 tons of c. The cost per day for running mine A is \$ 3000 and for running mine B \$ 2000. Each day A will produce 60 tons of a, 20 tons of b, and 40 tons of c. The corresponding figures for B are 20, 20, and 80. Construct the most economical production plan.

Solve the following transport problems:

12.

	3	3	4	5
4	13	11	15	20
6	17	14	12	13
9	18	18	15	12

13.

	75	75	75	75
100	19	15	19	20
100	20	23	17	31
100	14	25	20	18

14.

	10	5	7	6
18	2	11	6	3
9	12	10	15	5
11	4	9	13	10

(Give at least two different solutions.)

15.

	10	5	5
8	3	2	2
3	2	4	2
6	4	2	2

16.

	20	40	30	10	50	25
30	1	2	1	4	5	2
50	3	3	2	1	4	3
75	4	2	5	9	6	2
20	3	1	7	3	4	6

17. For three different services, a, b, and c, 100, 60, and 30 people, respectively, are needed. Three different categories, A, B, and C, are available with 90 of A, 70 of B, and 50 of C. The following table displays the "suitability numbers":

	a	b	c
A	6	4	4
B	6	10	8
C	9	8	8

Make a choice among the 210 people available for the 190 places so that the sum of the "suitability numbers" becomes as large as possible.

18. The following transport problem has a degenerate solution with only five transports. Find this solution and the minimum value!

	5	25	70
10	2	2	6
60	5	6	10
20	2	4	8
10	6	3	4

Answers to exercises

Answers to exercises

Chapter 1

1. 11 259 375   2. $0.2406631213\ldots$ 3. $3\%_0$

4. The error in the latter case is $14(2 + \sqrt{3})^5 = 10136$ times greater than in the former case.

5. $y \sim \dfrac{1}{2x} + \dfrac{1}{2^2 x^3} + \dfrac{1 \cdot 3}{2^3 x^5} + \dfrac{1 \cdot 3 \cdot 5}{2^4 x^7} + \cdots$

6. $a_{2k} = 0;\ a_{2k+1} = (-1)^k (2k)!$
   $b_{2k} = (-1)^{k+1}(2k-1)!;\ b_{2k+1} = 0$

Chapter 2

1. $x = 1.403602$ 2. $\begin{cases} x_1 = 1.7684 \\ x_2 = 2.2410 \end{cases}$ 3. $x = 1.30296$

4. $x = 2.36502$ 5. $x = 2.55245$

6. $x = 0.94775$ 7. $x = 1.28565$ 8. $x = 2.8305$

9. $y_{\max} = -y_{\min} = y(0) = -y(1) = 0.00158$ 10. $a = 0.85124$

11. $K = 0.98429$ 12. $\alpha = 0.804743$ 13. 0.23611

14. $ab < \frac{3}{4}$ 15. $\xi \simeq \dfrac{x_{n-1} x_{n+1} - x_n^2}{x_{n-1} - 2x_n + x_{n+1}}$ 16. $x = 1.4458$

17. $x = y = 1.1462$ 18. $\alpha_1 = \frac{1}{20},\ \alpha_2 = \frac{2}{525},\ \alpha_3 = \frac{13}{37800};\ x = \pm 0.551909.$

19. $\xi = 0.268$

20. $\begin{cases} x_1 = 7.017 \\ x_2 = -2.974 \\ x_3 = 0.958 \end{cases}$ 21. $\begin{cases} c = 0.05693 \\ x = 7.78975 \\ y = 1.99588 \end{cases}$ 22. $\begin{cases} \alpha_0 = e \\ \alpha = 3.03 \end{cases}$ 23. $\begin{cases} a = 0.3619 \\ b = 1.3093 \\ x = 0.8858 \end{cases}$

24. 0.278465. This value is a root of the equation $\log \alpha + \alpha + 1 = 0$.

25. $\begin{cases} x = 0.9727 \\ y = 1.3146 \end{cases}$ 26. $A = 1,\ B = \frac{2}{3},\ C = \frac{13}{15},\ x = 20.371303$

27. $A = -\frac{5}{6},\ B = \frac{169}{120}$ 28. $x_{1,2} = 1 \pm i,\ x_{3,4} = 3 \pm 4i$

29. $\begin{cases} x = 1.086 \\ y = 1.944 \end{cases}$ 30. The error is proportional to the fourth power of ε.

                        31. 0

Chapter 3

3. $\begin{pmatrix} -430 & 512 & 332 \\ -516 & 614 & 396 \\ 234 & -278 & -177 \end{pmatrix}$ 5. $\begin{cases} a = 0.469182 \\ b = -2.53037 \end{cases}$

6. $\begin{pmatrix} \cos\alpha & \sin\alpha \\ -\sin\alpha & \cos\alpha \end{pmatrix}$ 7. $\begin{pmatrix} \frac{6}{7} & \frac{2}{7} \\ \frac{3}{7} & \frac{1}{7} \end{pmatrix}$ 10. $\alpha = \dfrac{d - a \pm \sqrt{(d-a)^2 + 4bc}}{2b}$

                                           $p = \pm(\alpha\alpha^* + 1)^{-1/2}$
                                           $a' = 7 + 4i,\ d' = 7 - 4i$

13. For instance, $U = \dfrac{1}{\sqrt{2}} \begin{pmatrix} 1 & i \\ i & 1 \end{pmatrix}$; $B = \begin{pmatrix} a + bi & 0 \\ 0 & a - bi \end{pmatrix}$.

14. P_r has at least one characteristic value 0; hence det $P_r = 0$.

$$P_2 = \frac{1}{20} \begin{pmatrix} 7 & -3 & -9 & -1 \\ -3 & 7 & 1 & 9 \\ -9 & 1 & 13 & -3 \\ -1 & 9 & -3 & 13 \end{pmatrix}$$

15. One solution is $X = A^T(AA^T)^{-1}$ (provided AA^T is regular). The general solution of the special example is

$$\begin{pmatrix} a & b & c \\ 1-a & 1-b & 1-c \\ 2a-3 & 2b-3 & 2c-4 \\ a & b+1 & c+1 \end{pmatrix},$$

where a, b, and c are arbitrary.

Chapter 4

1. $x = 3$, $y = -2$, $z = 1$, $v = 5$ 2. $x = 4$, $y = 3$, $z = 2$, $v = 1$

3. $x = 0$, $y = 1$, $z = -1$, $u = 2$, $v = -2$

4. $x_1 = 0.433$, $x_2 = 0.911$, $x_3 = 0.460$, $x_4 = -0.058$, $x_5 = -0.115$, $x_6 = 0.244$

5. $x = 0.660$, $y = 0.441$, $z = 1.093$

6. $\begin{cases} x = 2.999\,966 \\ y = -1.999\,982 \end{cases}$

   General solution (cf. Chapter 13): $\begin{cases} x_{k+1} = 3 - \frac{7}{13}\left(\frac{49}{338}\right)^k \\ y_{k+1} = -2 + 2\left(\frac{49}{338}\right)^{k+1} \end{cases}$

Chapter 5

1. $\dfrac{15}{64} \begin{pmatrix} 15 & -70 & 63 \\ -70 & 588 & -630 \\ 63 & -630 & 735 \end{pmatrix}$

2. $\begin{pmatrix} 25 & -41 & 16 & -6 \\ -16 & 27 & -11 & 4 \\ 16 & -27 & 13 & -5 \\ -6 & 10 & -5 & 2 \end{pmatrix}$

3. $\begin{pmatrix} 6.1 & -2.3 & 0.1 & -0.1 \\ 8.6 & -2.8 & -0.4 & -0.6 \\ -4.4 & 1.2 & 0.6 & 0.4 \\ -7.1 & 3.3 & -1.1 & 0.1 \end{pmatrix}$

4. $\begin{pmatrix} 5 & -10 & 10 & -5 & 1 \\ -10 & 30 & -35 & 19 & -4 \\ 10 & -35 & 46 & -27 & 6 \\ -5 & 19 & -27 & 17 & -4 \\ 1 & -4 & 6 & -4 & 1 \end{pmatrix}$

5. $\begin{pmatrix} 13.5 & -6 & 2 & -1.5 \\ -6 & 3 & -2 & 1 \\ 2 & -2 & 10 & -3 \\ -1.5 & 1 & -3 & 1 \end{pmatrix}$

6. $\begin{pmatrix} 4.1 & -0.8 & -1 \\ -0.2 & 0.1 & 0 \\ -1.25 & 0.25 & 0.25 \end{pmatrix}$

7. $\alpha = 1/(1-n)$; $n \geq 2$

8. $M = 42{,}000$; $N = 5500$

9. $\dfrac{1}{6} \begin{pmatrix} 6 - 8i & -2 + 4i \\ -3 + 10i & 1 - 5i \end{pmatrix}$

10. $\begin{pmatrix} 7 & -3 & 0 & -5 \\ 8 & 1 & -2 & -11 \\ -5 & 0 & 1 & 6 \\ 19 & 5 & -6 & -28 \end{pmatrix}$

11. $B - \delta B E_{ik} B$; $\begin{pmatrix} 12.76 & 13.72 & 5.88 & 3.92 \\ 7.46 & -1.63 & 12.73 & 8.82 \\ 5.88 & 6.86 & 2.94 & 1.96 \\ 8.34 & 4.23 & 15.67 & 10.78 \end{pmatrix}$

12.
$$L = \begin{pmatrix} 1 & 0 & 0 & 0 & 0 \\ 1 & 1 & 0 & 0 & 0 \\ 0 & 1 & 1 & 0 & 0 \\ 0 & 0 & \frac{1}{2} & 1 & 0 \\ 0 & 0 & 0 & \frac{2}{7} & 1 \end{pmatrix}; \quad D = \begin{pmatrix} 1 & 0 & 0 & 0 & 0 \\ 0 & 1 & 0 & 0 & 0 \\ 0 & 0 & 2 & 0 & 0 \\ 0 & 0 & 0 & \frac{7}{2} & 0 \\ 0 & 0 & 0 & 0 & \frac{33}{7} \end{pmatrix}$$

14. $\alpha(n\varepsilon)^{p+1}/(1 - n\varepsilon)$

15.
$$10^{-3} \begin{pmatrix} -18.895 & -1.791 & 12.150 & 21.900 \\ -2.524 & 20.476 & 33.939 & 18.775 \\ 11.744 & -7.854 & 1.949 & -3.695 \\ -24.755 & -10.918 & -3.422 & 1.998 \end{pmatrix}; \quad \text{maximum error} = 0.18 \cdot 10^{-3}.$$

17. $\begin{cases} \alpha_i = -a^T C_i / a^T C_r, & i \neq r \\ \alpha_r = 1/a^T C_r \end{cases}$

Chapter 6

1. 98.522

2. 0.0122056; $\begin{pmatrix} -110.595 \\ 24.957 \\ -27.665 \\ 1 \end{pmatrix}$

3. 12.054; $\begin{pmatrix} 1 \\ 0.5522i \\ 0.0995(3 + 2i) \end{pmatrix}$

4. 19.29; -7.08 5. 4.040129   6. 8.00   7. 70.21

8. $\lambda_1 = a_0 + a_1 + a_2 + a_3$
$\lambda_2 = a_0 - a_1 + a_2 - a_3$
$\lambda_3 = a_0 + ia_1 - a_2 - ia_3$
$\lambda_4 = a_0 - ia_1 - a_2 + ia_3$
$x_1 = \begin{pmatrix} 1 \\ 1 \\ 1 \\ 1 \end{pmatrix}$; $x_2 = \begin{pmatrix} 1 \\ -1 \\ 1 \\ -1 \end{pmatrix}$; $x_3 = \begin{pmatrix} 1 \\ i \\ -1 \\ -i \end{pmatrix}$; $x_4 = \begin{pmatrix} 1 \\ -i \\ -1 \\ i \end{pmatrix}$

9. $p = 2a + (n - 2)b$; $q = (a - b)[a + (n - 1)b]$
$\lambda_1 = a + (n - 1)b$ (simple); $\lambda_2 = a - b$ $[(n - 1)$-fold$]$.

10. $A_1 = \begin{pmatrix} 9.5714 & 6 \\ 0.6122 & 1.4286 \end{pmatrix}$; $A_2 = \begin{pmatrix} 9.9552 & 6 \\ 0.0668 & 1.0448 \end{pmatrix}$; $A_3 = \begin{pmatrix} 9.9955 & 6 \\ 0.0067 & 1.0045 \end{pmatrix}$
Exact eigenvalues: 10 and 1.

11. An arbitrary vector v can be written as a linear combination of the eigenvectors.

12. $\lambda_1 = \dfrac{R_1 R_3 - R_2^2}{R_1 - 2R_2 + R_3} = R_3 + \dfrac{R_3 - R_2}{(R_2 - R_1)/(R_3 - R_2) - 1}$.

Chapter 7

6. $\delta^2(x^2) = 2$; $\delta^2(x^3) = 6x$; $\delta^2(x^4) = 12x^2 + 2$
$f(x) = 2x^4 + 4x^3$ (To this we can add an arbitrary solution of the equation $\mu f(x) = 0$.)

7. $a = r/4$; $b = r(r + 3)/32$

9. $y = 1 - \dfrac{p^2}{2!}x^2 - \dfrac{p^2(1 - p^2)}{4!}x^4 - \dfrac{p^2(1 - p^2)(4 - p^2)}{6!}x^6 - \cdots$

10. $\delta y_0 = \frac{1}{2}(y_1 - y_{-1}) - \frac{1}{16}(\delta^2 y_1 - \delta^2 y_{-1}) + \frac{2}{256}(\delta^4 y_1 - \delta^4 y_{-1}) - \cdots$

11. b) $c_0 = 0$; $c_1 = 1$; $c_2 = 30$; $c_3 = 150$; $c_4 = 240$; $c_5 = 120$; $N = 63$.

12. $J_0 y_0 = h(y_0 + \frac{1}{24}\delta^2 y_0 - \frac{17}{5760}\delta^4 y_0 + \cdots)$

13. $\delta'^2 = n^2\delta^2 + \dfrac{n^2(n^2-1)}{12}\,\delta^4 + \dfrac{n^2(n^2-1)(n^2-4)}{360}\,\delta^6 + \cdots$

14. Use operator technique! The new series is $2, 1, 3, 4, 7, 11, \ldots$

Chapter 8

1. $f(3.63) = 0.136482$ 2. 3.625 3. $0.55247\ 22945$

4. 3.4159 5. $0.46163\ 21441$ (correct value   6. 0.000034
                                       $0.46163\ 21450$)

7. 0.267949 9. The remainder term is of fourth order.
                             $f(2.74) = 0.0182128$

10. $y_p \simeq py_1 + \dfrac{p(p^2-1)}{6}\,h^2 y_1'' + \dfrac{p(p^2-1)(3p^2-7)}{360}\,h^4 y_1^{IV}$

$\qquad + \dfrac{p(p^2-1)(3p^4-18p^2+31)}{15120}\,h^6 y_1^{VI} + $ (similar terms in q and y_0).

11. $a = q$; $c = -\dfrac{pq}{6}(q+1)$;   12. $\begin{cases} x = 0.46996 \\ y = 1.56250 \end{cases}$

$\quad b = p$; $d = -\dfrac{pq}{6}(p+1)$; $Ai(1.1) = 0.120052$.

13. $\begin{cases} x = 0.191 \\ y = 0.525 \end{cases}$ 14. 12.95052 15. 27.7718 16. 2.7357

19. Differences of the rounded values   20. 3.1415 21. $c = 0.33275$

Chapter 9

1. -0.43658 2. 0.9416 3. $a = 9, b = 30$ 4. 0.061215

5. $a = 8, b = 6$ 6. $n = 33$

Chapter 10

1. 20.066 (exact value 21)   2. 0.7834 3. 0.6736

4. 1.46746 5. Speed 3087 m/sec; height 112.75 km   6. 0.54003

7. 3.1044 8. 1.9049

9. $C_2 = 0.6565 \left(\text{correct value} = \prod_{(\text{odd primes})} \left\{ 1 - \dfrac{1}{(p-1)^2} \right\} = 0.6601618 \right)$.

10. 1.8521 11. -0.94608 12. 1.3503 13. 9.688448

14. $\dfrac{4h}{3}(2y_1 - y_0 + 2y_{-1})$ 15. $A = \frac{19}{15}$; $B = \frac{17}{45}$; $C = -\frac{1}{90}$

16. $\begin{cases} x_1 = -0.2899 \\ x_2 = 0.5266 \end{cases}$ or $\begin{cases} x_1 = 0.6899 \\ x_2 = -0.1266 \end{cases}$ 17. $a = \frac{1}{12}$; $R = O(h^5)$

18. $y_{\max} = 1.01494$ for $x = \frac{1}{3}\pi$ 19. 0.6716 20. 1.816

21. -0.57722 (exact value $= -C$, where $C = $ Euler's constant).

22. 1.644934 and 0.822467 (exact values $\frac{1}{6}\pi^2$ and $\frac{1}{12}\pi^2$)   23. 0.2797

24. $\begin{cases} a_1 = 0.389111 \\ a_2 = 0.277556 \end{cases}$ $\begin{cases} x_1 = 0.821162 \\ x_2 = 0.289949 \end{cases}$ 25. $\begin{cases} A_1 = 0.718539 \\ A_2 = 0.281461 \end{cases}$ $\begin{cases} x_1 = 0.1120088 \\ x_2 = 0.6022769 \end{cases}$

26. $k_1 = \dfrac{1 + 3\alpha}{3(1 + \alpha)}$; $k_2 = \dfrac{1 - 3\alpha}{3(1 - \alpha)}$; $k_3 = \dfrac{4}{3(1 - \alpha^2)}$ · 5.06734 (exact value $\frac{76}{15} = 5.06667$).

27. $a = \frac{1}{10}$; $b = \frac{49}{90}$; $c = \frac{32}{45}$; $\alpha = \sqrt{\frac{3}{7}}$

28. $A = \dfrac{3}{20}$; $B = \dfrac{7}{20}$; $C = \dfrac{1}{30}$; $D = -\dfrac{1}{20}$; $R = \dfrac{h^6}{2 \cdot 6!} f^{IV}(x_0 + \xi)$; $0 < \xi < h$.

29. $k = \frac{2}{3}$; $\begin{cases} x_1 = 0.0711 \\ x_2 = 0.1785 \\ x_3 = 0.7504 \end{cases}$ 30. 0.1089

31.
x	y_1	y_2	y_3	y_4	y_5
0	1	1	1	1	1
0.25	1.1224	1.1299	1.1305	1.1303	1.1303
0.50	1.2398	1.2685	1.2709	1.2710	1.2710
0.75	1.3522	1.4134	1.4213	1.4220	1.4221
1.00	1.4597	1.5654	1.5816	1.5835	1.5836

Chapter 11

1. 0.27768   2. 0.6677   3. 0.92430   4. 0.91596 55942

5. 0.63201   6. 0.58905   7. 0.2257   8. 1.20206

9. $5e$ 10. 0.6557   11. 1.40587   12. 1.341487

13. $0.6719 + 1.0767i$ 14. 1.7168   15. 9.20090   16. 0.8225

17. $S = f(x) - \frac{1}{2}f'(x) - \frac{1}{4}f''(x) + \frac{1}{6}f'''(x) + \frac{5}{48}f^{IV}(x) - \frac{1}{15}f^{V}(x) + \cdots$

18. 1.782   20. $\frac{1}{10}$ (Incidentally, the answer is exact.)   21. $\pi^2/4$ and $\pi^2/6$

22. $\pi^2/6$ and $\pi^2/12$ 23. 0.309017 $(= \sin(\pi/10))$ 24. 777564

Chapter 12

1. 0.2313   2. Numerically: 0.722; exactly: $\pi^{3/2}/8 = 0.696$.

3. $a = \frac{2}{3}$; $b = \frac{1}{12}$. The value of the integral is 2.241.

4. The side of the square is a.

Chapter 13

1. $f(x) = \omega_1(x)\lambda_1^x + \omega_2(x)\lambda_2^x$, where $\lambda_{1,2} = \frac{1}{2}(\alpha + 2 \pm \sqrt{\alpha^2 + 4\alpha})$.
   Special case: $f = \omega_1 \cos x + \omega_2 \sin x$ and $f = \omega_1 e^x + \omega_2 e^{-x}$

2. $\begin{cases} x_1 = 19.912 \\ x_2 = -2.895 \\ x_3 = 1.020 \\ x_4 = -1.038 \end{cases}$ 3. e 4. $A = \frac{1}{6}$; $B = \frac{1}{3}$; $C = \frac{1}{2}$

5. (a) $u_n = \cos nx$; (b) $\sin nx/\sin x$

7. $y_0 = \pi\sqrt{3}/9$; $y_1 = \frac{1}{2}(\log 3 - \pi\sqrt{3}/9)$

k	y_k	k	y_k	k	y_k	k	y_k
0	0.604600	4	0.080339	7	0.047006	10	0.033117
1	0.247006	5	0.065061	8	0.041251	11	0.030140
2	0.148394	6	0.054600	9	0.036743	12	0.027652
3	0.104600						

$$N = 8.255$$

8. 3.359886

9. $f_n(\lambda) = \dfrac{\sin(n+1)\varphi}{\sin\varphi}$ where $2\cos\varphi = \lambda$.

Eigenvalues: $\lambda = 2\cos\dfrac{k\pi}{n+1}$; $k = 1, 2, \ldots, n$.

10. $k = N\tan\dfrac{p\pi}{N}$, $p = 1, 2, 3, \ldots, p < \dfrac{N}{2}$. $y_n = \left|\cos\dfrac{p\pi}{N}\right|^{-n}\sin\dfrac{pn\pi}{N}$.

11. $\tan(n\arctan x)$ 12. $\begin{cases} x_n = 5\cdot 9^n - 2\cdot 2^n \\ y_n = 9^n + 2^n \end{cases}$

Chapter 14

1. $\begin{cases} x & 0.2 \quad\quad 0.4 \quad\quad 0.6 \quad\quad 0.8 \quad\quad 1 \\ y & 0.8512 \quad 0.7798 \quad 0.7620 \quad 0.7834 \quad 0.8334 \end{cases}$ Minimum: $(0.58, 0.76)$.

2. $\begin{cases} x & 0.5 \quad\quad 1.0 \quad\quad 1.5 \quad\quad 2.0 \\ y & 1.3571 \quad 1.5837 \quad 1.7555 \quad 1.8956 \end{cases}$

3. $\begin{cases} x & 0.5 \quad\quad 1.0 \\ y(\text{R.-K}) & 0.50521 \quad 1.08508 \\ y(\text{Scr.}) & 0.50522 \quad 1.08533 \end{cases}$

4. $A(h) = 1 - h + h^2/2! - h^3/3! + h^4/4!$

h	$A(h)$	$A(h) - e^{-h}$
0.1	0.905	$8.20\cdot 10^{-7}$
0.2	0.819	$2.58\cdot 10^{-6}$
0.5	0.607	$2.40\cdot 10^{-4}$
1	0.375	$7.12\cdot 10^{-3}$
2	0.333	$1.98\cdot 10^{-1}$
5	13.71	13.71

$y_{100} - e^{-10} = 4.112\cdot 10^{-10}$

5. $\begin{cases} x & 2.4 \quad\quad 2.6 \quad\quad 2.8 \quad\quad 3.0 \\ y & 0.6343 \quad 0.5189 \quad 0.4687 \quad 0.4944 \end{cases}$ Minimum: $(2.84, 0.47)$.

6. $\begin{cases} x & 0.5 \quad\quad 1.0 \\ y & 1.2604 \quad 2.2799 \end{cases}$ Exact solution: $y = \dfrac{3x^2 + 4}{4 - x^2}$

7. $y = \dfrac{k\sin n\varphi}{\sin\varphi}$ where $\cos\varphi = \dfrac{12 - 5h^2}{12 + h^2}$, $y_6 = -0.0005$.   8. $\xi = 1.455$.

9. $z_n = (-1)^n n!\, a^n b$.   10. Try $y = \sin(\alpha x^n)$; $n > 1$.

11. $\begin{cases} x & 0.25 \quad\quad 0.50 \quad\quad 0.75 \\ y_a & 0.2617 \quad 0.5223 \quad 0.7748 \\ y_b & 0.2629 \quad 0.5243 \quad 0.7764 \end{cases}$ 12. $y(\tfrac{1}{2}) = 0.21729$; $z(\tfrac{1}{2}) = 0.92044$

13. $y(\tfrac{1}{2}) = 0.496$, $z(0) = 0.547$ 14. $p = 0.54369$; $q = 1.83928$

15. $h \le \sqrt{6}$ 16. $h \le a/b$

17. Weak stability; we get the approximate values

$$\left(1 + \frac{h}{8}\right)\left(-\frac{1}{2} \pm i\frac{\sqrt{3}}{2}\right) \quad \text{and} \quad 1 - h.$$

18. $y(0) = 0.0883$; $y(\tfrac{1}{2}) = 0.0828$ 19. $y(0) = 0.0856$

20. $a_n(\lambda) = 1 + \dfrac{2(1-\lambda)}{3!} + \dfrac{2^2(1-\lambda)(3-\lambda)}{5!} + \dfrac{2^3(1-\lambda)(3-\lambda)(5-\lambda)}{7!} + \cdots$

$$\lambda = 4.58$$

21. 17.9   22. 8.60   23. 12.362   24. 2.40   25. 31

Chapter 15

1.

	$x=0.25$	$x=0.5$	$x=0.75$
$y=0.75$	-0.176	-0.067	0.035
$y=0.5$	-0.379	-0.200	-0.057
$y=0.25$	-0.538	-0.297	-0.120

2.

	$x=1.5$	$x=2$	$x=2.5$
$y=1.5$	-0.84	-0.57	0.05
$y=1$	-0.46	0.00	0.78
$y=0.5$	-0.32	0.27	1.14

3. -0.21 4. $u(\frac{1}{3},\frac{1}{3}) = 0.71$; $u(\frac{2}{3},\frac{1}{3}) = 1.04$; $u(\frac{1}{3},\frac{2}{3}) = 1.05$; $u(\frac{2}{3},\frac{2}{3}) = 1.38$

5. $u = J_0(px)\exp(-p^2 t)$ where $J_0(p) = 0$ (smallest value of $p \simeq 2.4048$).

More general: $u = \sum\limits_{r=1}^{\infty} a_r J_0(p_r x)\exp(-p_r^2 t)$ where $J_0(p_r) = 0$.

7. $\lambda = 12.27$ (exact value $5\pi^2/4 = 12.34$)

8. $h = \frac{1}{4}$ gives $\lambda_1 = 41.3726$
   $h = \frac{1}{5}$ gives $\lambda_1 = 44.0983$
   Extrapolated value: $\lambda_1 = 48.94$
   (exact value: $5\pi^2 = 49.35$)

9. $h = \frac{1}{2}$ gives $\lambda_1 = 64$
   $h = \frac{1}{3}$ gives $\lambda_1 = 71.85$
   Extrapolated value: 78.13

10. $\lambda_1 \simeq 3.4$; $\lambda_2 \simeq 12$

11. $-\alpha h^2 \dfrac{\partial^4 u}{\partial x^2 \, \partial t^2} + O(h^4)$

Chapter 16

1. $D(\lambda) = 1 - \lambda/3$;
   $D(x, t; \lambda) = \lambda x t$

2. $y = x^{-1/2} + x$

3. $y = \dfrac{(45 - 15\lambda)x^2 + 9\lambda}{45 - 30\lambda - 4\lambda^2}$

4. $\lambda = -8 \pm \sqrt{76}$ 5. $y = 1.486x - 0.483x^3 + 0.001725x^5$

7. $\lambda = \pm 4/\sqrt{\pi^2 - 4}$.—Note that $\cos(x + t)$ can be written as a degenerate kernel.

8. $y = x\exp(x^4/4)$

9. $y = x + \dfrac{x^3}{1\cdot 3} + \dfrac{x^5}{1\cdot 3\cdot 5} + \dfrac{x^7}{1\cdot 3\cdot 5\cdot 7} + \cdots = \exp\left(\dfrac{x^2}{2}\right)\int_0^x \exp\left(-\dfrac{t^2}{2}\right) dt$

10. $\lambda_n = (2n+1)^2\pi^2/4$; $y_n = \sin(n + \frac{1}{2})\pi x$. The equation can be transformed to the differential equation $y'' + \lambda y = 0$.

11. $y = 6x + 2$ 12.

x	0	0.2	0.4	0.6	0.8	1.0
y	1	1.167	1.289	1.384	1.461	1.525

13. $y(x) = 1 + \displaystyle\int_0^x y(t)\, dt$ 14. $y(x) = 1 - 2x - 4x^2 + \displaystyle\int_0^x [3 + 6(x - t) - 4(x - t)^2] y(t)\, dt$

Chapter 17

1. $y = 0.59x + 2.22$ 2. (a) $y = 0.9431x - 0.0244$; (b) $y = 0.9537x - 0.0944$

3. $a = 0.96103$; $b = 0.63212$ 4. $a = 1$; $b = \frac{2}{3}$; $c = \frac{1}{15}$

5. $a_1 = -1$; $a_2 = \frac{1}{2}$; $a_3 = \frac{1}{3}$; $a_4 = \frac{3}{8}$; $a_5 = \frac{1}{5}$; $a_6 = \frac{13}{72}$

6. $a = -0.33824$; $b = -0.13313$; $c = 1.01937$ 7. $\alpha = 2.89$; $I_0 = 5.63$

8. $a = 1.619$; $b = 0.729$; $A = 3.003$; $B = 0.996$; y satisfies the difference equation $\Delta^2 y + (\alpha + \beta)\,\Delta y + \alpha\beta y = 0$, where $\alpha = 1 - e^{-ah}$; $\beta = 1 - e^{-bh}$. $\xi = \alpha + \beta$ and $\eta = \alpha\beta$ are determined with the least-squares method.

9. $a = 2.415$; $b = 1.258$. y satisfies the same difference equation as in Ex. 8 with $\alpha = 1 - e^{-ah}e^{ibh}$ and $\beta = 1 - e^{-ah}e^{-ibh}$. We find $e^{-2ah} = 1 - \xi + \eta$; $\cos bh = (1 - \frac{1}{2}\xi)/\sqrt{1 - \xi + \eta}$.

10. $a = 4.00$; $b = 120$ 11. $a = 8(3 - \sqrt{8}) = 1.3726$; $\varepsilon = (3 - \sqrt{8})/8 = 0.02145$

12. $2\sqrt{2} - 1$ 13. $0.99992 - 0.49878x + 0.08090x^2$

14. $P(x) = 4x^3 - 6x^2 + 1$; $\varepsilon = 0.0200$ 15. $a = 0.862$; $b = 0.995$; $c = 0.305$

16. $P(x) = 0.999736x - 0.164497x^3 + 0.020443x^5$ 17. $\dfrac{2}{\pi} - \dfrac{4}{\pi}\displaystyle\sum_{k=1}^{\infty}\dfrac{\cos 2kx}{4k^2 - 1}$

18. $\cos x = J_0(1) - 2J_2(1)T_2(x) + 2J_4(1)T_4(x) - \cdots$ (For definition of $J_\nu(x)$, see Ch. 18.)

Chapter 18

2. $\frac{4}{3}$ 3. $2\pi\sqrt{3}/9$ 4. $(\pi/n)/\sin(\pi/n)$ 6. $27\pi/1729$

7. $\sin(\pi/10) = (\sqrt{5} - 1)/4 = 0.309$ 8. $2n!\,(2n + 1)!!$

9. $(-1)^{(n-1)/2} \cdot \dfrac{2^n(n - 1)!}{((n - 1)/2)!\,((n + 1)/2)!}$ 10. $y = \displaystyle\int_0^{\infty} \exp\left(-t - \frac{x}{t}\right)dt$; $x > 0$

11. $y = \sqrt{4x}\{AI_1(\sqrt{4x}) + BK_1(\sqrt{4x})\}$; $\displaystyle\int_0^{\infty} \exp\left(-t - \frac{x}{t}\right)dt = \sqrt{4x}K_1(\sqrt{4x})$.

From the definition of K_1 we have $\lim_{x\to 0} xK_1(x) = 1$ and further the integral approaches zero when $x \to \infty$, and hence we find $A = 0$, $B = 1$.

12. $\cos(x \sin\theta) = J_0(x) + 2J_2(x)\cos 2\theta + 2J_4(x)\cos(4\theta) + \cdots$
$\sin(x \sin\theta) = 2J_1(x)\sin\theta + 2J_3(x)\sin 3\theta + 2J_5(x)\sin 5\theta + \cdots$
For $\theta = 0$ and $\theta = \pi/2$, we obtain the relations (18.5.11). Necessary trigonometric integrals can be computed by the same technique as was used in Section 17.5.

13. $[\Gamma(1 + 1/p)]^2/\Gamma(1 + 2/p)$; $\pi^2/6$

Chapter 19

1. $x = -\lambda^{-1}\log(1 - \xi)$ 2. $m + \sigma\xi$ 3. 0.688

4. Theoretical values: $0.75, 0.50, 0.25$ 5. (a) $\frac{1}{3}$ (b) $\frac{11}{18}, \frac{5}{18}, \frac{2}{18}$

Chapter 20

1. $y_{\max} = 4.4$ for $x^T = (4.4, 0, 0, 0.6)$ 2. $y_{\min} = -2$ for $x^T = (6, 8, 0)$

3. $\lambda \le -1$: $f_{\min} = 6\lambda$; $-1 \le \lambda \le -\frac{1}{2}$: $f_{\min} = 2\lambda - 4$; $-\frac{1}{2} \le \lambda$: $f_{\min} = -5$

4. $f_{\max} = 1$ for $x^T = (2, 2, 0, 0, 0)$

5. $f_{min} = 0$ for $x^T = (0, 3 - c, c, 0)$; $0 \leq c \leq 3$

6. $y_{min} = \frac{67}{12}$ for $x^T = (0, \frac{11}{12}, \frac{5}{4})$

7. $0 \leq \alpha \leq 2$: $f_{max} = 2\alpha$
   $2 \leq \alpha \leq 8$; $f_{max} = 2 + \alpha$
   $8 \leq \alpha \leq 12$: $f_{max} = 10$

8. $f_{min} = 3$ for $x = 3, y = 3, z = -3$.
   z can be written $z = u - v$ where $u \geq 0$,
   $v \geq 0$.

9. $z_{max} = \frac{857}{16}$ for $x^T = (\frac{43}{16}, 1, 3, \frac{11}{2})$

10. 0 lb of A, 17.5 lb of B, and 15 lb of C give a maximum profit of $1,375.

11. Mine A should be running 2 days, and B 6 days. Minimum cost: $18,000 per week.

12.

3	1	
	2	4
		5

Min = 186

13.

	75		25
25		75	
50			50

Min = 5000

14.

5		7	6
5	5		

or

10		7	1
			5
	5		

or any combination of these two solutions. Min = 135.

15. Several optimal solutions exist, for example,

4		4	
3			
	5	1	

or

4	4	
3		
	1	5

Min = 38

16. There are several optimal solutions, for example,

20		10			
		20	10	20	
	40			10	25
			20		

or

10		20			
		10	10	30	
10	40				25
			20		

Min = 430

17.

70		
	60	10
30		20

18.

	10	
		60
5	15	
		10

Min = 730

Index

ABCDE69